Food Colloids
Interactions, Microstructure and Processing

Food Colloids
Interactions, Microstructure and Processing

Edited by

Eric Dickinson
Procter Department of Food Science, University of Leeds, UK

RS•C
advancing the chemical sciences

Sep/re

Chem

The proceedings of Food Colloids 2004: Interactions, Microstructure and Processing held on 18–21 April 2004 in Harrogate, UK.

Chemistry Library

Special Publication No. 298

ISBN 0-85404-638-0

A catalogue record for this book is available from the British Library

Published by The Royal Society of Chemistry,
Thomas Graham House, Science Park, Milton Road,
Cambridge CB4 0WF, UK

Registered Charity Number 207890

For further information see our web site at www.rsc.org

Typeset by the Charlesworth Group, Wakefield, UK
Printed by Athenaeum Press Ltd, Gateshead, Tyne and Wear, UK

7/23/05

Preface

There can surely never have been a time when the field of colloid and surface science was radiating more excitement or confidence than it is in the early 21st century. Under the fashionable—and as yet relatively untarnished—brand name of 'nanoscience' (and its sibling subject 'nanotechnology'), a traditional discipline that a decade or so ago had become physical chemistry's poor relation seems to have been relaunched into the scientific mainstream. As a consequence, food researchers working on the nanometre scale are deriving a technical payout from a dynamic growth in activity. They are gaining access to powerful modern instrumental and computational techniques. And they are deriving lively spin-off from the collaboration and recently discovered interest of other scientists whose previous 'involvement' with food was mainly limited to the kitchen or local supermarket.

This volume is based on a conference entitled 'Food Colloids 2004' held in Harrogate (UK) on 18–21 April 2004, the tenth in a series of European conferences on the subject of food colloids organized under the auspices of the Food (Chemistry) Group of the Royal Society of Chemistry. The programme was sub-titled 'Interactions, Microstructure and Processing' and it was arranged under six main themes: (i) interfacial characteristics of food emulsifiers and proteins; (ii) microstructure and image analysis; (iii) processing issues; (iv) phase transitions; (v) interactions of macromolecules, particles, droplets and bubbles; and (vi) perception of taste, texture and appearance.

The Harrogate conference was attended by 192 delegates from five continents. The technical programme consisted of 41 lectures and 97 posters. Most of the invited and contributed lectures are recorded in full in this book. In addition to being displayed at the formal poster sessions, over half of the poster contributions were displayed electronically on the conference web-site—before, during, and after the conference. Research papers based on selected poster presentations are to appear separately in a special issue of the journal *Food Hydrocolloids*.

The editor of this book acknowledges enthusiastic assistance from the other members of the International Organizing Committee in assembling the scientific programme: Prof. Björn Bergenståhl (University of Lund), Prof. David Horne (Hannah Research Institute), Dr Reinhard Miller (Max-Planck-Institute, Golm), Prof. Juan Rodríguez Patino (University of Seville), Prof. Ton van Vliet (Wageningen University) and Prof. Pieter Walstra (Wageningen University). In addition, for their spirited commitment and hard work, he would like to express his personal and sincere thanks to the members of the Local Organizing Committee: Caroline Eliot-Laizé (Secretary), Jianshe Chen, Rammile Ettelaie, Jonathan James, Brent Murray, Malcolm Povey, and Luis Pugnaloni.

Eric Dickinson (Leeds)
July 2004

Contents

Foams and Emulsions

Sensory Perception

Structure Control and Processing

Gels and Gelation

Rheology, Structure and Texture Perception in Food Protein Gels

By E. Allen Foegeding

DEPARTMENT OF FOOD SCIENCE, NORTH CAROLINA
STATE UNIVERSITY, RALEIGH, NC 27695, USA

1 Introduction

One could propose that mankind's original encounter with a protein gel occurred when the first human accidentally dropped an egg into a fire or hot water and discovered the hard cooked egg. If we accept this humble origin of a culinary art, then the presence of current research[1] dealing with the gelation of ovalbumin demonstrates that making protein gels is a simple process while understanding the molecular mechanisms continues to present scientific challenges.

The seminal review by Ferry[2] defines some general properties of protein gels as "...systems in which small proportions of solid are dispersed in relatively large proportions of liquid by the property of mechanical rigidity ... the characteristic property common to all gels." The rheological characteristics of gels were further refined[3] to include viscoelastic properties and to distinguish between 'solid' and 'solid-like' gels. The key elements of a solid-like gel are a storage modulus, $G'(\omega)$, which has a plateau extending to times at least of the order of seconds, and a loss modulus, $G''(\omega)$, that is much smaller than the storage modulus in the plateau region. For gels formed by denatured proteins, Ferry[2] went on to propose: (1) they are the result of forming a three-dimensional network of solute; (2) they arise from interactions that act along the entire solute molecule, and therefore require a proper balance of attractive and repulsive forces regulated by pH and ionic strength; and (3) their macroscopic appearance and water-holding properties reflect microscopic networks that have extremes of coarse or fine structures. The following mechanism for gelation of denatured proteins was suggested: native protein (corpuscular) → denatured protein (long chains) → association network.

It is amazing that the description for gelation of denatured proteins proposed by Ferry in 1948, based on primarily 4 research articles, is generally valid today. While it has since been proven that proteins do not have to unfold into long chains to form gel networks,[4,5] there have been, and continue to be, numerous investigations into how factors (pH, salts) regulate the association of denatured molecules, the resulting microstructure, and the links between microstructure and macroscopic properties (texture and water-holding).

Research on protein gelation can be generally categorized into two areas, the first being the nature of the gelation process. This involves the various factors that transform proteins from non-aggregating to aggregating structures; the definition and determination of the gel point; the factors responsible for development of gel elasticity; and theories capable of explaining all aspects of the process. The second area concerns the physical and chemical properties of gels. To that end, the microstructure of gels along with the rheological and fracture properties have received the greatest attention, but appearance, intra-gel diffusion, and water-holding attributes have also been investigated.[6,7] While the nature of the gelation process can be considered a general scientific curiosity, the appearance, water-holding and textural properties of gels have direct applications to food quality. Some of the foods that rely on the gelation of denatured proteins to produce desirable characteristics are cooked eggs, processed meats, and some cheeses. In theory, if one understands the physical and chemical basis for the quality characteristics of a food protein gel, such as a cooked egg, then the potential exists to form the same structure with another protein. Indeed, this was pointed out[8] in 1930 as a motivation for developing whey protein ingredients to replace egg proteins in foods. The many soy protein-based imitation meat products on the market today serve as examples of this application. It is also possible to augment a product by adding proteins as gelling agents. An example is the addition of milk proteins to water-added hams as a means of increasing water-holding and rheological properties.[9]

The fundamental science of globular protein gelation took a step forward with the publication of three articles in the early 1980s.[10–12] These articles were the first to focus on molecular and microstructural changes associated with heat-induced gelation of a variety of globular proteins. The protein gelation process was reviewed by Clark and Lee-Tuffnell[5] and by Clark.[13] In these reviews, the theoretical foundation of the gelation process is discussed, along with comprehensive coverage of protein aggregation, changes in protein secondary structure, and microstructure. These areas will be addressed in the present article only from the standpoint of how they help to explain differences in gel structure type and textural properties. This review will focus on gelation from the perspective of producing food structure and textural properties.

2 Gel Structure Type

In a recent article[14] entitled 'From food structure to texture', Wilkinson and coworkers state that advances in the understanding of texture perception will depend on advances in sensory analysis of texture, physiological perception, and food structure. Included under the heading of food structure are

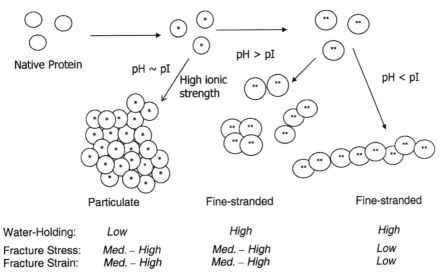

	Particulate	Fine-stranded	Fine-stranded
Water-Holding:	*Low*	*High*	*High*
Fracture Stress:	*Med. – High*	*Med. – High*	*Low*
Fracture Strain:	*Med. – High*	*Med. – High*	*Low*

Figure 1 *Model of globular protein gelation depicting changes from native protein to slightly unfolded (*) and more extensively unfolded (**) states, and resultant gel types. Characteristic physical properties are listed below the gel types.*

"rheological parameters, microstructure and other relevant food characteristics and their relation to sensory perception." This begs the question: are there specific structures in food protein gels? If so, what are the distinguishing characteristics that separate one structure from another? In many ways this issue has already been addressed. Heat-induced globular protein gels can be transparent, turbid or opaque depending on the gelation conditions (see Figure 1).[2,12,15,16] Gels that are transparent or translucent are generally classified as *fine-stranded* (also called fibrillar, 'string of beads', or 'true' gels). They are formed under conditions of pH greater or less than the p*I*, and at low ionic strength. The name "fine-stranded" derives from transmission electron micrographs showing microstructures consisting of strands of various lengths, but with diameters corresponding to the length of one to several molecules.[4,12,16] Fine-stranded gels formed at low and high pH have different textural properties. The gels formed at low pH are weak and brittle, and those formed at high pH are strong and elastic.[17–19] Gels that are opaque are called *particulate* (also called aggregated, random aggregated, or coagulated).[12,16,17] Particulate gels are formed under conditions where there is minimal charge repulsion, such as at a pH close to p*I* or at high ionic strength. The microstructure of particulate gels is composed of spherical particles with diameters in the micrometre range that are associated into a gel network.[20,21] The large particles and low water-holding properties of particulate gels have led to the proposal[12,22] that they are formed by phase separation. Needless to say, there are conditions that lie between the two limiting cases that produce gels with structures that can be considered either (i) a combination of stranded and particulate or (ii) distinct and intermediate between these extremes.

3 Gelation of Fine-Stranded and Particulate Gels

The networks formed in fine-stranded gels at low pH (generally ≤ 3.5) are different from those formed at high pH (generally ≥ 6.5). β-Lactoglobulin gel networks formed at pH 7.0 are composed of strands that are slightly thicker, longer and more curled than those formed at pH 3.5.[21] Heating a β-lactoglobulin solution at pH 2 at a concentration below the critical gelation concentration ($< C_0$) produces fibrils rather than a gel network. Fibrils of β-lactoglobulin of length of 1–7 μm are observed.[23,24] The fibril length is independent of ionic strength in the range 0.01–0.05 M.[24] Static and dynamic light scattering measurements support[25] the formation of β-lactoglobulin fibrils at pH 2.0. Recent atomic force microscopy images confirm[26] the formation of β-lactoglobulin strands composed of dimers or partially unfolded monomers. For β-lactoglobulin, an increase in pH (to 2.5 or 3.0) or ionic strength (> 50–90 mM) causes the fibrils to shorten.[23,24] The formation of fibrils and their association into a gel network seems to be a general property of gels formed at low pH and low ionic strength. This type of gelation can be explained[24] by an adjusted random contact model that takes into account the charge and semi-flexibility of the fibrils. This model has been shown[27] to apply to fibril network gels formed by bovine serum albumin, β-lactoglobulin and ovalbumin.

The type of β-lactoglobulin aggregates formed at pH 7.0 varies among investigations. Kavanagh *et al.* reported[23] the formation of short (< 0.05 μm) linear aggregates based on transmission electron microscopy. Using size-exclusion chromatography and light scattering, Le Bon *et al.* found[28] that, when β-lactoglobulin was heated at 60 °C at pH 7.0 in a 0.1 M salt solution, self-similar aggregates are formed composed of 85 protein molecules and having a radius of 15 nm. A similarly sized aggregate (18 ± 4 nm) was observed[26] by atomic force microscopy for β-lactoglobulin solutions that were heated at 80 °C in pH 7.0 solutions containing 0.1 M salt. Although differences in aggregates have been observed among studies conducted at neutral pH, possibly due to variations in heating conditions and methods of analysis, the types of aggregates formed at high pH appear to be consistently different from those formed at low pH. This suggests that, at some level, the fundamental structural element is different for low pH and high pH fine-stranded gels.

Particulate gels can be formed in solutions where the pH is close to the pI of the protein (or proteins), or where the pH is not close to the pI but there is a relatively high ionic strength (usually > 0.1 M).[12,21] Once the pH and ionic strength are sufficient to form a particulate gel network, increasing ionic strength causes a decrease in rheological parameters (G' or G^*) and water-holding properties, accompanied by changes in the microstructure described as a coarsening of the network or an increase in voids.[6,29,30]

Molecular changes associated with fine-stranded and particulate gels have been investigated using several spectroscopic techniques. Using Fourier transform infrared spectroscopy or Raman scattering spectroscopy, a more extensive unfolding was seen[31,32] when β-lactoglobulin formed fine-stranded gels as compared to particulate gels. While there were apparent differences between fine-stranded and particulate gels, no major differences were seen when these types of gels were formed under different gelation conditions. Therefore, differences in rheological properties associated with fine-stranded gels formed at low and high pH, as well as particulate gels formed close to the pI or at high ionic strength, cannot be explained based on protein structural transitions alone.

A series of studies has established[30,33,34] how the incorporation of protein into the gel network at and after the gel point determines whey protein gel microstructure, permeability and rheology. These experiments were conducted under conditions where gelation was relatively slow (20–24 h at 68.5 °C), and mainly particulate gels were formed. The amount of aggregated protein forming the incipient gel network determines the gel network type (microstructure) and the permeability (a function of microstructure and other properties). Gel permeability decreases only slightly upon further heating after the gel point, suggesting that the network type is 'fixed' at the gel point. In contrast, the elasticity (G') is close to zero at the gel point, and it increases with the total amount of protein incorporated into the established gel network.[3,33,34]

4 Fracture Properties

Fracture behaviour can be investigated by compressing, extending or twisting a sample to the point of fracture.[35] While compression is most often used, due to the simplicity of sample preparation, it is limited to conditions where there is a uniform expansion of the sample when compressed (*i.e.*, where friction between sample and testing apparatus is very low), and for fracture strains occurring at compression levels above $\sim$80% of the original height h (*i.e.*, $\Delta h/h \leq 0.8$). Materials that fracture at high strains must be deformed in tension or torsion. Torsional deformation is pure shear, and there are minimal shape changing forces. So samples that exude fluids in compression and tension will not do so in torsion. Also, since the tensile and shear forces are of equal magnitude, the sample fracture pattern in torsion indicates the weakest mode. The angle of the fracture plane relative to the long axis is 45° for tension fracture and 90° for shear fracture.[35]

While the rheological properties determined in the linear viscoelastic region are associated with small deformations of the gel network, the fracture properties are related to weaknesses in the structure due to defects and/or inhomogeneous elements in the network.[36–38] The rheological and fracture properties of heat-induced protein gels can be altered by adjusting four general parameters: the protein concentration, the heating temperature/time, the solvent conditions/quality, and the addition of other macromolecules or filler particles.

The development of fracture stress and strain in heat-induced whey protein and egg white gels has been investigated by several approaches. When a 10% w/v solution (pH 7.0, 0.1 M NaCl) of whey protein isolate is heated at 80 °C, the development of fracture stress and strain can be observed by sampling over various heating times.[39] After 10 minutes of heating, a self-supporting gel is formed with a low fracture stress (13 kPa) and high fracture strain (1.58).

As heating progresses, the value of the fracture stress increases and levels out at ~ 22 kPa, and the fracture strain decreases to 1.05. This trend of increasing fracture stress and a coinciding decrease in fracture strain is similar to that observed with increasing protein concentration.[39,40] The solvent conditions for gelation of each type of protein can be altered such that whey protein isolate and egg white gels have the same fracture stress at a given protein concentration. Fracture stress values can range from 12 kPa to 130 kPa, as the protein concentration increases from 6 to 19% w/v.[40] While the fracture stress is similar for egg white and whey protein isolate gels, the fracture strain is different. Fracture strain values remain around 2.0 for egg white gels formed at various protein concentrations or heating times.[40] Therefore, gel deformability is one of the properties differentiating whey and egg white protein gels.

Solvent conditions, such as pH and salt type and concentration, play a major role in determining all of the distinguishing physical properties associated with heat-induced protein gels. In whey protein isolate gels formed at pH 7.0 by heating at 80 °C for 30 min, the addition of NaCl (20–500 mM) or $CaCl_2$ (5–500 mM) causes an increase in fracture stress from ~ 3 kPa up to 30 kPa when the ionic strength goes from 20 to 100 mM.[41] Fracture strain follows cation-specific trends, with values starting at ~ 2.8 at low NaCl concentrations, and then decreasing and ultimately increasing as NaCl concentration is increased. In comparison, a higher concentration of $CaCl_2$ causes a progressive increase in fracture strain to a final level of 2.5.[41] These trends are general for chloride salts with monovalent (Li, Na and K) and divalent (Mg, Ca and Ba) cations. The changes in fracture properties coincide with transformations from fine-stranded to particulate gels. As stated previously, the fracture properties of low pH and high pH stranded gels vary greatly. Gels formed at pH 7.0 (14% w/v protein, heated at 80 °C for 30 min) had measured fracture stress and strain values of 74.5 kPa and 1.25, in contrast to gels formed at pH 3.0 with respective values of 18.5 kPa and 0.39 (Figure 2).[18] It is interesting to note that, when fracture stress and strain are combined into a fracture modulus, the respective values for pH 7 and pH 3 gels are 60 and 48 kPa. This agrees with small-deformation rheological investigations that show similar G' values for stranded gels made under low and high pH conditions.[17,19]

The separate behaviour of fracture stress and fracture strain show the relatively weak and brittle nature of stranded gels formed at low pH. It is possible, however, to alter the fracture strain of gels formed at low pH using a two-step gelation process. Gels can be formed by initially heating at pH 7.0 with no salt present, such that gelation does not occur, and then lowering the pH to 3.1 followed by heating to induce gelation. The resultant gels have an increased fracture strain but no increase in fracture stress (Figure 2). Evidence points[18] to the reduced extent of formation of disulfide bonds being associated with the brittle nature of gels formed at low pH. An alternative explanation,[42,43] that does not rule out the participation of disulfide bonds, is that the fracture strain is determined by the shape of strands in the gel network. According to this

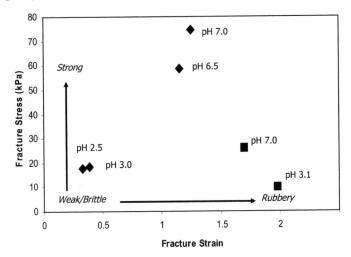

Figure 2 *Fracture stress and strain for whey protein isolate gels formed at various pH values from solutions of 14% protein (◆) or 10% protein (■).* (Data taken from ref. 18.)

argument, at low pH the straight rigid strands deform by stretching and this is associated with low fracture strain. The curved strands formed at high pH deform by bending, and this leads to a greater fracture strain. Therefore, it is possible that the aggregates formed at pH 7.0 remain curved when gelled at pH 3.1 and thereby increase the fracture strain.[18]

5 Non-linear Properties and Rate Dependence

There is a major information gap between rheological tests in the linear viscoelastic region, with strain values usually <0.1, and tests for fracture at strains from 0.3 to 2.8. When testing at a level of strain past the linear viscoelastic region but prior to fracture, the stress–strain relationship can be linear, or it can show strain hardening or strain weakening.[44] In general, egg white and whey protein isolated gels have a high level of strain hardening at low protein concentration, and decreasing degrees of strain hardening as the protein concentration increases.[40]

In the previous discussion on fracture properties, all the reported tests were conducted under a constant rate of compression or twisting. However, heat-induced protein gels are viscoelastic, and so any variations associated with deformation rate were not determined. The effect of strain rate on non-linear and fracture properties of fine-stranded whey protein isolate gels (pH 7.0, 50 mM NaCl; 8, 10 or 18% w/v protein) has been investigated.[45] The gels were characterized over a shear strain-rate range of 0.014–0.69 s^{-1}. Only at the lowest strain rate of 0.014 s^{-1} did the fracture properties vary. The fracture strain increased such that for the lowest protein concentration the values were beyond the sensitivity of the instrument (>3.0). However, at strain rates of

0.02–0.69 s^{-1}, the fracture stress and strain remained constant for all protein concentrations. The increased fracture strain at a strain rate of 0.014 s^{-1} may be of interest from a material science perspective; but as it implies up to 10 min to fracture, it is not on the time-scale relevant to normal chewing. There was a general trend observed for all gels involving the relationship between apparent modulus (stress/strain at any strain level) and strain rate. At strain rates of 0.014–$0.021 s^{-1}$, the apparent modulus decreased, became linear, and then showed strain hardening as the strain increased. It was also observed that the critical level of strain corresponding to the onset of strain hardening decreased as strain rate increased.[45] An overall conclusion is that the large-strain region, and the associated level of strain hardening, is more strain-rate dependent than is the fracture behaviour.

6 Defining the Characteristics of Gels

It is clear from the previous discussion that one can classify the gel type as fine-stranded or particulate based on visible appearance and microstructural considerations. It is also apparent that one can extensively characterize the physical properties of a gel based on rheology and fracture testing. However, if we match a variety of physical properties between two gels of one type, are the gels indistinguishable? In other words, can one produce 'identical' gels from different processes with one protein, or use different proteins to produce the same gel? Since pH and salts both contribute to electrostatic effects on denaturation and aggregation, can they be exchanged to produce identical gels?

In the case of fracture stress and fracture strain the answer is yes. Particulate whey protein isolate gels (10 % w/v protein) can be made to have similar fracture stress and strain when solution conditions are balanced by either holding pH constant at 7.0 and adding NaCl, or by adjusting the pH without adding NaCl.[46] Two pairs of gels were formed. Gels with a fracture stress of 24 ± 1 kPa were formed by adding 0.25 M NaCl to pH 7.0 solutions, or by adjusting the pH to 5.47. A second pair with a fracture stress of 13 ± 1 kPa was made by adding 0.6 M NaCl to pH 7.0 solutions or adjusting the pH to 5.68. All four gels had the same fracture strain. But the gels were different in the degree of strain hardening, in terms of microstructure and permeability. Changing gelation pH changed the level of strain hardening whereas adding NaCl at pH 7.0 had minimal effect on strain hardening. Since fracture strain was associated with disulfide bonding,[18] the effect of covalently blocking a portion of the sulfhydryls was tested. A small amount of sulfhydryl blocking caused by adding 2 mM n-ethylmaleimide was found to alter the strain hardening of gels made at pH 7.0, but it had no effect on fracture properties. The slight adjustment, presumably in disulfide bonding, produced two gels formed under different solution conditions with similar values for fracture stress, strain and strain hardening. This implies similar force–deformation relationships between gels at one deformation rate. Even though these gels had identical fracture and strain-hardening properties, there were differences in permeability, suggesting that microstructural differences remained between the gels.

7 Rheological Properties, Fracture Properties and Sensory Texture

The relationship between mechanical breakdown (fracture) properties and sensory texture is complex and varies amongst food materials.[47,48] The perception of physical properties in the mouth can be visualized by considering the compression of a piece of cooked egg white between the molar teeth. The gel is compressed to the point of fracture with continual sensing of the force–deformation relationship up to fracture. After fracture there are several properties that can be evaluated, such as the number of particles, the cohesiveness of the particles, and the tactile sensation between the gel and the mouth. Corresponding mechanical properties of the said material would start with those determined in the linear viscoelastic region, and then progress to measurement with strain increasing through the non-linear region (strain hardening or weakening), ending at fracture. Note that mechanical evaluation generally stops once the gel is fractured, while sensory perception evaluates the result of the initial fracture and then most likely continues to fracture several additional times until the food is swallowed. Note also that it is likely that the levels of stresses and strains associated with the linear viscoelastic region are below the sensory perception threshold.

The goal of relating mechanical to sensory properties has at least two paths with different assumptions and approaches. The first is imitative. With this approach, the goal is to develop one or more tests that will imitate the chewing process such that mechanical properties are correlated with sensory properties. These tests are seldom conducted such that the force and deformation measurements can be converted to fundamental stresses and strains; therefore, their main purpose is to serve as substitutes for sensory analysis. The other extreme is to measure fundamental properties in the linear viscoelastic region, in the non-linear region, and at fracture—and then see what combination of these best correlates with sensory analysis. Mechanical properties can be augmented with microstructural and water-holding data. The advantage of the latter approach is that it allows one to use theories relating gel structure to mechanical properties, and so to develop scientific principles that can be used to engineer precise structures into foods. This overall approach is outlined in Figure 3.

8 Sensory Texture of Some Gels

The defining test for similarities and differences between food gels is sensory analysis. The complex process of chewing food, and thereby evaluating its texture, occurs in three stages.[14] The preparatory first stage involves movement of the food from the front of the mouth to the back and initial jaw movement

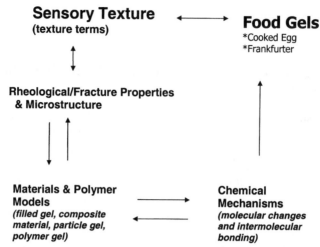

Figure 3 *Relationships among sensory texture, material and polymer models, and chemical mechanisms for food gels.*

with the teeth not reaching occlusion. In this stage, the tactile response of the food touching the interior of the mouth is evaluated. Food particle size is decreased in the second stage, and in the third stage the food is put into a form that is ready for swallowing.[14] Therefore, physical properties of gels that are evaluated at each stage of mastication will encompass sensory texture. Also, the deformation rate used in determining rheological and fracture properties should be within a range where viscoelastic effects are similar between mechanical and sensory analysis.[49]

The effect of gel structure and filler volume on gel texture has been investigated[50] by comparing the fracture properties and sensory texture of particulate and fine-stranded whey protein isolate gels containing 0–20% v/v sunflower oil. Gels were formulated such that fracture stress (evaluated at a shear-rate of 0.126 s^{-1}) was similar between gel type, and ranged from 11 to 54 kPa depending on oil content. Increasing the oil content increased the fracture stress and decreased the fracture strain for fine-stranded gels. These trends are similar to those observed for heat-induced whey protein emulsion gels.[51,52] In contrast, increasing the lipid content of particulate gels was found to cause an increase in the fracture stress, but it had no effect on the fracture strain. It is not clear if this is associated with the particulate structure or is due to other factors. In soy protein isolate gels containing 10–30% v/v soybean oil, increased phase volume or particle size increases the fracture stress.[53] In contrast to whey protein gels, the fracture strain of soy protein gels did not change or showed just a slight increase with increased oil phase volume.

Eighteen textural attributes were evaluated[50] to characterize the gel from initial tactile sensation, through first bite, mastication and preparation for swallowing (Figure 4). The first-bite property of firmness, and the pre-swallowing properties of 'number of chews' and 'time to swallow' were the only properties that increased with lipid content and were independent of gel structure type. The increase in firmness coincides with an increase in the

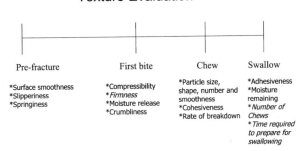

Figure 4 *Terms commonly used to describe the texture of whey protein isolate emulsion gels. The terms are listed according to the stage of evaluation. Properties shown in italics are significantly increased with increasing amount of oil.* (Data taken from ref. 50.)

fracture stress, and known effects of lipid filler particles in heat-induced protein gels.[51,52] The remaining 15 sensory texture terms, including those describing the breakdown process of the gel during mastication, were dependent primarily on gel structure type (Figure 4).[50] Clearly, in this case, the addition of lipid stiffened the gel network, but it did nothing to change the defining characteristics associated with gel type. This suggests that gel type is the primary determinant of sensory texture and that filler lipids just add firmness to the network without changing the primary texture type.

9 Future Research Trends and Needs

Heat-induced gelation of proteins is one of the key mechanisms for forming texture in foods, and therefore it warrants a comprehensive understanding. As depicted in Figure 4, the design of specific textures requires not only an understanding of how gel structures and physical properties arise, but also how they are sensed during consumption. There have been many advances[5,13] in understanding the scientific principles regulating critical gelation time, gelation concentration, and G'. Likewise, with the advent of proteomics and associated research tools, the ability to identify the structural changes and interactions associated with protein gelation should progress. Furthermore, there has been some recent progress in understanding gel fracture properties. But much work is still needed to determine the link between physical properties and sensory texture.

References

1. M. Weijers, L. M. C. Sagis, C. Veerman, B. Sperber, and E. van der Linden, *Food Hydrocoll.*, 2002, **16**, 269.
2. J. D. Ferry, *Adv. Protein Chem.*, 1948, **4**, 1.
3. K. Almdal, J. Dyre, S. Hvidt, and O. Kramer, *Polymer Gels and Networks*, 1993, **1**, 5.

4. M. P. Tombs, in 'Proteins as Human Food', ed. R. A. Lawrie, AVI Publishing, Westport, CT, 1970, p. 126.
5. A. H. Clark and C. D. Lee-Tuffnell, in 'Functional Properties of Food Macromolecules', eds J. R. Mitchell and D. A. Ledward, Elsevier, Essex, 1985, p. 203.
6. W. Chantrapornchia and D. J. McClements, *Food Hydrocoll.*, 2002, **16**, 467.
7. E. L. Bowland and E. A. Foegeding, *Food Hydrocoll.*, 1995, **9**, 47.
8. P. N. Peter and R. W. Bell, *Ind. Eng. Chem.*, 1930, **22**, 1124.
9. S. J. Haylock and W. B. Sanderson, in 'Interactions of Food Proteins', eds N. Parris and R. Barford, ACS Symposium Series, American Chemical Society, Washington, DC, 1991, vol. 454, p. 59.
10. A. H. Clark and C. D. Tuffnell, *Int. J. Peptide Protein Res.*, 1980, **16**, 339.
11. A. H. Clark, D. H. P. Saunderson, and A. Suggett, *Int. J. Peptide Protein Res.*, 1981, **17**, 353.
12. A. H. Clark, F. J. Judge, J. B. Richards, J. M. Stubbs, and A. Suggett, *Int. J. Peptide Protein Res.*, 1981, **17**, 380.
13. A. H. Clark, in 'Functional Properties of Food Macromolecules', 2nd edn, eds S. E. Hill, D. A. Ledward and J. R. Mitchell, Aspen, Gaithersburg, MD, 1998, p. 77.
14. C. Wilkinson, G. B. Dijksterhuis, and M. Minekus, *Trends Food Sci. Technol.*, 2000, **11**, 442.
15. R. K. Richardson and S. B. Ross-Murphy, *British Polym. J.*, 1981, **13**, 11.
16. E. Doi, *Trends Food Sci. Technol.*, 1993, **4**, 1.
17. M. Stading and A.-M. Hermansson, *Food Hydrocoll.*, 1991, **5**, 339.
18. A. D. Errington and E. A. Foegeding, *J. Agric. Food Chem.*, 1998, **46**, 2963.
19. Q. Tang, O. J. McCarthy, and P. A. Munro, *J. Texture Stud.*, 1995, **26**, 255.
20. I. Heertje and F. S. M. van Kleef, *Food Microstruct.*, 1986, **5**, 91.
21. M. Langton and A.-M. Hermansson, *Food Hydrocoll.*, 1992, **5**, 532.
22. A. H. Clark, in 'Biopolymer Mixtures', eds S. E. Harding, S. E. Hill and J. R. Mitchell, Nottingham University Press, Nottingham, 1995, p. 37.
23. G. M. Kavanagh, A. H. Clark, and S. B. Ross-Murphy, *Int. J. Biol. Macromol.*, 2000, **28**, 41.
24. C. Veerman, H. Ruis, L. M. C. Sagis, and E. van der Linden, *Biomacromolecules*, 2002, **3**, 869.
25. P. Aymard, T. Nicolai, D. Durand, and A. Clark, *Macromolecules*, 1999, **32**, 2542.
26. S. Ikeda and V. J. Morris, *Biomacromolecules*, 2002, **3**, 382.
27. L. M. C. Sagis, C. Veerman, and E. van der Linden, *Langmuir*, 2004, **20**, 924.
28. C. Le Bon, T. Nicolai, and D. Durand, *Int. J. Food Sci. Technol.*, 1999, **34**, 451.
29. C. M. M. Lakemond, H. H. J. de Jongh, M. Pâques, T. van Vliet, H. Gruppen, and A. G. J. Voragen, *Food Hydrocoll.*, 2003, **17**, 365.
30. M. Verhuels and S. P. F. M. Roefs, *Food Hydrocoll.*, 1998, **12**, 17.
31. T. Lefèvre and M. Subirade, *Biopolymers*, 2000, **54**, 578.
32. S. Ikeda and E. C. Y. Li-Chan, *Food Hydrocoll.*, 2004, **18**, 489.
33. M. Verheul and S. P. F. M. Roefs, *J. Agric. Food Chem.*, 1998, **46**, 4909.
34. M. Verheul, S. P. F. M. Roefs, J. Mellema, and C. G. de Kruif, *Langmuir*, 1998, **14**, 2263.
35. D. D. Hamann, in 'Physical Properties of Foods', eds E. B. Bagley and M. Peleg, AVI Publishing, Westport, CT, 1983, p. 351.
36. H. McEvoy, S. B. Ross-Murphy, and A. H. Clark, in 'Gums and Stabilisers for the Food Industry', eds G. O. Phillips, D. J. Wedlock and P. A. Williams, Pergamon, Oxford, 1984, vol. 2, p. 111.
37. H. McEvoy, S. B. Ross-Murphy, and A. H. Clark, *Polymer*, 1985, **26**, 1483.

38. T. van Vliet, H. Luyten, and P. Walstra, in 'Food Polymers, Gels and Colloids', ed. E. Dickinson, Royal Society of Chemistry, Cambridge, 1991, p. 392.
39. E. A. Foegeding, *J. Texture Stud.*, 1992, **23**, 337.
40. H. Li, A. D. Errington, and E. A. Foegeding, *J. Food Sci.*, 1999, **64**, 893.
41. P. R. Kuhn and E. A. Foegeding, *J. Agric. Food Chem.*, 1991, **39**, 1013.
42. M. Mellema, J. H. J. van Opheusden, and T. van Vliet, *J. Rheol.*, 2002, **46**, 11.
43. J. M. S. Renkema, *Food Hydrocoll.*, 2004, **18**, 39.
44. A. Bot, I. A. van Amerongen, R. D. Groot, N. L. Hoekstra, and W. G. M. Agterof, *Polymer Gels and Networks*, 1996, **4**, 189.
45. L. L. Lowe, E. A. Foegeding, and C. R. Daubert, *Food Hydrocoll.*, 2003, **17**, 515.
46. M. K. McGuffey and E. A. Foegeding, *J. Texture Stud.*, 2001, **32**, 285.
47. P. J. Lillford, *J. Texture Stud.*, 2001, **32**, 397.
48. T. van Vliet, *Food Qual. Pref.*, 2002, **13**, 227.
49. P. Sherman, in 'Food Structure—Its Creation and Evaluation', eds J. M. Blanshard and J. R. Mitchell, Butterworths, London, 1988, p. 417.
50. E. A. Gwartney, D. K. Larick, and E. A. Foegeding, *J. Food Sci.*, 2004, **69**, S333.
51. K. R. Langley and M. L. Green, *J. Texture Stud.*, 1989, **20**, 191.
52. Y. Mor, C. F. Shoemaker, and M. Rosenberg, *J. Food Sci.*, 1999, **64**, 1978.
53. K. H. Kim, J. M. S. Renkema, and T. van Vliet, *Food Hydrocoll.*, 2001, **15**, 295.

Mechanism of Acid Coagulation of Milk Studied by a Multi-Technique Approach

By Douglas G. Dalgleish, Marcela Alexander, and Milena Corredig

DEPARTMENT OF FOOD SCIENCE, UNIVERSITY OF GUELPH, GUELPH, ONTARIO N1G 2W1, CANADA

1 Introduction

The acid coagulation of milk, especially after heat treatment, is a complex process involving the dissolution of the calcium phosphate from the casein micelles,[1-3] followed by the coagulation of the casein micelles, and, if they are present, the soluble complexes formed between whey proteins and κ-casein on heating.[4,5] It is not yet clearly established whether the changing pH alters the internal structure of the casein micelles during acidification, because of the diminishing content of calcium phosphate. Finally, it is also not clearly established how the coagulation process occurs,[6] nor what is the particular mechanism of action of the denatured serum proteins in the gel formation in heated milk.

Casein micelles are believed to be surrounded by a 'hairy' layer consisting of the caseinomacropeptide portions of the surface-bound κ-casein,[7,8] which extends several nanometres into solution,[9] and is sufficiently charged to give stabilization of the casein micelles by both steric and charge-repulsion mechanisms.[10] There is some evidence that the collapse of this layer is partly responsible for the destabilization of casein micelles by acidification, but it is not clearly established at what point the collapse occurs. A decrease in micellar size, which may be interpreted as a collapse of the layer, has been shown to occur at a relatively high pH (6 or above) in diluted milks,[11] but in another study, using undiluted skim milk, the collapse of the hairs appeared to occur at pH values below 6, just before gelation starts.[12]

Studies of the rheology of acidifying milks suggest that the acid gelation of heated milk may not occur via a single-stage aggregation. The presence of denatured whey proteins,[13] in solution and bound to the casein micelles,[14] offers the opportunity for a multi-step aggregation process. The modification of the charge of the hairy layer and its changed geometry may well be the

16

causative factor in the coagulation of the unheated milk. However, in heated milks the structure of the micellar surface is likely to be more complex and it seems likely that simple collapse is insufficient to initiate the aggregation. It appears possible that there are already interactions between the whey protein complexes and the micellar surface before true micellar coagulation occurs.[12,15]

This paper describes studies of the acid coagulation of milk using the two techniques of ultrasound spectroscopy and transmission diffusing wave spectroscopy (DWS), and compares them with rheological measurements. The objective is to combine the results from the different physical techniques to elucidate the mechanism of the aggregation reaction, and to establish whether changes occur in the structure of the micelles during acidification. In principle, rheology describes the bulk properties of the material, while ultrasound and DWS relate more to the molecular and particulate properties of the system.[16-18] In fact, as will be seen, the new techniques appear to measure aspects of the reactions which have not previously been considered.

2 Experimental

Fresh milk from the University farm was collected, treated with sodium azide (0.02% w/v), skimmed at 6000 *g* for 20 min at 5 °C, and filtered three times through glass fibre filters. Heat treatment was performed on samples of milk in glass vials in a water bath at 85 °C for 20 min, followed by immediate cooling in an ice bath. The milk samples were acidified by the addition of solid glucono-δ-lactone (GDL) at concentrations of 0.8, 1.0, 1.5 or 2.0% w/w at 30 °C. The pH of the acidifying milk was measured regularly over a period of 3.5 hours. Sub-samples of this milk were used in the ultrasonic and DWS equipment, so that the measurement of pH against time could be used in the analysis of results from both experiments.

Ultrasonic Measurements

The ultrasonic spectrometer was an HR-US 102 instrument (Ultrasonic Scientific, Dublin, Ireland); its principles of operation have been described.[19] The instrument measures the velocity and the attenuation of ultrasound at up to 8 selected frequencies in sample and reference cells, each containing approximately 1 ml of sample or reference solution. The temperature of the instrument was controlled by an external programmable water bath. Samples were thoroughly degassed before use. Normally, the sample cell contained acidifying milk, and the reference cell contained untreated milk. Continuous measurements of both ultrasonic velocity and attenuation were made at ultrasonic frequencies of 5.099, 7.836 and 14.665 MHz, starting as soon as possible after the addition of GDL to the milk, and continuing for at least 3.5 hours.

Diffusing Wave Spectroscopy

Measurements were made using an apparatus constructed in our laboratory. Light from a solid-state 50 mW laser of wavelength 488 nm was passed through a milk sample contained in a rectangular glass cuvette (path length $L = 4$ mm) immersed in a tank of water maintained at 30 °C. The tank was fitted with two glass windows to allow passage of the laser beam and the scattered light. Scattered light was collected by a single mode optical fibre placed directly behind the exit window of the water bath. The fibre optic was bifurcated at its outlet end, and the 50/50 split light signal was fed into two matched photomultipliers, the signals from which were amplified and fed to a correlator which performed a cross-correlation analysis. The output from the instrument was therefore a time-correlation function $g_{(1)}(t)$ which can be expressed in the simple form for light passing through a sufficiently long cuvette:[16]

$$g_{(1)}(t) \approx \frac{\left(\dfrac{L}{l^*} + \dfrac{4}{3}\right)\sqrt{\dfrac{6t}{\tau}}}{\left(1 + \dfrac{8t}{3\tau}\right)\sinh\left[\dfrac{L}{l^*}\sqrt{\dfrac{6t}{\tau}}\right] + \dfrac{4}{3}\sqrt{\dfrac{6t}{\tau}}\cosh\left[\dfrac{L}{l^*}\sqrt{\dfrac{6t}{\tau}}\right]}. \tag{1}$$

Here, the parameter τ is defined as $\tau = (Dk^2)^{-1}$, where D is the particle diffusion coefficient, $k = 2\pi n/\lambda$ is the wave vector of the light, and n is the refractive index of the milk serum. The length L is the thickness of the sample being measured, and the factor l^* is the photon transport mean free path. The correlation function in equation (1) is very nearly exponential in time, with a characteristic decay time of τ $(l^*/L)^2$. The equation is valid for samples with $L \gg l^*$ (*i.e.*, $L/l^* > 10$) and $t \ll \tau$.

Equation (1) contains the two unknowns—τ and l^*. To obtain τ from the measured correlation function, l^* was calculated from the transmission of light at any time,[16]

$$T_i = \frac{I}{I_0} = \frac{5l^*\big/3L}{1 + 4l^*\big/3L}, \tag{2}$$

where I and I_0 are the initial and transmitted intensities of the laser light. The scattering of a colloid of known τ (monodisperse latex spheres of diameter 269 nm) was used to calibrate the system[20] and to determine I_0 from equation (2). For milks, the measurement of the transmitted light intensity, together with the determined value of I_0, allowed the value of l^* to be calculated at all times. The measured correlation functions at any time gave the values of τ and hence the particle radius, using values of 1.34 and 1.13×10^{-3} Pa s for the refractive index and viscosity of milk serum, respectively.[20] This full analysis allowed the determination of the apparent particle radii in milk without any assumptions being made, except that the casein micelles and the aggregates should be approximately spherical.

Measurement of Diffusible Calcium during Acidification

The release of ionic calcium (Ca^{2+}) from the casein micelles during acidification was followed using an ultrafiltration cartridge with a nominal molecular weight cut-off of 10^4 Da, collecting samples of permeate every 15 min throughout the period of acidification. The concentrations of Ca^{2+} in the permeates were measured either by a calibrated calcium electrode or by the technique of inductively coupled plasma–optical emission spectrometry.

3 Results

Rheological measurements of acidifying milks are not discussed here: our results were essentially in agreement with many published studies of the acid gelation of milk.[4,6,13,14] In this article we consider only our ultrasound and light scattering results.

Diffusing Wave Spectroscopy

The values of the apparent particle radius and of the quantity $1/l^*$ for unheated and heated milks acidified by 1% GDL are shown in Figure 1. In unheated milk, little change was observed in the value of l^* until the pH had decreased to about 5.4, after which point there was a sharp increase in $1/l^*$ until the value peaked at about pH = 5.1 (Figure 1A). The values of the apparent particle radius (Figure 1B) followed a different pattern. They decreased steadily by about 10–15 nm as the pH fell from 6.7 to about 5.4, after which they started to rise slowly. However, the rapid change in particle size ('gelation') did not occur until the pH had dropped below a value of 5.0. This is also around the point where changes in viscoelastic behaviour become evident. These results show that the process of acid-induced gelation, even for unheated milk, begins, albeit inefficiently, at a considerably higher pH than the formation of the acid gel (*i.e.*, the explosive growth of the apparent molecular size). It should be noted also that the value of $1/l^*$ started to change considerably before any significant increase in particle size; indeed, it started to increase at pH ≈ 5.4, where only the first indications of change in particle size were apparent.

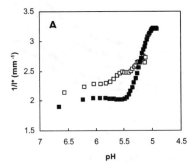 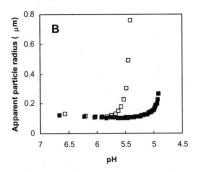

Figure 1 *DWS measurements of acid-induced gelation of heated and unheated milks. Plotted against pH are (A) the reciprocal of the photon transport mean free path, $1/l^*$, and (B) the apparent particle radius for heated (□) and unheated (■) milks during acidification with 1.5% GDL.*

The heated milks behaved differently from the unheated milks. It is known that heated milk coagulates at a higher pH, and Figure 1B confirms this by showing that the particle size has an explosive growth at pH ≈ 5.4. Prior to the aggregation there was a decrease in the particle radius, similarly to that in unheated milk, leading to a minimum of particle size at pH ≈ 5.8. The value of 1/l* increased steadily from the moment that the acidulant was added until the milk gelled, and there was no sharp increase preceding the pH of gelation, in contrast to the case of unheated milk (Figure 1A). However, the plot of 1/l* against pH reproducibly showed an inflection at pH ≈ 5.5, preceding the rapid gelation. Nevertheless, the fact that 1/l* was increasing at all times throughout the acidification suggests that the casein micelles in heated milk have a tendency to form weak structures even at pH values far from the gelation point. It has often been noted that the elastic modulus of heated milk shows an increase at pH ≈ 5.3, but DWS suggests that the interactions become important at pH values well in excess of this.

Ultrasonic Spectroscopy

Measurements of changes in ultrasonic velocity and attenuation are plotted in Figure 2 as functions of pH for milk acidified with 1.5% GDL. The measured velocity did not depend on the frequency of the ultrasound. As acidification proceeded, the velocity increased smoothly, after an initial short stage where the increase was slow, until it levelled off towards the end of the acidification, the net result being a smooth sigmoidal curve (Figure 2A).

The ultrasonic attenuation also started to increase slowly with pH, and then increased steadily and apparently smoothly until it also levelled off towards the end of the reaction (Figure 2B). As expected, the attenuation was frequency dependent, the values being larger at the higher ultrasonic

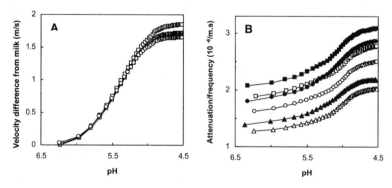

Figure 2 *The change in ultrasonic velocity (A) and attenuation (B) of skim milk caused by acidification with 1.5% GDL. Results are for heated milks (filled symbols) and unheated milks (open symbols) at frequencies of 14.665 MHz (squares), 7.836 MHz (circles) and 5.099 MHz (triangles). Results for the velocities of heated milks overlie those for unheated milks and are not shown. Attenuation values are divided by the frequency.*

frequencies. A power-law dependence has been demonstrated[21] for casein micelles; and from measurements made at 8 different frequencies we have found that the log–log plots of attenuation against frequency for skim milks are linear, with a power-law exponent of 1.25–1.30. The same overall behaviour patterns were seen in heated and unheated milks (Figure 2B), but it was noticed that the heated milk gave a higher attenuation at a given frequency than did the unheated milk.

It is possible to emphasize changes of the slopes of the curves in Figure 2 by calculating their gradients and plotting them against pH. These gradients were calculated numerically from the original data by a running 15-point average (with measurements recorded at each frequency at approximately 40-second intervals). The plots of the gradient of ultrasonic velocity change are shown in Figure 3. For unheated milk, the main curve from the ultrasonic velocity (Figure 3A) shows an inverted bell-shape, suggesting that only a single process was being measured over all the pH values above 5.0. This appears consistent with the dissolution of the micellar calcium phosphate. However, at pH 5.0, a small peak was seen, showing that there had been a sharp, short-lived and small increase in the slope of the velocity *versus* pH plot.

In contrast, the curve for heated milk shows a similar bell-shaped curve, but lacking the contribution at pH 5.0, and having a small shoulder starting at pH ≈ 5.5. Evidently, the onset of gelation in either heated or unheated milk causes only a very small change in the behaviour of the ultrasonic velocity, although the use of gradients allows the onsets of the gelation processes to be distinguished. The additional components on the basic curves in Figure 3A appear at the same points at which the rapid increase in apparent particle radius is measured by DWS. It seems that in milks the elasticity of the gel network has very little effect on the ultrasonic velocity of the transverse waves.

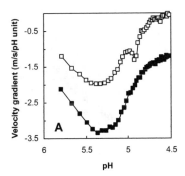 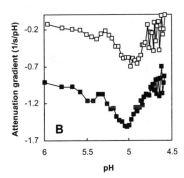

Figure 3 *The gradients of the curves in Figure 2 as calculated from the original measurements at 14.665 MHz: (A) velocity, (B) attenuation. Results for other frequencies are very close to those shown. In both figures the curves for heated milk are displaced downwards for clarity; in actuality, the curves for heated and unheated milks almost superimpose. Note that gradients are negative because pH is decreasing during acidification.*

This is in contrast to results for ultrasonic shear waves where the formation of gels can be seen.[22,23]

The instrument measures attenuation with less accuracy than it does velocity, and the gradient method just described gives somewhat less clear results, especially at low pH where the noise is greatest. However, it does show (Figure 3B) for both heated and unheated milk that the gradient decreases smoothly, with a small effect at pH ≈ 5.5, until it reaches a minimum (*i.e.*, a maximum negative slope) at pH ≈ 5.0. The overall behaviour pattern was approximately independent of frequency, and of heating, and it appears that the measurement of ultrasonic attenuation was not capable of clearly detecting the gelation of the milks. Rather it would seem that the attenuation depends on factors in the milk that are not directly affected by the gelation process.

The changes in ultrasonic velocity occurring during acidification are probably caused by the dissolution of the micellar calcium phosphate, which alters the ultrasonic properties of the milk serum because of the high degree of hydration of the released calcium ions.[22] It has not been established if the changes in attenuation can be explained in this way. We have investigated this hypothesis by plotting the measured increases in ultrasonic velocity during acidification against the measured concentrations of Ca^{2+} which had been dissolved from the casein micelles, and found that, although the curves were close to one another, they did not totally coincide for all of the GDL concentrations which were used (Figure 4A); the observed ultrasonic velocity at a given value of free Ca^{2+} was greatest for the lowest GDL concentration used. That is, although there was a strong correlation between Ca^{2+} release and the ultrasonic velocity, the dependence on the GDL concentration (and hence the kinetics of acidification) suggested that the release of Ca^{2+} was not quite sufficient to explain the increase in velocity during acidification. Because of

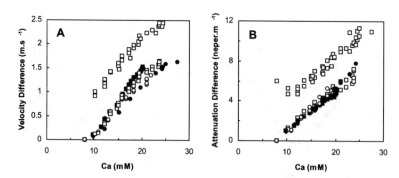

Figure 4 *Plots of the changes in (A) velocity and (B) attenuation against the concentration of ionic calcium in the milk serum, during acidification with different concentrations of GDL. The top sets of points (□) in each graph are from milk permeates, and the lower sets from milks. For clarity, the measurements for the permeates are displaced; in reality, the sets of points for milk and for permeate superimpose almost completely. For the milks, the symbols refer to measurements made at different GDL contents: ■, 0.8%; □, 1.0%; ●, 1.5%; o, 2%.*

the gelation of the milk, it was not possible to prepare permeate to allow measurement of Ca^{2+} liberated during the last stages of acidification, and so the results shown in Figure 4A describe the pre-gelation behaviour. However, it was shown above that the gelation of the protein makes little apparent effect on the evolution of ultrasonic velocity.[24] We found little difference (not shown) in the overall behaviour of heated and unheated milks, which gel at different pH values, but show very similar plots of velocity *versus* calcium content.

To confirm further the importance of the soluble Ca^{2+} on the ultrasonic velocity, samples of milk permeates were collected by ultrafiltration at intervals during the acidification process and their ultrasonic parameters were measured immediately. The increases in the ultrasonic velocity of the permeate during acidification were found to parallel exactly those in the milk (Figure 4A). Thus, the change in ultrasonic velocity of milk is to a great extent defined by the dissolution of the colloidal calcium phosphate from the casein micelles, and the consequent increase in the velocity of ultrasound through the milk serum. Therefore, although it is very difficult to measure the very small ultrasonic effects arising from changes in the casein micelles, it may be possible to measure the release of Ca^{2+} from the micelles by measuring the ultrasonic velocity quite accurately.

The changes in attenuation of skim milk during acidification over most of the reaction time are linearly related to the extent of dissociation of the Ca^{2+} (Figure 4B). In contrast to the measurements of ultrasonic velocity, they seem to show no dependence at all on the rate of acidification (*i.e.*, the concentration of GDL). The results for heated milk (data not shown) and unheated milk appear virtually superimposable. The changes in attenuation for the milk samples were compared with those for the milk sera. The attenuation of milk permeate isolated at the natural pH of the milk (6.7) is approximately half of that of intact skim milk.[21] However, permeates produced at intervals during the acidification of milk and measured immediately showed significant increases in their ultrasonic attenuation as the pH decreased. These increases were very similar to the increase of attenuation of the acidifying milks (Figure 4B). Thus, the changes in the attenuation of the skim milk can be completely explained by the changes in the properties of the serum. This observation, like that for the velocity, explains why there is so little effect of gelation on the ultrasonic properties. Although the casein micelles contribute significantly to the overall attenuation of the skim milk, their properties in respect of ultrasound scattering appear to remain unaltered despite the changes in the composition of the micelles (loss of calcium phosphate) and their colloidal state (gelled or ungelled). This was confirmed by the observation that the contribution to the attenuation from the milk proteins (*i.e.*, the attenuation for the milk minus the attenuation for the permeate) changed very little during acidification.

4 Discussion

Qualitatively, the results for the particle sizes measured by DWS are in agreement with previous measurements using other techniques. Rheology has shown[24] that unheated and heated milks begin to coagulate at pH values around 5.0 and 5.3, respectively, and the apparent particle sizes that we have measured bear out these results, in respect of the main aggregation. However, it is evident from our DWS results that, even in unheated milks, there is some aggregation of the casein micelles starting at pH 5.4 or just below, which has not been clearly detected before. Similarly, aggregation of heated milk was seen to start at around pH 5.8. Although there is general agreement about the point of gelation, our results show that the decrease of particle radius is more gradual than suggested by the results of Vasbinder *et al.*,[12] and the pH values at which we observe the onset of aggregation are greater than those reported by these authors. Our results suggest that the collapse of the κ-casein layer is a gradual process (assuming that the collapse in the surface layer is the cause of the decrease in micellar size), and that it occurs equally in heated and unheated milks.

The measurement of $l*$ not only allows calculation of the particle sizes but it is in itself indicative of changes or interactions within the system, even in the absence of particle aggregation.[16] The considerable increase in $1/l*$ starting at pH 5.4 in unheated milk is not correlated with a significant change in the particle size. It must therefore indicate the existence of long-range interparticle forces (so that the spatial positions of the particles in effect become more correlated) in advance of the formal aggregation of the system. This effect is more pronounced in heated milk, where these long-range interactions seem to be important even at pH > 6.0, where no changes in particle size during acidification are detected. We have no detailed explanation of the increase in $1/l*$ at high pH values, although we have observed it in heated milks acidified with concentrations of GDL between 0.8 and 2.0% (unpublished observations).

The measurement of $l*$ during acidification confirms the differences between heated and unheated milks. The aggregation also shows differences, as may be expected from previously published results. The significant increase in the particle size of the heated milk occurs at a higher pH, and much more sharply, than is the case for unheated milk, reflecting the results seen by rheology,[4,14,24] and by backscattering DWS.[12,18] In general, the results (especially for $l*$) show that the acid precipitation of milk, and especially of heated milk, should not be regarded as a simple phenomenon occurring over a narrow pH range. Rather it is the result of a gradually increasing tendency of the particles to interact as the pH is reduced.

In contrast, the measurements of ultrasonic velocity and attenuation tell us little about the physical condition of the particles in the system. The results clearly demonstrate that, in acidifying milks, it is only the transport of calcium and phosphate from the casein micelles to the milk serum which significantly affects the ultrasonic parameters. Although it would be expected in principle that the changes in hydration of the casein micelles would modify the ultrasonic parameters, this was found not to be the case either for velocity or attenuation. Even gelation did not cause significant changes, and we can only assume therefore that the types of bonds formed between micelles must be very similar to those which hold the micelles together. As far as the behaviour of acidified milks is concerned, it seems that the use of ultrasound is valuable principally as a means of studying the acidification process itself, rather than its effects on the casein micelles.

Acknowledgements

The authors acknowledge the Ontario Dairy Council and the Natural Sciences and Engineering Research Council of Canada for their continuing financial support. We thank Professor Ross Hallett, University of Guelph, for his help in the construction of the DWS apparatus.

References

1. D. G. Dalgleish and A. J. R. Law, *J. Dairy Res.*,1989, **56**, 727.
2. Y. Le Graët and G. Brulé, *Lait*, 1993, **73**, 57.
3. A. Laligant, M.-H. Famelart, G. Brulé, M. Piot, and D. Paquet, *Lait*, 2003, **83**, 181.
4. J. A. Lucey and H. Singh, *Food Res. Int.*, 1998, **30**, 529.
5. F. Guyomarc'h, A. J. R. Law, and D. G. Dalgleish, *J. Agric. Food Chem.*, 2003, **51**, 4652.
6. D. S. Horne, *Int. Dairy J.*, 1999, **9**, 261.
7. P. Walstra, V. A. Bloomfield, G. J. Wei, and R. Jenness, *Biochim. Biophys. Acta*, 1981, **669**, 258.
8. C. G. de Kruif and E. B. Zhulina, *Colloids Surf. A*, 1996, **117**, 151.
9. D. G. Dalgleish and C. Holt, *J. Colloid Interface Sci.*, 1988, **123**, 80.
10. C. G. de Kruif and C. Holt, in 'Advanced Dairy Chemistry—1. Proteins', 3rd edn, eds P. F. Fox and P. L. H. McSweeney, Kluwer/Plenum, New York, 2003, part A, p. 233.
11. D. S. Horne and C. M. Davidson, *Colloid Polym. Sci.*, 1986, **264**, 727.
12. A. J. Vasbinder, A. C. Alting, and C. G. de Kruif, *Colloids Surf. B*, 2003, **31**, 115.
13. J. A. Lucey, M. Tamehana, H. Singh, and P. A. Munro, *J. Dairy Res.*, 1998, **65**, 555.
14. F. Guyomarc'h, C. Queguiner, A. J. R. Law, D. S. Horne, and D. G. Dalgleish, *J. Agric. Food Chem.*, 2003, **51**, 7743.
15. J. A. Lucey, C. T. Teo, P. A. Munro, and H. Singh, *J. Dairy Res.*, 1997, **64**, 591.
16. D. A. Weitz and D. J. Pine, in 'Dynamic Light Scattering: The Method and Some Applications', ed. W. Brown, Oxford University Press, Oxford, 1993, p. 652.
17. D. S. Horne and C. M. Davidson, *Colloids Surf.*, 1993, **77**, 1.
18. A. J. Vasbinder, P. J. J. M. van Mil, A. Bot, and C. G. de Kruif, *Colloids Surf. B*, 2001, **21**, 245.
19. V. Buckin and C. Smyth, *Sem. Food Anal.*, 1999, **4**, 113.
20. M. Alexander, L. F. Rojas-Ochoa, M. Leser, and P. Schurtenberger, *J. Colloid Interface Sci.*, 2000, **253**, 35.
21. W. G. Griffin and M. C. A. Griffin, *J. Acoust. Soc. Am.*, 1990, **87**, 2541.
22. E. Kudryashov, C. Smyth, and V. Buckin, *Prog. Colloid Polym. Sci.*, 2000, **115**, 287.
23. C. Smyth, E. D. Kudryashov, and V. Buckin, *Colloids Surf. A*, 2001, **183–185**, 517.
24. J. A. Lucey, P. A. Munro, and H. Singh, *Int. Dairy J.*, 1999, **9**, 275.

Rheology of Acid Skim Milk Gels

By Catriona M. M. Lakemond and Ton van Vliet[1]

DEPARTMENT OF AGROTECHNOLOGY AND FOOD SCIENCES,
LABORATORY OF FOOD PHYSICS, WAGENINGEN UNIVERSITY,
P.O. BOX 8129, 6700 EV WAGENINGEN, THE NETHERLANDS
[1]WAGENINGEN CENTRE FOR FOOD SCIENCES, P.O. BOX 557,
6700 AN WAGENINGEN, THE NETHERLANDS

1 Introduction

The microstructure of traditional products as yoghurt and cheese is largely determined by aggregated milk proteins. Nowadays milk proteins are also used for structuring many other food products. One of the most important ways to get a continuous network structure of milk proteins is by acidification. In this study we use acidification and heating in different combinations to obtain a variety of microstructures. Preheating of milk followed by acidification will be referred to as the *T–gel* route throughout this article. Heating the gel after it has been formed will be referred to as the *gel–T* route. The latter route is of industrial relevance because in many cases the pasteurization or sterilization of a product is required after the structure has been formed.

The T–Gel Route

It has frequently been observed[1-7] that the heating of milk leads to association of whey proteins with casein micelles. However, not all the whey proteins are bound to the casein micelles on heating. A considerable fraction stays in the serum.[3] This fraction has been found[2,5-7] to depend on the pH of heating. Vasbinder *et al.*[8] proposed a qualitative model to describe the effect of pH adjustment of milk in the range 6.35–6.9 prior to heat treatment. According to this model, heating at $pH \geq 6.6$ leads to a partial coverage of the casein micelle with single whey protein molecules and the formation of separate whey protein aggregates. But heating at $pH < 6.6$ causes attachment of all the whey proteins to the casein micelles.

For gel formation by acidification, the preheating of milk at its natural pH has been shown[4,9,10] to have a distinct effect on the properties of acid-induced milk gels. However, the effect of preheating at pH values slightly below the

26

natural pH of milk on the rheological properties of acid milk gels has not been studied so far. In the first part of this paper, we investigate to what extent the type of aggregates formed by heating milk at different pH values influences the gel characteristics of acid-induced milk gels.

The Gel–T Route

An alternative way to modify gel properties is by heating the gel after it has been formed. Although many authors have already reported on gel formation via the T–gel route, research performed on the gel–T route is scarce. Acidification in the cold followed by heating has been studied by Roefs[11] and acidification at 20 °C followed by heating by Vasbinder et al.[12] In these studies the gel was not formed during acidification, but as a result of the heating step. In industry it is known that heating of acid milk gels may significantly affect texture, often even leading to collapse of the protein network. In this study we investigate under what conditions (acidification and heating temperature) the collapse of the gel network occurs, and whether the occurrence of such collapse can be related to the initial network structure.

2 Materials and Methods

Adjustment of pH and Heat Treatment of Skimmed Milk

Skimmed milk was prepared by dissolving 10.45 g of low-heat skim milk powder (Nilac; NIZO Food Research Ede, The Netherlands) in 100 g distilled water. The milk contained 2.8% w/w casein and 0.5% w/w whey protein. To prevent bacterial growth 0.02% of sodium azide was added. The milk was stirred for 16 hours at 32 °C and used for rheological experiments afterwards.

The skimmed milk (pH = 6.65) was acidified by addition of 0, 0.05, 0.10 and 0.15% w/w glucono-δ-lactone (GDL, Sigma) in order to reach final pH values of 6.65, 6.5, 6.35 and 6.2, respectively, after overnight incubation at 20 °C. A sample of pH 6.9 was obtained by addition of 1 M NaOH.

The skimmed milk or its gel was heated in glass tubes (volume 8 ml, diameter 1 cm) for 10 min at 80 °C and cooled with tap water to room temperature. The samples required approximately 2 min to warm up to 80 °C, giving a total heating time of 12 min. Gel stability was checked by visual examination.

Rheological Properties

Rheological measurements were performed with a Bohlin VOR Rheometer (Bohlin Instruments, Cirencester, UK). A concentration of 1.05–1.2% w/w GDL was added to the skimmed milk samples, depending on the amount of GDL that had been used for acidifying to pH 6.2–6.65 before heating, in order to keep the total amount of GDL added constant. For the milk heated at pH 6.9, first the pH was lowered to pH 6.65 using 1 M HCl and then 1.2% of GDL was added. After GDL addition, the mixture was stirred vigorously for

approximately 2 min before it was transferred into the rheometer, and the pH
of the milk was followed separately and continuously. For incubation at 32 °C,
the final pH was reached about 16 hours after GDL addition.

For dynamic measurements the samples were oscillated at a frequency of
0.1 Hz and measurements were taken every 5 min for 20 hours. The amplitude
of oscillation (maximum strain 0.0103) was kept sufficiently low to ensure
linear behaviour. Gel ageing was followed at 32 °C. All experiments were
performed in duplicate. In this work we consider a gel to be formed when the
value of the storage modulus G' is larger than 1 Pa (except for the unheated
sample, where a value of 0.1 Pa was used).

Large-deformation properties of the gels were determined 20 hours after
addition of GDL. In separate experiments, the gels were subjected to a
constant shear rate of 0.00185 s^{-1} up to yielding (or fracture), defined as the
point where the shear stress starts to decrease, as described by Luyten et al.[13]

Permeability Measurements

The permeability coefficient B was determined by measuring the flow rate Q of
a 0.12 M salt solution through the acid skim milk gel. The quantities B and Q
are related to each other according to the Darcy equation,

$$Q = \frac{BA_c \Delta P}{\eta l},$$ (1)

where A_c is the cross-sectional area of the gel through which the liquid is
permeating, ΔP is the applied pressure difference over the distance l, and η the
viscosity of the salt solution. Measurements were performed according the
method first described by van Dijk and Walstra.[14]

3 Results and Discussion: The T–Gel Route

Results are only shown for milk heated at pH 6.2 and pH 6.9. The dynamic
moduli, the fracture properties and the gel coarseness were also determined for
pH values in between. Generally, a consistent trend was observed with pH,
although the results obtained for gel stiffness and fracture stress showed no
significant differences between heating at pH 6.65 and at 6.9.[15]

Small-Deformation Rheology

Figure 1 shows the development of the storage modulus G' as a function of
(decreasing) pH for milk heated at pH 6.2 and 6.9 and for unheated milk. The
results show that around pH 5.2–4.9 the milk starts to gel. Thereafter, the
value of G' for all samples increases linearly with pH between pH = 4.95 and
pH ≈ 4.6. The non-linearity observed around pH 4.5–4.6 is caused by the
ongoing increase in G' due to rearrangements within the gel after reaching the

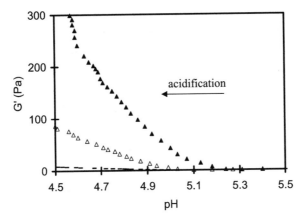

Figure 1 *Storage modulus G' of T–gel skim milk gels acidified with 1.2% GDL at 32 °C for milk heated for 10 min at 80 °C at pH 6.2 (△), pH 6.9 (▲) and for unheated milk (–).*

final pH (4.5–4.6). The development of the loss modulus G'' as a function of pH (results not shown) was similar to that of G', although the absolute values were smaller. We assume that the kinetics of the (rearrangement) processes in the gels are similar for milk heated at pH 6.2 and 6.9, and for unheated milk when the pH drops below pH 4.95, since all the curves of G' against pH were linear in pH in that region. A similar linearity can be deduced from the results of Roefs et al.[11]

At an ageing time of 1200 min after GDL addition, which was about 4 hours after a pH value of 4.5–4.6 was reached, the value of G' of milk heated at pH 6.9 was about 3 times higher than that of milk heated at pH 6.2. Compared to unheated milk, G' was even a factor 10 or 30 higher for milk heated at pH 6.2 or 6.9, respectively. Indeed, it is known[4,9,10] that heating itself has a distinct effect on the rheological properties.

So, for the same amount of protein, large differences in gel stiffness can be obtained depending on heat treatment and the pH at which the milk is heated. The stiffness of gels measured by applying small deformations is determined both by the geometrical structure of the gel and by the interaction forces between the relevant structural elements.[16,17] A general expression for the shear modulus G is

$$G = NC\frac{\mathrm{d}^2 F}{\mathrm{d}x^2},\qquad(2)$$

where N is the number of effective strands per unit area bearing the stress (or the effective strand length per unit volume), C is a characteristic chain length relating the local deformation to the macroscopic strain, and $\mathrm{d}F$ is the small change in Gibbs free energy due to a small distance change $\mathrm{d}x$ between the cross-links. The factors N and C are determined by the geometrical structure of the network.[16,17] For gels that can be described as consisting of an ensemble of fractal clusters, the quantity C is related to the way the clusters are linked to each other (*e.g.*, straight links, hinged links, or links with a more tortuous shape)[18] and to the curvature of the strands within the clusters.[19] The role of the geometrical structure is discussed further below. From equation (2) it is clear that the modulus is also determined by the change in Gibbs free energy upon deformation. Since the bond energy for covalent bonds is a factor 20 to 200 higher than for non-covalent bonds,[20] it is expected that the shear modulus will be higher when relatively more covalent bonds are present (see below).

Behaviour of Tan δ Directly Following Gel Formation

The quantity tan δ (= G''/G') is a measure of gel viscoelasticity. It was found that the major differences in the time development of tan δ between acid-induced milk gels made from milk heated at pH 6.2 or 6.9, or from unheated milk, were observed just after the onset of gelation (*i.e.*, 100–300 min after addition of GDL, corresponding roughly to pH values in the range 5.2–4.7). Therefore Figure 2 focuses on the changes in tan δ in this range. (A very sensitive torsion bar was used to register these tan δ values). For milk heated at pH 6.65, results show that tan δ, after an initial decrease, first increased to a maximum around pH = 5.1, and then decreased with pH; in contrast, for pH 6.2 milk, tan δ decreased with time from the point at which gelation took place (pH = 5.05). We note that below pH 5.05 the development of tan δ as a function of pH was similar for both gels. The horizontal offset is only 0.03 pH units, which is within the experimental error. For unheated milk the maximum was not observed.[21–23]

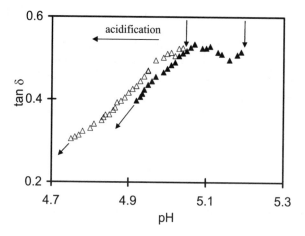

Figure 2 *The development of tan δ as a function of pH for acidified milk (1.2% GDL) heated for 10 min at 80 °C at pH 6.2 (△) and pH 6.65 (▲). The points of gelation are indicated with vertical arrows.*

Values of tan δ measured 1200 min after addition of GDL were also similar for both gels (0.22–0.23). The measured value for unheated milk is somewhat higher (0.25), but the difference is so small that it is not considered important.

Comparison of the data in Figures 1 and 2 shows that the higher gel stiffness observed for heating at pH 6.9, compared to heating at pH 6.2, goes along with differences at and after the onset of gelation (the pH of gelation, and the change in tan δ with pH). The observed maximum for milk heated at pH 6.65 (Figure 2) has been reported before.[21-24] Similar to our results, Lucey *et al.*[22] observed that the maximum in tan δ is lower and eventually absent as the pH of gelation is lowered. The pH of gelation is higher for heated milk than for unheated milk, probably because the apparent isoelectric point (p*I*) of the casein micelles is raised by attachment of β-lactoglobulin molecules on the surface of the casein micelle.[25-27] The difference in pH of gelation for heating at pH 6.2 and pH 6.65 is probably not linked to differences in the apparent p*I* of the whey protein–casein aggregates. We hypothesize that the structure of the aggregates formed from whey proteins and casein micelles upon heating is one of the factors that explains the altered pH of gelation. Because the surface of the casein micelle becomes smoother on heating at pH 6.9 than on heating at a lower pH,[8] the resulting casein micelles could possibly approach each other more closely upon acidification, which should then result in easier bond formation. A second factor that could explain the altered pH of gelation is a difference in the extent of S–S interactions. From the results of Lucey *et al.*,[22] we conclude that the higher pH of gelation possibly correlates with the maximum in tan δ, and with increased development of S–S interactions during the onset of gelation. This effect could be the direct result of a higher pH of gelation, because at higher pH values disulfide bond formation is favoured.[28]

Effect of Blocking of Free Thiol Groups

One of the possibilities to explain the differences observed in the pH of gelation, and possibly also in the final G' value, is a difference in the extent of S–S interactions. To test this possibility, the gelation of milk heated at pH 6.2 or 6.9 was followed in the presence of 5 mM N-ethylmaleimide (NEM) to block the accessible free SH groups. (The NEM was added after the heating of the milk.) Without NEM added, we found that the pH at which a gel is formed is higher at the higher pH of heating (Table 1), and for both gels formed from heated milk higher than the gel formed from unheated milk. Addition of 5 mM NEM did not affect the pH at which gelation started for the milk heated at pH 6.2, but it did for the milk heated at pH 6.9. Adding 20 mM NEM gave similar results. For G' (1200 min after GDL addition) similar trends were found upon NEM addition as were observed for the effect on the gelation pH. A higher pH of gelation goes along with a higher value of G'.

In our opinion the results in Table 1 provide direct evidence for a greater role of S–S bridges in gels made from milk heated at pH 6.9 as compared to milk heated at pH 6.2. However, a difference in S–S interactions cannot explain fully the higher pH of gelation for milk heated at pH 6.9 as compared

Table 1 *Gel stiffness (1200 min after GDL addition) and pH of gelation for milk heated for 10 min at 80 °C at pH 6.2 and pH 6.9 with and without NEM added after heating (to block the accessible free SH groups) and for unheated milk.*

pH of heating	5 mM NEM	G' (Pa)	pH of gelation
6.2	−	101 ± 4	5.05 ± 0.02
	+	103 ± 10	5.06 ± 0.01
6.9	−	323 ± 32	5.25 ± 0
	+	194 ± 16	5.15 ± 0.01
Unheated	−	9 ± 1	4.94 ± 0.01

with pH 6.2, since even when the free thiol groups are blocked there is still a significant difference (0.09 units) in the pH of gelation. Similarly, the differences in G' values are also not fully explained by differences in S–S interactions.

It has been reported previously[29-31] that gelation of heated milk at its natural pH involves disulfide bonding, which is in line with our results obtained for milk heated at pH 6.9. Heating of milk at its natural pH probably leads to an increase in the amount of free thiol groups.[29,31] We assume that heating milk at pH 6.2 does not lead to such an increase, thereby explaining the smaller involvement of S–S interactions in the gelation process. However, a difference in the extent of S–S interactions could also be explained by higher reactivity of the free thiol groups at higher pH.[28]

Curvature and Fracture Properties

It is obvious that it is easier to extend a curved strand than a similar straight strand. Whether the strands in a particular gel will tend to be deformed by bending or by stretching is determined by the curvature of the strands and their connectivity.[19] In general, for strands that are not too brittle, those gels in which the strands are more curved will exhibit a higher fracture strain than gels containing straighter strands; this is because the curved strands will first be straightened during deformation before they are stretched.[32] Thus, measuring the fracture strain gives information about the degree of curvature of strands in the gel network.

The results for the fracture properties of gels formed from milk heated at pH 6.2 and pH 6.9, and from unheated milk, are shown in Figure 3. Typically, as the strain increases, so does the stress (up to a maximum). The maximum in the stress is designated as the point where fracture occurs (indicated by the arrow in Figure 3). The fracture strain was found to decrease from 1.1 to 0.9 as the pH of heating was increased from 6.2 to 6.9. The fracture strain for unheated milk was 1.4–1.7 times larger than that for heated milk. From these results we conclude that the curvature of the strands within the gels increases in the following order of treatment: milk heated at pH 6.9 < milk heated at pH 6.2 < unheated milk. It is expected that gels with more curved strands have a

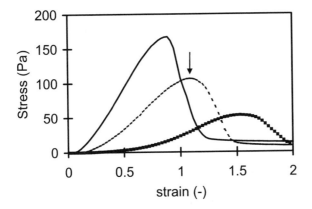

Figure 3 *Stress as a function of applied strain at a constant shear rate (0.00185 s⁻¹) for milk heated prior to acidification for 10 min at 80 °C at pH 6.2 (dashed line), at pH 6.9 (solid line) and for unheated milk (■). The assumed fracture point for the system heated at pH 6.2 is indicated by the arrow.*

lower gel stiffness because the resistance against bending is lower than against stretching.[19] Therefore we can conclude that the curvature of the strands is one of the factors explaining the differences in gel stiffness observed in Figure 1.

Figure 3 also shows that the fracture stress is a factor of 1.7 higher for heating at pH 6.9 as compared with heating at pH 6.2. For unheated milk the fracture stress is 2 to 3 times smaller than for heated milk. Lower fracture stresses and larger fracture strains for gels prepared from unheated milk, as compared to heated milk (≥ 80 °C), have also been reported by others.[4]

It is known that the fracture stress σ_f is partly determined by the type of bonds present in the gel network. Since covalent bonds have a higher bond energy,[20] the force necessary to break a strand in which the particles are joined together by covalent bonds is higher than for connections made of non-covalent bonds. This effect impacts more on σ_f than on G' because in fracture experiments the bonds need to be broken, whereas in small-deformation experiments they only need to be deformed. We conclude that the higher σ_f values observed for gels made from milk heated at pH 6.9 (compared to pH 6.2) can be (partly) explained by the increased contribution of S–S interactions to the gel structure for milk heated at pH 6.9 (see Table 1). Furthermore, it is known that large-deformation properties depend more on the physical structure of the gel—especially the presence of large inhomogeneities—than do the small deformation properties. In general, the value of σ_f will tend to be much lower when large inhomogeneities are present;[33] their presence has been studied by permeability measurements, which provide information particularly on the large pores in a gel.

Gel Coarseness

Permeability measurements were performed on acid-induced milk gels made from milk heated at pH 6.2 and pH 6.9 and from unheated milk. The results

Table 2 *Permeability coefficient B of skim milk gels 20 hours after addition of 1.2% GDL after heating at 80 °C for 10 min at pH 6.2, pH 6.9 and for unheated milk.*

	Heated at pH 6.2	Heated at pH 6.9	Unheated
$B/10^{-13}$ m^2	7.5	18.6	10.6

in Table 2 show that the permeability coefficient of the gel prepared from milk heated at pH 6.9 was more than a factor 2 higher than that for milk heated at pH 6.2. Since the size of the relevant pores scales quadratically with the permeability,[11] it is expected that the mean pore size for gels made from milk heated at pH 6.9 is about factor 1.5 larger than that for heating at pH 6.2. The permeability coefficient of the gels made from unheated milk lies in between the values measured for the gels prepared from milk heated at pH 6.2 and pH 6.9. So the sizes of the inhomogeneities alone cannot explain the observed differences in fracture stress.

It was shown above that the curvature of the strands in the gel network varies amongst the samples (fracture strain between 0.9 and 1.5). However, the fracture strains are all clearly larger than for gels containing straight strands (~ 0.5).[15,18] So it can be concluded that all gels contain strands that are curved to some extent. For gels containing curved strands, it is expected that the modulus will be higher if the strands are thicker, because the bending force for a thick strand is higher than for a thin strand, since it is known[34] that the bending load of a cylinder of diameter d scales as d^4. We conclude that the higher gel stiffness found in Figure 1 for gels made from milk heated at pH 6.9 is (partly) explained by the higher coarseness, which implies the presence of thicker strands in the gel network. Additionally, we conclude that the higher gel coarseness of those gels cannot explain the relatively high values for σ_f observed for heating at pH 6.9 (Figure 3) compared to heating at pH 6.2.

4 Results and Discussion: The Gel–T Route

We have investigated whether the gel structure remains stable or collapses after heating for 10 min at 80 or 90°C (step 2) for different incubation temperatures (20–55 °C) during step 1. A stable gel was found to be formed for the incubation temperatures 20, 32 and 40 °C, but not for 55 °C (Table 3). Permeability measurements showed that the coarseness of these gels became higher as the incubation temperature increased, in agreement with the results reported by Lucey *et al.*[35] for acid milk gels. At higher temperatures the rearrangements can take place more readily,[36] which probably results in a coarser network with straighter strands.[37] It is assumed that the probability of breaking bonds/strands is so high at an incubation temperature of 55 °C that no stable gel structure can be formed.

Heating for 10 min at 90 °C (step 2) was found to cause all the stable gels to collapse (Table 3). However, the gel structure remained intact for gels

Table 3 *Gel stability and permeability coefficient (B) after step 1 (20 h after addition of GDL; incubation temperature 20–55 °C) and gel stability after step 2 (10 min at 80 °C or 90 °C).*

T (°C) of step 1	Gel stability after step 1	B / 10^{-13} m^2 after step 1	Gel stability after step 2 (10 min, 80 °C)	Gel stability after step 2 (10 min, 90 °C)
20	+	1.1	−	−
32	+	10.6	+	−
40	+	62	+	−
55	−	n.d.[a]	n.d.[a]	n.d.[a]

[a] n.d. = not determined.

incubated at 32 and 40 °C (step 1) and then heated for 10 min at 80 °C (step 2). For these two conditions the gel structure is coarser (with a higher value of *B*) than for an incubation temperature of 20 °C (Table 3). In general, a gel structure will tend to collapse when all the protein–protein bonds in each strand break at about the same time. In a coarser gel, the strands are thicker and therefore the probability that all bonds will break at the same moment is lower than for a thin strand.[36] This may explain the higher resistance towards heating of the coarser network structure. A gel formed at 32 °C according the gel–T route has a higher stiffness than a gel formed according the T–gel route after a similar total ageing time: for example, after ageing for 40 h, the values of *G'* were 740 and 330 N m^{-2}, respectively.[15]

5 Conclusions

Following the T–gel route, we have found that the preheating of skimmed milk at pH 6.9 compared to heating at pH 6.2 before acidification results in gels that are about 3 times stiffer. The higher gel stiffness observed for heating at pH 6.9 can be explained only on the basis of a combination of effects: S–S interactions, strand curvature, and gel coarseness. Probably the higher observed stiffness is the result of aggregation and rearrangement processes before and just after gel formation. These differences are probably related to the different types of aggregates formed between casein micelles and whey proteins during heating at pH 6.2 and pH 6.9.

Our study of the gel–T route has shown that gels can be made that are stable against heat treatment. They were found to be more stable when the gel microstructure before heat treatment was coarser.

References

1. H. Singh and P. F. Fox, *J. Dairy Res.*, 1987, **54**, 509.
2. A. J. R. Law, *J. Dairy Res.*, 1996, **63**, 35.
3. H. Singh, M. S. Roberts, P. A. Munro, and C. T. Teo, *J. Dairy Sci.*, 1996, **79**, 1340.
4. J. A. Lucey, C. T. Teo, P. A. Munro, and H. Singh, *J. Dairy Res.*, 1997, **64**, 591.
5. S. G. Anema and Y. Li, *J. Agric. Food Chem.*, 2003, **51**, 1640.

6. M. Corredig and D. G. Dalgleish, *Food Res. Int.*, 1996, **29**, 49.
7. D. J. Oldfield, H. Singh, M. W. Taylor, and K. N. Pearce, *Int. Dairy J.*, 2000, **10**, 509.
8. A. J. Vasbinder and C. G. de Kruif, *Int. Dairy J.*, 2003, **13**, 669.
9. M. Kalab and D. B. Emmons, *Milchwissenschaft*, 1976, **31**, 402.
10. T. van Vliet and C. J. A. M. Keetels, *Neth. Milk Dairy J.*, 1995, **49**, 27.
11. S. P. F. M. Roefs, 'Structure of acid casein gels', Ph.D. Thesis, Wageningen Agricultural University, the Netherlands, 1986.
12. A. J. Vasbinder, H. S. Rollema, A. Bot, and C. G. de Kruif, *J. Dairy Sci.*, 2003, **86**, 1556.
13. H. Luyten, W. Kloek, and T. van Vliet, *Food Hydrocoll.*, 1994, **8**, 431.
14. H. J. M. van Dijk and P. Walstra, *Neth. Milk Dairy J.*, 1986, **40**, 3.
15. C. M. M. Lakemond and T. van Vliet, submitted for publication.
16. T. van Vliet and P. Walstra, *Neth. Milk Dairy J.*, 1985, **39**, 115.
17. T. van Vliet, in 'Food Emulsions and Foams: Interfaces, Interactions and Stability', eds E. Dickinson and J. M. Rodriguez Patino, Royal Society of Chemistry, Cambridge, 1999, p. 307.
18. L. G. B. Bremer, B. H. Bijsterbosch, R. Schrijvers, T. van Vliet, and P. Walstra, *Colloids Surf.*, 1990, **51**, 159.
19. M. Mellema, J. H. J. van Opheusden, and T. van Vliet, *J. Rheol.*, 2002, **46**, 11.
20. P. Walstra, 'Physical Chemistry of Foods', Marcel Dekker, New York, 2003.
21. B. Y. Kim and J. E. Kinsella, *J. Food Sci.*, 1989, **54**, 894.
22. J. A. Lucey, M. Tamehana, H. Singh, and P. A. Munro, *J. Dairy Res.*, 1998, **65**, 555.
23. J. A. Lucey, M. Tamehana, H. Singh, and P. A. Munro, *J. Dairy Res.*, 2000, **67**, 415.
24. M. H. Famelart, J. Tomazewski, M. Piot, and S. Pezennec, *Int. Dairy J.*, 2003, **13**, 123.
25. A. J. Vasbinder, P. J. J. M. van Mil, A. Bot, and C. G. de Kruif, *Colloids Surf. B*, 2001, **21**, 245.
26. A. J. Vasbinder, A. C. Alting, and C. G. de Kruif, *Colloids Surf. B*, 2003, **31**, 115.
27. J. F. Graveland-Bikker and S. G. Anema, *Int. Dairy J.*, 2003, **13**, 401.
28. K. Wüthrich, 'NMR of Proteins and Nucleic Acids', Wiley, New York, 1986.
29. K. Hashizume and T. Sato, *J. Dairy Sci.*, 1988, **71**, 1439.
30. S. J. Goddard, *J. Dairy Res.*, 1996, **63**, 639.
31. A. J. Vasbinder, A. C. Alting, R. W. Visschers, and C. G. de Kruif, *Int. Dairy J.*, 2003, **13**, 29.
32. J. M. S. Renkema, *Food Hydrocoll.*, 2004, **18**, 39.
33. T. van Vliet and P. Walstra, *Faraday Discuss.*, 1995, **101**, 359.
34. W. C. Young, 'Roark's Formulas for Stress and Strain', McGraw-Hill, New York, 1989.
35. J. A. Lucey, T. van Vliet, K. Grolle, T. Geurts, and P. Walstra, *Int. Dairy J.*, 1997, **7**, 389.
36. T. van Vliet, H. J. M. van Dijk, P. Zoon, and P. Walstra, *Colloid Polym. Sci.*, 1991, **269**, 620.
37. T. van Vliet, in 'Hydrocolloids—Part 1. Physical Chemistry and Industrial Application of Gels, Polysaccharides and Proteins', ed. K. Nishinari, Elsevier, Amsterdam, 2000, p. 367.

Linear and Non-linear Rheological Properties of β-Lactoglobulin Gels in Relation to their Microstructure

By Matthieu Pouzot, Lazhar Benyahia, Dominique Durand, and Taco Nicolai

POLYMÈRES, COLLOÏDES, INTERFACES, UMR CNRS, UNIVERSITÉ DU MAINE, 72085 LE MANS CEDEX 9, FRANCE

1 Introduction

Due to the importance of the applications in the food industry, the rheological properties of heat-set globular protein gels have been studied extensively in the past.[1-3] Nevertheless, we are still far from fully understanding the relation between the texture of these gels and their linear and non-linear rheological properties.

Two main classes of globular protein gels can be distinguished: (i) gels formed from linear and more-or-less rigid aggregates, and (ii) gels formed from densely cross-linked aggregates. The first type is generally formed in the presence of strong repulsive electrostatic interactions, *i.e.*, at pH values away from the isoelectric point pI and at low ionic strength. The second type is formed when electrostatic interactions are weak. In some cases, when the electrostatic interactions become very weak, the protein aggregates show a tendency to phase separate, and the gels become heterogeneous on the microscopic scale with inclusions of dense protein domains. Of course, these different types of gels exhibit different rheological behaviour.

In the present work, we restrict our investigation to the particular kind of turbid gel formed by heating β-lactoglobulin (β-lg) at pH 7 and in 0.1 M NaCl at protein concentrations above 15 g L^{-1}. This milk protein, the main component of whey, is probably the most widely studied globular food protein. It has a molecular radius of about 2 nm and a molar mass of 19 kg mol^{-1}.[4] Our main purpose here is to correlate the morphology of the β-lg gels to their linear and non-linear rheological properties.

2 Material and Methods

Materials

The β-lg used in this study was a gift from Lactalis (Laval, France). Solutions were extensively dialysed against salt-free Milli-Q water at pH = 7, with 200 ppm NaN_3 added to prevent bacterial growth. After dialysis the ionic strength was set to 0.1 M with NaCl. The protein concentration was measured after filtration by UV absorption at 278 nm assuming an extinction coefficient of 0.96 L g^{-1} cm^{-1}. Solutions contained in air-tight light-scattering cells were heated for 24 h in a thermostat bath at 80 °C and then rapidly cooled to 20 °C. For rheology measurements the samples were degassed and heated at 80 °C directly in the rheometer cell.

Light Scattering

Dynamic light-scattering measurements were carried out at 20 °C using a commercial version of the cross-correlation instrument.[5] The light source was a diode laser with wavelength 685 nm. Photon correlation was performed with a digital correlator (ALV-5000E, ALV). The relative excess scattering intensity (I_r) was taken as the intensity minus the solvent scattering divided by the scattering intensity of toluene at 20 °C. The values of I_r were corrected for multiple scattering, and were measured as a function of the scattering wave vector q for each sample in its light scattering cell (inner tube diameter 4 mm). Turbidity was determined at 685 nm using a spectrometer with cells of varying thickness.

After dilution of aggregates, the value of I_r is related to the weight-average molar mass M_w and the structure factor $S(q)$ of the solute:[6]

$$I_r = HCM_wS(q). \tag{1}$$

Here, C is the concentration and H is an optical constant. We have used $\partial n/\partial C = 0.189$ mL g^{-1} for the refractive index increment, and a toluene standard with Rayleigh constant of 1.02×10^{-5} cm^{-1}. The structure factor $S(q)$ may be written in terms of the z-average radius of gyration (R_{gz}) of the solute for $qR_{gz} < 1$:

$$S(q) = (1 + q^2R_{gz}^2/3)^{-1}. \tag{2}$$

At higher concentrations, M_w and R_{gz} should be replaced by an apparent molar mass M_a and apparent radius R_a.

Rheology

Linear elastic properties of the gels were evaluated by measuring the shear modulus as a function of heating time with an ARES strain-controlled rheometer. We used either a Couette geometry (cup diameter 34 mm, bob diameter 32 mm) or a plate–plate geometry (diameter 50 mm, gap 1 mm). The temperature was controlled with a Peltier system or circulating water from a thermostat tank. The samples were covered with silicone oil to avoid evaporation. The shear strain of 5% was shown to be in the linear regime. Non-linear elastic behaviour of the gels was studied by measuring the shear modulus as a function of strain up to fracture using an ARES strain-controlled rheometer with Couette geometry (cup diameter 34 mm, bob diameter 32 mm) or a cone-and-plate geometry (diameter 25 mm or 50 mm).

3 Results and Discussion

Gel Structure

Heat-induced denaturation of globular proteins leads to aggregation, and eventually a gel is formed above a certain characteristic concentration C_g.[1] The β-lg aggregates are stable to cooling and dilution and can be characterized in dilute solution at room temperature. Static light scattering (SLS) has proven to be a powerful technique to characterize the growth and the structure of the aggregates and gels.[7]

Figure 1 shows the structure factor of highly diluted β-lg aggregates in a solution of pH 7 and 0.1 M NaCl obtained after heating at 80 °C for 24 h. (The fraction of unaggregated protein is about 5% over the whole range of concentrations investigated). This master curve shows that the aggregates are self-similar with a fractal dimension $d_f = 2$. By analogy with the highly diluted system, we can determine an apparent molar mass M_a and an apparent radius R_a for the undiluted systems. After dividing $I_r(q)$ by M_a, and plotting the result as a function of qR_a, we obtain values of the structure factor that are the same as for the diluted aggregates. This is consistent with Figure 2, where the dependence of M_a on R_a is compared with the dependence of M_w on

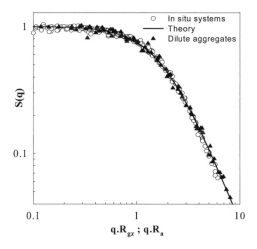

Figure 1 *Structure factor $S(q)$ of highly diluted β-lg aggregates (filled triangles) and undiluted systems (open circles) obtained after heating for 24 h at 80 °C. The solid curve represents equation (2) with $R = R_{gz}$ or R_a*

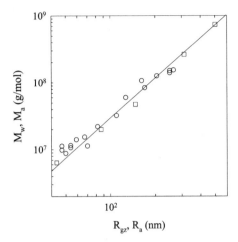

Figure 2 *Comparison of the dependence of M_a on R_a (○) with that of M_w on R_{gz} (□). The straight line represents the relationship $M_a = 3 \times 10^3 \, R_a^2$.*

R_{gz} for the diluted aggregates. This power relation with exponent 2 confirms independently the value of $d_f = 2$.

The concentration dependence of R_a and R_{gz} is shown in Figure 3. For the very dilute solutions, interactions between the aggregates are negligible, and so we have $R_a \approx R_{gz}$. With increasing concentration of protein, the aggregates become bigger and R_{gz} diverges at C_g. However, R_a remains finite due to the excluded volume interaction that increases with increasing concentration.

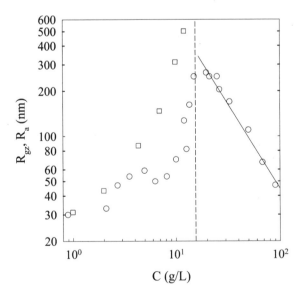

Figure 3 *Comparison of the concentration dependence of R_a (○) and R_{gz} (□). The dashed line indicates the gel concentration C_g and the solid line has a slope of −1.1.*

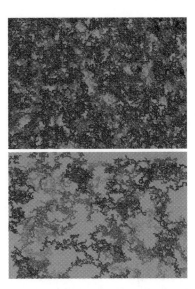

Figure 4 *Images from computer simulation of a diffusion-limited cluster aggregation process at two volume fractions of 2% (top) and 0.5% (bottom).*

Above $C \approx 20$ g L^{-1}, the interactions become predominant, and R_a decreases with C as a power law ($R_a \propto C^{-1.1}$). The system may be described as a collection of randomly close-packed blobs with radius R_a and apparent molar mass M_a, with $R_a \propto C^{1/(d_f-3)}$ and $M_a \propto R_a^{d_f}$. These relations are consistent with the experimental results for $C > 20$ g L^{-1} with $d_f = 2.0 \pm 0.1$. To illustrate this structure visually, we show in Figure 4 representations of gels obtained from off-lattice computer simulations of a diffusion-limited cluster aggregation process of spheres at two different volume fractions.[9] The gels have a self-similar fractal structure on intermediate length scales, and are homogeneous on length scales above the so-called correlation length.

Linear Elasticity of β-Lactoglobulin Gels

It was already shown elsewhere[10] for this system that, while the temperature influences strongly the kinetics, it does not influence the self-similar structure of the aggregates. Figure 5 shows the evolution of G' and G'' at different temperatures for $C = 50$ g L^{-1}. It is clear that the gelation rate increases strongly with increasing temperature. However, there is not a significant effect of temperature on the shape of $G'(t)$ and $G''(t)$. This is demonstrated in Figure 6 where we show sets of master curves obtained by time–temperature superposition at different protein concentrations. As expected, at a given temperature, both the rate of gelation and the long-time modulus increase with increasing concentration.[11]

The frequency dependence of the linear elasticity was determined by performing frequency sweeps during the gelation process. The values of G' and G'' at equal times were obtained by interpolation.

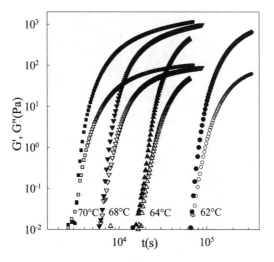

Figure 5 *Evolution of the storage and loss moduli at 0.1 Hz as a function of heating time t for β-lg solutions at different temperatures (C = 50 g L^{-1}).*

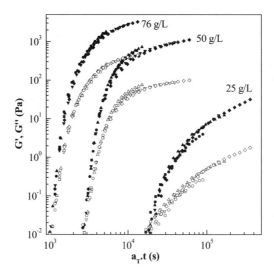

Figure 6 *Master curves of the evolution of G' and G'' at reference temperature 70 °C obtained by time–temperature superposition for different β-lg concentrations.*

Figure 7 shows the frequency dependence of G' at different heating times at 70 °C for $C = 25$ g L^{-1}. It is clear that, as soon as G' rises above the noise level (0.01 Pa), it has a very weak frequency dependence at low frequencies followed by a strong frequency dependence at high frequencies. With increasing heating time, G' increases and the transition point (F_c) between the weak and the strong frequency dependence shifts to higher frequencies. The fast relaxation process in the high frequency range is caused by the response on length scales

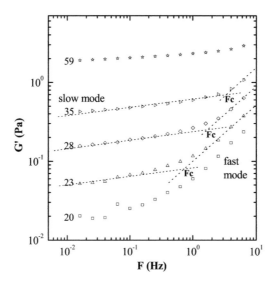

Figure 7 *Frequency dependence of the storage modulus of a β-lg solution at different heating times (10^3 s) at 70 °C (C=25 g L^{-1}). F_c is the transition frequency between fast and slow modes. Dashed lines represent the asymptotic frequency behaviour of the two modes.*

smaller than the mesh size of the gel. The longest timescale of the internal dynamics of the blobs (*i.e.*, $1/F_c$) is determined by the motion of the blobs over the characteristic distance R_a.[11] The value of G' continues to decrease weakly below F_c, which implies that the stress-bearing strands are not fully elastic.

As the storage shear modulus G' ($\gg G''$) is almost independent of frequency in the range 0.01–10 Hz, we can take G' at 0.1 Hz to be the elastic modulus G_0. The concentration dependence of G' after 24 h of heating is shown in Figure 8. After 24 h of heating at 80 °C, almost all of the native protein molecules have aggregated, so that the sol fraction is negligible (except close to C_g). For $C > 20$ g L^{-1} the concentration dependence can be approximated by the power-law $G' \propto C^{4.5}$. At lower concentrations G' decreases strongly and the sol–gel transition occurs at $C_g = 15$ g L^{-1}.[8] Light scattering has shown that at this concentration the aggregates are already strongly interpenetrated. The sol–gel transition may therefore be described as a bond percolation process of close-packed blobs of size R_a. Whereas light scattering is sensitive to the structure of the whole system, the elastic modulus is determined by the structure of the gel fraction. Percolation of blobs would explain the initial rapid increase of the modulus above t_g for a given concentration, and above C_g after prolonged heating.

For $C > 20$ g L^{-1}, where C has a power-law dependence, the sol fraction is negligible and all the blobs are connected; so the structural unit of the gel may be identified with R_a. Therefore, the results for $C > 20$ g L^{-1} at long heating times may be interpreted in terms of the so-called fractal colloidal gel model,[12,13] where the gel is modelled as a collection of interconnected,

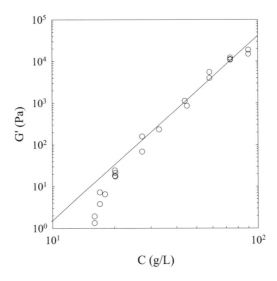

Figure 8 *Concentration dependence of the storage shear modulus G' at 0.1 Hz for β-lg solutions after heating for 24 h at 80 °C. The straight line has a slope of 4.5.*

space-filling, monodisperse fractal blobs. Each constitutive blob acts as a spring with elastic constant $K = K_0 N_b^{-1} R_a^{-2}$, where K_0 is the bending constant of each of N_b links constituting the stress-bearing backbone. Due to the assumed self-similarity of the network, we have $N_b \propto R_a^{d_b}$, where d_b is the fractal dimension of the elastic backbone, which lies between 1 and d_f. We showed previously[8] that the power-law concentration dependence of the shear modulus of the β-lg gels (see Figure 8) is consistent with this model. The combination of results obtained from light scattering and rheology provides strong evidence that the linear elasticity of heat-induced β-lg gels formed at pH 7 and in 0.1 M NaCl can be well understood within the framework of the colloidal fractal gel theory.

Non-linear Elasticity of β-Lactoglobulin Gels

Oscillatory measurements of the shear modulus at large strain γ indicate a strain-hardening effect followed by macroscopic failure of the gel (see Figure 9), as has been reported previously[14-17] for different globular protein gels. The effect of strain hardening decreases with increasing gel strength and disappears for gels with a shear modulus above 5 kPa. The results obtained at 0.1 Hz and 1 Hz are the same within the experimental error.

Gisler *et al.*[19] proposed a model that predicts the non-linear dynamic response of a network of interconnected fractal clusters with randomly oriented backbones. Assuming that strain hardening is caused by anisotropic unbending of the stress-bearing strands, they derived the following relation for the strain dependence of the shear modulus:

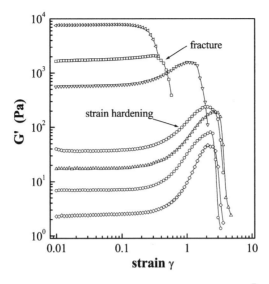

Figure 9 *Dependence of the storage shear modulus G′ on the shear strain for β-lg gels with different zero-strain moduli.*

$$G'(\gamma) = G'_0 \int_0^\pi \int_0^\pi \sin 3\theta \, \sin 2\varphi \left[\sum_{i=1}^n \prod_{j=0}^{i-1} \frac{(A/2-j)}{i!} \lambda^{2i} \right] d\theta \, d\varphi \qquad (4)$$

where

$$\lambda\,(\theta,\phi,\gamma) = (1 + \gamma \, \sin^2\theta \, \sin 2\phi + \gamma^2 \sin^2\theta \, \cos^2\phi)^{1/2} \qquad (5)$$

is the mean longitudinal extension ratio of the strands. The quantity A is defined by

$$A = (1 + d_b)/(d_b - 1), \qquad (6)$$

and G'_0 is the shear modulus of the gel in the linear regime. The model predicts that $G'(\gamma)/G'(0)$ is a universal function of the strain and of d_b.

Figure 10 shows that the strain hardening of β-lg gels at pH 7 in 0.1 M NaCl is well described by equation (4) with $d_b = 1.27$. The deviation from the model at higher strains is attributed to the onset of a microscopic yielding of the gel before macroscopic failure. Macroscopic yielding causes a strong increase in tan δ at higher strains.[18]

4 Conclusions

This study demonstrates the close relationship between the structural and rheological properties of heat-induced β-lg gels at pH = 7 and in 0.1M NaCl. The protein network is composed of an assembly of rather monodisperse

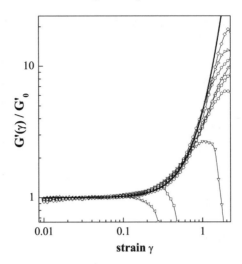

Figure 10 *Master curve of the strain hardening behaviour for the heat-induced β-lg gels shown in Figure 9. The solid line represents equation (4) with $d_b = 1.27$.*

fractal blobs and is homogeneous on larger length scales. The linear elastic behaviour of the gels is well described by the fractal colloidal gel model where the enthalpic elasticity of the network stems from the bending of the stress-bearing backbone of the blobs. Non-linear mechanical properties at large deformations reveal strain hardening followed by failure. The strain hardening is well explained by a model that allows for anisotropic un-bending of the stress-bearing strands under shear.

References

1. A. H. Clark, in 'Functional Properties of Food Macromolecules', 2nd edn, eds S. E. Hill, D. A. Ledward, and J. R. Mitchell, Aspen Publishers, Gaithersburg, ME, 1998, p. 77.
2. D. Renard, A. Axelos, and J. Lefebvre, in 'Food Macromolecules and Colloids', eds E. Dickinson and D. Lorient, Royal Society of Chemistry, Cambridge, 1995, p. 390.
3. M. Verheul and S. P. F. M. Roefs, *Food Hydrocoll.*, 1998, **12**, 17.
4. H. A. McKenzie, in 'Milk Proteins: Chemistry and Molecular Biology', ed. H. A. McKenzie, Academic Press, New York, 1971, vol. 2, p. 257.
5. C. Urban and P. Schurtenberger, *J. Colloid Interface Sci.*, 1998, **207**, 150.
6. W. Brown, in 'Light Scattering: Principles and Developments', Clarendon Press, Oxford, 1996.
7. D. Durand, J. C. Gimel, and T. Nicolai, *Physica A*, 2002, **304**, 253.
8. M. Pouzot, T. Nicolai, D. Durand, and L. Benyahia, *Macromolecules*, 2004, **37**, 614.
9. M. Rottereau, J.C. Gimel, T. Nicolai, and D. Durand, unpublished results.
10. C. Le Bon, T. Nicolai, and D. Durand, *Int. J. Food Sci. Technol.*, 1999, **34**, 451.
11. M. Pouzot, L. Benyahia, and T. Nicolai, *J. Rheol.*, 2004, **48**, 1123.

12. W. Shih, W. Y. Shih, S. Kim, J. Liu, and I. A. Aksay, *Phys. Rev. A*, 1990, **42**, 4772.
13. M. Mellema, J. H. J. van Opheusden, and T. van Vliet, *J. Rheol.*, 2002, **46**, 11.
14. T. Hagiwara, H. Kumagai, and K. Nakamura, *Food Hydrocoll.*, 1998, **12**, 29.
15. A. Bot, R. D. Groot, and W. G. M. Agterof, in 'Gums and Stabilisers for the Food Industry', eds G. O. Phillips, P. A. Williams, and D. J. Wedlock, IRL Press, Oxford, 1996, vol. 8, p. 117.
16. R. D. Groot, A. Bot, and W. G. M. Agterof, *J. Chem. Phys.*, 1996, **104**, 9202.
17. L. L. Lowe, E. A. Foegeding, and C. R. Daubert, *Food Hydrocoll.*, 2002, **17**, 512.
18. M. Pouzot, T. Nicolai, D. Durand, and L. Benyahia, unpublished results.
19. T. C. Gisler, R. C. Ball, and D. A. Weitz, *Phys. Rev. Lett.*, 1999, **82**, 1064.

Gelation of Bovine Serum Albumin in the Presence of Low-methoxyl Pectin: Effects of Na$^+$ and Ca^{2+} on Rheology and Microstructure

By L. Donato, C. Garnier, B. Novales, and J.-L. Doublier

UNITE DE PHYSICOCHIMIE DES MACROMOLECULES, INRA, RUE DE LA GERAUDIERE, BP 71642, 44316 NANTES CEDEX 03, FRANCE

1 Introduction

Globular proteins and polysaccharides are two kinds of gelling biopolymers used in the food industry for their wide range of textural properties.[1] When mixed together, a phase separation process often occurs because of the mechanisms of thermodynamic incompatibility and depletion flocculation.[2,3] If one or both of the biopolymers is capable of gelation, mixed gels can be formed and there is a kinetic competition between gelation and phase separation. The control and understanding of the relative rates of these processes can lead to a wide range of microstructures and a large variety of textures.[4]

The present study aims to evaluate the influence of solvent conditions (the presence of sodium and calcium ions) on the properties of a model system composed of a globular protein, bovine serum albumin (BSA), and an anionic polysaccharide, low-methoxyl pectin (LM pectin). BSA is one of the most widely studied proteins, and it is the most abundant protein in blood plasma contributing to the colloidal osmotic pressure. It has an isoelectric point of p$I \approx 5.2$.[5] BSA denatures on heating, and a gel can be formed if the medium conditions are favourable.[6] Pectins are complex polysaccharides that belong to the cell walls of plant materials. The main chain of the pectin molecule is a polygalacturonic acid partially esterified by methoxyl groups and interrupted by insertion of rhamnose residues carrying neutral sugars as side-chains.[7] LM pectin, with a degree of methoxylation below 50%, shows a high affinity for calcium ions leading to the formation of a thermoreversible gel.[8] Mixtures of BSA + LM pectin at 20 °C remain homogeneous and translucent even after centrifugation. These biopolymers are therefore considered as compatible.[9,10]

The influence of heat-induced protein gelation on anionic polysacharide + globular protein mixtures has been the subject of many studies.[3,11-17] However, fewer studies have been reported[13-18] on the heat-induced gelation of a globular protein in a mixture with a gelling polysaccharide. Results available so far show a variety of mechanical and structural properties depending on the nature of the biopolymers and on solvent conditions such as ionic strength, *e.g.*, as reported by Neiser *et al.*[16,17] on BSA + alginate systems. In order to understand better the influence of solution conditions on the mechanical and structural properties of this kind of mixed system, the objective of the present investigation was, first, to understand the influence of ionic strength (NaCl) on the heat-induced gelation of BSA + LM pectin at pH = 6.8. As a second stage, we consider the influence of the combination of NaCl and $CaCl_2$ on the structural and mechanical properties of the same mixed system. Ultrastructural information on the organization of the biopolymers was obtained by confocal laser scanning microscopy (CLSM), and viscoelastic properties of the mixed gels were studied by means of oscillatory shear measurements.

2 Materials and Methods

Materials

The LM pectin sample, kindly given by Degussa Food Ingredients (Baupte, France), had a degree of esterification of 28.1%. The pectin powder was purified by washing with acidic ethanol in order to eliminate ions in excess and to obtain the polysaccharide in an acidic form. Polysaccharide solutions were prepared by dissolving the LM pectin powder in deionized water. The BSA (purity 98–99 wt%) was purchased from ICN Biomedicals (Aurora, OH). The protein powder was defatted with *n*-pentane. Any insoluble matter was eliminated by centrifugation ($1.6 \times 10^4 g$ for 20 min). Protein solutions were prepared by adding the BSA powder to deionized water under gentle magnetic stirring overnight at 4 °C. Calcium contents determined by atomic absorption spectroscopy were 2.3 µmol g^{-1} for the defatted BSA powder and 34.1 µmol g^{-1} for the purified LM pectin powder. The pH of each solution was adjusted to pH = 6.8 with NaOH. Sodium azide (0.02 wt%) was added to the biopolymer solutions to prevent bacterial contamination.

Preparation of Biopolymer Mixtures

Protein and LM pectin solutions were mixed at room temperature. It was verified that the pH after mixing was still 6.8. Blends were then stirred at 50 °C before addition of a warm solution of NaCl and/or $CaCl_2$. The final concentrations in the mixture were 8 wt% BSA, 0.85 wt% LM pectin, 9–100 mM NaCl, and 3 mM $CaCl_2$. The intrinsic ionic strength of the biopolymers has to be taken into consideration in the calculation of total ionic strength. For LM pectin, the ionic strength is given by $I = \alpha\gamma C$,[19] where α is the fraction of ionized groups ($\alpha = 1$ at pH 6.8), γ is the monovalent counterion activity coefficient ($\gamma_{Na} = 0.507$), and C is the polymer concentration in equivalents per litre. The contribution to the ionic strength from the BSA (7.6×10^{-2} mM), calculated from the known ion content of the powder, was considered to be negligible as compared with the ionic strength contribution from the LM pectin (12.5 mM).

Dynamic Oscillatory Measurements

Time-sweep oscillatory measurements were performed at a frequency of 1 rad s⁻¹ for a strain amplitude of 1% using a controlled-strain rheometer (AR2000, TA Instruments) equipped with a Peltier temperature controller and a cone-and-plate geometry (40 mm diameter, 4° angle). Warm protein + LM pectin solutions were poured into the rheometer cell at 50 °C. The temperature was increased from 50 to 80 °C at 6 °C min⁻¹, kept at 80 °C for 30 min, and decreased from 80 to 20 °C at 10 °C min⁻¹. The temperature was then maintained for one hour at 20 °C, and a frequency sweep test was performed.

Confocal Microscopy

CLSM was used in the fluorescence mode. Observations were made with a Carl Zeiss LSM 410 Axiovert (Le Pecq, France) with a laser at a wavelength of 543 nm. As proteins do not exhibit intrinsic fluorescence at this wavelength, the BSA was stained by adding rhodamine B isothyocyanate (RITC) to the protein solution under magnetic stirring for 1 hour. Mixtures of RITC–BSA + LM pectin were prepared as described above, poured between a concave slide and a coverslip, and then hermetically sealed. The same heat treatment as for rheological measurements was applied using a thermostated stage (Linkam PE 60). It was verified that labelling did not change the rheological behaviour of the systems. For each system, 10 images were taken after one hour at 20 °C after the end of heat treatment. Each image was composed into 512×512 pixels with grey levels ranging from 0 (black) to 255 (white).

Texture Image Analysis

A method of texture image analysis, the grey level spatial interdependence method (also refered to as the co-occurrence method), was applied to the images taken by CLSM.[20,21] A textured area in an image is characterized by a non-uniform, or varying, spatial distribution of grey level intensity. This image analysis method is based on the observation of grey level values of pixels separated by a given distance in a given direction. The result is a two-dimensional histogram describing the probability that pairs of grey values occur in a given spatial relationship. For this study, only adjacent pixels were considered. The number of times that a pixel of grey level i was found adjacent to a pixel of grey level j was computed in each direction. The frequencies of occurrence were arranged into square tables called co-occurrence matrices of size equal to the number of grey level classes. From each co-occurrence matrix, 10 parameters were extracted.[22] Each parameter gives information on the structure of the system (homogeneity, complexity, presence of aggregates, pore sizes, *etc.*). The co-occurrence parameters were then analysed using principal component analysis (PCA). In the similarity maps, two points close to each other correspond to images of similar texture.[23]

3 Results and Discussion

System Without Added Salts

Variations of the storage modulus G' as a function of time for BSA alone and for the BSA + LM pectin mixture are shown in Figure 1a. For pure BSA, the modulus increased weakly during the thermal processing (final $G' < 1$ Pa); and, as shown in Figure 1b, a weak gel was formed. Addition of LM pectin resulted in a rapid rise in G' during the plateau at 80 °C, and a smoother increase during the cooling step. After one hour at 20 °C, the value of G' did not increase significantly and the system was stable, a strong gel having been formed (Figure 1b).

The corresponding microstructures obtained for BSA and BSA + LM pectin after the thermal processing are shown in Figures 2a and 2b. For BSA only, the fluorescence was regularly distributed in the medium. As weak gelation has been shown by rheology, it could be concluded that there is a protein network, but it cannot be detected at this scale of observation. For BSA + LM pectin mixtures, inhomogeneities in the distribution of fluorescence were observed: bright zones in Figure 2b correspond to regions enriched in protein, and dark zones correspond to regions devoid of BSA (and containing LM pectin). These observations confirm that the heat-induced BSA gelation becomes modified by the presence of LM pectin as a result of a phase separation process occurring during heating. This results in a modification of the structure of the protein network, and an enhanced gel strength due to the

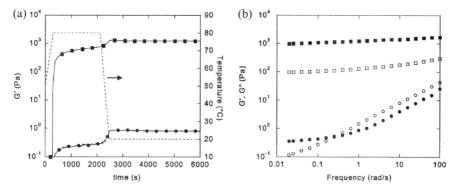

Figure 1 *Heat-induced gelation of 8 wt% BSA (●) and 8 wt% BSA + 0.85 wt% LM pectin (■) in water at pH = 6.8. (a) Change in storage modulus G′ during the thermal process. Dashed line corresponds to the temperature profile. (b) The corresponding frequency-sweep test: G′, solid symbols; G″, open symbols.*

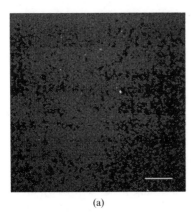
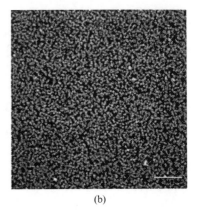

(a)   (b)

Figure 2 *CLSM micrographs of (a) 8 wt% BSA gel and (b) 8 wt% BSA+0.85 wt% LM pectin gel in water at pH=6.8, after one hour at 20 °C following the thermal treatment. Protein appears bright. Scale bar=25 µm.*

increase in the local protein concentration. The same effect has been shown[15] by Beaulieu *et al.* for 8 wt% whey protein + 1 wt% LM pectin at pH=6.

Effect of NaCl on Gelation of BSA

The kinetics of gel formation of BSA with different concentrations of added NaCl is shown in Figure 3a. The same evolution of G' as described for the BSA+LM pectin system in water was observed. Increasing the ionic strength reinforces the gel in comparison with the results obtained in water. A strong gel was formed as shown for instance by the mechanical spectra of BSA in 100 mM NaCl (Figure 3b). The corresponding CLSM observations (not shown)

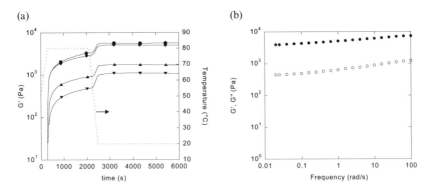

Figure 3 *Heat-induced gelation of 8 wt% BSA at pH=6.8 with added NaCl. (a) Change in storage modulus G' during the thermal process for various ionic strengths:* ▼, *12 mM;* ▲, *25 mM;* ◆, *50 mM;* ●, *100 mM. Dashed line corresponds to the temperature profile. (b) Frequency-sweep test in 100 mM NaCl:* ●, G'; *O,* G''.

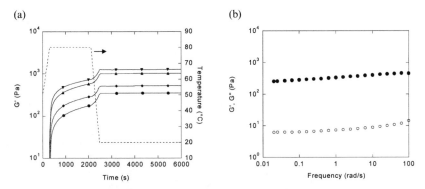

Figure 4 *Heat-induced gelation of 8 wt% BSA+0.85 wt% LM pectin at pH=6.8 with added NaCl. (a) Change in storage modulus G' during the thermal process for various ionic strengths:* ▼, *24.5 mM;* ▲, *37.5 mM;* ◆, *62.5 mM;* ●, *112.5 mM. Dashed line corresponds to the temperature profile. (b) Frequency-sweep test in 112.5 mM NaCl:* ●, *G';* O, *G".*

gave the same result as in water. The reinforcement of the gel can be related to the type and the strength of electrostatic interactions between the protein molecules. At pH=6.8, the BSA is negatively charged. Electrostatic repulsion between protein molecules is reduced by screening of charges brought about by the presence of ions in the vicinity of the surface of the protein molecules. A 'salting-in' process takes place.[24]

Effect of NaCl on Gelation of BSA + LM Pectin

Addition of LM pectin in the presence of different concentrations of NaCl (Figure 4a) does not change the overall shape of the gelation curves as compared with BSA alone (Figure 3a) and the mixture in water (Figure 1a). In contrast to pure BSA, however, increasing the concentration of NaCl in the system reduced the final gel strength. Overall, the G' values were all found to be lower than for the corresponding pure BSA gel at the same NaCl concentration. The corresponding microscopic observations for different ionic strengths showed that phase separation takes place, the final microstructure of the mixtures showing dense fluorescent particles concentrated in protein and dark zones containing LM pectin.

To evaluate how the increase in sodium ion concentration modifies the microstructure of the BSA+LM pectin gel, texture image analysis has been performed on pictures obtained for all ionic strengths. The similarity map obtained after texture image analysis of all images is presented in Figure 5. The first axis separates the samples into three groups (12.5 mM $\leq I \leq$ 42.5 mM, $I \geq$ 52.5 mM, and $I >$ 87.5 mM). For $I \leq$ 42.5 mM, the parameters describing the images are related to complex and less homogeneous systems with dense protein aggregates and small size of pores, in contrast to parameters describing systems for higher ionic strength conditions related to more homogenous

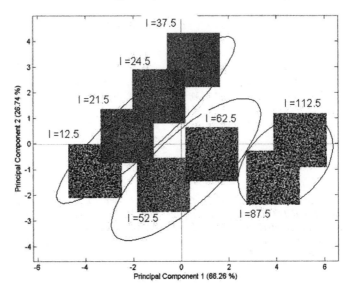

Figure 5 *Similarity map for 8 wt% BSA+0.85 wt% LM pectin gels made at pH=6.8 with different total ionic strength varying from 12.5 to 112.5 mM (in NaCl).*

images with large aggregates separated from dark zones corresponding to the LM pectin phase. The second axis does not provide any additional information. Axis 1 also shows the decrease in G' as the ionic strength increases. This result can be compared with a similar study on 8 wt% BSA + 1 wt% alginate at pH 6.8 by Neiser *et al.*,[16] where it was found that increasing ionic strength caused the gel strength to increase (up to ~20 mM NaCl) and then to decrease. They suggested[16] that this behaviour may be due to a balance between the long-range repulsion of net negatively charged BSA molecules and alginate chains and a short-range attraction between local positive sites on the BSA molecules and the alginate. Another explanation can be proposed. By assuming that the protein aggregates behave as spherical particles, phase separation could be the result of a depletion flocculation of the BSA aggregates, caused by the presence of the LM pectin chains in the medium.[25] Increasing ionic strength seems to enhance the phase separation process.

Effect of NaCl/CaCl₂ on Gelation of LM pectin

Changes in G' during thermal processing for LM pectin in the presence of 3 mM CaCl₂ for different NaCl concentrations are shown in Figure 6a. Whatever the ionic strength, no modulus was measured at 80 °C. The value of G' began to increase only during the cooling step at a temperature below 60 °C, and it continued to increase as a function of time at 20 °C. A gel is clearly formed (see Figure 6b). As already known,[26] increasing the ionic strength increases the gel strength due to the diminution of electrostatic repulsion between LM pectin chains.

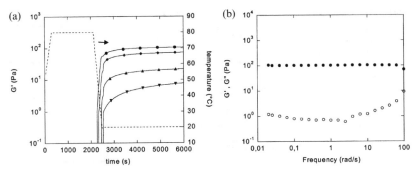

Figure 6 *Thermal processing of 0.85 wt% LM pectin at pH=6.8 in NaCl/CaCl₂ with CaCl₂=3 mM. (a) Change in storage modulus G' during the thermal process for various ionic strengths: ▼, 24.5 mM; ▲, 42.5 mM; ◆, 62.5 mM; ●, 112.5 mM. Dashed line corresponds to the temperature profile. (b) Frequency-sweep test for I=112.5 mM: ●, G'; O, G".*

Effect of NaCl/CaCl₂ on Gelation of BSA

Similar G' profiles to those in Figure 3a were obtained, and addition of calcium ions led to an increase in the gel strength with increasing ionic strength. The same conclusions as for the systems without calcium could be inferred. However, it also could be envisaged that there is an affinity between BSA and ionic calcium, as suggested by Powell Baker and Saroff,[27] or an increase in the reactivity of sulfhydryl groups and increased hydrophobic interactions as was reported for β-lactoglobulin.[28]

Effect of NaCl/CaCl₂ on Gelation of BSA + LM Pectin

Gel formation data of the BSA+LM pectin mixture for different ionic strengths in the presence of calcium are reported in Figure 7. The profiles for the mixture are similar to the one for pure BSA for ionic strengths ranging from 24.5 mM to 42.5 mM. For higher ionic strengths, different kinetics of gelation from that for BSA alone were observed. Increasing ionic strength yielded a slight increase of G' during the step at 80 °C. A sharp G' increase was found to occur during the cooling step below 60 °C, followed by a levelling out at 20 °C. This result suggests that protein aggregation at 80 °C is hindered by the presence of both LM pectin and ionic calcium for $I > 42.5$ mM. In addition, the final values of G' for BSA+LM pectin gels were much lower than for the BSA gels (result not shown) and higher than for the corresponding LM pectin gel (see Figure 6a), except at $I = 112$ mM, where the final gel strength was of the same order of magnitude as for the pure LM pectin gel. The mixed gel formed at $I = 24.5$ mM showed a microstructure of the same type as for BSA alone, whereas at $I = 112$ mM the structure was closer to that of the LM pectin gel.

As LM pectin gelation occurs on cooling, it can be supposed that it is the polysaccharide gelation that governs the gelation of the mixed systems when

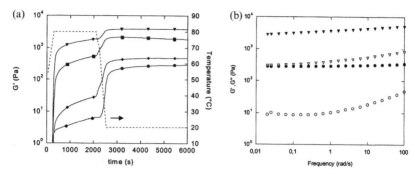

Figure 7 *Thermal processing of 8 wt% BSA+0.85 wt% LM pectin at pH=6.8 in NaCl/CaCl₂ with CaCl₂=3 mM. (a) Change in storage modulus G' during the thermal process for various ionic strengths: ▼, 24.5 mM; ■, 42.5 mM; ◆, 62.5 mM; ●, 112.5 mM. Dashed line corresponds to the temperature profile. (b) Frequency-sweep tests for I=24.5 mM (▼,▽)and I=112.5 mM (●,O): ▼,●, G'; ▽,O, G".*

the ionic strength is increased. The corresponding microscopic observations of BSA + LM pectin gels in the presence of calcium ions also show evidence of a phase separation process.

Texture image analysis has been performed on pictures obtained for the mixed system at different ionic strengths. The similarity map obtained is presented in Figure 8. The first axis describes the systems as a function of

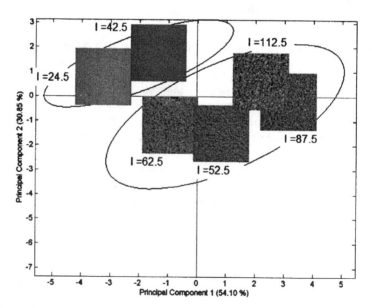

Figure 8 *Similarity map for 8 wt% BSA + 0.85 wt% LM pectin gels made at pH=6.8 with different total ionic strength varying from 24.5 to 112.5 mM (in NaCl/CaCl₂ with CaCl₂=3 mM).*

ionic strength, separating into two groups ($I \leq 42.5$ mM and $I \geq 52.5$ mM). For $I \leq 42.5$ mM, the parameters describing the images are related to complex systems with dense protein aggregates and small sizes of pores, in contrast to parameters describing systems at $I \geq 52.5$ mM, whose final microstructure shows protein aggregates separated from the LM pectin phase. Again the second axis does not provide any additional information. Increasing ionic strength results in an increase in the degree of phase separation, as for systems without calcium—but different microstructures and mechanical properties are observed.

From the combination of rheology and microscopy, the final gel microstructure can be interpreted as an interpenetrated structure.[29] For $I > 42.5$ mM, the protein aggregates form a very weak gel that is visible by microscopic observation, interpenetrated by the LM pectin network, as deduced from rheological results. For $I \leq 42.5$ mM, the protein aggregates form a strong gel network (visible by microscopic observation and identified rheologically) interpenetrated by the LM pectin weak gel.

4 Conclusions

The nature of the ions in the system plays a major role in determining the textural and structural properties during heat-induced gelation of BSA in the absence and presence of LM pectin. For the protein alone, increasing ionic strength increases the gel strength and addition of ionic calcium reinforces the gel. No variation in the microstructure could be found for the different ionic strengths studied. Addition of LM pectin to the protein was found to induce a phase separation process that is kinetically trapped by the protein gelation. This causes a weakening effect in the mixed gel as compared to the pure BSA gel due to a modification of the protein aggregation. Addition of Ca^{2+} to the mixture induces pectin gelation and the formation of a mixed gel. A balance between pectin gelation and protein gelation seems to govern the overall behaviour of the system depending on solvent conditions.

Texture image analysis allows more precise classification of all the different kinds of microstructures observed by CLSM, in cases where visual observation was not sufficient to distinguish the images from one to another. Hence, mixed gelled systems can yield a variety of structures and textures depending on the medium conditions. This could lead to interesting new properties in real food systems.

References

1. E. Dickinson and D. J. McClements, 'Advances in Food Colloids', Blackie, Glasgow, 1995, p. 81.
2. J.-L. Doublier, C. Garnier, D. Renard, and C. Sanchez, *Curr. Opin. Colloid Interface Sci.*, 2000, **5**, 1.
3. V. B. Tolstoguzov, *Food Hydrocoll.*, 1991, **4**, 429.
4. V. B. Tolstoguzov, *Food Hydrocoll.*, 2003, **17**, 1.

5. T. Peters, *Adv. Protein Chem.*, 1985, **37**, 161.
6. A. H. Clark and C. D. Lee-Tuffnell, 'Functional Properties of Food Macromolecules', J. R. Mitchell and D. A. Ledward, Elsevier Applied Science, London, 1986, p. 203.
7. M. C. Jarvis, *Plant Cell Environ.*, 1984, **7**, 153.
8. A. G. J. Voragen, J.-F. Thibault, W. Pilnik, M. A. V. Axelos, and C. M. G. C. Renard, 'Food Polysaccharides and their Applications', ed. A. M. Stephen, Marcel Dekker, New York, 1995, p. 287.
9. N. Takada and P. E. Nelson, *J. Food Sci.*, 1983, **48**, 1408.
10. M. G. Semenova, V. S. Bolotina, A. P. Dmitrochenko, A. L. Leontiev, V. I. Polyakov, and E. E. Braudo, *Carbohydrate Polym.*, 1991, **15**, 367.
11. S. Neiser, K. I. Draget, and O.Smidsrød, *Food Hydrocoll.*, 1998, **12**, 127.
12. S. Turgeon and M. Beaulieu, *Food Hydrocoll.*, 2001, **15**, 583.
13. E. E. Ndi, B. G. Swason, J. L. Smith, and D. W. Stanley, *J. Agric. Food Chem.*, 1996, **44**, 86.
14. V. M. Bernal, C. H. Smajda, J. L. Smith, and D. W. Stanley, *J. Food Sci.*, 1987, **5**, 1121.
15. M. Beaulieu, S. Turgeon, and J.-L. Doublier, *Int. Dairy J.*, 2001, **11**, 961.
16. S. Neiser, K. I. Draget, and O. Smidsrød, *Food Hydrocoll.*, 1998, **12**, 127.
17. S. Neiser, K. I. Draget, and O. Smidsrød, *Food Hydrocoll.*, 1999, **13**, 445.
18. L. Donato, C. Garnier, B. Novales, S. Durand, J.-L. Doublier, *Biomacromolecules*, in press.
19. M. Rinaudo and A. Domard, *C. R. Acad. Sci.*, 1973, **177**, 339.
20. G. Lohmann, *Comput. Graphics*, 1995, **19**, 29.
21. B. Novales, S. Guillaume, M. F. Devaux, and M. Chaurand, *J. Sci. Food Agric.*, 1998, **78**, 187.
22. R. Haralick, *IEEE Trans. Syst. Man. Cybernet.*, 1979, **3**, 610.
23. I. T. Jolliffe, 'Principal Component Analysis', Springer, New York, 1986.
24. P. Relkin, *CRC Crit. Rev. Food Sci. Nutr.*, 1996, **36**, 565.
25. R. Tuinier, J. K. G. Dhont, and C. G. de Kruif, *Langmuir*, 2000, **16**, 1497.
26. C. Garnier, M. A. V. Axelos, and J.-F. Thibault, *Carbohydr. Res.*, 1993, **240**, 219.
27. H. P. Powell Baker and H. A. Saroff, *Biochemistry*, 1965, **4**, 1670.
28 S. Jeyarajah and J. C. Allen, *J. Agric. Food Chem.*, 1994, **42**, 80.
29. V. B. Tolstoguzov, *Food Hydrocoll.*, 1995, **9**, 317.

Colloidal Interactions

Stabilization of Food Colloids by Polymers

By C. G. de Kruif[1,2] and R. Tuinier[3]

[1]NIZO FOOD RESEARCH, KERNHEMSWEG 2, 6718 ZB EDE,
THE NETHERLANDS
[2]UNIVERSITEIT UTRECHT, VAN'T HOFF LABORATORIUM,
PADUALAAN 8, 3584 CH UTRECHT, THE NETHERLANDS
[3]FORSCHUNGSZENTRUM JÜLICH, INSTITUT FÜR
FESTKÖRPERFORSCHUNG, 52425 JÜLICH, GERMANY

1 Introduction

In a famous lecture series on colloids delivered at the Massachusetts Institute of Technology,[1] Overbeek discussed the properties of lyophilic and lyophobic colloids. Lyophobic or 'solvent-fearing' colloids are thermodynamically unstable, but are kinetically stable because of a high energy barrier that prevents flocculation in an energy minimum. The classical example is the so-called DLVO colloid or charge-stabilized colloid, where the kinetic stability derives from a strong electrostatic repulsion.[2] Destroying the repulsion, by adding salt, makes the particles flocculate. Lyophilic or 'solvent-loving' colloids are thermodynamically stable. Unlike lyophobic colloids, they form or redisperse spontaneously. Examples are *micro*emulsions, soap micelles and many colloids stabilized with a steric (or polymeric) layer on the surface. In food products we would call casein micelles from milk a typical example of a lyophilic colloid. The solvency of the colloid derives from a free energy gain in 'dissolving' the steric stabilizing polymer in the solvent. Thus, the continuous phase must be a good solvent for the polymer chains. A clear example here is the case of silica particles (radius 10–1000 nm), which can be stabilized with octadecyl chains (C18) grafted to their surface.[3] In a good solvent such as cyclohexane, these particles redisperse spontaneously—after drying, for instance. In a marginal or poor solvent, the polymer layer or brush collapses and steric stabilization is lost. The fundamental principles of steric stabilization are now well understood, and described in the classic book by Napper.[4]

Small proteins are lyophilic as well, because of the entropy gain on dispersion, although at too high or low salt concentrations they may lose their solvency. The group of polymers of the polysaccharide category are often collectively named hydrocolloids. They are colloids as well, because their size is in the colloidal range. Almost without exception food hydrocolloids are polysaccharides. They are called *hydro*colloids because they are water-loving polymers and they disperse spontaneously in water. In that sense the casein proteins could be called hydrocolloids as well—but that is usually not done.

The most important colloids in foods are the *macro*emulsions. They are not stable thermodynamically and coalescence and Ostwald ripening will lead, eventually, to bulk water and oil phases. Nevertheless, for all practical purposes, most emulsions are as stable as they can be. Usually the kinetic stability of emulsions derives from a combination of both electrostatic and polymeric repulsion between the emulsion droplets.

Virtually all food products can be classified as food colloids. This means that food consists of particles, proteins, polysaccharides, all finely dispersed in a continuous phase. If we would know the number, size and interaction potential for all the particles, we would be able to describe the product properties in great detail. For selected systems such descriptions are indeed available. *e.g.*, pure proteins and polysaccharides. A major challenge to the food scientist is to control physical product stability; this typically means no flocculation, no syneresis, no creaming, and no phase separation. Biological, chemical and enzymatic stability must be achieved as well, but such aspects are not considered here.

The physical stability of food colloids can be understood from their inter-particle interactions. In order to illustrate this theme, we will discuss the stability of casein micelles as found in milk. We will show that their stabilization is not due to electrostatic repulsion, as has often been suggested, but due to steric stabilization. That is, casein micelles are stabilized by a brush of casein polymers on the surface. Admittedly, there is a small contribution from electrostatic repulsion, but it is only of secondary importance. It must be, of course, because milk has a relatively high ionic strength of 80 mM, and adding more salt does not destabilize it either. There is also the special case of polymeric (de)stabilization, namely the situation arising when different polymers and colloids are mixed. For instance, mixing gelatin and dextran leads to a clear phase separation with a variety of phase morphologies.[5] Adding a polymeric component to a particle system may lead to so-called depletion-induced phase separation and flocculation, possibly followed by a restabilization at high polymer concentrations. In this paper we will discuss the various contributions to the interaction potential: van der Waals attraction, electrostatic repulsion, steric or polymeric repulsion/attraction, and the excluded volume interaction. For this we will use the casein micelle as a model colloid to illustrate the phenomena and to calculate the various contributions. In doing so, it is not implied that casein micelles are a perfect model system for which theory can be proven correct or incorrect. Rather, the aim is to use accepted theoretical models and theories to obtain a better understanding of practical systems. Using these concepts it is easier to understand the factors affecting food product properties and (in)stability.

Figure 1 *Cartoon of a casein micelle.*

Casein micelles are present in milk and are real association colloids. A casein micelle (CM) can be dissociated (and reformed) by, for instance, applying high pressure or in a warm ethanol solution. The CM structure has long been a subject of study and debate—see the reviews of Holt,[6] Rollema,[7] Holt and Horne,[8] Horne,[9] Walstra,[10] and, most recently, de Kruif and Holt.[11] Figure 1 shows a schematic representation. Basically there is agreement on the composition and main features of the CM. So, there is consensus that the CM contains small calcium phosphate clusters (size 4–5 nm) and that it is sterically stabilized by κ-casein. There is, however, considerable dispute on the mechanisms involved in the formation of the CM and on its detailed structure (as discussed by Horne[9] and Walstra[10]). Here we will not dwell on such details, but consider casein micelles simply as colloidal particles that are roughly monodisperse in size. In the description used here we neglect the fact that the integrity of the CM may change in time. It is assumed that the phenomena described herewith are occurring on a timescale much shorter than the time for rearrangement of the CM when environmental conditions are changed. This point is emphasized here, since critics may say that one cannot apply colloid theory to this type of system because it is changing in time. However, it is not the aim of this paper to validate theory, but rather to use theory to classify and interpret colloidal phenomena.

The relevance of understanding the behaviour of casein micelles for food science can hardly be underestimated. Much of the world food production relies on caseins, maybe more so than for any other protein. Nevertheless, research on caseins is declining strongly on a relative scale with respect to other proteins. The casein micelle is the irreplaceable basis of the dairy industry and therefore it deserves the keen interest of food colloid scientists. The world production of milk is estimated at 560×10^9 kg year^{-1}. Only 2% of that production is from the 1.3 million cows in the Netherlands. Every cow in the Netherlands produces 4 drops of milk per second. A litre of milk contains on average 2.7% w/w of casein micelles. The CM radius is 100 nm and the colloid is rather monodisperse (in contrast to what is said in the literature). Skim milk contains about 12 vol% of casein micelles. Thus, in numbers per year, the world CM production is ~2×10^{27}.

2 Stability of Casein Micelles

From a colloid physics point of view, casein micelles can be regarded as colloidal particles stabilized by a coat or brush of κ-casein (and probably some β-casein) on the outside. From this point of view they resemble latex particles, and the sort of experiments done by physicists and colloid chemists on latex particles can also be done on casein micelles. Recently, we considered[12] the colloidal stability of casein micelles using a DLVO-like approach, *i.e.*, summing the various contributions to the interaction potential between a pair of casein micelles. We distinguish separate steric, electrostatic and van der Waals contributions as indicated below.

Steric Interactions

We consider the κ-casein hairs on the outside as a polyelectrolyte brush whose interaction with a similar brush must be calculated. For the interaction potential between hard plates at a distance h carrying brushes, we use the theory of Alexander and de Gennes,[13,14] which leads to the following expression for the *force* between two parallel flat plates:

$$\frac{K_{brush}(h)}{kT} = -\begin{cases} \sigma^{3/2}\left[\left(\dfrac{2H}{h}\right)^{9/4} - \left(\dfrac{h}{2H}\right)^{3/4}\right] & \text{for } h < 2H \\ 0 & \text{for } h > 2H \end{cases} \tag{1}$$

The first term in the square brackets in equation (1) represents the increase in osmotic pressure (leading to repulsion) due to the increase of the brush concentration between the plates, and the second term represents the change in elastic free energy of the brush upon compression. The parameter H is the brush height, h is the distance between the plates, and σ is the grafting density.

By applying the Derjaguin approximation to the potential between two flat surfaces, the following expression for the potential between two spheres of radius a is obtained:[15,16]

$$\frac{W_{brush}(h)}{kT} = \begin{cases} \infty & \text{for } h < 0 \\ \dfrac{16\pi aH2\sigma^{\frac{3}{2}}}{35}\left[\left[28\left(\dfrac{2H}{h}\right)^{1/4} - 1\right] + \dfrac{20}{11}\left(1 - \left(\dfrac{h}{2H}\right)^{\frac{11}{4}}\right)\right] & \\ \qquad + 12\left(\dfrac{h}{2H} - 1\right)\Bigg] & \text{for } 0 < h < 2H \\ 0 & \text{for } h > 2H \end{cases} \tag{2}$$

Equation (2) is supposed to be valid for $a \gg H$. In order to calculate $W_{brush}(h)$, we need to calculate the brush height H as a function of pH and chain density σ. As discussed by de Kruif and Zhulina,[17] κ-casein molecules can be described as charged brushes in the 'salted brush' regime. In this regime the ionic strength is such that it penetrates the brush and screens the electrostatic interactions between the charged polyacid groups (Debye length $\kappa^{-1} \ll H$). The polyacid brush is then quasi-neutral; that is, its characteristics are identical to those of neutral brushes. Consequently, in the strong-stretching approximation, the brush height is[18]

$$H = Nb \left(\frac{8 v^{eff} \theta}{\pi^2} \right)^{\frac{1}{3}},$$ (3)

where N is the number of segments of the brush, each of length b. For the case considered here, we have $N = 15$ (coinciding with the number of charged groups) and $b = 0.6$ nm. The quantity θ is the grafting density—the fraction of 'sites' occupied by the brushes at the surface. The parameter σ (in m^{-2}) is the grafting density θ divided by the surface area occupied by a brush. It is calculated from the known composition of milk and casein micelles and their size distribution. The molar mass of κ-casein is 19032 g mol^{-1}, and so the brush density corresponds to approximately 7500 chains per casein micelle, and the average distance between the chains is 4 nm. The area occupied by one chain at the surface is at least the squared length of an amino acid group in the κ-casein brush, where the length equals *ca.* 0.32 nm. We therefore estimate σ as 0.006. The above data give a brush height of 7.0 nm, which is consistent with dynamic light-scattering experiments.[19] The effective excluded volume is denoted by v^{eff}, and is defined as[17]

$$v^{eff} = v + \frac{\alpha^2}{\phi_s}$$ (4)

where v is the (dimensionless) excluded volume per segment (normalized with respect to the segment volume) and ϕ_s ($\cong 0.01$ in milk) is the salt volume fraction. The parameter v is given by $v = 1 - 2\chi$, where χ is the Flory–Huggins interaction parameter, taken as $\chi = 0.5$ (a theta solvent).

We take a very simple model for the dissociation of the polyacid brush in which we assume that the various dissociating groups do not affect one another. This leads to the well-known relation between the pH and the degree of dissociation α,

$$\alpha = \frac{K_a}{K_a + \left[H^+ \right]}$$ (5)

where $[H^+]$ is the proton concentration ($\equiv 10^{-pH}$) and K_a the mean value of the dissociation constant of the carboxylic groups along the chain. We assume

$pK_a = 4.9$, a value given by Swaisgood.[20] The above treatment offers us a simplified model for the brush height as a function of pH. It should be realized that equation (3) loses its applicability at low pH where there is no longer a strong-stretching regime.

Electrostatic Repulsion

Casein micelles have a clearly measurable electrophoretic mobility, which becomes very small, however, near pH = 5. Under physiological conditions (pH = 6.7), a zeta potential of 8 mV has been reported by Schmidt.[21] The expression for the interaction potential between two spheres of radius a due to electrostatic repulsion, $W_{er}(h)$, can be derived from the linear approximation to the Boltzmann charge distribution around a pair of spheres:[22]

$$\frac{W_{er}(h)}{kT} = 2\pi\varepsilon_0\varepsilon_r\psi^2\ln(1+\exp(-\kappa h)) \tag{6}$$

Here, Ψ is the surface potential (taken to be equal to 20 mV at pH = 7 and to fall linearly to zero at pH = 4.9), ε_0 is the permittivity in vacuum, ε_r is the dielectric constant (taken as 80), and $a = 100$ nm. The Debye length κ^{-1} is found from:

$$\kappa^{-1} = \sqrt{\frac{\varepsilon_r\varepsilon_0 RT}{F^2 I}} \rightarrow \kappa \cong 3\sqrt{I}\,(\text{nm}^{-1}) \tag{7}$$

The ionic strength I is about 0.08 M in milk, giving a Debye length of the order of 1 nm. The electrostatic repulsion between casein micelles in milk is therefore short-ranged.

Van der Waals Attraction

The the van der Waals attraction potential, $W_{vdw}(h)$, between two spheres of radius a separated by a distance h is given by[23]

$$\frac{W_{Vdw}(h)}{kT} = -\frac{A}{6}\left(\frac{2a^2}{(h+2a)^2-4a^2}+\frac{2a^2}{(h+2a)^2}+\ln\left\{\frac{(h+2a)^2-4a^2}{(h+2a)^2}\right\}\right) \tag{8}$$

where A is the Hamaker constant. We may use values of the Hamaker constant quoted for proteins. Values of 5 and $7\,kT$ are reported[24,25] for β-lactoglobulin and lysozyme, respectively. The CM protein density is about 6 times less than that of a globular protein, and therefore we estimate A as $1.0\,kT$.

3 Interaction Potential between Two Casein Micelles

We are now in a position to calculate the total interaction potential between two casein micelles by adding the various contributions:

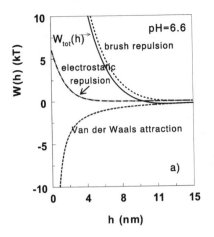
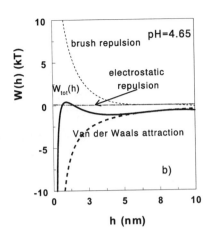

Figure 2 *Total interaction potential and its various contributions as a function of the distance between two casein micelles at (a) pH=6.6 (fresh skim milk) and pH=4.65 (acidified skim milk).*

$$\frac{W_{tot}(h)}{kT} = \frac{W_{vdW}(h)}{kT} + \frac{W_{er}(h)}{kT} + \frac{W_{brush}(h)}{kT}. \tag{9}$$

In Figure 2 we can see the contributions of the various terms to $W_{tot}(h)/kT$ at two different pH values. Clearly, the contribution of the electrostatic repulsion is small, if not negligible, which is consistent with the insensitivity of the system to salt. Changing the ionic strength (0.08 M in milk) by a factor of 10 up or down hardly influences the calculated potential.

From statistical mechanics,[26] we can calculate the second osmotic virial coefficient B_2 from $W_{tot}(r)$ using

$$B_2 = \frac{2\pi}{V_c} \int_0^\infty r^2 \left[1 - \exp\left(-\frac{W_{tot}(r)}{kT} \right) \right] dr \tag{10}$$

where $r = h + 2a$, and V_c is the volume of the casein micelle. The second osmotic virial coefficient is a parameter that monitors the stability of a colloidal dispersion.

Attractive interactions in colloid and interface science can be simplified considerably by modelling the interaction in terms of a square-well potential. This potential has the main features of $W_{tot}(h)/kT$ for casein micelles in that its attractive well is very short-ranged (a few nm) as compared to the particle size (100 nm). The square-well potential is more realistic in that its depth is finite, at incomplete renneting, and at pH values slightly above clotting. Thus we observe a *reversible* attractive interaction rather than a probability to get stuck in the primary minimum, as is often discussed. The square-well potential is defined in equation (11) and depicted schematically in Figure 3.

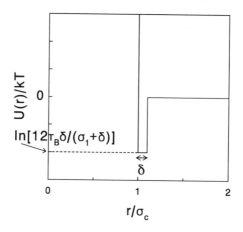

Figure 3 *Schematic representation of the square-well potential of depth* $-\varepsilon$ *and width* δ.

$$\frac{V(r)}{kT} = \begin{cases} \infty & \text{for } r < 2a \\ -\varepsilon & \text{for } 2a < r < 2a + \delta \\ 0 & \text{for } h > 2a + \delta \end{cases} \tag{11}$$

The second osmotic virial coefficient for this potential is

$$B_2 = 4 - \frac{\delta}{2a}[\exp(\varepsilon) - 1]. \tag{12}$$

Even simpler is the adhesive hard-sphere theory of Baxter,[27] which can be used to describe both equilibrium and transport properties of colloidal dispersions. In Baxter's approach the well-width is negligibly small ($\delta \to 0$). The effective attraction in this model is expressed by the Baxter parameter τ_B, and the second osmotic virial coefficient becomes:

$$B_2 = 4 - \frac{1}{\tau_B}. \tag{13}$$

Since the square-well potential is easily mapped on the Baxter potential, we have also used this potential by relating the well-depth to an observable parameter directly. For the relative viscosity Cichocki and Felderhof derived[29] the following equation for adhesive hard spheres:

$$\eta_r = 1 + \frac{5}{2}\phi + \left(5.9 + \frac{1.9}{\tau_B}\right)\phi^2 \tag{14}$$

where ϕ is the volume fraction of the spheres.

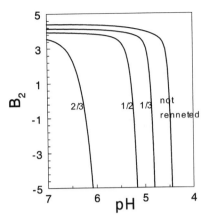

Figure 4 *Calculated second virial coefficient B_2 as a function of pH for various degrees of CM renneting, i.e., brush density. The volume fraction of micelles is 0.11.*

The second virial coefficient can be measured, and equations (10) and (12) can than be used to calculate the strength of the interaction (see Figure 4). Values of the second virial coefficient can be obtained from small-angle scattering, but it is even more convenient to obtain from measuring turbidity. Plotting ln(turbidity/concentration) against volume fraction gives a curve whose slope contains τ_B.[29]

4 Cheese-Making

An enzyme preparation called rennet (extracted from the stomach of calves/cows) is added to the milk during cheese-making. The enzyme chymosin present in the rennet cleaves off very specifically the κ-casein brush on the surface of the casein micelles. Actually, chymosin cuts off 64 amino acids from the C-terminal end. Pictorially speaking, we can say that the κ-casein fur is cut off. We do not use the word 'shave' here because the enzymatic reaction occurs at random. As the reaction obeys simple first-order reaction kinetics, the brush density as a function of time is given by

$$\frac{P_\infty - P(t)}{P_\infty} = e^{-\frac{k[E]t}{[E_0]}} = \frac{\sigma}{\sigma(t)} \qquad (15)$$

where k is a rate constant $(8 \times 10^{-4}\,\mathrm{s}^{-1})$ and $[E]=[E_0]$ corresponds to an enzyme concentration used as standard addition in practice.

The well-depth is simply given by[30]

$$\varepsilon/kT = \gamma kt[E]/[E_0] \qquad (16)$$

where γ is a proportionality constant $(=1.63)$. Thus, in this model the interaction potential and consequently the static and dynamic properties are simple functions of enzyme kinetics in a given experiment. We checked this and found out that turbidity, viscosity and diffusivity can be described with a single set of

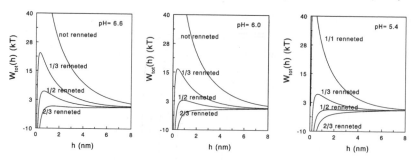

Figure 5 *Total interaction potential $W_{tot}(h)$ between two casein micelles at pH 6.6, 6.0 and 5.4 for various degrees of renneting.*

parameters using the adhesive hard-sphere model. Figure 5 shows calculations of the interaction potential at three pH values as a function of degree of renneting.

In Figure 6 we give an example of the viscosity as a function of the amount of κ-casein split off from the surface of the micelles. The dashed line in Figure 6 represents the behaviour obtained using $W_{tot}(h)$ at pH=6.7. The result is qualitatively correct. According to these calculations, the casein micelles lose their stability if 60% of the steric layer of κ-casein is cut off. In practice the value is as high as 90%.

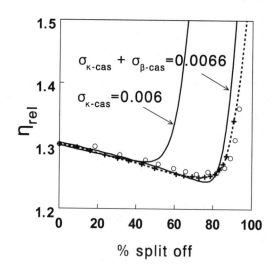

Figure 6 *Viscosity of CM dispersion as a function of the percentage of κ-casein cleaved off during renneting. The experimental data (2 independent measurements) are given as the data points from refs. 30 and 31. One solid curve is calculated using equations (9), (10), (13) and (14), and the other solid curve represents the influence of adding 10% β-casein to the brush which is not renneted.*

The discrepancy may arise from several effects not accounted for in the calculations. The two most important are probably the following. First, the outside layer will also contain some β-casein. It is known that β-casein leaves the micelle on lowering the temperature to near 0 °C. This extra polymer is not renneted, but it does contribute to the steric stabilization. We therefore artificially added 10% of extra β-casein to the brush. At the same time we can adjust the brush height so as to allow the viscosity to decrease, as was observed experimentally. Secondly, it is known that the strength of steric stabilization at low coverage may be underestimated by the theory, leading to lower stability than found experimentally after partial renneting.

5 Yoghurt-Making

In yoghurt-making, the pH of the milk is gradually lowered by the activity of lactic acid bacteria, which convert the milk sugar (lactose) into lactic acid. Depending on the conditions and the type of bacteria used, the acidity may reach values as low as pH = 4.

The experimental results in Figure 7 were obtained by partially renneting the casein micelles. The partial renneting lowers the pH stability as was shown by de Kruif and Zhulina.[17] The results in Figure 7 were fitted to the adhesive hard-sphere model by setting the well depth equal to 1/(pH − pCrit) where pCrit is a critical value of the pH where the brush collapses.[31] Figure 7 is consistent with the results in Figure 6 if we assume that the increase in apparent particle radius is comparable to the viscosity increase.

6 Conclusions

Casein micelles in milk are sterically stabilized association colloids. The stabilization derives from a short brush of κ-casein 'hairs' on the surface of the

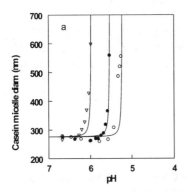
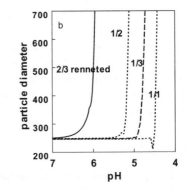

Figure 7 *Stokes diameter of casein micelles as a function of pH for various degrees of renneting: (a) experimental, (b) calculated.*

Stabilization of Food Colloids by Polymers

Table 1 *List of symbols and parameter values used in the calculations.*

Parameter	Value	Symbol [dimension]
Chain density	0.006	σ [m^{-2}]
Volume fraction	0.11	ϕ
Volume fraction of salt	0.01	ϕ_s
Flory–Huggins parameter	0.5	χ
Number of charged groups	15	N_g
Kuhn length of segment	0.6	LK [nm]
Radius of colloid	100	R, a [nm]
Radius of gyration	$(5/3R)^{-1/2}$	R_g [nm]
Distance between surfaces		h [nm]
Height of brush		H [nm]
Debye–Hückel length	1	κ [nm^{-1}]
Acidity	4–74	pH
– Log dissociation constant	4.9	pK_a
Surface potential	0.020	ξ [V]
Hamaker constant	1 kT	A [J]
Boltzmann constant	1.38×10^{-23}	k [JK^{-2}]
Temperature	298	T [K]
Dielectric constant of water	$80 \times 8.854 \times 10^{-12}$	ε [Fm^{-2}]
Elementary charge	7.056×10^{-10}	e [C]
Refractive index	1.333	n
Wavelength of light	650×10^{-9}	λ_0 [m]
Second virial coefficient		B_2 [m^3]
Reduced second virial coef.		$B_2/(4/3\pi R^3)$
Baxter parameter	$B_2 = 4 - 1/\tau$	τ

casein micelles. The physical behaviour of casein micelles in fresh skimmed milk can be described by the hard-sphere model. During the production of dairy products the stabilizing brush is changed in either of two ways.

In cheese-making the brush is cut off by an enzymatic reaction. As a result the casein micelles lose their steric coat and an attractive interaction between the micelles eventually leads to flocculation and gelation (the cheese curd). The behaviour of the micelles can be understood quantitatively using the adhesive hard-sphere model with parameter values given in Table 1.

In yoghurt-making, lowering the pH diminishes the brush stability. The brush hairs can be considered as polyelectrolyte chains. Their solvency depends on the charge density along the chain. On lowering the pH, the carboxylic acid groups become less dissociated and the κ-casein hairs lose their stability. As a result the brush collapses, and the casein micelles flocculate at a lowered pH.

References

1. J. Th. G. Overbeek, 'Colloid and Surface Chemistry: Lecture Series', Massachusetts Institute of Technology, Cambridge, MA, 1971.
2. J. Th. G. Overbeek, *Adv. Colloid Interface Sci.*, 1982, **16**, 17.
3. J. W. Jansen, C. G. de Kruif, and A. Vrij, *J. Colloid Interface Sci.*, 1986, **114** , 471.

4. D. H. Napper, 'Polymeric Stabilization of Colloidal Dispersions', Academic Press, London, 1983.
5. R. H. Tromp, A. R. Rennie, and R. A. L. Jones, *Macromolecules*, 1995, **28**, 4129.
6. C. Holt, *Adv. Protein Chem.*, 1992, **43**, 63.
7. H. S. Rollema, in 'Advanced Dairy Chemistry', ed. P. F. Fox, Elsevier Applied Science, Barking, Essex, 1992, vol. 1, p. 111.
8. C. Holt and D. S. Horne, *Neth. Milk Dairy J.*, 1996, **50**, 1.
9. D. S. Horne, *Int. Dairy J.*, 1998, **8**, 171.
10. P. Walstra, *Int. Dairy J.*, 1999, **9**, 189.
11. C. G. de Kruif and C. Holt, in 'Advanced Dairy Chemistry—1. Proteins', 3rd edn, eds P. F. Fox and P. L. H. McSweeney, Kluwer/Plenum, New York, 2003, part A, p. 223.
12. R. Tuinier and C. G. de Kruif, *J. Chem. Phys.*, 2002, **117**, 1290.
13. S. Alexander, *J. Phys. (France)*, 1977, **38**, 983.
14. P. G. de Gennes, *Adv. Colloid Interface Sci.*, 1987, **27**, 189.
15. G. J. C. Braithwaite, P. F. Luckham, and A. M. Howe, *J. Colloid Interface Sci.*, 1999, **213**, 525.
16. C. N. Likos, K. A. Vaynberg, H. Löwen, and N. J. Wagner, *Langmuir*, 2000, **16**, 4100.
17. C. G. de Kruif and E. B. Zhulina, *Colloids Surf. A*, 1996, **117**, 151.
18. R. Israels, F. A. M. Leermakers, and G. J. Fleer, *Macromolecules*, 1994, **27**, 3087.
19. C. G. de Kruif, *Langmuir*, 1992, **8**, 2932.
20. H. E. Swaisgood, in 'Advances in Dairy Chemistry—1. Proteins', ed. P.F. Fox, Elsevier Applied Science, London, 1992, p. 63.
21. D. G. Schmidt and J. K. Poll, *Neth. Milk Dairy J.*, 1986, **40**, 269.
22. R. Hogg, T. W. Healy, and D. W. Fuerstenau, *Trans. Faraday Soc.*, 1996, **62**, 1638.
23. E. J. W. Verwey and J. Th. G. Overbeek, 'Theory of the Stability of Lyophobic Colloids', Elsevier, Amsterdam, 1948.
24. H. M. Schaink and J. A. M Smit, *Phys. Chem. Chem. Phys.*, 2000, **2**, 1537.
25. C. Gripon, L. Legrand, I. Rosenman, O. Vidal, M. C. Robert, and F. Boué, *J. Cryst. Growth*, 1997, **178**, 575.
26. D. A. McQuarrie, *Statistical Mechanics*, Harper & Row, New York, 1976.
27. R. J. Baxter, *J. Chem. Phys.*, 1968, **49**, 2770.
28. M. Penders and A. Vrij, *J. Chem. Phys.*, 1990, **93**, 3704.
29. B. Cichocki and B. U. Felderhof, *J. Chem. Phys.*, 1988, **89**, 3705.
30. C. G. de Kruif, Th. J. M. Jeurnink, and P. Zoon, *Neth. Milk Dairy J.*, 1992, **46**, 123.
31. C. G. de Kruif, *J. Dairy Sci.*, 1998, **81**, 3019.

Self-Consistent-Field Studies of Mediated Steric Interactions in Mixed Protein + Polysaccharide Solutions

By Rammile Ettelaie, Eric Dickinson, and Brent S. Murray

PROCTER DEPARTMENT OF FOOD SCIENCE, UNIVERSITY OF LEEDS, LEEDS LS2 9JT, UK

1 Introduction

The functional properties of many individual food ingredients, including food macromolecules, have been extensively studied in the literature and are now relatively well understood.[1,2] However, foods are generally complex systems in which the behaviour of any added functional molecules are strongly influenced by the presence of other such species in the system. Thus, in recent years the centre of research activity has shifted towards understanding the more complicated behaviour exhibited by mixtures of such ingredients in foods. In particular, much attention has been directed towards mixed biopolymer systems, such as protein + protein and polysaccharide + protein aqueous solutions, as well as systems containing various combinations of macromolecules and surfactants.[3–6]

In many instances the interactions between different functional ingredients are not desirable. For example, the presence of small surface-active molecules in emulsion-based systems can often cause the displacement of larger amphiphilic biopolymers from the surface of the droplets.[7,8] Such biopolymers often provide the droplets with a certain degree of surface charge,[9] which in turn leads to the interdroplet electrostatic repulsive force necessary to stabilize the droplets against aggregation and coalescence. In other cases adsorbed biopolymers can lead to steric stabilization, or indeed to the formation of covalently cross-linked elastic films, which can also make a contribution towards preventing droplets from coalescing. Therefore, it is generally accepted that the displacement of these macromolecules by smaller surface-active species is detrimental to the colloidal stability and so may result in the eventual breakdown of the emulsion.[10] Additionally, in the presence of small amounts of certain types of polysaccharide gums, protein-stabilized emulsion droplets can also become flocculated. This happens when there is repulsion (or no net interaction) between protein and polysaccharide molecules, and the aggregation is attributed to the phenomenon of depletion attraction,[9] mediated by non-adsorbed polysaccharide. Yet other examples where the combination of two functional macromolecules may be undesirable include the well-studied cases involving mixed solutions of polysaccharides and proteins. When the unlike interactions are sufficiently strong, they can induce bulk phase separation behaviour in these systems. Depending on whether the interactions are favourable or unfavourable, this phase separation takes place, respectively, due to coacervation or incompatibility.[4,11,12]

In contrast to the situations mentioned above, there are other instances where the simultaneous presence of several functional food molecules can lead to enhanced functionality, or indeed to the appearance of new functionalities not possible with the individual ingredients separately. For example, the addition of non-adsorbing polysaccharide can increase the emulsifying power of a food protein by increasing the thermodynamic activity of the latter in solution,[5] whilst complexation of surfactant with polysaccharide can assist in the gel-forming behaviour of the food hydrocolloid.[6] Similarly, gelation in mixtures of polysaccharides can provide novel textures involving interpenetrating but independent gel networks, as well as mechanically enhanced coupled gels due to synergistic effects.[13] Such novel structures in biopolymer mixtures typically arise as a result of the competition between the kinetics of gelation and the thermodynamic phase separation, involving the mechanisms of nucleation or spinodal decomposition.[14]

In this article we are interested in the behaviour of protein + polysaccharide mixtures at interfaces. In particular, we wish to examine theoretically the proposition that, under favourable conditions, such mixtures can provide steric interactions that are sufficiently repulsive to stabilize emulsions or dispersions.[12] As will be shown, each of these two constituent biopolymers, on its own, may not be capable of sterically stabilizing colloidal particles. But an interacting mixture of the two may provide the required stability. Conjugated polymers formed from covalent bonding of protein and polysaccharides will not be discussed here. Such molecules, whether occurring naturally, such as in the high-molecular-mass fraction in gum arabic,[12] or formed deliberately, e.g., via Maillard reactions,[15,16] are effectively *single* species possessing the macromolecular structure one expects for good steric stabilizers. Similarly, chemically modified polysaccharides, such as hydrophobically modified starch, are also excluded from the current discussion.

Our calculations here of mediated steric forces in polysaccharide + protein mixtures are based on the so-called self-consistent-field (SCF) theory. This theory has previously been used in the study of food biopolymer systems to model adsorbed casein layers.[17,18] It is an equilibrium theory which is now well-established in the field of colloid science.[19] There are a number of simplifications involved in our approach, in particular the simplified way in which the structure of the various biopolymers are represented. Despite these assumptions, we do believe that the theoretical approach presented here provides a useful preliminary indication of the power of the technique for explaining and predicting the nature and occurrence of steric interactions arising from the combined presence of two or more biopolymers.

The methodology and underlying theory of SCF calculations have been described extensively in the literature.[19,20] We therefore confine ourselves to just a brief description of the method.

2 Methodology

There are many different ways of formulating the SCF theory. Essentially, in these calculations one wishes to determine the density profiles of the various species existing in the solution between a pair of colloidal particles that minimizes the free energy of the system. It is an inherent assumption of SCF theory that these profiles—the most probable density profiles—dominate all the thermodynamic behaviour of the system, in so far as calculations of the steric interactions between the particles are concerned. As such, the theory ignores the contribution of any fluctuations that there might be around this most probable set of profiles.

The most convenient and wide-spread implementation scheme of the SCF theory is that due to Scheutjens and Fleer,[20] in which the solution between a pair of particles (or flat walls, as the case might be) is represented in terms of a lattice space. Each lattice site is occupied by either a monomer, belonging to one of the polymer species in the solution, or a solvent molecule. In such lattice theories, as a matter of convenience, the sizes of the various monomeric groups and the solvent molecules are assumed to be the same. Also, since the most probable density profile is expected to vary only in the single direction perpendicular to the walls, all the sites at the same distance away from a wall are considered together as a single layer. The density of each monomer type making up the macromolecules appears in the theory only as its averaged value within that layer. The enthalpic interactions between various groups, and between those groups and the solvent, are represented in the free energy expression by terms of the form $\sum_z \chi_{\alpha\beta} \phi^\alpha(z) <\phi^\beta(z)>$, where $\phi^\alpha(z)$ is the mean density of type α monomers in the zth layer. The quantity $<\phi^\beta(z)>$ is a further averaging of $\phi^\beta(z)$ over three consecutive layers ($z-1$, z, $z+1$), in a manner which depends on the topology of the underlying lattice used, reflecting the fact that species in one layer also interact with those residing in the neighbouring ones. The interaction between different monomeric groups, α and β, is therefore assumed to be of short range, with its strength characterized as usual by the Flory–Huggins parameter $\chi_{\alpha\beta}$.[4,20]

SCF calculations of the type mentioned above have already been applied successfully to a number of food-related systems. In particular, their use in the study of the adsorption of a simple model of the disordered milk protein β-casein has produced results[17,18] which are in good agreement with neutron reflectivity experiments on the same protein.[21] In these calculations a relatively detailed representation of the primary structure of β-casein was used.

Furthermore, the long-range nature of the electrostatic interactions, between various charged amino-acid groups, was explicitly taken into account. The present work, involving a mixture of biopolymers, adopts an even simpler model for both the polysaccharide and the protein, treating all the interactions as being of short range along the lines described above. Once again, however, we believe that these simplifications do not detract from the main conclusions of the study.

3 Steric Interactions Mediated by Pure Protein or Polysaccharide

Polysaccharide hydrocolloids, unless covalently bonded to proteins or to other surface-active molecules, generally have little affinity for oil–water or air–water interfaces.[12] The presence of such macromolecules in aqueous solution, far from stabilizing colloidal particles, is expected to result in attractive depletion forces,[1,9] and consequently to flocculation of the particles. Figure 1 shows the nature of the colloidal interaction mediated by a non-adsorbing biopolymer, in a solution containing 0.9% by volume of such chains, as calculated using SCF theory. The interaction energy is displayed as a function of the separation between a pair of approaching surfaces. The polysaccharide molecules in these calculations are represented in a simplistic manner, each containing five linear chains of size $N = 200$ monomers, with all five chains joined together at one end through a single cross-linking point. The monomers are assumed to reside in an athermal solution, *i.e.*, with the χ parameter for the solvent–monomer interaction set to zero. It is clear that no significant interactions are mediated by these macromolecules until the separation distance between the particles becomes comparable to the radius of gyration (or, more

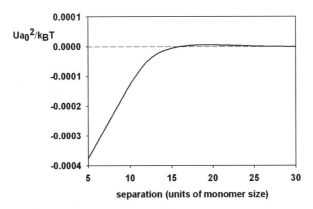

Figure 1 *Interaction potential per unit area (in units of kT per monomer-sized area a_0^2) is plotted against the separation distance between two solid surfaces, mediated in a 0.9% solution of biopolymer, representing polysaccharide in the SCF calculations. The overall size of molecules is $N = 1000$, with one cross-link point, and all the groups comprising the molecules are identical.*

precisely, to the screening length[9]) of the molecules. At this close separation distance, the macromolecules leave the gap between the approaching surfaces, and an attractive depletion force develops.

The above result is to be contrasted with that for solutions of biopolymers comprised of both hydrophobic and hydrophilic groups. The amphiphilicity alone, however, is not a sufficient requirement to give good steric stabilizing properties. Work on synthetic polymers has shown[19] that good steric stabilizers tend to have a diblock-type structure, with one block having a strong affinity for the hydrophobic surface, and the other distinctly preferring to remain in the aqueous solvent. With the almost unique exception of β-casein,[21] for which most of the hydrophobic amino-acid groups tend to be towards one end of the molecule and the hydrophilic ones towards the other, most food proteins do not posses this kind of idealized amphiphilic structure. Instead they tend to have their hydrophobic groups distributed in short blocks along the entire backbone. Thus, at least in their unfolded state, they more closely resemble multi-block synthetic polymers, consisting of alternating randomly distributed hydrophobic and hydrophilic sections, rather than simple diblock copolymers.

Figure 2 presents some results of our SCF calculations on the steric forces mediated by multi-block polymers of the type mentioned above. For both the curves shown, the total size of the chains is set to $N = 100$, with half of the monomers in each molecule taken to be hydrophobic and the other half hydrophilic. While the hydrophilic groups have no affinity for the surface of the particles, the hydrophobic ones are assumed to have an adsorption energy of $-1\ kT$ per monomer, quite typical of hydrophobic interactions. The strengths of the hydrophilic and hydrophobic interactions with the solvent molecules

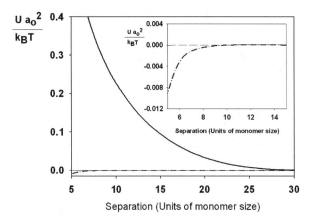

Figure 2 *The same as Figure 1, but now involving a 0.1% solution of an amphiphilic biopolymer. The solid line is for a diblock structure, with one block of size 50 units being hydrophobic while the other (also of size 50 units) is assumed to be hydrophilic. The dashed line is for a model of an unfolded globular protein, involving alternating small hydrophobic and hydrophilic blocks of size 5 units, but the same overall size of $N = 100$. The inset shows a graph on an expanded scale of the interaction potential per unit area for the latter case.*

are, respectively, taken to be $\chi = 0$ (a good solvent) and $\chi = 1\,kT$ (a poor solvent). The only difference between the structures of the macromolecules represented by the two curves in Figure 2 is that the solid curve is for a diblock chain, while the dashed line is calculated for a chain with 20 alternating hydrophobic/hydrophilic blocks of size 5 units each. In the former case, the polymer gives a strongly repulsive interaction potential; for the latter case the repulsive interaction between the particles is no longer present. More significantly, as can be seen from the inset of Figure 2, for the molecules with the small blocks, there is also a separation range over which the mediated colloidal interactions are actually attractive.

It is interesting to compare the surface affinities of the two molecular structures considered above. Figure 3 shows the density profile for the adsorbed multi-block polymer on a single isolated flat solid wall, plotted as a function of the perpendicular distance away from the wall. The concentration of chains in the bulk solution is fixed at 1%. There is clearly a high degree of adsorption onto the surface, with the density of the chains near the wall attaining a value at least an order of magnitude higher than its bulk concentration. In Figure 4 the same calculation is repeated at the same total polymer concentration, except now for a mixed solution with a tiny fraction (1%) of the chains assumed to be of the diblock type. Despite the fact that the diblock polymers have the same overall size and composition as the multi-block chains, and are present at a very much lower concentration, they are seen completely to displace the multi-block chains from the vicinity of the interface. This result dramatically confirms the experimental finding that the amphiphilic β-casein, at rather small concentrations, is capable of removing other food biopolymers from air–water or oil–water interfaces.

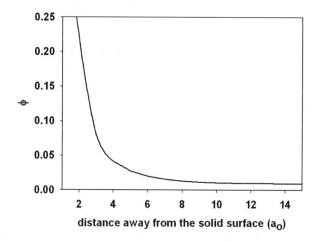

Figure 3 *Variation of the polymer concentration profile ϕ for adsorption of the multi-block amphiphilic biopolymer of Figure 2, in a 1% solution, plotted against the distance away from a hydrophobic solid surface (in units of monomer size a_0).*

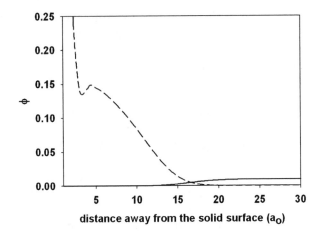

Figure 4 *The same as Figure 3, but now a tiny fraction (1/100) of the multi-block amphiphilic biopolymer molecules is replaced with chains having a diblock struc-ture, i.e., to form a mixed solution. The dashed line represents the concentration profile of the diblock macromolecules, while the solid line is for the multi-block ones.*

4 Steric Interactions in the Presence of Both Polysaccharide and Protein

Polysaccharides and proteins are capable of interacting favourably with each other through a variety of molecular mechanisms, including hydrogen bond-ing, and, at least in the pH range intermediate between the isoelectric points of the two biopolymers, *via* electrostatic interactions. In Figure 5 we present the calculated effective potential that is mediated between two surfaces in a solution containing a mixture of 0.9% polysaccharide + 0.1% protein. For each biopolymer the same simplistic representation, exactly as described in section 3, was adopted. All segments on the polysaccharide chains are assumed to interact with the hydrophilic groups on the protein molecules, with the χ parameter for such interactions set to $-3\ kT$. This is quite close to the limit at which the system is predicted to phase separate due to coacervation, accord-ing to simple Flory–Huggins calculations. As can be seen from Figure 5, there is no evidence for any predicted repulsive steric forces in such a system. Instead the forces remain attractive and of rather short range. In Figure 6 we display the calculated density profile of the adsorbed layer, formed at an isolated single interface. The solid curve shows the variation in concentration of the protein component, plotted as a function of distance away from the surface, while the dashed curve represents the same result for the polysaccharide component. Although the presence of the polysaccharide molecules near the interface is quite evident, the adsorbed layer formed is not much thicker than that formed by the protein chains on their own. Indeed, it is hard to distinguish a clear secondary polysaccharide layer. Instead, what emerges is a mixed single interfacial layer formed from the two biopolymers. The possibility of attractive

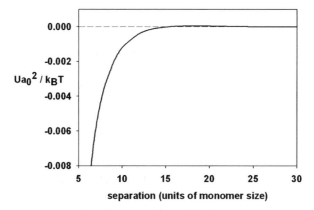

Figure 5 *Interactions mediated between two solid surfaces in a mixed solution involving 0.9% polysaccharide + 0.1% unfolded globular protein. All monomer groups of the polysaccharide molecule are assumed to be the same and to interact favourably with the hydrophilic parts of the protein with an interaction parameter $\chi = -3$ kT. The structures modelled are identical to those considered in Figures 1 and 2.*

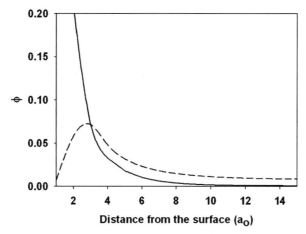

Figure 6 *Density profile of the adsorbed layer comprising the protein (the solid line) and the polysaccharide (the dashed line) plotted as a function of distance from an isolated solid–solution interface, in a mixed solution of 0.9% polysaccharide + 0.1% protein, identical to that in Figure 5.*

colloidal interactions between emulsion droplets in a mixed solution of bovine serum albumin + dextran sulfate was suggested by the experimental study of Dickinson and Galazka.[22]

In the above calculations we have simplistically assumed that each part of a polysaccharide molecule is equally capable of interacting favourably with a hydrophilic group on the protein chain. This clearly may not be the case in practice. For example, only certain sections of a polysaccharide might be sulfated, carboxylated or hydroxylated, and therefore possess sufficient charge or favourable polarity to allow attractive interaction with proteins. This possibility is explored in an extension of our model, in which we allow only 100 of the units (from a total of 1000) on each polysaccharide to interact with the appropriate sections of the protein, but with a rather higher energy ($\chi = -7\,kT$). The remaining 900 units, while still treated as hydrophilic, have no particular affinity for the protein molecules. As Figure 7 shows, a drastically different picture of the adsorbed film (and resulting interactions) now emerges. The mediated interactions between the colloidal particles in such a mixed solution are found to be strongly repulsive over a relatively broad range of surface separations. In particular, both the range and magnitude of the mediated force should be contrasted with those seen in Figures 5. The density profile for the adsorbed layer in the refined model, as displayed in Figure 8, is also equally different to that of the uniformly interacting polysaccharide shown in Figure 6. The most pronounced feature seen in Figure 8 is the presence of a broad, but distinct, secondary polysaccharide layer extending for a considerable distance into the aqueous phase, clearly beyond that formed by the adsorbed protein molecules. It is indeed the presence of such an extended hydrophilic adsorbed layer that is the main feature expected of a good steric stabilizer.[9,12]

5 Conclusions

We have considered, from a theoretical point of view, whether the existence of sufficiently favourable interactions between polysaccharide chains and proteins can give rise to weak complexes that have sufficient affinity for a hydrophobic surface and also have the ability to provide steric stability to a food colloidal system. Using a simplified representation of the typical

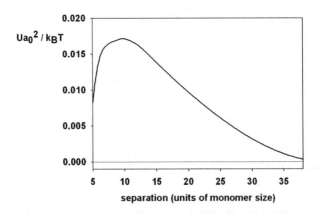

Figure 7 *The same as Figure 5, but now only a fraction (10%) of the monomers on the polysaccharide molecule are assumed to interact with the hydrophilic parts of the amphiphilic protein. For the fraction that still interacts with the protein chains, the interaction energy is $\chi = -7\,kT$.*

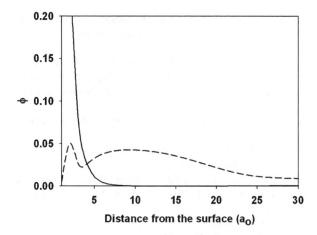

Figure 8 *Concentration profile ϕ of the two biopolymers in the mixed solution considered in Figure 7. The curves should to be compared to those in Figure 6.*

structures of these two kinds of biopolymers, in which only the most prominent and essential features are preserved, we have performed self-consistent-field (SCF) calculations to explore this possibility. We have also studied the structure of the adsorbed layers that are formed from these mixed solutions.

The results demonstrate that, under certain conditions, where only parts of the polysaccharide chains are able to interact with the protein, such strong steric repulsive forces can arise. In contrast, when every part of the polysaccharide molecule is free to associate with hydrophilic sections of the protein, the repulsive interparticle interaction disappears, and indeed it becomes net attractive. The calculated polymer concentration profiles of the adsorbed layers in these two contrasting cases also reflect these differences. Where all polysaccharide groups can interact favourably with the protein chains, it can be inferred that the polysaccharide molecule lies nearly flat on top of the adsorbed protein layer, so that no clear secondary layer is detectable. On the other hand, when more specific interaction is assumed, a much broader secondary adsorbed layer is generated. While some pieces of experimental evidence for both of these scenarios are available in the literature,[12,22,23] we believe that a more systematic experimental investigation of the theoretical predictions presented here is an interesting opportunity for future research.

References

1. E. Dickinson and G. Stainsby, 'Colloids in Food', Applied Science, London, 1982.
2. A. G. Gaonkar, 'Ingredient Interactions: Effect on Food Quality', Marcel Dekker, New York, 1995.
3. V. B. Tolstoguzov, in 'Functional Properties of Food Macromolecules', 2nd edn, eds S. E. Hill, D. A. Ledward, and J. R. Mitchell, Aspen, Maryland, 1998, p. 252.
4. E. Dickinson, in 'Biopolymer Mixtures', eds S. E. Harding, S. E. Hill, and J. R. Mitchell, Nottingham University Press, Nottingham, 1995, p. 349.

5. E. Dickinson, in 'Food Colloids and Polymers: Stability and Mechanical Properties', eds E. Dickinson and P. Walstra, Royal Society of Chemistry, Cambridge, 1993, p. 77.
6. B. Lindman, A. Carlsson, S. Gerdes, G. Karlström, L. Piculell, K. Thalberg, and K. Zhang, in 'Food Colloids and Polymers: Stability and Mechanical Properties', eds E. Dickinson and P. Walstra, Royal Society of Chemistry, Cambridge, 1993, p. 113.
7. A. R. Mackie, A. P. Gunning, P. J. Wilde, and V. J. Morris, *Langmuir*, 2000, **16**, 2242.
8. A. R. Mackie, A. P. Gunning, L. A. Pugnaloni, E. Dickinson, P. J. Wilde, and V. J. Morris, *Langmuir*, 2003, **19**, 6032.
9. P. Walstra, 'Physical Chemistry of Foods', Marcel Dekker, New York, 2003.
10. L. A. Pugnaloni, E. Dickinson, R. Ettelaie, A. R. Mackie, and P. J. Wilde, *Adv. Colloid Interace Sci.*, 2004, **107**, 27.
11. V. B. Tolstoguzov, *Food Hydrocoll.*, 2003, **17**, 1.
12. E. Dickinson, *Food Hydrocoll.*, 2003, **17**, 25.
13. V. J. Morris, in 'Functional Properties of Food Macromolecules', 2nd edn, eds S. E. Hill, D. A. Ledward, and J. R. Mitchell, Aspen, Maryland, 1998, p. 143.
14. I. T. Norton and W. J. Frith, in 'Food Colloids, Biopolymers and Materials', eds E. Dickinson and T. van Vliet, Royal Society of Chemistry, Cambridge, 2003, p. 282.
15. E. Dickinson and M. G. Semenova, *Colloids Surf.*, 1992, **64**, 299.
16. M. Akhtar and E. Dickinson, *Colloids Surf. B*, 2003, **31**, 125.
17. F. A. M. Leermakers, P. J. Atkinson, E. Dickinson, and D. S. Horne, *J. Colloid Interface Sci.*, 1996, **178**, 681.
18. P. J. Atkinson, E. Dickinson, D. S. Horne, and F. A. M. Leermakers, *Ber. Bunsen. Phys. Chem.*, 1999, **100**, 994.
19. G. J. Fleer, M. A. Cohen Stuart, J. M. H. M. Scheutjens, T. Cosgrove, and B. Vincent, 'Polymers at Interfaces', Chapman & Hall, London, 1993.
20. O. A. Evers, J. M. H. M. Scheutjens, and G. J. Fleer, *Macromolecules*, 1990, **23**, 5221.
21. P. J. Atkinson, E. Dickinson, D. S. Horne, and R. M. Richardson, *J. Chem. Soc. Faraday Trans.*, 1995, **91**, 2847.
22. E. Dickinson and V. B. Galazka, in 'Gums and Stabilisers for Food Industry', eds G. O. Phillips, D. J. Wedlock, and P. A. Williams, Oxford University Press, Oxford, 1992, vol. 6, p. 351.
23. M. G. Semenova, A. S. Antipova, A. S. Belyakova, E. Dickinson, R. Brown, E. G. Pelan, and I. T. Norton, in 'Food Emulsions and Foams: Interfaces, Interactions and Stability', eds E. Dickinson and J. M. Rodriguez Patino, Royal Society of Chemistry, Cambridge, 1999, p. 163.

Probing Emulsion Droplet Interactions at the Single Droplet Level

By A. Patrick Gunning, Alan R. Mackie, Peter J. Wilde, Robert A. Penfold, and Victor J. Morris

INSTITUTE OF FOOD RESEARCH, NORWICH RESEARCH PARK, COLNEY, NORWICH NR4 7UA, UK

1 Introduction

The colloidal interactions that occur between the droplets of an emulsion are crucial in controlling many important characteristics of the system.[1] Examples include factors such as the processing behaviour of the emulsion,[2] its textural and sensory properties,[3] and its stability against coalescence,[4] which ultimately determines the shelf-life. Traditionally these interactions are measured indirectly by rheological methods.[5] These measurements are made on bulk samples and so they reveal only an average picture of events. Furthermore, food emulsions are typically complex multi-component heterogeneous systems,[6] which limits the usefulness of much of the data obtained by traditional methods.

In order to overcome such limitations, we have developed[7] a method for attaching individual oil droplets to the end of an AFM cantilever which allows the measurement of droplet–droplet interactions in aqueous solution at the level of a single pair of droplets. Coating of droplet surfaces with added proteins or surfactants allows the production of model emulsions. This approach gives rise to a plethora of new measurement possibilities over various length scales. For example, we can demonstrate that AFM measurements of emulsion droplet deformability are sensitive to interfacial rheology by modifying the interfacial film on a pair of droplets *in situ*. On droplets coated with the anionic surfactant SDS, we have been able to monitor coalescence after screening of the double-layer interactions by the addition of counter-ions. We also demonstrate in this paper that the presence of a non-adsorbing polymer (polyethylene oxide) in the aqueous phase alters the droplet–droplet interaction between surfactant coated droplets, suggesting that our approach could be used to monitor depletion interactions.

2 Experimental

Materials

n-Tetradecane (T-4633, purity ≥ 99%) was obtained from Sigma Chemicals (Poole, UK) and was used without further purification. Ultra-pure water was used throughout this study (surface tension 72.6 N m^{-1} at 20 °C, conductivity 18 MΩ, Elga, High Wycombe, UK). The non-ionic water-soluble surfactants, Tween 20 (polyoxyethylene sorbitan monolaurate) and Brij 35 (polyoxyethyleneglycol dodecyl ether), were obtained from Pierce (Surfact-Amps 20 & 35, Pierce, Rockford, USA) as 10% w/v solutions in water. Aliquots of the relevant surfactant solution were added to the liquid cell in order to achieve the concentrations described in the text. The whey protein β-lactoglobulin (L-0130, Sigma Chemicals) was dissolved to a concentration of 100 μM in water, and an aliquot of this solution was added to the liquid cell of the AFM in order to achieve the desired concentration. Sodium dodecyl sulfate (SDS) and sodium chloride were obtained from Sigma Chemicals (Poole, UK) and dissolved to form stock concentrations of 10% w/w. Aliquots of these solutions were added to the liquid cell in order to achieve the desired concentrations for the coalescence experiments, *i.e.*, 1 mM SDS and 40 mM NaCl.

Instrumentation

The AFM apparatus used in this study was a TM Lumina instrument (Veeco, USA). This consists of an AFM head mounted above an inverted optical microscope (Olympus IX-70). The AFM head is fitted with a small CCD camera that views the sample from one side, which, in combination with the liquid scanner, permits observation of the sample under liquid. For the studies of the variation of cantilever deflection with interdroplet distance, the oil droplets were attached to cantilevers without tips (200 μm long thin variety of Nanoprobe NP0, Veeco, USA) in order to ensure that the measurements represented droplet–droplet interactions and not interactions between a droplet and the protruding apex of a silicon nitride tip. The AFM was operated in contact mode and the scan rates were typically 1–2 Hz. Approach and retract speeds during force *versus* distance cycles were 0.5 μm s^{-1}, except in the depletion experiments which were carried out at the much lower speed of 0.05 μm s^{-1} in order to minimize drag effects. Both cantilever types have typical quoted force constants of 0.06 N m^{-1}. All the AFM data presented were captured under liquid (water or buffer solution).

Preparation of Oil Droplets

An *ad hoc* 'sprayer' was used to prepare oil droplets in the following manner. The *n*-tetradecane was sucked from a small reservoir into a 200 μL pipette (Gilson). A pipette tip, connected to an air line, was held at 90° to the aperture of the oil-filled pipette tip in order to create a fine spray of *n*-tetradecane droplets which was directed over a clean, dry, glass microscope slide (BDH

SuperFrost). It was found that the best method was to direct the mist over the top of the slide, which was placed approximately 10 cm from the 'sprayer' tip, and then to allow some of the oil droplets to fall onto the slide. This resulted in a fairly uniform coverage of the slide with discrete 10–50 μm diameter droplets. Pointing the sprayer tip directly at the glass slide simply led to coalescence of the droplets on the glass slide, producing a thin film of oil. Although such a continuous oil film results in the formation of a larger number of oil droplets attached to the glass slide upon the addition of water, it also results in a greater amount of free oil on the surface of the water, which is inconvenient.

Glass surfaces are hydrophilic, and so the oil droplets were easily displaced when water was placed on top of them, prior to AFM measurement. However, the technique used to wet the glass was found to be critical. A small volume of water was squeezed out of the tip of the Gilson pipette, held approximately 3–4 mm above the glass surface, in order to form a drop hanging on the end of the pipette tip. The drop was then made to fall onto the glass slide by squeezing a fraction more water out of the pipette. The majority of the oil drops on the slide were displaced during the addition of water, and they formed a film of *n*-tetradecane on the water surface, together with some free floating oil droplets. However, some oil droplets (typically 10–15) did remain attached to the glass slide. The adhering droplets were generally found to be positioned in the middle of the wetted region of the glass slide. These droplets could be distinguished from floating oil patches on the water surface by their more spherical shape, and by their lack of mobility in the visual field when the microscope was tapped. The AFM head was then lowered down onto the slide in order to sandwich the water between the liquid scanner and the glass slide. The remainder of the measured volume of water in the pipette was added drop-wise onto the side of the liquid cell. This avoided the flow and shear effects associated with discharging the pipette directly into the sandwiched drop of water, with consequent risks of displacing the attached oil drops from the glass slide. Addition of the various surfactant and polymer solutions was carried out in the same manner at appropriate times during the course of the experiments.

Attachment of Oil Droplets onto the AFM Cantilever

The AFM head was positioned using lateral micrometers, so that the end of the tip-less cantilever sat directly above a droplet on the surface of the glass slide. The cantilever was then driven down into the target droplet using the fine approach motor on the AFM. Engulfment of the lever into the drop was observed via the instrument's onboard camera. Once engulfment had been achieved, the AFM head and cantilever were retracted away from the surface of the slide, whereupon the oil droplet was found to detach itself from the slide surface. This method allowed droplets to be placed with high precision at the end of the cantilever. Droplets attached to the cantilevers were typically in the size range 20–60 μm.

Once a droplet had been attached to the cantilever, then the surfactant or protein, required to stabilize the droplet, was added to the liquid cell of the

AFM. It was found that this had to be done afterwards, otherwise droplet attachment to the cantilever was found to be impossible, as the cantilever could not be made to penetrate the droplets. Droplet transfer to cantilevers was found to be only possible for oil droplets that had been deposited onto glass slides. The cantilever consists of silicon nitride which is nominally hydrophilic, although presumably less so than the glass surface. This explains why the oil droplets can be transferred onto the cantilevers from a glass surface. In fact there is a delicate balance involved: when the cantilever was too hydrophobic, it was found that oil droplets would simply wet onto the lever rather than form a drop.

Monitoring 'Force–Distance' Interactions

The cantilever with its attached oil droplet was positioned directly over a second droplet, which was still attached to the glass slide, and the variation of the interaction force with separation was monitored and recorded. Pairs of droplets of approximately equal size were chosen for the measurements described in this study. Cantilever deflection was quantified by the photodiode detector of the AFM in terms of electrical output. This was converted into real distances by determination of the inverse optical lever sensitivity (INVOLS) of the particular lever used in each measurement. This involved determining a 'force–distance' curve on the *rigid* surface of the glass slide using the bare cantilever. Since there is no sample deformation in this case, the gradient for the resulting cantilever deflection *versus* distance graph provides a scale factor for the conversion. Finally the data were scaled to a nominal point of zero cantilever deflection, this being defined as the point of furthest droplet separation distance on each graph.

3 Results and Discussion

Figure 1 shows a video image* captured through the $10 \times$ objective of the Olympus microscope, illustrating an AFM cantilever with an oil drop attached. Such images demonstrate the relative scale of droplet and cantilever, and allow the size of the droplets to be measured with a reasonable degree of accuracy. The image also illustrates that the V-shaped cantilever used in this study effectively acts as a three-point cradle holding the droplet firmly in place. We carried out experiments where we attached oil droplets to rectangular cantilevers, since these have the advantage of being easier to calibrate in terms of their spring constants, but we found that the single-point attachment of the oil droplets to these levers was much less stable. The resulting oil droplets were seen to move from side-to-side when the droplet–lever assemblies were scanned across a glass surface, and so the risk of oil droplet movement during the interaction measurements seemed unacceptably high.

---

*A video of the entire attachment process is available from the authors on request.

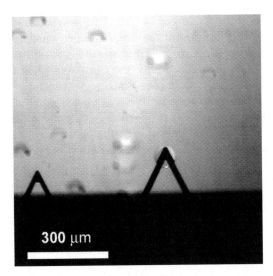

Figure 1 *Optical micrograph of an AFM cantilever with an attached droplet under water. The length of the cantilever is 200 μm.*

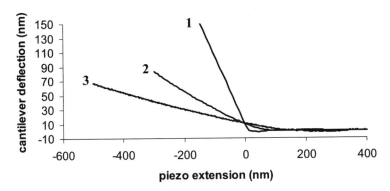

Figure 2 *Deflection versus distance graphs: (1) the tip-less cantilever driven towards a glass surface; (2) the same cantilever with an n-tetradecane droplet attached driven towards a glass surface; (3) the same cantilever with droplet attached driven towards a second n-tetradecane droplet attached to a glass surface.*

The graph presented in Figure 2 depicts the variation in the deflection of the AFM cantilever as it was pressed against various surfaces. Curve 1 was obtained by pressing the bare cantilever against the surface of a clean glass slide under water. It can be seen that the deflection of the cantilever is directly proportional to the extension of the piezo scanner (driving the surfaces together) with a unit coefficient. This indicates, as one might expect, that the slide surface was perfectly rigid, and that no indentation or deformation occurred. After this measurement a droplet of *n*-tetradecane was attached to the cantilever. The droplet was 'stabilized' by the addition of β-lactoglobulin (β-lg) to the water in the liquid cell of the AFM at a concentration of 24 μM. This was done in order to prevent the droplet either sticking to, or coalescing with, surfaces during the carrying out of the subsequent measurements. Curve 2 is a record of the deflection of the cantilever when this combined oil droplet–cantilever assembly was pressed into a glass surface. It is immediately obvious that the relation between the piezo extension and the subsequent lever deflection remains approximately proportional, but with a coefficient less than unity, indicating that some deformation of the oil droplet has occurred during the measurement, reducing the degree of deflection of the cantilever. Curve 3 reveals that, when this oil droplet–cantilever assembly was pressed into a second oil droplet located on the glass surface, the deflection of the cantilever was lower still, indicating an even higher degree of deformation during the measurement. This data series tells us that the gradient in the deflection *versus* distance curves shown in Figure 2 obtained for the oil droplet–cantilever assembly is sensitive to droplet deformation—but it does not reveal the origins of this effect.

There are two possible sources of droplet deformability in this type of experiment. The first is the excess internal Laplace pressure ΔP of the droplets, defined by

$$\Delta P = \frac{2\gamma}{r},\tag{1}$$

where γ is the interfacial tension and r is the droplet radius. The second source of droplet deformability comes from the interfacial elasticity, and is given by

$$E_{mod} = \frac{\mathrm{d}\gamma}{\mathrm{d}\ln A}.\tag{2}$$

Equation (2) defines the interfacial elastic modulus, E_{mod}, as the rate of change of surface tension γ with area A. There is a crucial distinction between these two contributions, which enables testing of the origin of the deformability seen in the AFM experiments. In the first case the deformability of the droplets is inextricably linked to the interfacial tension. A reduction in interfacial tension will produce a corresponding reduction in the Laplace pressure of the droplet. Interfacial elasticity has no such direct connection to interfacial tension, but it can be increased or decreased for the droplet systems simply by the addition of proteins or surfactants. The following experiment was carried out to exploit this fact in order to determine which of the two effects predominates in the AFM measurements shown in Figure 2.

An oil droplet attached to a cantilever was first coated with β-lg by adding it to the liquid cell at a concentration of 2 μM. After 30 minutes the deflection *versus* distance measurements were commenced by pressing the oil drop–cantilever assembly against a second oil droplet attached to the glass substrate. The gradient in the region of constant compliance from the resulting curves was determined by fitting a linear function to the data, and the moduli were plotted against time (Figure 3). Examination of the data shown in Figure 3

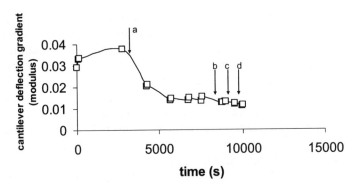

Figure 3 *Effect of interfacial rheology on droplet deformation. Each data point represents the modulus of the gradient of the constant compliant region for AFM deflection versus distance measurements with time, for a pair of droplets of varying interfacial composition. The droplets were initially coated with β-lg, and then Tween 20 was added at the times indicated by the arrows: (a) 2 μM, (b) 4 μM, (c) 6 μM, and (d) 9 μM. (This set of data was not scaled using the INVOLS conversion factor since only changes in cantilever deflection gradient were needed.)*

reveals that initially the modulus increases. This is consistent with the fact that the interfacial film strengthens, as the molecules are given time to unfold and to increase the degree of interaction with neighbouring proteins.[8] It is also counter to the behaviour of the interfacial tension, which will tend to reduce with time, indicating that the interfacial elasticity is the origin of the deformability measured in the AFM data. This conclusion is reinforced when one considers the data obtained from the next part of the experiment. After the initial increase in the gradient values seen in Figure 3, when the oil droplets were exposed to the pure protein solution, the interfacial protein film was exchanged for an interfacial surfactant film. This was accomplished by adding 2 μM Tween 20 (at time labelled 'a' on the graph) to the liquid cell; as reported previously,[9] this will completely displace the protein from the oil–water interface given sufficient time. The first data point following the addition of Tween 20 shows that the modulus of the deflection *versus* distance gradient has reduced significantly and thereafter seems to reach a plateau, despite further additions of Tween 20 (see labels 'b–d' on the graph). It is known[10] that Tween 20 forms interfacial films with much lower elasticity than proteins, consistent with the decrease seen in Figure 3 for the AFM data. Furthermore, each addition of Tween 20 would have reduced the value of the interfacial tension, since the concentration was well below the cmc value for this surfactant (59 μM), and yet these additions do not appear to affect the deflection *versus* distance gradient value in any significant way. Taken together, the data from the two stages of the experiment shown in Figure 3 provide compelling evidence that the droplet deformability, seen in the gradient of the AFM cantilever deflection *versus* distance curves, originates from the interfacial elasticity of the droplets, and is not due to their internal pressure.

So far we have considered what may be described as short-range effects—namely the interactions that occur as a pair of emulsion oil droplets are squeezed against one another. Another interesting aspect to examine is the interactions which occur at longer range, and for the purposes of this study these have been grouped into two categories. The first type of interaction of interest here occurs over the intermediate distance range—it is usually dictated by the charge on the droplets and the counter-ion concentration in the aqueous solution that surrounds them. n-Tetradecane droplets have a significant negative zeta potential in water, and the surface charge on the droplets is modified by the nature of the stabilizing interfacial film.[11] In an oil-in-water emulsion the surface charge influences the stability of the aqueous film between the droplets, and hence the coalescence phenomena upon droplet approach. Figure 4 shows the results of an experiment designed to explore the effect of altering the charge distribution between two approaching n-tetradecane droplets, one droplet being attached to the cantilever and the other attached to the surface of the glass slide. The droplets were initially stabilized against coalescence using the anionic surfactant SDS, added to the liquid cell of the

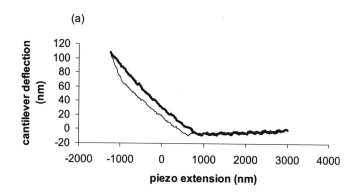

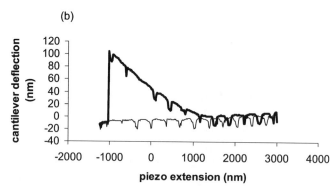

Figure 4 *Evidence for droplet coalescence and the effect of charge screening. Deflection versus distance data representing the interaction between two droplets. Approach curves (thick lines) and retraction curves (thin lines) were obtained in the presence of (a) 1 mM SDS and (b) 1 mM SDS +40 mM NaCl.*

AFM at a concentration of 1 mM. Under these conditions, the Debye screening length is 5.6 nm, meaning that the thin aqueous film between the approaching droplets is unlikely to thin down to less than this distance, since the repulsive interaction will act to deform the approaching droplet surfaces, making the chance of coalescence very low.

The cantilever deflection *versus* piezo extension in Figure 4a shows that the two droplets are simply squeezed against one another on approach (thick line), as indicated by the positive deflection of the cantilever to which one droplet was attached, and that the cantilever then relaxes along virtually the same path on subsequently pulling apart (thin line). The charge on the droplets was further screened by the addition of 40 mM NaCl. This reduces the Debye length to 1.5 nm, and so raises the probability of coalescence. The resultant deflection *versus* distance graph is shown in Figure 4b. Initially the positive deflection of the droplet-bearing cantilever looks the same as for the pure SDS case; but, at a piezo extension distance of ~1 μm, the deflection drops abruptly to the baseline and remains there for the rest of the approach (some 250 nm). Upon retraction of the piezo scanner, the value for the cantilever deflection remains fixed to the baseline value for the entire retraction distance. The abrupt reduction in the cantilever deflection value during approach indicates that a catastrophic event has occurred during the measurement cycle—this event is, in fact, the coalescence of the droplet attached to the cantilever with the droplet attached to the glass surface. This was confirmed by visual inspection of the droplets after the measurement using the inverted microscope: upon complete retraction of the cantilever away from the glass slide, the droplet that was previously attached to the glass surface was no longer present, and the one attached to the cantilever had grown to around twice its original volume.

The second type of long-range interaction which influences the stability of emulsion droplets is caused by the addition of non-adsorbing polymers or particles to the continuous phase. The presence of such entities generates an attractive force between the droplets known as a depletion force.[12-14] This is due to the osmotic gradient set up when two approaching emulsion droplets exclude polymer molecules or particles from the thin aqueous film as the gap between them closes. The magnitude of this force is related to the ratio of sizes of the droplet and the polymer molecules (or particles) and also to the volume fraction that each occupies in solution.[15]

We have attempted to observe the depletion effect experimentally for a system comprising two approaching *n*-tetradecane droplets stabilized by the non-ionic surfactant Brij 35. The measurement was initially carried out in a 10 mM KCl solution containing 225 μM Brij 35 (Figure 5a). The salt was added to the imaging solution in order to reduce the Debye length of the electrostatic repulsion, so that it did not overlap with the expected depletion region. Afterwards, a non-adsorbing polymer, polyethylene oxide (PEO), was added (Figure 5b) to the liquid cell at a concentration of 1% w/w, which approximates to the overlap concentration $C_p{}^*$ for this polymer. The PEO was also dissolved in 10 mM KCl to maintain the ionic strength of the imaging solution. Once again the droplets were squeezed together (the approach stage)

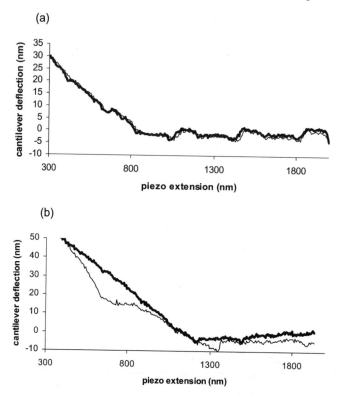

Figure 5 *Direct evidence for the depletion interaction. Deflection versus distance data representing the interaction between two droplets. Approach curves (thick lines) and retraction curves (thin lines) were obtained for (a) 10 mM KCl + 225 μM Brij 35 and (b) 10 mM KCl + 225 μM Brij 35 + 1% w/w PEO.*

and then pulled apart (the retraction stage) during the measurement cycle. The data in Figure 5a illustrate that there was no measurable adhesion between the droplets in the absence of polymer, since there was no hysteresis between the deflection of the droplet-bearing cantilever upon approach (thick line) and retraction (thin line) from the second droplet. (Indeed it is difficult to visualize the retraction curve for most regions of the data in Figure 5a.) The data obtained after the addition of polymer (see Figure 5b) appear quite different: there are two deviations in the deflection of the cantilever upon retraction (thin line), indicative of an adhesive interaction between the droplets.

It is possible to convert the cantilever deflection into a force provided that the spring constant k of the cantilever is known. In this experiment the value of k was determined to be 0.05 N m^{-1}. In such AFM deflection *versus* distance measurements, the adhesive force is defined as the magnitude of negative deflection of the cantilever relative to the baseline value obtained (or null point) when the cantilever is far away from the surface. For the data presented in Figure 5b, this corresponds to a distance of 4.7 nm and therefore a force of 234 pN. It should be noted, however, that the deviations seen upon retraction of the droplets were quite variable on repeated measurement, due to the significant viscosity imparted to the solution by the relatively high polymer concentration.

The maximum force at contact due to depletion between a pair of oil droplets can be estimated from

$$F \approx -3\left(\frac{C_p}{C_p^*}\right)\frac{k_B T \sigma}{\sigma_p^2}, \qquad (3)$$

where C_p is the concentration of the polymer solution, k_B is Boltzmann's constant, T is the absolute temperature, σ is the diameter of the droplets, and σ_p is the diameter of the polymer molecules. This latter value was estimated from published hydrodynamic radius data.[16] Equation (3) is valid provided that the ratio of droplet size to the polymer size is large, as is the case in the present experiments. If we insert typical values into equation (3) (*i.e.*, $\sigma = 50$ µm, $\sigma_p = 50$ nm, $T = 300$ K), then the maximum depletion force at contact is $F \approx 60$ pN. This is about one quarter of the observed adhesion force of 234 pN determined in the AFM measurement. When one considers the effect that viscous drag might have on the movement of the drop-bearing cantilever in the polymer solution, it is not surprising to find a relatively large difference between the calculated figure and the value determined experimentally. Nevertheless, the fact that the values do agree to within an order of magnitude seems encouraging. It suggests that the AFM approach used here may be capable of measuring long-range depletion effects if the system can be fine-tuned with respect to maximizing the size ratio between the emulsion droplets and polymer molecules, and if the solution viscosity can be kept within reasonable bounds.

4 Conclusions

A method has been devised for attaching an oil droplet to an AFM cantilever. This enables the monitoring of the interaction between a protein-coated droplet attached to the cantilever and a protein-coated droplet attached to a flat glass surface. Differences in the interactions between two droplets, and the interactions between a droplet and the flat glass surface, appear to be due to deformation of the droplets. Based on experiments in which the interfacial film on a pair of droplets was changed *in situ*, the main origin of droplet deformability has been shown to be due to the interfacial rheology.

The procedures developed for measuring droplet interactions have been used to monitor coalescence of droplets. For droplets coated with the ionic surfactant SDS, screening of the electrical double-layer by the addition of salt has been found to promote coalescence.

Finally, data have been obtained illustrating the effect of a non-adsorbing polymer (PEO) on the interaction between a pair of surfactant-coated droplets, suggesting that this technique may also be of value for monitoring long-range interactions between emulsion droplets, such as depletion effects.

Acknowledgements

The authors thank David Hibberd, Gary Barker and Andrew Watson for useful discussions. This work was funded by the BBSRC through its core strategic research grant to IFR.

References

1. B. Bergenståhl and P. M. Claesson, in 'Food Emulsions', 3rd edn, eds S. E. Friberg and K. Larsson, Marcel Dekker, New York, 1997, p. 57.
2. H. D. Goff, *J. Dairy Sci.*, 1997, **80**, 2620.
3. J.-L. Gelin, L. Poyen, R. Rizzotti, C. Dacremont, M. Le Meste, and D. Lorient, *J. Texture Stud.*, 1996, **27**, 199.
4. G. A. van Aken, F. D. Zoet, and J. Diederen, *Colloids Surf. B*, 2002, **26**, 269.
5. P. Manoj, A. D. Watson, D. J. Hibberd, A. J. Fillery-Travis, and M. M. Robins, *J. Colloid Interface Sci.*, 1998, **207**, 294.
6. H. D. Goff, *Int. Dairy J.*, 1997, **7**, 363.
7. A. P. Gunning, A. R. Mackie, P. J. Wilde, and V. J. Morris, *Langmuir*, 2004, **20**, 116.
8. J. Benjamins and E. H. Lucassen-Reynders, in 'Proteins at Liquid Interfaces', Studies in Interface Science Series—vol. **7**, eds D. Möbius and R. Miller, Elsevier, Amsterdam, 1998, p. 341.
9. A. R. Mackie, A. P. Gunning, P. J. Wilde, and V. J. Morris, *J. Colloid Interface Sci.*, 1999, **210**, 157.
10. A. R. Mackie, A. P. Gunning, M. J. Ridout, P. J. Wilde, and J. M. Rodriguez Patino, *Biomacromolecules*, 2001, **2**, 1001.
11. P. J. Anderson, *Trans. Faraday Soc.*, 1959, **55**, 1421.
12. S. Asakura and F. Oosawa, *J. Chem. Phys.*, 1954, **22**, 1255.
13. S. Asakura and F. Oosawa, *J. Polymer Sci.*, 1958, **33**, 183.
14. A. Vrij, *Pure Appl. Chem.*, 1976, **48**, 471.
15. H. N. W. Lekkerkerker, W. C. K. Poon, P. N. Pusey, A. Stroobants, and P. B. Warren, *Europhys. Lett.*, 1992, **20**, 559.
16. K. Devanand and J. C. Selser, *Macromolecules*, 1991, **24**, 5943.

Adsorbed Layers

Formation and Properties of Adsorbed Protein Films: Importance of Conformational Stability

By Martien A. Cohen Stuart, Willem Norde, Mieke Kleijn, and George A. van Aken

LABORATORY OF PHYSICAL CHEMISTRY AND COLLOID
SCIENCE, WAGENINGEN UNIVERSITY, DREIJENPLEIN 6,
6703 HB WAGENINGEN, THE NETHERLANDS

1 Introduction

When a solution of protein molecules is exposed to an interface, a complex sequence of events takes place which ultimately results in the formation of a layer of adsorbed molecules. These adsorbed molecules have often lost a good deal of their original shape—or more precisely, their conformation—as well as the properties that depend on that shape, and the interface has acquired new properties. For food systems the mechanical properties of the interface are often highly relevant, as they play a role in the structure of the food product on mesoscopic length scales, *e.g.*, in foams and emulsions. The rates of the processes contributing to the formation of protein films are also important, as they may determine the fate of a dispersion during various processing operations.

In this paper we review the formation of protein films from a dynamic point of view, attempting to develop a coherent picture. This is not a simple task, since phenomena like diffusion, attachment and molecular deformation must find a natural place in that picture, and each of these is influenced by a variety of parameters in the experimental situation. Quantities like diffusion coefficient, protein concentration, and adsorption rate constant have at least to be considered, and also the tendency for the protein to be deformed should somehow be quantified. Our leading idea is that to some extent protein molecules can be considered as organic nanoparticles with a variable degree of 'softness'. As such they respond in a characteristic way to forces exerted by surfaces with which they make contact.

2 Macromolecules as 'Soft' Particles

Perhaps the most typical characteristic of a macromolecule as compared to a small molecule is the large number of internal degrees of freedom—in other words, the ability to change shape. Polymer chains, of which proteins are a sub-class, can indeed assume many different shapes or conformations, and these shape changes are crucial to their interfacial behaviour.

An appropriate starting point for discussing the conformations of simple homopolymers in solution is the concept of the random coil, *i.e.*, the shape taken by a linear chain with effectively non-interacting segments. The random coil must be described in terms of statistical averages rather than in terms of a well-defined map of atomic positions. It is a shape of high entropy, and any perturbation of the random coil is always accompanied by an entropic penalty.

Macromolecular shape is strongly influenced by effective (*i.e.*, solvent-dependent) interactions between monomers. In good solvents there is an effective repulsion between the monomers which leads to larger average dimensions of the molecules (swelling), whereas in poor solvents the monomers attract each other and the macromolecule tends to assume a compact globular shape. The size R of a simple homopolymer, consisting of N identical monomers, can be generally expressed by the power law $R \sim N^{\alpha}$. Compact coils have an N-independent density, so that $R^3 \sim N$, or $\alpha = 1/3$; swollen coils have a density *decreasing* with increasing N, corresponding to higher α values, usually in the range 0.5–0.6. When the effective interaction between the monomers changes (*e.g.*, due to changes in temperature or solvent composition), a macromolecule responds by a change in its compactness.[1]

While proteins are no exception to this general pattern, their behaviour in aqueous solution is more complicated. Firstly, the monomer units (except glycine) are strictly chiral, which is not the case for many synthetic polymers. This introduces a tendency for the chains to form secondary structures—like helices and sheets. Secondly, proteins are not simple homopolymers. Some of the amino-acid monomers are typically hydrophobic and some are hydrophilic. Proteins with a large number of hydrophilic monomers tend to form swollen random coils in water, whereas proteins with many hydrophobic units must remain compact. The size of the latter satisfies the $\alpha = 1/3$ scaling fairly well, and they are accordingly referred to as 'globular proteins'. The fact that globular proteins are still soluble in water is due to an appropriate number of hydrophilic units that populate the molecular surface in the native conformation. As for simple homopolymers, the changes in effective interactions between the protein monomers lead to changes in compactness; and such changes can be induced by changing temperature or solvent composition. For example, additives like the organic salt, guadininium chloride, or urea have the effect of weakening the hydrophobic monomer–monomer attractions in water, whereas addition of sucrose or glucose has a strengthening effect on these interactions.

A convenient parameter to quantify the strength of monomer–monomer interactions is the excluded volume parameter v. This parameter is the effective

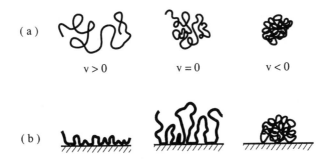

Figure 1 *Conformations of a chain molecule at various values of the monomer–monomer excluded volume corresponding to v>0 ('good solvent'), v=0, or v<0 ('poor solvent'): (a) in solution; (b) on a surface.*

(interaction weighted) probability that one monomer excludes a certain volume from access by another monomer. The parameter takes positive values in good solvents, and negative values in poor solvents; and, when it is just zero, the monomers behave ideally, as if they had no volume (see Figure 1).

When a polymer is subjected to a deforming force in compression or in shear, its response depends strongly on whether it is swollen or collapsed. The monomer units in a swollen polymer molecule repel each other, and further separation is not counteracted by monomer–monomer interactions. The only restoring force is of an entropic nature, being derived from the reduced number of configurations possessed by a stretched or compressed coil as compared to a spherical one. Compact macromolecules behave differently. Due to the attraction between their monomers, they not only tend to keep most of the poor solvent out of their interiors, but they also try to minimize their external contact area. In a certain way, they may be considered as tiny droplets of insoluble liquid, having a certain interfacial tension. Any deformation of such a droplet will then not only be counteracted by an entropic force, but also by the increased surface area and concomitant unfavourable interactions with the solvent. For a simple homopolymer in the globular state, the 'interfacial tension' γ_{ps} has been estimated[2] to be of order kTv^2. The larger the tension, the higher the Gibbs free energy associated with particle deformation—in other words, the 'harder' the particle.

Generalizing these ideas to globular proteins, we propose that these molecules can also be assigned an (effective) interfacial tension and a related 'hardness' characterizing the ease with which they can be deformed by external forces. As one increases the temperature or adds a denaturant, the molecules gradually become 'softer' (their interfacial tension decreases) and unfolding becomes increasingly easier. Thus we come to an (admittedly oversimplified) picture of a globular protein molecule in water as a collapsed copolymer with a certain number of hydrophilic units on its surface. An important refinement is that this 'interfacial tension' is not constant—it may change as the molecule is being deformed. The presence of the hydrophilic units on the surface has an important influence; in particular, the charged units will exert a certain surface pressure that counteracts the 'interfacial tension' due to the more hydrophobic units (as is the case for any charged surface). Hence, the charging up of a protein (moving away from its isoelectric point) will tend to lower its interfacial tension and to increase the tendency for it to yield under deformation.

3 Macromolecular Adsorption Considered as Droplet Spreading

How will such a deformable object respond to contact with an interface? Most likely, it will first attach somehow to the surface, and then spread out to develop a significant contact area; this latter process stops as soon as further spreading costs more than it yields. Hence, the natural approach is to think of a balance between a driving force, namely the formation of a protein–substrate (ps) contact at the expense of a substrate–water (sw) contact, and a counteracting effect due to deformation, mainly coming from increased contact of the hydrophobic units of the protein with water (pw). The free energy G of the protein/substrate system will change as the area A of the protein–substrate contact increases. We can cast this in terms of a 'force' by taking the derivative of G with respect to A; then, the first effect may be expressed in terms of a difference in interfacial tensions, $\gamma_{sw} - \gamma_{ps}$, whereas the second effect involves the protein–water interfacial tension, γ_{pw}. Hence, we suggest that the extent of spreading may be discussed in terms of a 'spreading parameter', $S = -\mathrm{d}G/\mathrm{d}A = \gamma_{sw} - \gamma_{ps} - \gamma_{pw}$, much in the same way as for the spreading of a liquid droplet on a substrate. The more the protein resists deformation (*i.e.*, the higher is its 'interfacial tension' γ_{pw} or its 'hardness'), the less it will unfold; but also, the larger the free energy decrease per unit of protein–substrate contact (*i.e.*, the larger is the value of $\gamma_{sw} - \gamma_{ps}$), the larger the contact area it will tend to develop. One immediately sees that swollen polymer molecules (in a good solvent) will easily spread out because they have little resistance against deformation, whereas compact ones have a reduced tendency to do this (see Figure 1b).

It is not so easy to check these conjectures experimentally. In particular, for solid substrates, none of the interfacial tensions that make up S is experimentally accessible. We may, however, consider the number of adsorbed molecules per unit of surface area, Γ. Before doing so, we have to point out that there is an apparent paradox here: the higher the affinity of the protein for the substrate, the lower is the mass (given by ΓM, where M is the molar mass) of the saturated adsorbed layer! This is due to the fact that most macromolecules adsorb sufficiently strongly to attain a completely covered surface (saturation) even at very low concentrations. At saturation, the condition of having a filled surface corresponds to $\Gamma A = 1$, where A is the area per molecule. Hence, the more spreading, the larger is A and the smaller is Γ. Stronger interaction may lead to more spreading, and thus a stronger attraction between protein and substrate may imply a smaller mass per unit area. This is quite the opposite of what happens with small molecules, for which the adsorbed mass is determined by a balance between binding strength (surface affinity) and translational entropy, and the extent of spreading is not an issue.

It follows from the reasoning given so far that one way to increase the adsorbed amount is to *lower* the interfacial tension γ_{sw} between water and substrate, provided that we keep the other two interfacial tensions constant. This may seem an unusual approach—but a rather clean example exists when we have a solid substrate for which we can vary γ_{sw} by external control. A metal electrode can satisfy this requirement. As is well known (and extensively studied for mercury), the interfacial tension drops when the electrode is polarized, according to the classical Lippmann equation:[3]

$$\left(\frac{\partial\gamma_{sw}}{\partial E_{\pm}}\right)_{T,\mu} = -\sigma^0 = \int_{E_{\pm}(\sigma^0=0)}^{E_{\pm}} CdE_{\pm} \tag{1}$$

Here, σ^0 is the surface charge, $E_{\pm}$ is the applied potential, and C is the differential capacitance of the electrode/electrolyte interface. We recognize here a thermodynamic Maxwell relation, which is why the derivative at the left-hand side must be evaluated at constant temperature T and chemical potential μ. The implication of the Lippmann equation is that, as long as C is constant, the interfacial tension changes with increasing potential in a quadratic fashion; hence the plot of γ_{sw} against E is a parabola with its apex at the isoelectric point. What this means is that, if the wetting picture makes sense, a given protein, at fixed pH and ionic strength (and protein concentration), has its minimum adsorption at the isoelectric point (IEP) of the electrode, and, symmetrically around this minimum, the adsorption increases with increasing potential.

A beautiful illustration of this effect was recently found by Barten *et al.*[4] (see Figure 2a) for the adsorption of lysozyme (3 mg L^{-1}) at pH = 6.4 for various applied potentials on a gold electrode immersed in 1 mM KNO$_3$ (Figure 2a). Lysozyme has its own pI at pH $\approx$ 11, so that it is positively charged under all the measurement conditions. Yet the adsorption increases symmetrically around the IEP of the *gold surface*, despite the change in sign of the surface charge of the substrate. Moreover, the data can be fitted nicely to a parabola, which strongly suggests a connection to the Lippmann equation, and underpins the importance of interfacial tension-driven spreading. Somewhat similar data for insulin on a platinum electrode were reported by Razumas *et al.*[5] (see Figure 2b).

According to the 'droplet' idea explained above, the extent of spreading should also respond to changes in γ_{pw}. That is, a 'soft' protein should spread more than a 'hard' one, and charging a protein should tend to make it 'softer', and hence increase its tendency to spread. Examples of these effects can be found in many publications; in particular, the role of conformational stability of proteins on their adsorption has been amply demonstrated. We give one example. Most proteins have their stability optimum at or close to their pI, so that the adsorbed amount (other factors being constant) is also at a maximum at pH $\approx$ pI. An illustration is taken from recent work by van der Veen (see Figure 3) for adsorption of lysozyme (p$I \sim$ 10) and succinylated lysozyme

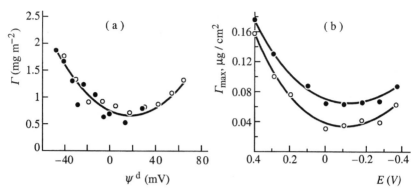

Figure 2 *Total adsorbed amount of protein as a function of the electrical potential. (a) Lysozyme on gold—the potential Ψ^d is the true double-layer potential with respect to the solution, as deduced from surface force data. Protein concentration $=3$ mg L^{-1}; background electrolyte $=1$ mM KNO_3, pH $=6.4$. (b) Insulin on platinum—the potential E is the applied electrode potential with respect to the reference: O, 2Zn insulin; •, 4Zn insulin. Protein concentration $= 0.3$ mM; background electrolyte $=0.15$ M NaCl, pH $=7.4$.*
(Redrawn from refs. 4 and 5.)

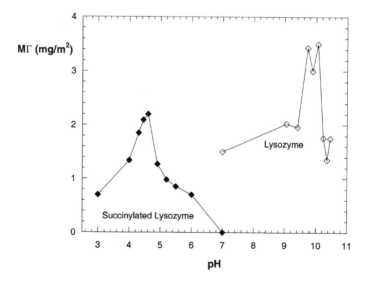

Figure 3 *The adsorbed amount of lysozyme $M\Gamma$ on silica as a function of pH. Protein concentration $=110$ mg L^{-1}; salt concentration $=50$ mM KNO_3. The adsorbed amount was determined from a kinetic curve at 1000 s after the start of the process.*
(Redrawn from ref. 6.)

(p$I \sim 4.5$) on negatively charged silica; for both proteins, the adsorption maximum is close to pI despite the rather different substrate charge.[6]

Protein molecules have a complex core/shell structure, and so one cannot make a simple guess at how the Gibbs free energy G or its derivative, the spreading parameter S, depends on the molecular area A. A constant value of the spreading parameter S obviously does not provide a satisfactory picture, because if S remains constant the spreading goes on indefinitely. As the protein molecule flattens out, counteracting effects clearly must build up, both due to the fact that the molecule consists of covalently bonded moieties, as well as because more and more of the available hydrophobic units get exposed to the solvent. Hence, S must eventually decrease with increasing molar area A. As the simplest conceivable form, one might propose a linear decrease,

$$S = S_0 - pA, \tag{2}$$

where p is a coefficient (N m^{-3} mol) characterizing that decrease. Spreading must stop at $S = 0$, which implies a maximum area:

$$A_{max} = S_0 / p. \tag{3}$$

If one assumes that the *rate* of spreading is also proportional to S this leads to a kinetic picture:

$$dA/dt = S/f, \tag{4}$$

where f (N mol s m^{-3}) is a friction factor. Inserting equation (2) into equation (4) we then obtain a simple equation for $A(t)$:

$$\frac{dA}{dt} = \frac{S_0 - pA}{f} = \frac{p}{f}(A_{max} - A) . \tag{5}$$

This implies an exponential spreading behaviour of the form

$$A(t) = A_0 + \Delta A[1 - \exp(-t/\tau_S)], \tag{6a}$$

where $\tau_S = f/p$ is the *spreading time*, $\Delta A = A_{max} - A_0$ is the maximum area increment due to spreading, and A_{max} is the largest area the molecule can attain.

A slightly different picture is that the spread and native conformations are separated by a barrier. In this case, the molecules primarily exist either in an almost 'fresh' (native) conformation or in a spread configuration; in this representation, a two-state model makes more sense. It suggests a first-order rate equation for the entire adsorbed population, converting molecules with initial contact area A_0 into molecules with area A_{max} and an overall increase of the average area per molecule given by

$$<A>(t) = A_0 + \Delta A[1 - \exp(-t/\tau_S)], \tag{6b}$$

where the spreading time is now related to a rate constant rather than to a friction factor, but the net effect on the *average* area per molecule is the same in equations (6a) and (6b).

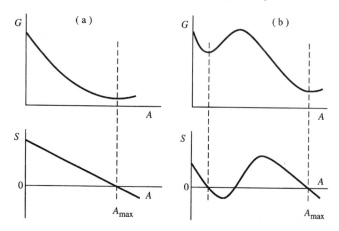

Figure 4 *The free energy G and the spreading parameter S as a function of the molecular area A: (a) linear dependence of S on A; (b) two-state model. The quantity A_{max} is the largest area that a molecule can attain.*

We sketch these simple scenarios in Figure 4. In the first case, G begins (at A_0) as a monotonically decreasing function of A which eventually reaches a minimum; for this case, the quantity $S = -\,dG/dA$ decreases linearly, becoming zero at A_{max} where equilibrium is reached and the spreading stops. In the second case, G has a shallow minimum at low A, but rising with increasing A, reaching a maximum, and decreasing until a second deeper minimum is reached. For this case, S crosses the zero axis three times; the first crossover is a metastable point, the second is unstable, and the third is the stable point A_{max}. This latter case would describe situations where the protein molecule does not unfold gradually, but tends to assume two distinct surface states, one modestly unfolded, and the other much more extensively unfolded.

It is known that the denaturation of many proteins in solution is a transition between two states rather than a gradual swelling. One is therefore inclined to expect that this also holds for surface-bound proteins. Some evidence supporting this idea comes from a study on adsorbed chymotrypsin in which differential scanning calorimetry was used[7] to determine the thermal behaviour of the molecules in the adsorbed film. It was found[7] that the adsorbed film produced an enthalpic peak at the same temperature as the native protein, but with an amplitude that decreased as the film became more dilute. This result is consistent with the existence of two populations, one loosely attached, with denaturation behaviour identical to that of the native molecules, and one considerably more deformed (but retaining much secondary structure), with no measurable heat of denaturation. One would expect that the fraction of this latter population would become smaller as the overall surface coverage increases, which is just what was observed,[7] with supporting evidence coming from measurements of the enzymatic activity on the same system.

4 The Supply-Rate Effect

Accepting that spreading on surfaces is a characteristic property of globular proteins, we must address the question of what it means for the overall kinetics of the adsorption process. For each adsorbing molecule, the adsorption occurs in three steps: (i) transport through the solution by convection and diffusion, (ii) attachment, and (iii) spreading.

The transport step is essentially trivial once the flow pattern near the surface is known. The flux of molecules is found by solving a diffusive–convective equation. Solutions are available for various types of flow.[8] We particularly emphasize here the fact that *solid* substrates are quite different from *liquid* ones, because the latter can change area and this is a factor in the flow pattern. The flux (supply rate) J towards the surface can be generally expressed as:

$$J(t) = c/R = c/(R_{tr} + R_{att}).$$ (7)

Here, c is the bulk concentration, and R is a resistance (s m^{-1}) that can be thought of as being built up from two components: the resistance R_{tr} to transport through the solution (given by the flow parameters and the diffusion constant of the adsorbate), and the resistance R_{att}, which accounts for barriers that must be overcome before a first stable contact is established, but that do *not* depend on surface coverage. This latter resistance can also be seen as an inverse adsorption rate constant: $R_{att} = k_{att}^{-1}$.

The value of J can be associated with a characteristic 'filling time' τ_f, *i.e.*, the time needed to reach a typical coverage Γ_{max} (assuming no desorption):

$$\tau_f = \Gamma_{max}/J(0).$$ (8)

The outcome of the adsorption process is now strongly dependent on the relative values of τ_f and τ_s. If the filling time is much shorter than the spreading time, the adsorbed layer is filled with molecules with a small contact area A, and the adsorbed amount at saturation (A^{-1}) is high. If the filling time is much longer than the spreading time, the majority of the adsorbed molecules occupy a much larger area (of order A_{max}), and the saturated adsorption will be much lower. In the general case, we have a mixed population, and the adsorbed amount has an intermediate value given by $\Sigma_A f_A/A$, where f_A is the fraction of molecules with area A (*i.e.*, $\Sigma_A f_A = 1$). Since we can manipulate τ_f by varying, for example, the concentration or the experimental flow-rate, we expect to see a saturated adsorbed amount which primarily depends on the kinetic conditions under which the adsorbed layer was formed. We call this the 'supply-rate effect'.

5 Adsorption on Solid Surfaces

For solid substrates, experimental examples of the supply-rate effect can easily be found. We present examples in Figures 5–7. In order to bring out clearly the differences, we plot the time-dependent adsorbed amount $\Gamma(t)$ as a function of the scaled variable $\tilde{t} = J(0)t$, which makes the initial slopes of the curves equal to unity by definition, and collapses the initial behaviour onto a single curve. Figure 5 is for savinase adsorbing on silica from a 10 mM salt solution at pH 8,[9] and Figure 6 is for an immunoglobulin (IgG 7B) adsorbing on silica at ionic strengths 0.1 M and 0.005 M.[10] As can be seen, the shapes of the curves are indeed dependent on the concentration of the bulk solution: the lower the concentration (the lower the flux J), the lower is the adsorbed mass where leveling off occurs. Differences in flow-rate have very similar effects. For uncharged flexible polymers like polystyrene, a supply-rate effect is not expected, since τ_s is probably very small. This is supported[9,10] by the experimental data in Figure 7.

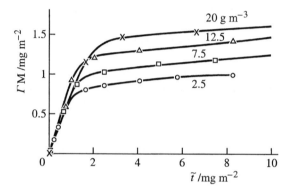

Figure 5 *Time-dependent adsorption $\Gamma(t)$ of savinase as a function of the scaled variable $\tilde{t}$ at four different concentrations (as indicated) on silica at pH 8 and 10 mM salt.*
(Redrawn from ref. 9.)

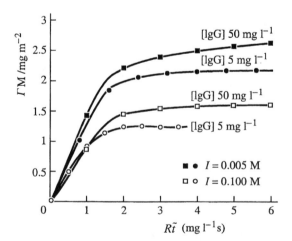

Figure 6 *Time-dependent adsorption $\Gamma(t)$ of IgG 7B on silica (pH 6) as a function of the scaled variable $R\tilde{t}$ for two IgG concentrations (5 and 50 mg L^{-1}, respectively) and two ionic strengths (5 and 100 mM).*
(Redrawn from ref. 10.)

Because the supply-rate effect is a consequence of the kinetics of spreading, it should be no surprise that it is affected not only by the type of macromolecule, but also by the factors controlling the spreading rate, and hence by the nature of the surface. An illustration is given in Figure 8 for the behaviour of lysozyme on two different surfaces, namely bare silica (hydrophilic) and polystyrene (hydrophobic). On silica there is a substantial supply-rate effect;

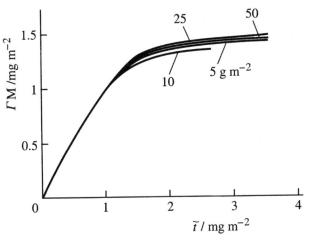

Figure 7 *Time-dependent adsorption $\Gamma(t)$ of polystyrene ($M=3000\ kDa$) from decalin (a good solvent) on silica as a function of the scaled variable $\tilde{t}$ for four polystyrene concentrations as indicated.*
(Data from ref. 11; redrawn from ref. 9.)

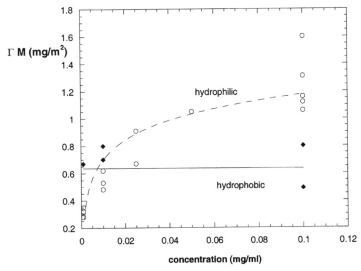

Figure 8 *Adsorbed amounts (at pH 10 and 50 mM KNO_3) for lysozyme on two different surfaces, a hydrophilic silica surface (open symbols) and a hydrophobic polystyrene surface (filled symbols).*
(Unpublished data; courtesy M. van der Veen, Wageningen University.)

the adsorbed amounts increase almost tenfold over the concentration range (1–100 mg L^{-1}). But on polystyrene the adsorbed amount is nearly independent of the concentration at an intermediate value. Presumably, the spreading time on the hydrophobic surface is too short to lead to a supply-rate effect, but the maximum extent of spreading is not so large as it is on bare silica.

6 Effect of Spreading Rate on Overall Adsorption Kinetics

In a rather transparent quantitative analysis of the supply-rate effect,[9] the exponential area relaxation of equation (6) (as introduced by Pefferkorn[12] as the 'growing disk model') can be combined with the assumption that attachment is only possible for the fraction β of the surface that remains unoccupied. Hence, the adsorption rate is given by

$$d\Gamma/dt = \beta J. \tag{9}$$

In equation (9), the quantity β is a decreasing function of time. Its value at time t is obtained by integrating over the entire distribution of molecules with various residence times $t-t'$:

$$\beta(t) = 1 - J \int_0^t dt' \beta(t')[A_0 + \Delta A(1 - e^{-(t-t')/\tau_s})]. \tag{10}$$

It is convenient to define dimensionless variables

$$t^* = t/\tau_s, \qquad J^* = JA_0\tau_s, \qquad a^* = \Delta A/A_0. \tag{11}$$

Equation (10) can then be rewritten as a second-order differential equation,

$$(d^2\beta/dt^{*2}) + (1 + J^*)(d\beta/dt^*) + J^*(1 + a^*)\beta = 0, \tag{12}$$

which has the same form as the equation describing the amplitude of a damped oscillator, containing an inertial term, a friction term, and a driving term. With the function $\beta(t^*)$ known, one easily obtains $\Gamma(t^*)$:

$$\Gamma(t^*) = J^* \int_0^{t^*} dt^{*'} \beta(t^{*'}). \tag{13}$$

We show an example of a set of kinetic curves obtained in this way by plotting adsorbed amount as a function of the dimensionless time $\tilde{t}$, for various values of the scaled variable J^* (Figure 9).

As can be seen from Figure 9 (dashed curves), the adsorption first rises, it reaches a maximum (at $\beta = 0$), and then it decreases again. This decrease is due to the fact that equation (10) allows for negative values of β, leading to desorption; in molecular terms this means that weakly adsorbed molecules may be displaced by more strongly adhering molecules when the latter spread to occupy a larger area. The differences in the heights of the maxima reflect the supply-rate effect, as expected. Adsorption–time curves of this kind, with a maximum, have indeed been observed experimentally.[13-15] However, the subsequent decrease may often not occur. In these cases, the experimental curve is better represented by a constant value of Γ as soon as the maximum has been passed, as indicated by the horizontal full lines in Figure 9. The data for savinase (see Figure 5) corresponds to this case. Fitting the growing disk model to the savinase data (Figure 10) allows the determination of a value of τ_s of 10^2 s. For the IgG data of Figure 6, the estimated spreading time is $\tau_s \sim 10^3$ s.

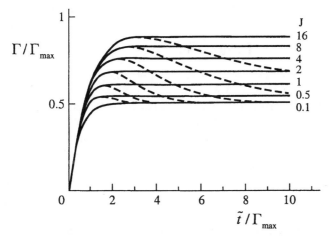

Figure 9 *Calculated adsorption kinetics according to equations (12) and (13), plotted using scaled variables and $a=1$. Both the adsorbed amount Γ and the scaled time $\tilde{t}$ are normalized by the value Γ_{max} for a saturated surface. Full curves: no desorption taken into account. Dashed curves: the solution of equation (12). (Redrawn from ref. 9.)*

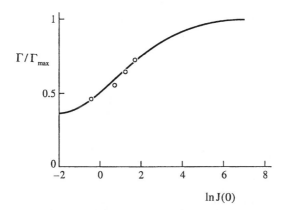

Figure 10 *Saturated adsorption Γ_{max} as a function of flux (logarithmic scale) for savinase on silica, as derived from the data in Figure 5 ($\Delta A/A_0=1.75$, $\tau_s=100$ s, $A_0/M=0.45$ m^2 mg^{-1}). (Redrawn from ref. 9.)*

A somewhat different experimental procedure for probing spreading processes and assessing the value of τ_s was followed by Wertz and Santore.[16] These authors used the supply-rate effect to deduce rates of spreading—or 'footprint' growth, as they called it—based on the time needed to cover 75% of the available area, taking the long-time limit as the saturation level. Figure 11 presents some of the data from this work. It turns out that the estimated spreading rates are very low indeed. For both albumin and fibrinogen, the footprint grows by a factor of about 5 over approximately 30 minutes; the growth seems to be linear in time, showing no signs of leveling off. Additional data indicate that the footprint can continue to grow for about 2 hours!

7 Adsorption on Liquid Surfaces

Proteins do also spread, of course, at liquid interfaces to form patches of a certain area. In particular, water–air interfaces have been extensively studied. The tendency of a protein to flatten, expressed as the area per unit mass A/M (where M is the molar mass), is easily determined for water surfaces by monitoring the surface pressure as a function of the surface density or adsorbed amount Γ. A typical surface pressure *versus* density isotherm of a protein at a water–air interface has the shape depicted in Figure 12. At low adsorbed amounts Γ, the protein film behaves as a dilute gas with a negligible surface pressure $\Pi = RT\Gamma$, but as soon as the molecules start to interact (at $\Gamma = \Gamma^* = A_{max}^{-1}$) the film rapidly builds up a substantial surface pressure.

We have assumed here that the surface compression experiment is done slowly, using spread protein monolayers which have sufficient time to relax before the experiment starts, so that all molecules have reached the maximum molar area A_{max}. One could take the value of area per unit mass, $A_{max}/M = (M\Gamma^*)^{-1}$, at the point where the surface pressure starts to increase as a measure of the ability of a molecule to spread: a large value of this parameter indicates that the molecules have a large area per unit mass and are very thin, whereas a small area per unit mass implies a more limited extent of spreading. Typical

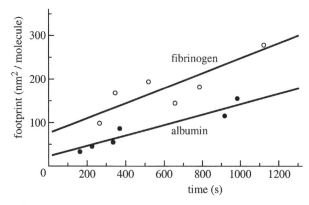

Figure 11 *Footprint (molecular area) of albumin and fibrinogen as a function of time.* (Redrawn using data from ref. 16.)

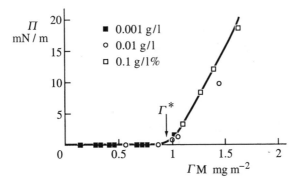

Figure 12 *Surface pressure Π at the water–air interface as a function of surface concentration Γ for ovalbumin at three bulk concentrations (as indicated). A sharp kink occurs at $\Gamma = \Gamma^* = A_{max}^{-1}$.*
(Redrawn from ref. 17.)

values of this parameter range from 1.3 m² mg⁻¹ for a 'hard' protein like lysozyme to 2.4 m² mg⁻¹ for a very 'soft' protein like β-casein. This range implies thicknesses of the molecular 'pancakes' varying between ~0.7 and 0.4 nm (if the densities are assumed to be the same). In experiments with proteins directly deposited at the interface (Langmuir films), the value of Γ is known, but in cases where the layer forms by adsorption it has to be measured, *e.g.*, by ellipsometry.

8 Steady State Measurements

Much less is known about how fast proteins spread on water. In order to get an idea, one would have to measure the ratio A/M as a function of the (average) residence time of the molecules at the interface. One way to tackle this may be to carry out measurements of the surface pressure and adsorbed amount under steady state conditions, *i.e.*, in an experiment where fresh protein is continuously supplied to the interface and removed, such that the adsorbed mass remains stationary. This can be done at a liquid surface, since we can simultaneously supply protein to a surface and remove it by expanding the surface. Experimental arrangements allowing implementation of this are the roller trough and the overflowing cylinder. In these instruments, the water–air surface expands at a well-defined (relative) expansion rate $\theta = \mathrm{d}\ln A/\mathrm{d}t$, independent of the distance from the stagnant zone at the centre. Under the expanding surface, a convective flow is present which transports protein molecules towards the surface; the flux of these molecules in the steady state is given by[9,18]

$$J^+ = \frac{c}{k_{att}^{-1} + \left(\dfrac{\pi}{2\theta D}\right)^{1/2}} \tag{14}$$

where D is the diffusion coefficient, and k_{att} is the rate constant for attachment. Due to the expansion of the surface, the adsorbed layer is also continuously diluted, leading to a loss term (a negative flux):

$$J^- = -\Gamma\theta . \qquad (15)$$

If no desorption occurs, the total flux is just given by the sum of these two terms: $J = J^+ + J^-$. In the steady state, the total flux is zero. It follows that the adsorbed amount in the steady state is given by:

$$\Gamma_{ss} = \frac{c}{\theta/k_{att} + \left(\dfrac{\pi\theta}{2D}\right)^{1/2}} . \qquad (16)$$

Hence, the surface is homogeneously covered with molecules at a coverage Γ_{ss}. We can now define an average residence time τ_r as

$$\tau_r = \Gamma_{ss}/J^+ = -\Gamma_{ss}/J^- = \theta^{-1}. \qquad (17)$$

While the *distribution* of residence times is a rather asymmetric function, the fraction of severely aged molecules is very small.

Studying adsorption at the surface of an overflowing cylinder can give us two pieces of kinetic information. First, we can get an idea of the rate of the attachment step. The idea is to measure the steady-state adsorbed amount Γ_{ss} as a function of concentration and expansion rate. Then, using equation (16), we can plot $c/(\Gamma_{ss}\theta)$ as a function of $\theta^{-1/2}$ which should give a linear plot, with slope $(\Pi/2D)^{1/2}$ and intercept k_{att}^{-1}. Secondly, we have access to the extent of spreading by measuring the surface pressure Π of a protein solution as a function of θ. When Π is plotted as a function of the expansion rate, at constant protein concentration, one obtains a line of negative slope which intersects the $\Pi = 0$ axis at a particular spreading rate θ^*. The residence time of the molecules at that spreading rate equals $(\theta^*)^{-1}$. Moreover, this is the point where the protein molecules are just touching, so that a measure of Γ_{ss}^* (by ellipsometry or fluorescence) *at this point* just gives the reciprocal molar area, which can then be expressed per unit mass to obtain a *dynamic* value of A/M corresponding to $(\theta^*)^{-1}$. Experiments of this sort have been carried out by van Aken and Merks,[19] and the data are reproduced in Figure 13. The authors determined surface pressures for several proteins at a constant protein concentration of 20 mg L^{-1} (50 mg L^{-1} for lysozyme). Clearly, the surface pressure decreases with increasing θ, and it eventually returns to zero at high expansion rates.

Finally, the kinetics of spreading can be studied by extending such measurements to a series in which the *protein concentration* is varied. At each concentration, the value of Γ_{ss}^* and the corresponding value of $(\theta^*)^{-1}$ are determined. A plot of $(\Gamma_{ss}^*)^{-1}$ as a function of $(\theta^*)^{-1}$ describes the spreading kinetics of the protein. As far as we know, such measurements have not yet been done.

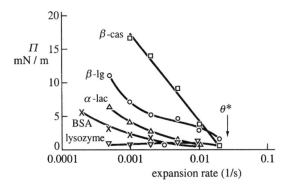

Figure 13 *Steady-state surface pressure Π at 25 °C as a function of expansion rate θ of an air–water interface, for several proteins at pH 7.0 and ionic strength 0.075 M. The protein concentration is 20 mg L⁻¹ (50 mg L⁻¹ for lysozyme). (Redrawn from ref. 19.)*

However, it is expected that, as spreading on water occurs much faster than on solids, the rate may even be too fast to measure by the methods indicated.

9 Mechanical Behaviour of Protein Films

The degree of 'softness' of a protein obviously affects the surface rheological properties of the protein film. Intuitively, one expects that a very 'soft' protein that spreads to a large extent will loose all internal cohesion (though inter-molecular physical or chemical bonds may compensate somewhat for this), and so will form an adsorbed layer that behaves predominantly in a liquid-like manner. Conversely, a 'hard' protein that retains much of its compactness will probably form a film with more elastic properties. An experiment with the potential to highlight such a difference is the shear deformation of a saturated protein layer. In such an experiment the shear strain on an adsorbed protein film is gradually increased and the resulting stress is measured. The stress pattern shows typical yielding behaviour somewhat analogous to that of concentrated polymer systems in three dimensions: there is an initial linear increase in stress corresponding to an elastic response, followed by yielding behaviour where the stress first decreases and then reaches a plateau value. We give an example in Figure 14 for glycinin.[20]

When one compares the plateau shear stress σ_{ss} for different proteins, it turns out that hard proteins with low A_{max}/M can sustain large stresses, whereas soft proteins yield very easily. Quantitatively, the differences are very pronounced, covering about three orders of magnitude. Plotting log σ_{ss} as a function of A_{max}/M for seven different proteins reveals a surprising correlation (Figure 15). The initial slope of the stress–strain curve (the modulus) and the drop in stress are substantially larger for hard proteins than for soft ones. Clearly, a soft protein like β-casein on a water–air surface behaves in a very liquid-like manner, whereas a hard proteins like glycinin forms films that exhibit more glass-like behaviour.

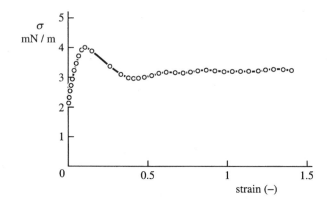

Figure 14 *Stress/strain curve in shear for a glycinin layer on the water–air surface at pH =3. The protein concentration in solution is 100 mg L⁻¹. The stress σ in the layer increases slightly upon ageing.*
(Redrawn from ref. 20.)

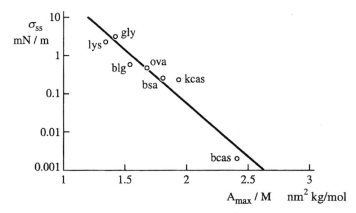

Figure 15 *The plateau shear stress σ_{ss} (on a logarithmic scale) for several proteins plotted as a function of the corresponding values of A_{max}/M (in nm² kg mol⁻¹). Each data point represents a particular protein: lysozyme (lys), glycinin at pH 3 (gly), ovalbumin (ova), β-lactoglobulin (blg), bovine serum albumin (bsa), κ-casein (kcas), and β-casein (bcas).*
(Redrawn from ref. 21.)

Another manifestation of these differences in mechanical behaviour appears when a protein film is neither sheared nor compressed, but is subjected to the growth of an oil film at the surface. The oil is chosen such that it can spread on water provided the pressure of the surrounding protein film does not exceed a given threshold value (in this experiment equal to 15 mN m⁻¹). If an oil reservoir is present under the surface in the form of small emulsion droplets, oil insertion begins when the protein film is diluted down to this critical surface pressure. From that point on, the area of the oil domains grows at the expense of the protein-covered domains. The difference between various proteins shows up in the way by which the oil domains grow: the time-dependent *size*

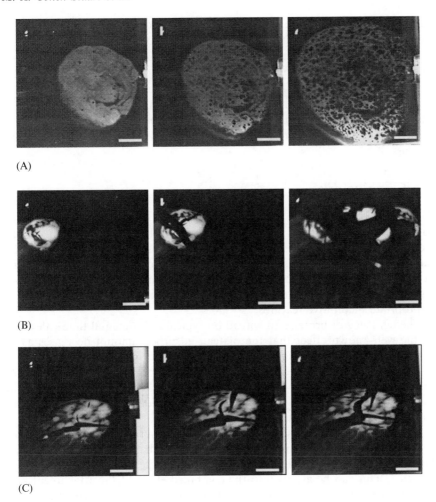

(A)

(B)

(C)

Figure 16 *Oil domains during the expansion of a protein film. The white areas correspond to regions where emulsion droplets are present under the surface; dark spots or cracks appear when these droplets insert into the surface to form an oil film. For each protein, three snapshots during the extension process are shown, from left to right, respectively, in order of increasing surface extension (with respect to the first picture): (A) β-casein (1.00, 1.045 and 1.09); (B) β-lacto-globulin (1.00, 1.024 and 1.156); (C) glycinin at pH 3 (1.00, 1.02 and 1.04). (Reproduced from ref. 22.)*

and *shape* of the domains is very different. This shape turns out to be easily visible. When an oil film appears, the emulsion droplets under the surface disappear, whereas they remain visible as a scattering milky cloud under regions of intact protein film. Figure 16 shows three examples—for β-casein, β-lactoglobulin and glycinin.[22]

In the film composed of the softest protein (β-casein) the oil film grows in the form of *circular islands* (case A). This is consistent with a liquid-like surface where domain shapes are determined by the line tension. The film of the significantly harder protein, β-lactoglobulin, responds very differently to oil invasion—it develops several nearly straight *fractures*, which suddenly appear one after another, and immediately fill up with oil (case B). Clearly, this is more like a brittle gel. The hardest protein of the three, glycinin (at pH = 3), forms a film which also disintegrates by fracturing (case C); but in this last case the rapidly propagating cracks form numerous branches, so that the protein film becomes entirely splintered, rather like the shattering of a plate of glass.

10 Conclusions

The internal cohesion of a protein molecule endows it with a certain resistance to deformation that bears analogy to the effect of a surface tension—as if it were a tiny liquid droplet. The adsorption process is therefore considered in terms of the spreading of such a droplet, and factors affecting the adsorption are reviewed from that perspective. A very simple model of a 'core–shell droplet' characterized by a time-dependent 'spreadability' parameter has all the appropriate qualitative features.

The tendency of proteins to spread on typical experimental timescales leads to the well-known effect that the plateau adsorbed amount decreases as the supply-rate diminishes. The 'growing disk' model, originally due to Pefferkorn *et al.*,[12] can be incorporated into a simple model of adsorption kinetics[9] that turns out to account for typical experimental data. Spreading times have been measured in several cases. They probably vary widely with protein type and substrate nature, but no general rules are known.

The kinetics of protein adsorption on water–air surfaces is quite often monitored by measuring the surface pressure rather than the adsorbed amount. Here, lag times are usually observed before a finite surface pressure can be detected. This can be ascribed to the combined effect of the attachment kinetics and the shape of the two-dimensional pressure–density isotherm, which has a region of negligible pressure as long as the adsorbed proteins are too dilute to interact.

The threshold area per unit mass at which the pressure first becomes measurable is a good indicator of the degree of protein hardness. Hard proteins like lysozyme have values of $\sim 0.8 \text{ m}^2 \text{ mg}^{-1}$, whereas for a soft protein like β-casein the value is $1.44 \text{ m}^2 \text{ mg}^{-1}$. Mechanical properties are found to correlate strongly with the spreadability. In particular, the steady-state shear stress, measured when a protein film is deformed at constant shear-rate, varies by over about three orders of magnitude when comparing soft and hard proteins. A linear correlation is found between the shear stress and the threshold area per unit mass. Finally, we note that the way in which the domains of an invading oil phase grow in a protein film is dramatically influenced by the mechanical properties of the film. For soft proteins, forming liquid-like films, the oil domains are circular, suggesting the action of a line tension. For hard proteins, the domains appear abruptly in the form of oil-filled fractures, indicating a more gel-like or glass-like character of the protein film.

Acknowledgements

We have used data obtained by several of our (former) Ph.D. students and coworkers: Desiree Barten, Marcel van Eijk, Renate Ganzevles, Natalie Hotrum, Anneke Martin, Mark Maste, Marjolein Merks, Marijn van der Veen, and Thierry Zoungrana.

References

1. P. G. de Gennes, *Scaling Concepts in Polymer Physics,* Cornell University Press, Ithaca, NY, 1979.
2. A. Johner and J. F. Joanny, *J. Phys. II*, 1991, **1**, 181.
3. G. Lippmann, *Compt. Rend.,* 1873, **76**, 1407; *Ann. Phys. Chem.*, 1873, **149**, 546.
4. J. M. Kleijn, D. Barten, and M. A. Cohen Stuart, *Langmuir*, 2004, **20**, 9703.
5. V. Razumas, J. Kulys, T. Arnebrant, and T. Nylander, *Elektrokhimiya*, 1988, **24**, 1518 (English translation).
6. M. van der Veen, W. Norde, and M. A. Cohen Stuart, *Colloids Surf. B*, 2004, **35**, 33.
7. T. Zoungrana, G. H. Findenegg, and W. Norde, *J. Colloid Interface Sci.*, 1997, **190**, 437.
8. Z. Adamczyk, T. Dabros, J. Czarnecki, and T. G. M. van de Ven, *Adv. Colloid Interface Sci.*, 1983, **19**, 183.
9. M. C. P. van Eijk and M. A. Cohen Stuart, *Langmuir*, 1997, **13**, 5447.
10. M. G. E. G. Bremer, 'Immunoglobulin adsorption on modified surfaces', Ph.D. thesis, Wageningen University, Netherlands, 2001.
11. J. C. Dijt, M. A. Cohen Stuart, and G. J. Fleer, *Macromolecules*, 1994, **27**, 3207.
12. E. Pefferkorn and A. Elaissari, *J. Colloid Interface Sci.*, 1990, **138**, 187.
13. J. Buijs, P. A. W. van den Berg, J. W. Th. Lichtenbelt, W. Norde, and J. Lyklema, *J. Colloid Interface Sci.*, 1996, **178**, 594.
14. M. C. L. Maste, 'Proteolytic stability in colloidal systems', Ph.D. thesis, Wageningen University, Netherlands, 1996, chap. 3.
15. W. Norde and C. E. Giacomelli, *Macromol. Symp.*, 1999, **145**, 125.
16. C. F. Wertz and M. M. Santore, *Langmuir*, 1999, **15**, 8884.
17. J. Benjamins, 'Static and dynamic properties of proteins adsorbed at liquid interfaces', Ph.D. thesis, Wageningen University, Netherlands, 2001.
18. F. van Voorst Vader, Th. F. Erkens, and M. van den Tempel, *Trans. Faraday Soc.*, 1964, **60**, 1170.
19. G. A. van Aken and M. T. E. Merks, *Colloids Surf. A*, 1996, **114**, 221.
20. A. Martin, M. Bos, M. A. Cohen Stuart, and T. van Vliet, *Langmuir,* 2002, **18**, 1238.
21. A. Martin, 'Mechanical and conformational aspects of protein layers on water', Ph.D. thesis, Wageningen University, Netherlands, 2003.
22. N. E. Hotrum, M. A. Cohen Stuart, T. van Vliet, and G. A. van Aken, *Langmuir*, 2003, **19**, 10210.

Thermodynamic and Adsorption Kinetic Studies of Protein + Surfactant Mixtures

By R. Miller, D. O. Grigoriev, E. V. Aksenenko,[1]
S. A. Zholob,[2] M. E. Leser,[3] M. Michel,[3] and V. B. Fainerman[2]

MPI KOLLOID- UND GRENZFLÄCHENFORSCHUNG, 14424
POTSDAM/GOLM, GERMANY
[1]INSTITUTE OF COLLOID CHEMISTRY AND CHEMISTRY OF
WATER, 42 VERNADSKY AV., 03680 KIEV, UKRAINE
[2]MEDICAL PHYSICOCHEM. CENTRE, DONETSK MEDICAL
UNIVERSITY, 16 ILYCH AVENUE, DONETSK 83003, UKRAINE
[3]NESTEC LTD, NESTLÉ RESEARCH CENTRE,
VERS-CHEZ-LES-BLANC, CH-1000 LAUSANNE, SWITZERLAND

1 Introduction

The practical significance of mixed adsorption layers of proteins with surfactants at fluid interfaces, for example in the food industry for the stabilization of emulsions and foams, is the driving force for the development of new theoretical models to describe their equilibrium and dynamic properties. The addition of surfactants can modify adsorbed protein layers resulting in a change of adsorption as well as rheological characteristics.[1-8] The mechanism of protein–surfactant interaction depends on the nature and concentration of the surfactant in the solution bulk.[4,9-13] With increasing surfactant concentration, hydrophobic interactions between non-polar groups of the surfactant and the protein become more important, leading first to the binding of individual surfactant molecules to the protein, and subsequently to the formation of micelle-like structures.

For mixed solutions of non-ionic surfactant + protein, the surface behaviour at low surfactant concentrations is governed mainly by competitive adsorption.[7,15-27] A theoretical analysis of competitive adsorption of large and small molecules was presented by Lucassen-Reynders.[15] This theoretical model was generalized[27] to take into account the ability of the protein molecule to occupy a variable area in the surface layer.[28] This theory was further extended[29] by accounting for Coulombic interactions between the protein and ionic

120

surfactant molecules. In the present work theoretical and experimental results will be discussed for the adsorption behaviour of mixed solutions of a non-ionic surfactant and a protein.

2 Thermodynamic Model

A model proposed recently[28] assumes that protein molecules can exist in a number of states with different molar areas, varying from a maximum value ω_{max} at very low surface coverage to a minimum value ω_{min} at high surface coverage. The molar areas of two 'neighbouring' conformations differ from each other by the molar area increment ω_0, chosen to be equal to the molar area of the solvent, or the area occupied by one segment of the protein molecule. If the total number of possible states of an adsorbed protein molecule is n, the molar area in the ith state is $\omega_i = \omega_1 + (i-1)\omega_0$, and the maximum area is $\omega_{max} = \omega_1 + (n-1)\omega_0$, for $1 \le i \le n$ and $\omega_1 = \omega_{min} \gg \omega_0$. An equation of state for the surface layer has been derived based on a first-order model for both the non-ideal entropy and the heat of mixing.[28]

Similar to that for pure protein solutions, an analysis of the chemical potentials of several components in the bulk solution and in the surface layer can be performed. Assuming the non-ideality of entropy and enthalpy in the surface layer, and using the approximation $\omega_0 \approx \omega_S$, an equation of state for mixtures of protein + non-ionic surfactant can be obtained:[27]

$$-\frac{\Pi\omega_0}{RT} = \ln(1-\theta_P-\theta_S) + \theta_P(1-\omega_0/\omega) + a_P\theta_P^2 + a_S\theta_S^2 + 2a_{PS}\theta_P\theta_S . \tag{1}$$

Here, Π is the surface pressure, R is the gas constant, T is the temperature, ω is the average molar area of the protein, ω_S is the molar area of the surfactant, a is the Frumkin-type intermolecular interaction parameter (a_P for protein, a_S for surfactant, and a_{PS} for protein–surfactant), Γ_i is the protein adsorption in the ith state, $\theta_P = \omega\Gamma = \sum_{i=1}^{n}\omega_i\Gamma_i$ is the total surface coverage by protein, $\Gamma = \sum_{i=1}^{n}\Gamma_i$ is the total adsorption of protein, $\theta_S = \Gamma_S\omega_S$ is the total surface coverage by surfactant, and Γ_S is the surfactant adsorption. For $\omega_0 \ne \omega_S$, equation (1) can be approximated by

$$-\frac{\Pi(\theta_P\omega_0 + \theta_S\omega_S)}{RT(\theta_P + \theta_S)} = \ln(1-\theta_P-\theta_S) + \theta_P(1-\omega_0/\omega) + a_P\theta_P^2 + a_S\theta_S^2 + 2a_{PS}\theta_P\theta_S. \tag{2}$$

The adsorption isotherms for the protein in state 1 and the surfactant are

$$b_P c_P = \frac{\omega_1 \Gamma_1}{(\omega_1 / \omega)(1-\theta_P-\theta_S)^{\omega_1/\omega}} \exp[-2a_P(\omega_1/\omega)\theta_P - 2a_{PS}\theta_S], \tag{3}$$

$$b_S c_S = \frac{\theta_S}{(1-\theta_P-\theta_S)} \exp[-2a_S\theta_S - 2a_{PS}\theta_P], \tag{4}$$

where c_P and c_S are the concentrations in the solution bulk, b_P is the adsorption equilibrium constant for the protein in the jth state ($j=1$ in equation (3)), and b_S is the adsorption equilibrium constant for the surfactant. The values of b_P for all states j are assumed to be identical,[28] and so the adsorption constant for the protein molecule as a whole is $\Sigma b_P = n b_P$. The distribution of adsorbed protein over all the states is given by

$$\Gamma_j = \Gamma \frac{(1-\theta_P-\theta_S)^{\frac{\omega_j - \omega_1}{\omega}} \exp[2a_P\theta_P(\omega_j - \omega_1)/\omega]}{\sum_{i=1}^{n} (1-\theta_P-\theta_S)^{\frac{\omega_i - \omega_1}{\omega}} \exp[2a_P\theta_P(\omega_i - \omega_1)/\omega]}. \tag{5}$$

The set of equations (2–5) is sufficient to describe the adsorption behaviour of mixed solutions of protein + non-ionic surfactant. The problem of the theoretical description of such a mixture is formulated as follows: given the known values of T, ω_0, ω_{min}, ω_{max}, a_S, a_P, a_{PS}, b_P, c_P and b_S, the dependencies of ω, Γ, Γ_S, θ_P, θ_S and Π as a function of the surfactant concentration c_S can be calculated. An algorithm for this calculation was recently presented.[27]

The set of equations (2–5) can be significantly simplified by assuming an ideal surface layer (with respect to enthalpy and entropy), *i.e.*, the set of equations transforms into a simple additive relationship which expresses the surface pressure of the mixed protein + surfactant solution:[30]

$$\exp\bar{\Pi} = \exp\bar{\Pi}_P + \exp\bar{\Pi}_S - 1. \tag{6}$$

Here the quantities $\bar{\Pi} = \Pi\omega_0/RT$ for $\omega_0 \cong \omega_s$ (or $\bar{\Pi} = \Pi(\omega_0 + \omega_S)/2RT$ for $\omega_0 \neq \omega_S$), $\bar{\Pi}_P = \Pi_P\omega_0/RT$ and $\bar{\Pi}_S = \Pi_S\omega_S/RT$ are the dimensionless surface pressures of the mixture and the pure solutions of protein and surfactant, respectively, the latter being taken at the same concentrations as in the mixture. Equation (6) is a generalization of a model derived recently from the analysis of ideal mixtures of two surfactants.[31] For a number of binary mixtures of even quite different surfactants (non-ionic + non-ionic or non-ionic + ionic) it has been shown[32] that equation (6) works very well. This validity can be ascribed primarily to the fact that some particular adsorption features of the components are accounted for 'automatically', because the surface pressures of the individual solutions enter into equation (6). It was also shown[30] that surface pressures of mixtures of some selected non-ionic surfactants and proteins agree well with estimates obtained from equation (6).

3 Dynamics of Adsorption and the Dynamic Surface Tension

Equations of state for surface layers and the respective adsorption isotherms are the basis for a quantitative analysis of adsorption kinetics and dynamic surface tensions under quasi-equilibrium conditions. The integral equation proposed by Ward and Tordai[33] represents a general relationship between the dynamic adsorption $\Gamma(t)$ and the sub-surface concentration $c_i(0,t)$. The corresponding equations for the separate time evolutions of adsorption of protein and surfactant, respectively, in a mixed solution have the following form[34]

$$\Gamma_P(t) = 2\sqrt{\frac{D_P}{\pi}}\left[c_{0P}\sqrt{t} - \int_0^{\sqrt{t}} c_P(0,t-t')\,d(\sqrt{t'})\right] \pm \frac{c_{0P}D_P}{r}t, \qquad (7)$$

and

$$\Gamma_S(t) = 2\sqrt{\frac{D_S}{\pi}}\left[c_{0S}\sqrt{t} - \int_0^{\sqrt{t}} c_S(0,t-t')\,d(\sqrt{t'})\right] \pm \frac{c_{0S}D_S}{r}t, \qquad (8)$$

where D_P and D_S are the diffusion coefficients of the protein and surfactant, respectively, t is the time, c_{0P} and c_{0S} are the corresponding bulk concentrations, and r is the radius of curvature. The signs '$-$' or '$+$' before the second term on the right of equation (8) correspond respectively to diffusion inside or outside a droplet (or bubble), and t' is a dummy integration variable. Using equations (7) and (8), and the isotherms (3) and (4) as boundary conditions, the functions $\Gamma_P(t)$ and $\Gamma_S(t)$ for a protein + non-ionic surfactant mixture can be calculated.

4 Results and Discussion

The surface tension isotherms for β-lactoglobulin (β-lg),[6,35] decyl dimethyl phosphine oxide ($C_{10}DMPO$),[36] and mixtures of β-lg+$C_{10}DMPO$ as a function of $C_{10}DMPO$ concentration at a fixed β-lg concentration of 10^{-6} mol L^{-1} are shown in Figure 1. The surface tensions of the pure and mixed β-lg solutions were measured with the profile analysis tensiometer.

The adsorption behaviour of the non-ionic surfactant is well described by the Frumkin equation of state and the corresponding adsorption isotherm:

$$-\frac{\Pi \omega_S}{RT} = \ln(1-\theta_S) + a_S\theta_S^2, \qquad (9)$$

$$b_S c_S = \frac{\theta_S}{(1-\theta_S)}\exp[-2a_S\theta_S]. \qquad (10)$$

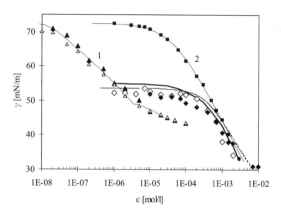

Figure 1 *Surface tension of β-lg solutions (△, data from ref. 6; ▲, data from ref. 35, curve 1), $C_{10}DMPO$ solutions (■, data from ref. 36, curve 2) and of mixed solutions of β-lg (10^{-6} mol L^{-1}) with $C_{10}DMPO$ (◇, ◆). Theoretical curves for pure $C_{10}DMPO$ solutions are calculated from equations (9) and (10), for pure β-lg solutions from the model given in ref. 28 using the parameters in the Table 1, and for mixtures from equations (2–5) with $a_{PS}=0$ (thin curve) and equation (6) (thick curve).*

Table 1 *Parameter values used in the theoretical analysis.*

Protein	ω_0 ($m^2\ mol^{-1}$)	ω_{min} ($m^2\ mol^{-1}$)	ω_{max} ($m^2\ mol^{-1}$)	a_P	b_P ($L\ mol^{-1}$)	$b_P n$ ($L\ mol^{-1}$)
HSA	2.5×10^5	3.0×10^7	7.5×10^7	1.0	3.0×10^5	5.43×10^7
β-lg	3.5×10^5	4.4×10^6	1.8×10^7	0.4	1.4×10^6	5.46×10^7

Surfactant	ω_S ($m^2\ mol^{-1}$)	a_S	b_S ($L\ mol^{-1}$)
$C_{10}DMPO$	2.5×10^5	-0.25	2.19×10^4

For $C_{10}DMPO$ the following parameters result:[36] $\omega_S = 2.5 \times 10^5$ m^2 mol^{-1}, $a_S = -0.25$, and $b_S = 2.19 \times 10^4$ L mol^{-1}. Two calculated curves for the β-lg + $C_{10}DMPO$ mixture are shown in Figure 1: one curve is calculated from equations (2–5) with $a_{PS} = 0$ using the parameters for β-lg as given in Table 1, and the second is from the simplified additive model of equation (6). Both theoretical curves agree well with the experimental data.

Figure 2 illustrates experimental surface tension isotherms for human serum albumin solutions (HSA),[37] $C_{10}DMPO$, and HSA + $C_{10}DMPO$ mixtures as a function of $C_{10}DMPO$ concentration at a fixed HSA concentration of 10^{-7} mol L^{-1}. Again two calculated curves are shown—one from equations (2–5) with $a_{PS} = 0$ using the parameters for HSA given in Table 1, and the other from the simplified equation (6).

The competitive nature of adsorption from a mixed protein + surfactant solution is illustrated in Figure 3, where the dependencies of adsorption of

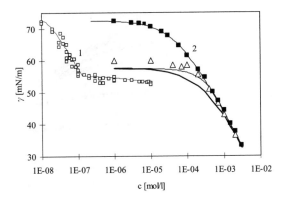

Figure 2 *Surface tension of HSA solutions (□, data from ref. 37, curve 1), $C_{10}DMPO$ solutions (■, data from ref. 36, curve 2), and of the mixture of HSA (10^{-7} mol L^{-1}) with $C_{10}DMPO$ (△). Theoretical curves for pure $C_{10}DMPO$ solutions are calculated from equations (9) and (10), for pure HSA solutions from the model given in ref. 28 using the parameters in Table 1, and for mixtures from equations (2–5) with $a_{PS}=0$ (thin curve) and equation (6) (thick curve).*

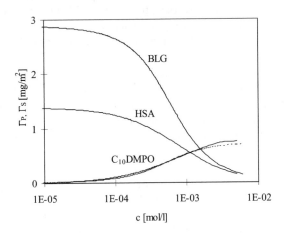

Figure 3 *Dependence of adsorption of β-lg, HSA and $C_{10}DMPO$ in HSA + $C_{10}DMPO$ and β-lg + $C_{10}DMPO$ mixtures on the $C_{10}DMPO$ concentration. The dotted line represents the adsorption of $C_{10}DMPO$ in β-lg + $C_{10}DMPO$ mixtures.*

protein (HSA and β-lg) and surfactant ($C_{10}DMPO$) on the surfactant concentration are shown. With increasing surfactant concentration, a significant decrease in the protein adsorption is observed due to the increased $C_{10}DMPO$ adsorption. This behaviour, first observed and discussed by Lucassen-Reynders,[9] arises from the non-ideality of the entropy of mixing. Mathematically, the contribution from the non-ideality of the entropy is expressed by the second term on the right of equation (2), and by the fact that the factor (ω_1/ω) and exponent (ω_1/ω) in the denominator of the protein adsorption isotherm (equation (3)) is lower than 1. The two proteins considered here are

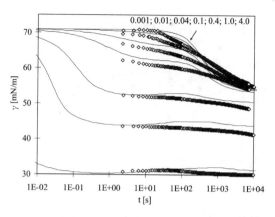

Figure 4 *Dynamic surface tension of a 10^{-6} mol L^{-1} β-lg solution mixed with different $C_{10}DMPO$ concentrations (in mmol L^{-1}), measured by profile analysis tensiometry ($\diamondsuit$). The theoretical curves are calculated from equations (2–5), (7) and (8).*

globular and have approximately the same ratio ω_1/ω. Therefore, the forms of the curves shown in Figure 3 (especially for C_{10}DMPO) are quite similar. It was shown earlier[27] that, since they have $\omega_1/\omega < 1$, flexible proteins are more subject to displacement from the adsorption layer than are globular proteins.

Figure 4 illustrates the dynamic surface tensions of mixed solutions of 10^{-6} mol L^{-1} β-lg at different C_{10}DMPO concentrations, as measured using the profile analysis tensiometer. The theoretical curves are calculated from the diffusion controlled adsorption kinetics model for mixtures using equations (2)–(5), (7) and (8) with the diffusion coefficients $D_P = 5 \times 10^{-11}$ m^2 s^{-1} and $D_S = 4 \times 10^{-10}$ m^2 s^{-1}, which are physically reasonable values. One can see that the lines fit the experimental data satisfactorily.

Considering the change in adsorption of the two components, as shown in Figures 5 and 6, one can see that β-lg starts to adsorb at longer times, and then it competes with the C_{10}DMPO molecules adsorbed at shorter times. Due to this competition, at small concentrations the adsorbed amount of C_{10}DMPO passes through a maximum, and then it decreases and levels off at the respective equilibrium value. The larger the C_{10}DMPO concentration, the less is the β-lg adsorption, and the less the β-lg is displaced by C_{10}DMPO from the surface layer. Hence, these model calculations allow a better understanding of the adsorption behaviour of mixed protein + surfactant solutions.

In Figure 7 the dynamic surface tension data of mixed solutions of 10^{-7} mol L^{-1} HSA at different C_{10}DMPO concentrations are shown. The theoretical curves were calculated from the same diffusion-controlled adsorption kinetics model with diffusion coefficients $D_P = 3 \times 10^{-11}$ m^2 s^{-1} and $D_S = 4 \times 10^{-10}$ m^2 s^{-1}. Again the lines provide a satisfactory fit to the experimental data.

In Figure 8 the change in HSA adsorption is compared to that for β-lg at several C_{10}DMPO concentrations. As the absolute values of the HSA and β-lg adsorption in the individual solutions at concentrations of 10^{-7} mol L^{-1} and 10^{-6} mol L^{-1}, respectively, are rather different (*cf.* Figure 4), the data are

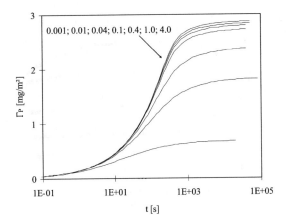

Figure 5 *Calculated adsorption of β-lg in a mixed 10^{-6} mol L^{-1} β-lg solution at different $C_{10}DMPO$ concentrations (in mmol L^{-1}).*

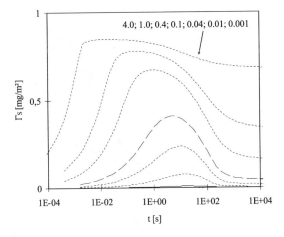

Figure 6 *Calculated adsorption of $C_{10}DMPO$ in mixed 10^{-6} mol L^{-1} β-lg solutions at different $C_{10}DMPO$ concentrations (in mmol L^{-1}).*

presented as relative adsorption values (ratio of dynamic adsorption in the mixture to the maximum adsorption for the individual protein solution at a chosen concentration). For the same $C_{10}DMPO$ concentration, the HSA is more intensively replaced from the adsorption layer than the β-lg. This is due mainly to the fact that the HSA concentration is lower. Another reason for this effect is the fact that the exponential term on the right hand side of equation (3) for $a_{PS} = 0$, *i.e.*, the value $\exp[-2a_P(\omega_1/\omega)\theta_P]$, is lower for HSA than for β-lg. This decrease in the exponential factor results from the fact that the a_P values for the compared proteins are different (see Table 1), which also leads to significantly lower θ_P values for HSA. Quite expectedly, the adsorption behaviour of $C_{10}DMPO$ in the two compared systems is quite different. It

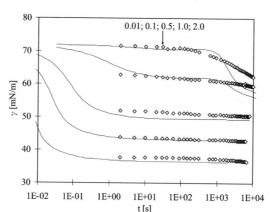

Figure 7 *Dynamic surface tension of a 10^{-7} mol L^{-1} HSA solution mixed with different $C_{10}DMPO$ concentrations (in mmol L^{-1}), measured by profile analysis tensiometry ($\diamond$). The theoretical curves are calculated from equations (2–5), (7) and (8).*

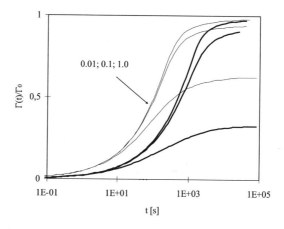

Figure 8 *Calculated relative adsorption of β-lg (10^{-6} mol L^{-1}, thin curve) and HSA (10^{-7} mol L^{-1}, thick curve) in a mixed solution at different $C_{10}DMPO$ concentrations (in mmol L^{-1}).*

follows from Figure 9 that the dynamic adsorption of $C_{10}DMPO$ is higher when HSA rather than β-lg is present in the solution. At the same time, over the short timescale when the extent of protein adsorption is insignificant, the $C_{10}DMPO$ adsorption dynamic curves for both systems are almost the same.

5 Conclusions

The set of thermodynamic equations presented here provides a good description of the equilibrium behaviour of mixtures of protein + non-ionic surfactant at liquid/fluid interfaces. The theoretical calculations of the adsorption

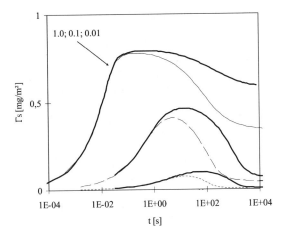

Figure 9 *Calculated adsorption of $C_{10}DMPO$ in mixed 10^{-6} mol L^{-1} β-lg solutions (thin curve) and 10^{-7} mol L^{-1} HSA solutions (thick curve) at different $C_{10}DMPO$ concentrations (in mmol L^{-1}).*

behaviour are based on the adsorption parameters of the individual protein and surfactant solutions. Analysis of the competitive character of adsorption from protein + surfactant mixtures shows that for increasing surfactant concentration the protein adsorption decreases. Experimental data for the mixtures agree satisfactorily with the theoretical estimates.

The dynamic surface tensions measured for binary mixtures of β-lg (10^{-6} mol L^{-1}) or HSA (10^{-7} mol L^{-1}) with $C_{10}DMPO$ are in satisfactory agreement with the predictions made from the diffusion model. The equation of state and adsorption isotherms derived from the thermodynamic analysis are used as boundary conditions in this kinetic model. The results present, for the first time, a quantitative analysis of experimental adsorption data for protein + surfactant mixtures.

References

1. R. Miller, V. B. Fainerman, A. V. Makievski, J. Krägel, D. O. Grigoriev, V. N. Kazakov, and O. V. Sinyachenko, *Adv. Colloid Interface Sci.*, 2000, **86**, 39.
2. A. Dussaud, G. B. Han, L. Ter Minassian-Saraga, and M. Vignes-Adler, *J. Colloid Interface Sci.*, 1994, **167**, 247.
3. J. Krägel, R. Wüstneck, D. Clark, P. Wilde, and R. Miller, *Colloids Surf. A*, 1995, **98**, 127.
4. N. J. Turro, X.-G Lei, K. P. Ananthapadmanabhan, and M. Aronson, *Langmuir*, 1995, **11**, 2525.
5. R. Wüstneck, J. Krägel, R. Miller, P. J. Wilde, and D. C. Clark, *Colloids Surf. A*, 1996, **114**, 255.
6. J. Krägel, R. Wüstneck, F. Husband, P. J. Wilde, A. V. Makievski, D. O. Grigoriev, and J. B. Li, *Colloids Surf. B*, 1999, **12**, 399.
7. E. Dickinson, *Colloids Surf. B*, 1999, **15**, 161.

8. R. Miller, V. B. Fainerman, A. V. Makievski, J. Krägel, and R. Wüstneck, *Colloids Surf. A*, 2000, **161**, 151.
9. J. Oakes, *J. Chem. Soc. Faraday Trans. 1*, 1974, **70**, 2200.
10. M. N. Jones, *Biochem. J.*, 1974, **151**, 109.
11. S. J. McClellan and E. I. Franses, *Colloids Surf. B*, 2003, **30**, 1.
12. S. F. Santos, D. Zanette, H. Fischer, and R. Itri, *J. Colloid Interface Sci.*, 2003, **162**, 400.
13. D. Kelley and D. J. McClements, *Food Hydrocoll.*, 2003, **17**, 73.
14. V. B. Fainerman and E. H. Lucassen-Reynders, *Adv. Colloid Interface Sci.*, 2002, **96**, 295.
15. E. H. Lucassen-Reynders, *Colloids Surf. A*, 1994, **91**,79.
16. M. A. Cohen Stuart, G. J. Fleer, J. Lyklema, W. Norde, and J. M. H. M. Scheutjens, *Adv. Colloid Interface Sci.*, 1991, **34**, 477.
17. P. Joos, 'Dynamic Surface Phenomena', VSP, Utrecht, 1999.
18. P. Joos and G. Serrien, *J. Colloid Interface Sci.*, 1991, **145**, 291.
19. M. A. Bos and T. van Vliet, *Adv. Colloid Interface Sci.*, 2001, **91**, 437.
20. J. Chen and E. Dickinson, *Food Hydrocoll.*, 1995, **9**, 35.
21. B. S. Murray, *Colloids Surf. A*, 1997, **125**, 73.
22. A. R. Mackie, A. P. Gunning, P. J. Wilde, and V. J. Morris, *Langmuir*, 2000, **16**, 8176.
23. A. R. Mackie, A. P. Gunning, P. J. Wilde, and V. J. Morris, *J. Colloid Interface Sci.*, 1999, **210**, 157.
24. C. M. Wijmans and E. Dickinson, *Langmuir*, 1999, **15**, 8344.
25. A. R. Mackie, A. P. Gunning, M. J. Ridout, P. J. Wilde, and V. J. Morris, *Langmuir*, 2001, **17**, 6593.
26. A. R. Mackie, A. P. Gunning, M. J. Ridout, P. J. Wilde, and J. M. Rodriguez Patino, *Biomacromolecules*, 2001, **2**, 1001.
27. V. B. Fainerman, S. A. Zholob, M. Leser, M. Michel, and R. Miller, *J. Colloid Interface Sci.*, 2004, **274**, 496.
28. V. B. Fainerman, E. H. Lucassen-Reynders, and R. Miller, *Adv. Colloid Interface Sci.*, 2003, **106**, 237.
29. V. B. Fainerman, S. A. Zholob, M. E. Leser, M. Michel, and R. Miller, *J. Phys. Chem. B*, in press.
30. R. Miller, V. B. Fainerman, M. E. Leser, and M. Michel, *Colloids Surf. A*, 2004, **233**, 39.
31. V. B. Fainerman and R. Miller, *J. Phys. Chem. B*, 2001, **105**, 11432.
32. V. B. Fainerman, R. Miller, and E. V. Aksenenko, *Adv. Colloid Interface Sci.*, 2002, **96**, 339.
33. A. F. H. Ward and L. Tordai, *J. Chem. Phys.*, 1946, **14**, 543.
34. R. Miller, V. B. Fainerman, E. V. Aksenenko, M. E. Leser, and M. Michel, *Langmuir*, 2004, **20**, 771.
35. J. Krägel, M. O'Neill, A. V. Makievski, M. Michel, M. E. Leser, and R. Miller, *Colloids Surf. A*, 2003, **31**, 107
36. V. B. Fainerman, R. Miller, E. V. Aksenenko, and A. V. Makievski, in 'Surfactants—Chemistry, Interfacial Properties and Application', Studies in Interface Science—vol. 13, eds V. B. Fainerman, D. Möbius, and R. Miller, Elsevier, Amsterdam, 2001, p. 189.
37. A. V. Makievski, V. B. Fainerman, M. Bree, R. Wüstneck, J. Krägel, and R. Miller, *J. Phys. Chem. B*, 1998, **102**, 417.

Computer Simulation of Interfacial Structure and Large-Deformation Rheology during Competitive Adsorption of Proteins and Surfactants

By Luis A. Pugnaloni, Rammile Ettelaie, and Eric Dickinson

PROCTER DEPARTMENT OF FOOD SCIENCE, UNIVERSITY OF LEEDS, LEEDS LS2 9JT, UK

1 Introduction

The process of adsorption at fluid interfaces is crucial to emulsion and foam formation as well as to the long-term stability of already-formed droplets and bubbles.[1] The competitive adsorption of different species is of particular interest in food systems where different surface-active components are commonly present. Proteins and low-molecular-weight surfactants have very different adsorption behaviour, and, most interestingly, they present complex interfacial structures when adsorbed from mixed bulk solutions. It has been shown by atomic force microscopy (AFM)[2] and Brewster angle microscopy[3] that the competitive adsorption of a protein + surfactant mixture occurs in a heterogeneous fashion, with the surfactant forming segregated domains in a quasi-two-dimensional protein matrix, at least over periods of time of a few days.

The large-deformation mechanical properties of these types of adsorbed films can be directly measured by using apparatus such as interfacial tensiometers.[4] Stress–strain curves can be obtained during the large-scale compression and expansion of the interfacial film. The mechanical properties of the adsorbed film determine its ability to support the stresses to which it may be subjected during droplet/bubble deformation or disproportionation.

In this paper we present some new results of the computer simulation of adsorbing binary mixtures. The molecules are modelled as spherical particles that adsorb at a planar interface by virtue of a steep external attracting force. The protein-like particles can form elastic bonds with each other; the idealized surfactant-like particles do not bond to other particles. The simulation is carried out using a Brownian dynamics (BD) algorithm. The simulated structure of the complex mixed adsorbed films has been shown[5] to compare very well with AFM images of adsorbed mixtures of globular protein + non-ionic surfactant.

We specifically investigate here the stress response to large uniaxial deformations of model single-protein films and protein + surfactant mixed films. We show how the breakability of the protein–protein bonds affects the general structure of a compressed protein film, beyond the collapse point at which the molecules are displaced from the interface to release stress. While unbreakable bonds lead to the formation of 'wrinkles' under compression, the presence of very easy-to-break bonds allows homogeneous displacement of the compressed film. The nature of these protein–protein bond parameters also affects the overall displacement behaviour of any co-adsorbed species during a large-deformation compression treatment.

2 The Model

For studying the structural implications of the adsorption of interacting surface-active components, a simple BD simulation approach was developed by Wijmans and Dickinson.[6,7] A detailed description of this model can be found in a recent review.[5] The model consists in essence of spherical particles interacting via a steeply repulsive spherical core potential. To mimic surface-active properties of the molecules, each particle interacts with an external potential, acting in the z-direction. This potential has a square-well-like shape with one infinite wall and one finite wall. The infinite wall prevents particles from escaping to the phase in which they are not soluble—typically air or oil in the case of proteins. Conversely, the finite wall allows for interchange of particles between the interface and the bulk phase in which they are soluble or dispersible—typically an aqueous solution in an experimental setup. A particle is said to be adsorbed if it lies within the external potential well.

In order to account for the intermolecular cross-linking behaviour of some proteins, the adsorbing particles are also allowed to interact with each other through flexible bonds. Nodes are created on the nominal surface of the spherical particles. The bond interaction acts along the straight line that joins the corresponding nodes, and it depends on the node-to-node distance b_{ij} only. As these bonding forces do not need to operate in purely radial directions, they can give rise to torques acting on each particle. An interparticle bond is created with a given probability $P_B^{\alpha\beta}$ when two particles approach within a distance b_1. Initially, nodes are created on the line that joins the particle centres. After bond formation, the nodes that define the ends of the bond are fixed within the corresponding particle reference system. That is, the nodes remain fixed at the initial position on the surface of each particle. If the particle moves such that the length of a bond exceeds b_{max}, the bond is deemed to be broken. By setting $b_{max} = b_{max} = \infty$, irreversible (unbreakable) bonds can be simulated. Throughout this article we consider bonds with an elastic constant of 200 $k_B T \sigma^{-2}$. Here, σ is the diameter of the particles, k_B is the Boltzmann factor, and T is the absolute temperature.

Particles cannot escape in the negative z-direction, as they are trapped by the infinite wall of the interface, but they can move away (and become "lost") in the positive z-direction. In order to avoid this happening, an additional "wall" has to be added at a given z-position, z_w.

The properties of the present model could, in principle, be studied by means of molecular dynamics, Monte Carlo, or BD simulation. However, the first two techniques would require the introduction of an extra component in the system to account for the solvent. Therefore, BD simulations are normally more efficient, as they introduce the solvent effect by just assuming a bulk solvent viscosity, η. The time is conventionally normalized with respect to the average time τ taken for a particle to diffuse a distance equal to its radius in an infinitely diluted system, *i.e.*,

$$\tau = \frac{(\sigma/2)^2}{6D_0} = \frac{\pi\sigma^3\eta}{8k_BT}, \tag{1}$$

where $D_0 = k_BT/(3\pi\eta\sigma)$ is the diffusion coefficient at infinite dilution.

3 Brownian Dynamics Simulation

The simulation is carried out using the BD technique to represent solvent effects. The dynamics of the particles are therefore assumed to be over-damped. As such, inertial terms are negligible and the velocity of each particle is directly proportional to the applied force on that particle. Consequently, one needs only to update the positions of the particles throughout the simulation. The diffusion coefficient at infinite dilution is set at $D_0 = 2.21 \times 10^{-4}\sigma\sqrt{k_BT/m}$, where m is the mass of the particles. In addition to the particle–particle and particle–interface interactions, a drag force and a random force consistent with D_0 are applied to each particle. The position update algorithm is

$$\mathbf{r}_i(t+\Delta t) = \mathbf{r}_i(t) + \mathbf{F}_i(t)\frac{\Delta t}{\xi} + \mathbf{R}_i(t,\Delta t), \tag{2}$$

where $\xi = k_BT/D_0$ is the Stokes friction coefficient, and $\mathbf{R}_i(t,\Delta t)$ is a Gaussian random displacement of zero mean and variance $\langle \mathbf{R}_i^2(t,\Delta t)\rangle = 2D_0\Delta t$.[8] Because of the torque applied by the bonds to the particles, there is also an analogous rotational update algorithm.[9]

Particles are normally allowed to spread at random at the interface ($z=0$) and then the system is allowed to equilibrate. The equilibration time depends on the particular set of parameters chosen; in some cases aggregation, coarsening, and/or desorption may require rather long equilibration times. The system is simulated within a cubic box of side length L. Due to the inhomogeneity introduced by the interfacial potential, periodic boundary conditions are applied in the x- and y-directions only. A neighbour list is used to speed up the calculation of the interparticle forces. The area fraction Γ is defined as the area occupied by the adsorbed particles divided by the total area of the interface in the simulation, *i.e.*, $\Gamma = \pi N^{ads}/(4L^2)$.

The response of the interface to an external perturbation can be measured through the interfacial stress tensor **S**. For a pairwise-additive interaction, its components are given by the Kirkwood formula[6,10]

$$S_{\alpha\beta} = \frac{N}{A} k_B T \delta_{\alpha\beta} - \frac{1}{A} \sum_{j>i}^{N} \sum_{i=1}^{N-1} r_{\alpha ij} F_{\beta ij},$$

(3)

where A is the area of the interface. Here, α and β indicate the different Cartesian components of the interfacial stress tensor **S**, the interparticle distance r_{ij}, and the interparticle force F_{ij}, respectively. We assume that the bulk stress tensor is negligible compared to the interfacial stress tensor, and therefore we sum over all particles in the system. Normally, the macroscopic rheological quantities of interest are related to the *symmetric* part of **S**, *i.e.*,

$$\bar{S}_{\alpha\beta} = \frac{1}{2}(S_{\alpha\beta} + S_{\beta\alpha}).$$

(4)

The dilatational rheology can be extracted by subjecting the system to changes in interfacial area.[6] We compress the interface in the x-direction at constant velocity, thereby simulating the type of compression experiment typically carried out in a Langmuir trough. The length of the interface is then given by

$$L_x(t) = L_{x,0} - v_x t,$$

(5)

where $L_{x,0}$ is the initial length, and v_x is the velocity of compression. In order to maintain a constant volume for the entire system, the simulation box has to be expanded in the z-direction according to

$$L_z(t) = L_{z,0} + \frac{L_{z,0} v_x t}{L_{x,0} - v_x t}.$$

(6)

4 Single-Component Monolayers

Three types of adsorbed layers consisting of a single component have been subjected to uniaxial compression and re-expansion: (a) a monolayer consisting of particles that interact through permanent bonds, (b) a monolayer consisting of particles that interact through transient bonds ($b_{max} = 0.5\sigma$), and (c) a monolayer consisting of particles that interact solely through the repulsive core potential.

Particles are initially spread at random at the interface and equilibrated over 10^6 time steps. During equilibration some particle–particle bonds are formed in cases (a) and (b), and eventually a cross-linked network of particles develops. The equilibrated interfacial structure for each system is shown in the first column of images in Figure 1. The monolayers are compressed in the

Initial   Compressed   Re-expanded

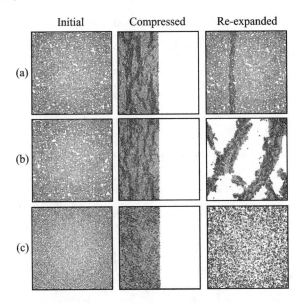

Figure 1 *Uniaxial compression and re-expansion of a model protein layer. From top to bottom, the type of particle–particle bonding is: (a) permanent bonds, (b) transient bonds, and (c) no bonds. Snapshots are shown of the system before compression, after compression (50% of the original area), and after full re-expansion. Dark spheres correspond to particles displaced from the interface.*

x-direction and re-expanded up to the original interfacial area. The velocity of compression–expansion has been set at $v_x = 0.132\sigma\tau^{-1}$. In different simulation runs, re-expansion has been started at various points during the compression process: these correspond to 15%, 30%, 50% and 70% reduction in area. The stress response curves obtained for the different monolayers are presented in Figure 2. Most cases present hysteresis in the stress curves, indicating an irreversible change in the system. An exception to this rule occurs for the smallest extent of deformation (15% compression) for the cases where the collapse point is not yet reached. However, for particles interacting through transient bonds (system type (b)), the irreversibility of the compression process is apparent even at such a relatively small deformation.

In case (a), there is neither the breaking of existing bonds nor the creation of new bonds during the compression–expansion process. During the compression phase, Figure 2a shows that the stress builds up rather fast up to the collapse point, whereupon some particles are forced out of the interface. At this point a few wrinkles are formed in the monolayer and some of the interfacial stress is released. These wrinkles are seen to lie perpendicular to the direction of compression. Upon re-expansion, the interfacial stress is progressively released, but it follows a different route to that of the compression route. The structure of the interface after 50% compression and after full re-expansion is shown in the second and third columns of Figure 1, respectively. Although a

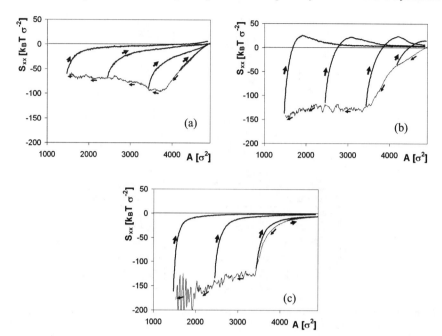

Figure 2 *Uniaxial compression and re-expansion of a model protein layer. The type of particle–particle bonding is: (a) permanent bonds, (b) transient bonds, and (c) no bonds. The stress response along the direction of compression is shown as a function of the interfacial area. The thin line refers to the compression, and the thick line refers to re-expansion at different stages of the compression (15%, 30%, 50% and 70% interfacial area reduction).*

rather wide single wrinkle still remains at this point, all the particles were observed to re-adsorb back into the monolayer shortly after the re-expansion had taken place. Therefore, the structure was fully recovered. The hysteresis in this case is a reflection of thermodynamic irreversibility, since part of the mechanical work done on the film is dissipated through the solvent, whereas the structural changes are fully reversible due to the maintenance of the bond connectivity of the monolayer network.

The monolayer formed by particles interacting through transient particle–particle bonds (case (b)) is able to break some existing bonds and to form some new bonds during the compression–expansion cycle. Initial compression induces a much faster stress build-up than for permanent bonds. The reason for this is that a large number of new bonds are created in case (b) as the particles are pushed together at the interface. Once the collapse point is reached, particles are forced out of the interface, and the stress remains roughly constant for the rest of the compression. When the interface is re-expanded, Figure 2b shows that a very fast stress release takes place. Following this, the stress overshoots, becoming positive (*i.e.* a negative interfacial pressure) due to the overstretching of many bonds created during the late stages of the compression. Finally, a large proportion of the bonds break up, the protein layer ruptures (see third column of Figure 1), and the interfacial stress is slowly relaxed.

When particles interact only through a repulsive core (case (c)), the stress increases very slowly during the initial stages of compression. As the monolayer becomes more close-packed, a faster stress build-up takes place, and suddenly the collapse point is reached. Just before the collapse point, small two-dimensional crystalline domains form at the interface. Beyond this point, particles desorb at the crystal boundaries and become dispersed in the bulk phase. The stress response in Figure 2c after the collapse point is characterized by large fluctuations associated with the sudden desorption of clusters of particles. Re-expansion of the interface leads to a fast stress release due to the absence of bonds.

We have also studied the effect of the degree of breakability, b_{max}, of the bonds in case (b). The lower the value of b_{max}, the more breakable the bonds are. Figure 3 shows the area fraction Γ, and the stress S_{xx}, in the direction of compression for different values of b_{max}. For particles that do not form bonds, the area fraction increases linearly with decreasing interfacial area as far as the collapse point. Further compression leads to desorption of particles. In this case, however, the area fraction keeps increasing beyond the collapse point, since particles at the interface rearrange in ever growing crystalline domains during compression. The area fraction goes beyond the limit of two-dimensional hexagonal close-packing ($\Gamma_{hex}^{max} = \pi/(2\sqrt{3})$) since adsorbed particles can accommodate themselves in different staggered planes within the narrow interfacial potential well. The introduction of transient bonds in the system was found not to affect the first part of the $\Gamma(A)$ curve, although the position of the collapse point may be slightly shifted. After the collapse

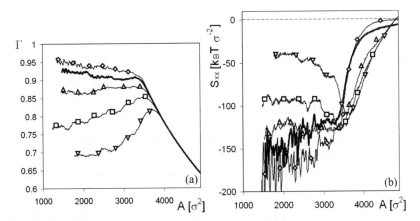

Figure 3 *Uniaxial compression of a model protein layer with transient particle–particle bonds for various degrees of breakability: no bonds (thick line), $b_{max}=0.3$ ($\Diamond$), $b_{max}=0.5$ ($\triangle$), $b_{max}=0.6$ ($\square$), $b_{max}=\infty$ (∇). The area fraction (a) and the interfacial stress (b) in the direction of compression are shown as a function of the interfacial area.*

point, the response of the simulated protein layer depends strongly on the degree of breakability of the bonds. The presence of very breakable bonds enhances crystal formation at the interface because it produces a more efficient kind of packing. This effect has been seen recently[11] for simulated nanoparticle monolayers interacting through short-range attractions. Less breakable bonds, however, promote a decrease in Γ after the collapse point. This is due to the fact that desorbing particles are able to drag their neighbours into the bulk phase if the elastic bonds last long enough. In general, wrinkle-type patterns appear in these cases as soon as the collapse point is passed.

The stress response of the protein layer is also very sensitive to the breakability of the bonds (see Figure 3b). When bonds are introduced between the particles, the stress under equilibrium conditions is reduced, which is evidenced by the difference in stress before compression. Bonds that are very easy to break do not significantly affect the stress build-up on compression, in comparison with the non-bond-forming case. However, the interface can take more stress at the collapse point, and the stress fluctuations beyond this point seem to be larger. On the other hand, the presence of bonds that are more permanent can promote a stress release after the collapse point, the extent of which grows with decreasing degree of breakability. Moreover, the initial stress build-up is faster in these cases, although it is not so sensitive to the degree of breakability.

5 Two-Component Monolayers

In order to assess the effect of compression on more complex layers, we have simulated four different binary systems. The mixtures are composed of two spherical species of the same size in a 1:1 molar ratio. In all cases, one of the species (type 1) interacts solely through the repulsive core potential, both with particles of its own type and with particles of the other type. The type 2 particles have the capacity to form bonds with particles of their own type. The four different cases correspond to different classes of bonding between the particles of type 2: (a) no bonds, (b) very-easy-to-break bonds ($b_{max} = 0.3$), (c) medium-strength breakable bonds ($b_{max} = 0.5$), and (d) permanent bonds ($b_{max} = \infty$).

The structure of the four different systems after 6×10^6 equilibration time steps is shown in Figure 4. Case (a) represents a perfect mixture since there is no asymmetry in the interactions of the two species. In case (b) a process of phase separation takes place due to the aggregation of type 2 particles led by the bonds being created and destroyed during the rearrangement of the clusters. The system is not fully equilibrated at this stage. In cases (c) and (d), the type 2 particles develop a cross-linked network, *i.e.*, a two-dimensional gel. Particles of type 1 sit in the gaps of this network. The main difference between cases (c) and (d) is the fact that in case (c) the particles of type 2 are able to rearrange the particle network by breaking and reforming bonds, so as gradually to reduce the system free energy by coarsening. However, this process is so slow that is not generally observed within the simulation

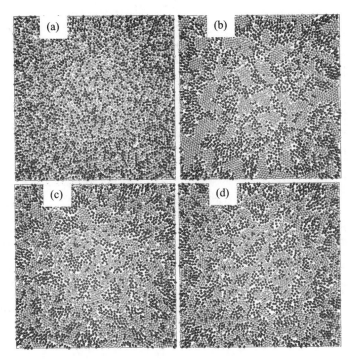

Figure 4 *Interfacial structure of complex mixed interfaces. In (a) two identical non-bond-forming species form a perfect random mixture. Mixtures with one species that does not form bonds (dark spheres) and another species that forms transient bonds (light spheres) are shown for different degrees of breakability: (b) $b_{max} = 0.3$, (c) $b_{max} = 0.5$, and (d) $b_{max} = \infty$. The system in case (b) is phase separating, but with very slow dynamics.*

timescale. The structure of this type of mixed monolayer has been discussed previously;[12] we focus here on the mechanical properties.

The response of these four systems to uniaxial compression ($v_x = 0.132\sigma\tau^{-1}$) is presented in Figure 5. The partial area fraction Γ_α of each species ($\alpha = 1$ or 2) is shown, and also the interfacial stress S_{xx} in the direction of compression. Interestingly, the stress response is not sensitive to the bond breakability with the exception of the extreme case (d) of fully permanent bonds. This shows that the core repulsion determines the stress response in the mixed systems—at least for a 1:1 composition. Contrasting with the stress response, the desorption of the competing species upon compression does depend on the nature of the bonds formed by the particles of type 2. Figure 5a shows that, in comparison with the symmetric case, the desorption beyond the collapse point for the case (b) of very-easy-to-break bonds is mainly due to the non-bond-forming particles of type 1. In case (c), however, the bond-forming particles of type 2 are desorbed preferentially. In this case, the type 2 particles tend to remain adsorbed in a secondary layer after displacement, by virtue of the bonds that connect them to the type 2 particles still remaining at the interface. The reason

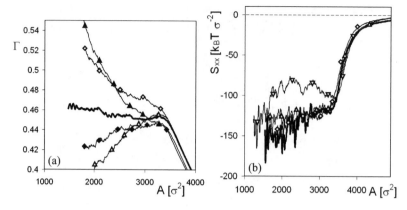

Figure 5 *Uniaxial compression of the complex mixed interface shown in Figure 4. The partial area fractions (a) and the interfacial stress (b) in the direction of compression are shown as a function of the interfacial area. Symbols represent the extent of breakability of the bond-forming species: no bonds (thick line), $b_{max}=0.3$ ($\Diamond$), $b_{max}=0.5$ ($\triangle$), $b_{max}=$ (∇). Filled symbols correspond to the partial area fraction of the non-bond-forming species (type 1 in the text). In (a), for clarity, only one of the partial area fractions is shown for the symmetric mixture, and no data are displayed for $b_{max}=\infty$.*

for this change in 'competitive desorption' behaviour lies in the balance of two contrasting effects. On the one hand, particle–particle bonds diminish the ability of an individual particle to desorb, since it will tend to be attracted to the interface not only by its individual adsorption energy but also by its connected neighbours. On the other hand, once a particle has been displaced from the interface, it will tend to drag its connected neighbours into the bulk solution enhancing the desorption of other particles of its own species. By varying the bond breakability, one is therefore able to change the balance between these two competing effects, since very breakable bonds are not able to contribute to the 'drag effect'.

6 Conclusions

We have shown by computer simulation that the response of a model protein monolayer to uniaxial compression is very sensitive to the nature of the particle–particle interactions. Indeed, the simulated structures of the compressed (and re-expanded) one-component monolayers present substantial differences when systems with various types of particle–particle bonding mechanisms are compared. The structural differences of the same systems at equilibrium are much less obvious. Similarly, the stress response on compression and re-expansion is qualitatively different between the various systems. These findings suggest good opportunities for validating simulation models against experiment by simply comparing their qualitative behaviour under large-deformation compression (expansion).

When particles interact through relatively permanent bonds, the displacement induced by compression leads to wrinkles that line up perpendicular to the direction of compression. This feature has been observed, for example, with proteins found in lung surfactants.[13] And desorption mechanisms induced by compression have been identified in some systems, *e.g.*, 2-hydroxytetracosanoic acid monolayers.[14] We believe that the same type of behaviour could also occur for adsorbed milk proteins, especially heat-treated whey proteins.

Finally, we have shown that the stress response of a mixed interfacial system to compression is not significantly affected by the nature of the bonding mechanism of the bond-forming species. In contrast, the 'competitive desorption' induced by compression favours the displacement of one or other of the species, depending on the extent of breakability of the bonds. This phenomenon seems to arise as a consequence of the delicate balance between a bond-enhanced adsorption effect and a drag-desorption mechanism.

The mechanical properties and collapse mechanisms studied here seem to merit further investigation both from the modelling perspective and also the experimental point of view. Understanding these large-deformation properties appears essential for controlling the response of interfaces during formation, stirring, and whipping of emulsions. Equally important, the response of fluid interfaces to slow compression (expansion) during disproportionation (and Ostwald ripening) is directly connected to the mechanical properties of monolayers[15] and their structural rearrangement upon collapse. The response of the simulated adsorbed layers to different compression (expansion) regimes is currently under investigation.

Acknowledgements

This research was supported by the BBSRC (UK). Computing was done on the Leeds Grid Node 1 facility, funded under the 2001 HEFCE Science Research Investment Fund initiative, at the University of Leeds, a partner in the White Rose Grid project.

References

1. E. Dickinson and D. J. McClements, 'Advances in Food Colloids', Blackie, Glasgow, 1995.
2. A. R. Mackie, A. P. Gunning, P. J. Wilde, and V. J. Morris, *J. Colloid Interface Sci.*, 1999, **210**, 157.
3. J. M. Rodríguez Patino, C. Carrera Sánchez, and M. R. Rodríguez Niño, *J. Agric. Food Chem.*, 1999, **47**, 4998.
4. D. B. Jones and A. P. J. Middelberg, *Langmuir*, 2002, **18**, 5585.
5. L. A. Pugnaloni, E. Dickinson, R. Ettelaie, A. R. Mackie, and P. J. Wilde, *Adv. Colloids Interface Sci.*, 2004, **107**, 27.
6. C. M. Wijmans and E. Dickinson, *Phys. Chem. Chem. Phys.*, 1999, **1**, 2141.
7. C. M. Wijmans and E. Dickinson, *Langmuir*, 1999, **15**, 8344.

8. M. P. Allen and D. J. Tildesley, 'Computer Simulation of Liquids', Clarendon Press, Oxford, 1987, chap. 9.
9. M. Whittle and E. Dickinson, *Mol. Phys.*, 1997, **90**, 739.
10. M. Doi and S. F. Edwards, 'The Theory of Polymer Dynamics', Oxford University Press, New York, 1986, chap. 3.
11. L. A. Pugnaloni, R. Ettelaie, and E. Dickinson, *Langmuir*, 2004, **20**, 6096.
12. L. A. Pugnaloni, R. Ettelaie, and E. Dickinson, *Langmuir*, 2003, **19**, 1923.
13. M. M. Lipp, K. Y. C. Lee, D. Y. Takamoto, J. A. Zasadzinski, and A. J. Waring, *Phys. Rev. Lett.*, 1998, **81**, 1650.
14. C. Ybert, W. Lu, G. Möller, and C. M. Knobler, *J. Phys. Chem. B*, 2002, **106**, 2004.
15. E. Dickinson, R. Ettelaie, B.S. Murray, and Z. Du, *J. Colloid Interface Sci.*, 2002, **252**, 202.

Molecular Interactions in Mixed Protein + Ionic Surfactant Interfaces

By Paul A. Gunning, Alan R. Mackie, A. Patrick Gunning, Peter J. Wilde, and Victor J. Morris

INSTITUTE OF FOOD RESEARCH, NORWICH RESEARCH PARK, COLNEY, NORWICH NR4 7UA, UK

1 Introduction

The nature of protein–surfactant interactions at interfaces can be critical for the functionality of food foams and emulsions.[1,2] It is well known that proteins and surfactants employ very different short-range mechanisms to stabilize against coalescence in foams and emulsions.[1,3] Proteins form viscoelastic networks at an interface which create a mechanical barrier to coalescence, whereas surfactants form a freely diffusing interface which maintains a continuous phase layer around the bubbles or droplets. Many food systems contain a mixture of proteins and surfactants which compete to adsorb at and stabilize the interface. Unfortunately, the competing stability mechanisms are mutually incompatible, and in many cases it results in a loss of stability.[1-3]

Various types of food surfactants and emulsifiers exist. They may differ in the degree of hydrophobicity. So they can be oil-soluble or water-soluble with a range of surface activities. The head-groups can be ionic or non-ionic, which influences surface packing and micelle formation, and which in turn affects surface activity. Nonetheless, however variable the chemistry of surfactants might be, their general surface-active behaviour is rather similar, going through the stages of adsorption, surface saturation and micelle formation. The main difference between the various types is the bulk concentration at which these stages occur, and the surfactant phase behaviour.

When the behaviour of surfactants is studied in the presence of protein, much more diversity is observed.[4] In studying a mixed system of a surfactant with an amphiphilic polyelectrolyte (protein), the interactions can be rather complex. The variation in behaviour between different systems is reflected in the use of a range of surfactant classes for different food applications. For example, in many dairy applications, the use of non-ionic emulsifiers such as Tweens and sugar esters is widespread, for the purpose of inducing instability and partial coalescence of fat globules. In contrast, in many baking applications, ionic surfactants such as sodium stearoyl lactylate (SSL) are used to improve gas-cell stability. Although all of the common surfactants and emulsifiers used in these applications are capable of stabilizing the colloidal systems in their own right, their special effects on functionality usually arise from particular interactions with the proteins also present in the system. Therefore, it is essential to understand the molecular basis for the observed differences in functionality between different types of surfactants used in protein-based foam and emulsion systems.

In this study we discuss the role of the surfactant headgroup on the interfacial and functional properties of some mixed protein + surfactant systems. Four classes of surfactant head-group are considered—non-ionic, anionic, cationic and zwitterionic. Although cationic surfactants are not actually used in the food industry, their study here is essential for understanding the role of specific charge interactions.

2 Materials and Methods

Two milk proteins were used in this study: β-lactoglobulin (β-lg, L-0130, lot no. 91H7005) and β-casein (C-6905, lot no. 12H9550), both obtained from Sigma Chemicals (Poole, UK). Initially, stock protein solutions were prepared at a concentration of 2 mg mL^{-1} in water. The water used for this study was obtained from an Elga Elgastat UHQ water purification system; it had a surface tension of $\gamma_0 = 72.6$ mN m^{-1} at 20 °C. The surfactants used had similar chain lengths, but differed in their headgroup composition. The anionic surfactant was sodium dodecyl sulfate (SDS), which was obtained as a 10% solution (L4522, lot no. 97H8505) from Sigma Chemicals. The cationic surfactant was cetyltrimethylammonium bromide (CTAB) which was obtained from Fluka (Buchs, Switzerland). The zwitterionic surfactant was lyso-phosphatidylcholine-lauroyl (LPC-L), and was obtained as a powder (lot no. 80F83551) from Sigma Chemicals. Finally, the non-ionic surfactant was Tween 20 (polyoxyethylene sorbitan monlaurate) which was obtained as a 10% solution (Surfact-Amps 20) from Pierce (Rockford, IL).

Surface tension measurements were made using a wetted, roughened, glass Wilhelmy plate and a PTFE Langmuir trough ($255 \times 112 \times 16$ mm), which was equipped with mobile and fixed PTFE barriers. The proteins (70 µg β-lg or 36 µg β-casein) were spread at the air–water interface from aqueous solution. The spread protein films were allowed to equilibrate for 30 minutes. Surface tension and surface rheological measurements have shown[5] that after this period of time the film has achieved a quasi-equilibrium state. Appropriate concentrations of surfactant were always added to the sub-phase behind the fixed barrier, in order to prevent the surfactants from spreading at the interface.

Langmuir–Blodgett (LB) dips were performed at a dipping speed of 8.4 mm min^{-1} on a hydrophilic substrate of freshly cleaved mica as described previously.[5] The surface pressure was closely monitored during each dip in order to confirm the efficient transfer of the adsorbed surface layer onto the mica. The LB monolayers were rinsed in water, and those containing LPC-L were further rinsed in 50:50 chloroform:methanol (BDH Chemicals, Poole, UK). The rinsing was carried out in order to remove surfactant, and thus to aid visualization of the protein remaining in the interfacial film by AFM. The excess solvent was evaporated by allowing the sample to stand in air for about 10 minutes.

Atomic force microscopy (AFM) images were produced using an East Coast Scientific AFM (ECS, Cambridge, UK). The probes used were Nanoprobe cantilevers (Digital Instruments, Santa Barbara, CA) with a quoted force constant of 0.38 N m⁻¹. Samples were imaged in dc contact mode under redistilled butanol. Images were subsequently analysed using Image Pro 4.5 image analysis software (Media Cybernetics, USA). Protein areas were calculated using a thresholding technique that determines the number of pixels above a threshold level as a percentage of the overall number of pixels in the image. This yields the percentage surface area covered by protein.

Surface shear rheological measurements of 1 µM solutions of β-lg, and of co-adsorbed mixtures of the protein with surfactant, were made in 10 mM sodium phosphate buffer at pH 7. Data were obtained using a Bohlin CS10 controlled stress rheometer (Bohlin Instruments, Cirencester, UK) equipped with an IFG10 interfacial geometry, which was a 6 cm diameter, 4° bicone. All measurements were made at 25 °C.

3 Results and Discussion

Foaming Properties

The foaming properties of protein + surfactant mixtures were studied. The effects of non-ionic surfactants, such as Tween 20, have been studied previously.[1-3] Like many other surfactants, they cause an initial destabilization of the foam due to the antagonistic competition between the protein and surfactant.[1-3] This behaviour is well known, and is thought to be the basis of the mechanism of action of many antifoaming agents. Similarly, in this study, the surfactants generally caused a destabilization of the foam, with the exception of the zwitterionic LPC-L.

Figure 1 shows the foam stability for β-lg in the presence of increasing concentrations of either Tween 20 or LPC-L. (The surfactant concentration is expressed in arbitrary units in order to aid clarity, due to the large difference in surface activity of the two surfactants). A similar foam stabilization effect has been observed previously with LPC and β-lg,[6] and with puroindolines,[7,8] but previous studies failed to explain the behaviour adequately. It was thought to be due to surfactant binding by the proteins, as all the proteins used in these studies were shown to bind to surfactants in solution. Such binding could lead to a more surface-active complex, but the physical mechanism underlying the improved foam stability could not be fully explained. It is possible that

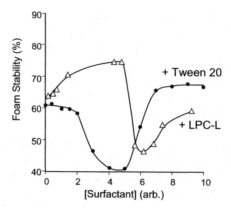

Figure 1 *Foam stability of 0.2 mg mL⁻¹ β-lg in the presence of increasing concentrations of the surfactants Tween 20 (•) and LPC-L (Δ). The surfactant concentrations have been normalized to account for the differences in absolute concentrations of each surfactant.*

the presence of the zwitterionic head group increases the number density of electrostatic interactions forming the interfacial protein network, particularly if the head-group is bound close to a protein molecule. Ionic surfactants have been found to increase the rheological parameters of polymer gels.[9] In the present study, surface rheology was used to investigate the impact of LPC-L on the strength of the interfacial protein network.

Surface Rheology

Previous studies with LPC-L + β-lg showed[6] that the surface dilatational elasticity was not sensitive to interfacial complexation. Surface shear rheology, on the other hand, is a direct measure of the mechanical stress–strain relationship, and as such is much more sensitive to the number and strength of molecular interactions at the interface. Figure 2 shows the surface shear elastic modulus G' of β-lg in the presence of increasing concentrations of surfactant. Upon addition of Tween 20, a general decrease in G' was observed, which agrees with previous findings.[10] However, in the presence of LPC-L, a small but significant increase in G' was observed. In fact, the presence of small concentrations of LPC-L showed a marked increase in the rate of adsorption, and consistently higher value of G' than for the protein alone. The anionic and cationic surfactants showed no evidence of an increase in G' for the mixed systems (results not shown), which is consistent with previous studies.[11] Therefore, although LPC-L has no net charge, the zwitterionic head-group appears to be able to interact with proteins at interfaces more strongly than the ionic surfactants, resulting in an increase in the shear elasticity of the interfacial protein network. This increase in the surface elasticity at low LPC-L concentrations may well be responsible for the observed increase in foam stability. It is possible that the protein may have to bind the LPC-L to produce this effect,

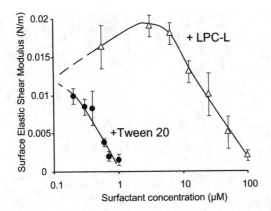

Figure 2 *Surface shear elastic modulus of 1 μM β-lg in the presence of increasing concentrations of Tween 20 (•) and LPC-L (Δ).*

since all the proteins studied to date have had strong binding affinities for surfactants. Further work is needed, however, to determine the role of surfactant binding in relation to the synergistic interactions between LPC and food proteins.

Interfacial Structure

Conventional surface techniques such as tensiometry and interfacial rheology give results which are average spatial values for the whole interface. Recent work has shown,[5,12,13] however, that it is important to understand the local interfacial structure of the interface on the sub-micron scale in order to explain the physical properties of mixed protein + surfactant interfaces. In particular, it has been demonstrated[5] that protein/surfactant interfaces are heterogeneous, with surfactants forming domains which can disrupt and destabilize the interfacial protein network. There also appear to be some differences in the structure of the interfaces when ionic surfactants are used instead of non-ionics.[14] Therefore, this aspect was investigated further as part of the present study.

Some AFM images of Langmuir–Blodgett films transferred from mixed protein + surfactant interfaces are shown in Figure 3. The images were taken at similar stages of the competitive displacement process, *i.e.*, the surfactants occupy a similar proportion of the interfacial area in all the images. There is a clear difference between the ionic surfactants and the non-ionic surfactant Tween 20, in that the latter forms much larger domains. (Note that the image for the Tween 20 mixed film is on a larger scale than the other images.) Studying the domain formation in more detail has shown that, in the case of Tween 20, the domains nucleate, possibly at defects within the protein film, and then begin to grow. The overall displacement process is dominated by the growth of individual domains.[15] In the case of the ionic surfactants, many more small domains were found to be formed. These domains did not appear to grow to the same extent as with the non-ionic surfactants. In fact, considering

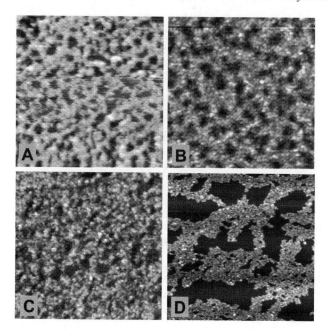

Figure 3 *AFM images of Langmuir–Blodgett films of mixed β-lg + surfactant interfaces. The various surfactants were as follows (image size in brackets): (A) SDS (1 μm × 1 μm); (B) CTAB (1 μm × 1 μm); (C) LPC-L (1 μm × 1 μm); (D) Tween 20 (3.7 μm × 3.7 μm).*

the number of domains, and hence the small distance between them, only a small amount of growth was required before the protein network was completely displaced. It is interesting to note that, despite the LPC-L molecule having no net charge, the appearance of its domains is similar to that found with anionic or cationic surfactants.

The presence of a small amount of LPC-L at the interface apparently increases the number and/or strength of interactions at the interface, as evidenced by the increase in surface shear modulus (Figure 2) as discussed above. Therefore, it would appear that the similarity in appearance of the AFM images in Figure 3, for the LPC-L and the ionic surfactants, can be attributed to increased interaction between the surfactants and the interfacial protein network. Although the proteins studied possess a net negative charge at neutral pH, they contain a range of charged and hydrophobic groups over the entire surface of the molecule. Therefore different parts of the protein's surface will be attractive to either anions or cations, and other parts will participate in hydrophobic interactions with surfactant alkyl chains. Thus the similarity between the different classes of ionic surfactants could be due to the fact that they all possess a certain balance of attractive and repulsive electrostatic interactions with the adsorbed protein. Also, one could argue that the interactions between surfactant head-groups may influence behaviour. Due to the increased electrostatic repulsion between head-groups, ionic surfactants have a much higher critical micelle concentration than non-ionics, with zwitterionics generally being in between. This will tend to affect the surface packing density and therefore make it less energetically favourable for an ionic surfactant to adsorb into surfactant domains. When the charges on the surfactants are screened at higher salt concentrations, it has been found[16] that the domain size increases, and the behaviour approaches that observed for the non-ionic surfactant.[16] In addition, Brownian dynamic simulations[17] have studied the effect of intermolecular interactions on the displacement of molecules from surfaces. This work has shown[17] that, if there is a long-range repulsion between displacer species (surfactant) and network-forming adsorbed particles (protein), then large domains are formed; but if there is only short-range repulsion, small domains are produced.[17] This appears to reflect the observed differences in the structure of the mixed films for non-ionic surfactants (long-range steric repulsion *via* the large hydrophilic head-group) and ionic surfactants (electrostatic interactions). The simulated behaviour highlights the importance of the details of the interactions between the surfactant and the protein.

Therefore, although the various ionic surfactants have very different electrostatic properties, the increased attraction to the protein, and the increased repulsion between surfactant head-groups, can result in superficially similar structures in the mixed protein + surfactant films. Certainly, the net effect is that it is it energetically favourable for the ionic surfactant to adsorb into the protein network and form small domains. For the non-ionic surfactant, however, the converse applies: it preferentially adsorbs into surfactant-rich regions, leading to the formation and growth of large domains.

Protein Displacement

AFM images acquired from mixed protein + surfactant films were analysed for the amount of protein remaining at the interface. The results are shown in Figure 4. The amount of protein remaining at the interface is given as a

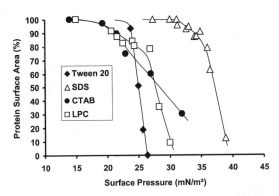

Figure 4 *Surface area occupied by protein on AFM images of Langmuir–Blodgett films of β-lg + surfactant interfaces, as a function of surface pressure for different surfactants.*

function of surface pressure; since the surface pressure is the driving force for displacement, this normalizes the data for any effects of time and concentration.

The displacement results in Figure 4 show that the non-ionic surfactant Tween 20 is the most effective at displacing the protein from the surface, and that the anionic surfactant SDS is the least effective. CTAB and LPC-L begin to displace the protein at lower surface pressures than Tween 20, but the overall displacement process appears more gradual. This suggests that the cationic and zwitterionic surfactants are more attracted to the interfacial protein film than is the anionic surfactant, which is possibly as a result of the overall net charge of the protein. This has been observed in previous interfacial studies,[11] where CTAB was found to have an influence on the surface properties of protein at lower concentrations than would be expected from the surface activity of the surfactant alone. Therefore, if the cationic/zwitterionic surfactants are more intimately mixed with the interfacial protein film, then the protein should be disrupted more easily because the adsorbed surfactant is inhibiting protein interactions from forming. However, the overall displacement may be influenced by the ability of the domains to expand, as the driving force for displacement is the surface pressure gradient between the expanding surfactant domains and the continuous protein network, which is counteracted by the elasticity of the protein film. Assuming that there is a surface tension gradient—it must exist for the domains to expand[5]—then there must be a line tension around the perimeter of each surfactant domain, in the same way that surface tension causes the Laplace pressure difference in gas bubbles. On this basis, smaller domains will be able to retard a higher surface pressure gradient than larger domains, and so the larger domains formed by non-ionic surfactant will be more effective at displacing protein from the interface than the smaller domains formed by ionic surfactant.

4 Conclusions

The influence of the surfactant head-group on protein displacement appears to be a complex interplay between molecular interactions and surface activity. Ionic surfactants exhibit a relatively higher affinity for adsorbing into the interfacial protein film. This can cause the protein network to be disrupted more easily, but it can also result in stronger electrostatic interactions. The relative attraction between the ionic surfactant and the protein film results in much smaller surfactant domains than those formed by a non-ionic surfactant. The smaller domains cannot displace the protein as effectively, possibly because they can support greater surface pressure gradients between the domains and the network.

Acknowledgements

The authors acknowledge the BBSRC (UK) for funding through the Core Strategic Grant to the Institute of Food Research and through research grant

D14067. We thank Eric Dickinson, Luis Pugnaloni and Rammile Ettelaie at the University of Leeds for their collaboration in the above research grant.

References

1. E. Dickinson, *Colloids Surf. B*, 1999, **15**, 161.
2. M. Coke, P. J. Wilde, E. J. Russell, and D. C. Clark, *J. Colloid Interface Sci.*, 1990, **138**, 489.
3. J. Krägel, R. Wüstneck, F. A. Husband, P. J. Wilde, A. V. Makievski, D. O. Grigoriev, and J. B. Li, *Colloids Surf. B*, 1999, **12**, 399.
4. A. R. Mackie, A. P. Gunning, P. J. Wilde, and V. J. Morris, *J. Colloid Interface Sci.*, 1999, **210**, 157.
5. D. K. Sarker, P. J. Wilde, and D. C. Clark, *Colloids Surf. B*, 1995, **3**, 349.
6. P. J. Wilde, D. C. Clark, and D. Marion, *J. Agric. Food Chem.*, 1993, **41**, 1570.
7. F. A. Husband, P. J. Wilde, D. Marion, and D. C. Clark, in 'Food Macromolecules and Colloids', eds E. Dickinson and D. Lorient, Royal Society of Chemistry, Cambridge, 1995, p. 283.
8. E. M. A. Bos and T. van Vliet, *Adv. Colloid Interface Sci.*, 2001, **91**, 437.
9. H. Lauer, A. Stark, H. Hoffmann, and R. Donges. *J. Surfactants Detergents*, 1999, **2**, 181.
10. J.-L. Courthaudon, E. Dickinson, Y. Matsumura, and D. C. Clark, *Colloids Surf.*, 1991, **56**, 293.
11. R. Wüstneck, J. Krägel, R. Miller, P. J. Wilde, and D. C. Clark, *Colloids Surf. A*, 1996, **114**, 255.
12. A. P. Gunning, A. R. Mackie, P. J. Wilde, and V. J. Morris, *Langmuir*, 1999, **15**, 4636.
13. A. R. Mackie, A. P. Gunning, M. J. Ridout, P. J. Wilde, and J. M. Rodriguez Patino, *Biomacromolecules*, 2001, **2**, 1001.
14. A. R. Mackie, A. P. Gunning, P. J. Wilde, and V. J. Morris, *Langmuir*, 2000, **16**, 8176.
15. A. R. Mackie, A. P. Gunning, L. A. Pugnaloni, E. Dickinson, P. J. Wilde, and V. J. Morris, *Langmuir*, 2003, **19**, 6032.
16. P. A. Gunning, A. R. Mackie, A. P. Gunning, P. J. Wilde, N. C. Woodward, and V. J. Morris, *Food Hydrocoll.*, 2004, **18**, 509.
17. C. M. Wijmans and E. Dickinson, *Langmuir*, 1999, **15**, 8344.

Conformational Changes of Ovalbumin Adsorbed at the Air–Water Interface and Properties of the Interfacial Film

By Stéphane Pezennec, Emmanuel Terriac,[1] Bernard Desbat,[2] Thomas Croguennec, Sylvie Beaufils,[1] and Anne Renault[1]

SCIENCE ET TECHNOLOGIE DU LAIT ET DE L'ŒUF, UMR 1253 INRA-AGROCAMPUS, 65 RUE DE SAINT-BRIEUC, 35042 RENNES CEDEX, FRANCE
[1]GROUPE MATIERE CONDENSEE ET MATERIAUX, UMR 6626 CNRS-UNIVERSITE DE RENNES 1, CAMPUS BEAULIEU, BATIMENT 11A, 35042 RENNES CEDEX, FRANCE
[2]LABORATOIRE DE PHYSICO-CHIMIE MOLECULAIRE, UMR 5803 CNRS-UNIVERSITE DE BORDEAUX 1, 351 COURS DE LA LIBERATION, 33405 TALENCE CEDEX, FRANCE

1 Introduction

Owing to their amphiphilicity, proteins are essential ingredients in the preparation and physical stabilization of multi-phase dispersed food systems. Like other surfactants, proteins are able to adsorb from aqueous solution to fluid interfaces and to lower the interfacial tension, thus helping thermodynamically the formation of large interfacial areas and partially hindering their spontaneous destabilization.

But the main specificity of proteins, as compared to low-molecular-weight surfactants, is their additional ability, under favourable physico-chemical conditions, to establish an interfacial network of intermolecular interactions, that contributes prominently to the macroscopic mechanical properties of the surface layer, thereby further stabilizing large interfacial areas. Examples of instability phenomena which may be counteracted by the mechanical resistance of the interfacial film are coalescence of emulsion droplets and film rupture between bubbles in foams.

In the case of globular proteins, the variable polarity of portions of the sequence partly determines the 'native' three-dimensional structure, in which the hydrophobic regions are protected, at least partly, from the solvent and so constitute the core of the globular structure. Adsorption of a globular protein at a hydrophobic interface may destabilize the 'native' structure, the interaction of hydrophobic regions of the protein molecule with the hydrophobic interface being entropically favoured. Such exposed regions may in turn become involved in further intermolecular interactions.

Aiming at further understanding of how the conformational rearrangements undergone by adsorbed protein may contribute to the properties of the interfacial film, we have studied various aspects of the behaviour of ovalbumin at the planar air–solution interface.

2 Ovalbumin and *S*-Ovalbumin

Ovalbumin (45 kDa) is a glycosylated, phosphorylated protein of 385 residues which accounts for about 60% of the protein in egg white.[1,2] It contains two phosphorylation sites and it is naturally present as a mixture of di-, mono- and non-phosphorylated forms in the approximate ratio 85:12:3. A large set of data about macroscopic aspects of the interfacial behaviour of ovalbumin is available.[3-7]

In order to avoid charge heterogeneity due to heterogeneity in phosphoserine content, we used throughout this study the purified di-phosphorylated (A1) form. This ovalbumin was purified by anion-exchange chromatography[8] or, for some of the preparations, by displacement anion-exchange chromatography.[9] The isoelectric pH of the A1 form is pI=4.75.[10]

S-Ovalbumin is a thermostable conformer of ovalbumin,[11,12] which is formed spontaneously during liquid egg-white storage, or which may be prepared by heating an ovalbumin solution at pH 10 for 16–24 h at 55 °C. The increase in thermostability is reflected in the change in denaturation temperature as measured by differential scanning calorimetry.[13] The structural differences between ovalbumin and *S*-ovalbumin have long been a matter of debate. The crystal structure of *S*-ovalbumin has been recently described.[14] Surprisingly, the tertiary structure of *S*-ovalbumin is quite similar to that of ovalbumin. *S*-ovalbumin mainly differs from native ovalbumin in the isomerization of some serine residues into D-serine and in changes in the conformation of lateral chains buried in the hydrophobic core.

The existence of these two ovalbumin conformers, with very similar structures, but clearly distinguished by properties such as the conformational stability (as reflected in the denaturation temperature) or the ability to form heat-set gels,[15,16] makes these proteins convenient models for studying the conformational aspects of interfacial phenomena.

For this study we prepared *S*-ovalbumin by heating ovalbumin in solution in an alkaline medium as already described.[17]

3 Methods

For both ovalbumin and *S*-ovalbumin, the interfacial behaviour was investigated under various conditions (variable protein charge) using complementary methods.

The adsorption kinetics were followed by surface tension measurements with a Wilhelmy plate and null ellipsometry.[8] The value of the ellipsometric angle Δ was considered faithfully to reflect the surface concentration Γ of the adsorbed protein.

The shear rheology of the adsorbed layer was characterized using an original apparatus,[18,19] in which a rotational strain is applied to the interface by a float carrying a small magnet, via periodic variations of a magnetic field. The amplitude and phase of the angular response were monitored using the reflection of a laser beam on a mirror carried by the float. The kinetics of interfacial rheological change during protein adsorption and ageing of the interfacial film could be followed with fixed-frequency oscillations of the magnetic field, while stable plateau values of the shear elastic constant μ were precisely measured through frequency spectra in the range 0.01–100 Hz.

Protein adsorption and conformational rearrangements at the air–solution interface were studied using polarization-modulated infrared reflection-absorption spectroscopy (PM-IRRAS).[20,21] This technique, through fast modulation of the polarization of the incident beam between directions parallel and perpendicular to the incidence plane, provides a differential reflectivity signal. The contribution of the interfacial layer is therefore extracted from the background generated by the sub-phase and the surrounding atmosphere. The conformation and orientation of various molecules at the air–water interface have been successfully studied by PM-IRRAS.[22–26] Here we have studied the conformational changes of ovalbumin and S-ovalbumin after adsorption from 0.625 g L^{-1} solutions in D$_2$O buffered at pD 3.9, 5.0, 6.4, and 10.4. For each of the eight protein–pD combinations, spectra were recorded at times ranging from 1–4 min to at least 4 h after the beginning of adsorption, which resulted in 9–12 spectra per combination.[17] After normalization, the whole data set was analysed by principal component analysis (PCA), in order to extract the most relevant features of the conformational changes.

4 Adsorbed Layers of Ovalbumin and *S*-Ovalbumin

As shown in Figure 1, with the exception of 0.625 g L^{-1} S-ovalbumin, the surface concentration as indicated by the ellipsometric angle Δ was only slightly affected by the protein net charge. In the case of 0.625 g L^{-1} S-ovalbumin, higher values of surface concentration were reached for low net charge of the protein (pH 3.5 and 4.6), as would be the case if adsorption of aggregates formed in the bulk or multi-layer adsorption had led to a greater thickness of the interfacial layer.

In contrast with the adsorbed protein concentration, for both forms of ovalbumin, the shear rheology of the interfacial layer was found to be strongly affected by the protein net charge. Peak plateau values were reached at a pH of the solution close to the protein's p*I* (4.75).

These results suggest that, while electrostatic repulsion did not limit the surface concentration of adsorbed protein, it strongly obstructed the development of the intermolecular interactions that determine the rheology of the interfacial

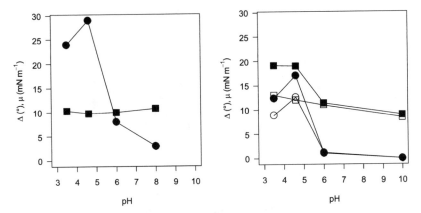

Figure 1 *Adsorption of ovalbumin and S-ovalbumin at the air–solution interface as a function of the pH of the solution. For ovalbumin (left) and S-ovalbumin (right),
plateau values of the ellipsometric angle Δ (■) and the shear elastic constant
μ (●) are given. The ovalbumin concentration is 1 g L^{-1}. The S-ovalbumin
concentrations are 0.1 g L^{-1} (open symbols) and 0.625 g L^{-1} (solid symbols).*

film. In addition, the changes of the plateau values of Δ and μ as a function
of ovalbumin concentration (not shown) reveal that, at low net charge, an
increase in concentration in the range 0.1–1 g L^{-1} did not affect the surface
concentration, but greatly increased the shear elastic constant μ. This suggests
that the rheology of the adsorbed layer also depends on the protein packing
density and on the extent of unfolding in the initial stage of adsorption. Thus,
together with a network of intermolecular interactions, it seems that the
intramolecular structure probably also contributes to the macroscopic
rheological properties of the interfacial layer.

5 Formation of an Intermolecular β-Sheet Network at the Air–Solution Interface

Using PM-IRRAS, we have investigated the conformational changes undergone by adsorbed ovalbumin and S-ovalbumin at different values of the net
charge. In order to draw the optimum amount of information from the spectral data set, it was analysed by principal component analysis. Successive
principal components (PCs), which are non-correlated, account for decreasing
amounts of the variance of the data set, *i.e.*, its initial 'information' content.
The principal components are associated with spectral patterns, which identify
the regions of the spectra which are the most variable. Each spectrum then can
be characterized by its score with respect to each of the PCs.

It was found that ~74 % of the variance was accounted for together by the
first two PCs (50.3 and 24.0 %, respectively). All the other PCs accounted each
for less than 10 % of the variance. In the following discussion, we focus on the
first principal component (PC1).

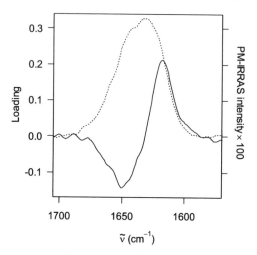

Figure 2 *Principal component analysis of PM-IRRAS spectra: the mean spectrum and the spectral pattern corresponding to principal component 1 (PC1). Adsorption of ovalbumin and S-ovalbumin from 0.625 g L⁻¹ solutions in D₂O at pD 3.9, 5.0, 6.4 and 10.4, and conformational changes in the interfacial layer were studied by PM-IRRAS. The whole data set consisting of spectra recorded for the two proteins at all pD values was analysed by PCA. The continuous line (left ordinate axis) represents the spectral pattern associated with PC1. The dotted line (right ordinate axis) represents the overall mean spectrum.*

The spectral pattern associated with PC1 is shown in Figure 2. PC1 represents the spectral variability and changes in the amide I band, which is sensitive to the protein backbone conformation and the secondary structure. PC1 exhibits a positive band at 1618 cm⁻¹ and a negative band at 1651 cm⁻¹. The former can be attributed to intermolecular β-sheets,[27] and the latter to α-helices. PC1 can therefore be associated with changes in the relative spectral contributions of these types of secondary structure (relatively to the mean spectrum), so that spectra with positive high scores on PC1 reveal a relatively high spectral contribution of intermolecular β-sheets in the corresponding interfacial layer.

Figure 3 shows, for ovalbumin and S-ovalbumin at different values of the net charge, the scores of the PM-IRRAS spectra as a function of time after the beginning of adsorption. For both these ovalbumin forms, the results clearly show that the score on PC1—namely the spectral contribution of intermolecular β-sheets—increases with time, and that both the extent and rate of increase are higher at low values of the protein net charge. The time effect and its modulation by pD have been shown to be significant by analysis of variance (not shown). It may also be noticed that, for equivalent times and pD values, the PC1 scores generally are higher for S-ovalbumin. This suggest that S-ovalbumin aggregates are present in solution at low charge, in agreement with the results obtained by ellipsometry. But the rate of development of intermolecular β-sheets for S-ovalbumin still appears to be as fast as for ovalbumin.

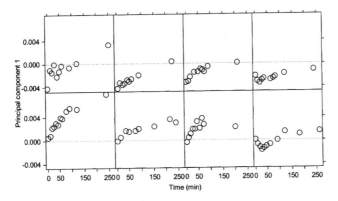

Figure 3 *Principal component analysis of PM-IRRAS spectra: the scores for PC1 as a function of time. Upper sequence of plots refers to ovalbumin. Lower sequence of plots refers to S-ovalbumin. From left to right: pD 3.9, 5.0, 6.4, 10.0.*

While it seems that the formation of intermolecular β-sheets upon adsorption at the air–water interface is not generally the case for all globular proteins, it has also been observed for glycinin.[28] In contrast to our results, it has recently been observed[29] with ovalbumin only upon compression of the surface. This apparent discrepancy might be explained by the fact that the latter study[29] was performed with a protein of relatively high net charge (at pH 7).

6 Conclusions

Our results indicate that a network of intermolecular β-sheets is formed in the interfacial film on adsorption of ovalbumin and *S*-ovalbumin and upon ageing of the interfacial film. Furthermore, this conformational rearrangement is correlated, in terms of charge effects, to the changes in interfacial rheology.

Together with the network of intermolecular interaction, the remaining internal structure—or, conversely, the extent of unfolding—are likely also to contribute to the surface rheological properties. The study of protein conformational changes upon adsorption as a function of the bulk concentration should give further insight into this contribution.

In addition to variations in the protein net charge, or the use of conformers such as ovalbumin and *S*-ovalbumin, numerous other factors may be varied in order to change the protein molecular properties and therefore the interfacial and foaming behaviour. One of these factors is the initial state of denaturation/aggregation state of the protein. Preliminary results, obtained in a study of the interfacial behaviour of ovalbumin samples heated in solution at 80 °C for different times, suggest that even the shortest heat treatment (5 min) results in an increase in the adsorbed amount, as reflected in the ellipsometric angle Δ. For all heated samples (5–40 min), it was found that foam density and stability had comparable values, and both were clearly higher than those obtained with the non-heated sample. Nevertheless, although marked changes were also observed in the shear elastic constant μ, the latter parameter seems insufficient to explain the variation in foaming ability.

With the aim of further investigating the contribution of adsorbed layer properties to the foaming behaviour, studies of the structure and dynamics of protein foams by methods such as phase-contrast X-ray tomography are currently in progress.

References

1. A. D. Nisbet, R. H. Saundry, A. J. G. Moir, L. A. Fothergill, and J. E. Fothergill, *Eur. J. Biochem.*, 1981, **115**, 335.
2. P. E. Stein, A. G. W. Leslie, J. T. Finch, and R. W. Carrell, *J. Mol. Biol.*, 1991, **221**, 941.
3. J. A. de Feijter and J. Benjamins, in 'Food Emulsions and Foams', ed. E. Dickinson, Royal Society of Chemistry, London, 1987, p. 72.
4. J. Benjamins and F. van Voorst Vader, *Colloids Surf.*, 1992, **65**, 161.
5. J. Benjamins and E. H. Lucassen-Reynders, in 'Proteins at Liquid Interfaces', eds D. Möbius and R. Miller, Elsevier, Amsterdam, 1998, p. 341.
6. E. H. Lucassen-Reynders and J. Benjamins, in 'Food Emulsions and Foams: Interfaces, Interactions and Stability', eds E. Dickinson and J. M. Rodríguez Patino, Royal Society of Chemistry, Cambridge, 1999, p. 195.
7. J. A. de Feijter, J. Benjamins, and F. A. Veer, *Biopolymers*, 1978, **17**, 1759.
8. S. Pezennec, F. Gauthier, C. Alonso, F. Graner, T. Croguennec, G. Brulé, and A. Renault, *Food Hydrocoll.*, 2000, **14**, 463.
9. T. Croguennec, F. Nau, S. Pezennec, and G. Brulé, *J. Agric. Food Chem.*, 2000, **48**, 4883.
10. N. Kitabatake, A. Ishida, and E. Doi, *Agric. Biol. Chem.*, 1988, **52**, 967.
11. M. B. Smith and J. F. Back, *Aust. J. Biol. Sci.*, 1968, **21**, 539.
12. M. B. Smith and J. F. Back, *Aust. J. Biol. Sci.*, 1965, **18**, 365.
13. J. W. Donovan and C. J. Mapes, *J. Sci. Food Agric.*, 1976, **27**, 197.
14. M. Yamasaki, N. Takahashi, and M. Hirose, *J. Biol. Chem.*, 2003, **278**, 35524.
15. B. Egelandsdal, *J. Food Sci.*, 1980, **45**, 570.
16. S. Shitamori, E. Kojima, and R. Nakamura, *Agric. Biol. Chem.*, 1984, **48**, 1539.
17. A. Renault, S. Pezennec, F. Gauthier, V. Vié, and B. Desbat, *Langmuir*, 2002, **18**, 6887.
18. C. Vénien-Bryan, P.-F. Lenne, C. Zakri, A. Renault, A. Brisson, J.-F. Legrand, and B. Berge, *Biophys. J.*, 1998, **74**, 2649.
19. C. Zakri, A. Renault, and B. Berge, *Physica B*, 1998, **248**, 208.
20. D. Blaudez, T. Buffeteau, J. C. Cornut, B. Desbat, N. Escafre, M. Pézolet, and J.-M. Turlet, *Appl. Spectr.*, 1993, **47**, 869.
21. D. Blaudez, J.-M. Turlet, J. Dufourcq, D. Bard, T. Buffeteau, and B. Desbat, *J. Chem. Soc., Faraday Trans.*, 1996, **92**, 525.
22. C. Alonso, D. Blaudez, B. Desbat, F. Artzner, B. Berge, and A. Renault, *Chem. Phys. Lett.*, 1998, **284**, 446.
23. S. Castano, B. Desbat, and J. Dufourcq, *Biochim. Biophys. Acta*, 2000, **1463**, 65.
24. S. Castano, B. Desbat, M. Laguerre, and J. Dufourcq, *Biochim. Biophys. Acta*, 1999, **1416**, 176.
25. E. Bellet-Amalric, D. Blaudez, B. Desbat, F. Graner, F. Gauthier, and A. Renault, *Biochim. Biophys. Acta*, 2000, **1467**, 131.

26. W. P. Ulrich and H. Vogel, *Biophys. J.*, 1999, **76**, 1639.
27. M. Jackson and H. H. Mantsch, *Crit. Rev. Biochem. Mol. Biol.*, 1995, **30**, 95.
28. A. H. Martin, M. B. J. Meinders, M. A. Bos, M. A. Cohen Stuart, and T. van Vliet, *Langmuir*, 2003, **19**, 2922.
29. E. V. Kudryashova, M. B. J. Meinders, A. J. W. G. Visser, A. van Hoek, and H. H. J. De Jongh, *Eur. Biophys. J.*, 2003, **32**, 553.

Displacement of β-Casein from the Air–Water Interface by Phospholipids

By Ana Lucero Caro, Mª. Rosario Rodríguez Niño,
Cecilio Carrera Sánchez, A. Patrick Gunning,[1]
Alan R. Mackie,[1] and Juan M. Rodríguez Patino

DEPARTAMENTO DE INGENIERÍA QUÍMICA, FACULTAD DE
QUÍMICA, UNIVERSIDAD DE SEVILLA, C/. PROF. GARCÍA
GONZÁLEZ, 1, 41012 SEVILLE, SPAIN
[1]INSTITUTE OF FOOD RESEARCH, NORWICH RESEARCH PARK,
COLNEY, NORWICH NR4 7UA, UK

1 Introduction

As phospholipids and proteins are the major components of biological membranes, the bulk of ongoing research concerning phospholipid–protein interactions is related to understanding biomembrane processes.[1,2] However, in this contribution we focus on phospholipid–protein interactions in relation to food dispersed systems.

Phospholipids and proteins, separately and in complex form, contribute significantly to the physical properties of many systems of technological importance, *i.e.*, emulsions and foams.[2-6] Phospholipids are often used as emulsifiers because of their amphiphilic characteristics and their high capacity to lower the interfacial tension.[3] Moreover, phospholipids have the capacity of spontaneous association in the aqueous bulk phase to form supramolecular structures that self-assemble. Thus, phospholipids can be used as raw materials for the preparation of vesicles (liposomes), which are the vehicles of choice in some drug delivery systems. This capacity to form physical barriers for the separation of two immiscibles phases is also of practical importance for functional foods and cosmetics formulation.[6] In addition, phospholipids have intrinsic properties (in nutrition and therapeutics), being a source of fatty acids, organic phosphate, and choline. This attractive synergy between technology and physiology justifies the interest in phospholipids for the formulation of functional foods.[6,7] In contrast, proteins form mechanically stable films but are unable to lower the surface tension below a certain value.[4,8] Milk

160

proteins have been utilized for traditional food formulations.[9,10] But, the demand for safe, high-quality, healthy foods with good nutritional value has increased interest in using vegetable proteins from cereals[11,12] and legumes[13,14] for the formation and stabilization of emulsions and foams.

Real foods contain mixtures of emulsifiers, with the potential to (de)stabilize fluid interfaces by antagonistic mechanisms,[3,15] and thus capable of contributing to the stability or instability of the dispersion.[16,17] The optimum use of emulsifiers depends on the knowledge of their interfacial physico-chemical characteristics, such as surface activity, structure, stability, interfacial rheological properties, interactions (compatibility) between film forming components, *etc.*, and the kinetics of film formation at fluid interfaces.[14]

The interactions between phospholipids and proteins at fluid interfaces are of great importance for the formation and stability of food colloid systems. In this work we analyse the interactions between dipalmitoyl-phosphatidylcholine (DPPC) and β-casein in mixed monolayers spread and adsorbed at the air–water interface by means of π–A isotherms (monolayer structure and miscibility), Brewster angle and atomic force microscopy (BAM and AFM), and interfacial rheology, as a function of surface pressure, pH, and phospholipid/protein interfacial ratio. The DPPC can be present in aqueous solution with zero net charge, as zwitterions or with a negative net charge, depending on the pH. Thus, the analysis of β-casein–DPPC mixed monolayers at different pH values can give insight into the importance of electrostatic interactions on the interfacial characteristics of mixed films at the air–water interface. The temperature (20 °C) and the ionic strength (0.05 M) were kept constant.

2 Experimental

Materials

The 99% pure β-casein was supplied and purified from bulk milk from the Hannah Research Institute (Ayr, Scotland). The DPPC was supplied by Sigma (>99%). To form the surface film, DPPC was spread in the form of a solution, using chloroform/ethanol (4:1, v:v) as a spreading solvent. Analytical grade chloroform (Sigma, 99%) and ethanol (Merck, >99.8%) were used without further purification. Samples for interfacial characteristics of protein and DPPC films were prepared using Milli-Q ultrapure water at pH 7. The water used as the sub-phase was purified by means of a Millipore filtration device (Milllipore, Milli Q™). To adjust the sub-phase pH, buffer solutions were used. Acetic acid/sodium acetate aqueous solution (CH_3COOH/CH_3COONa) was used to achieve pH = 5, and a commercial buffer solution called Trizma (($CH_2OH)_3CNH_2$ / ($CH_2OH)_3CNH_3Cl$) for pH = 7 and pH = 9. All these products were supplied by Sigma (>99.5%). The ionic strength was 0.05 M in all the experiments.

Methods

Measurements of surface pressure π *versus* average area per molecule A were performed on fully automated Langmuir-type and Wilhelmy-type film balances, as described elsewhere.[18–20] The sub-phase temperature was controlled at 20 °C by water circulation from a thermostat, within an error range of ±0.3 °C.

For microscopic observation of the monolayer structure, the Brewster angle microscope BAM 2 plus (NFT, Germany) was used as described elsewhere.[18,19] Using atomic force microscopy (AFM), images of β-casein + DPPC monolayers were taken of Langmuir–Blodgett films deposited onto hydrophilic mica substrates.[21]

To obtain surface dilatational rheological parameters of spread monolayers at the air–water interface, a modified Wilhelmy-type film balance (KSV 3000) was used.[20] The surface rheological parameters and the surface tension of β-casein adsorbing on a previously spread DPPC monolayer were measured in a ring trough according to the method of Kokelaar *et al.*[22] Measurements were made as a function of time and radial frequency, as described previously.[23]

3 Results and Discussion

Structural Characteristics of β-Casein + DPPC Mixed Films

Results derived from π–A isotherms at pH 5, 7, and 9 (Figure 1) measured in the Langmuir-type and Wilhelmy-type troughs are similar. These data are in good agreement with those obtained in the same troughs with spread monolayers of pure β-casein[19] and pure DPPC[24] on a similar sub-phase at pH 5 and 7. Briefly, the results confirm that β-casein monolayers have a liquid-expanded structure under these experimental conditions. The π–A isotherm is displaced towards the surface pressure axis and the monolayer structure is more condensed in the acidic sub-phase. It appears that a critical surface pressure and surface concentration exist at which the film properties change significantly.

With β-casein films we distinguish two different structures (1 and 2). The monolayer collapses at the highest surface pressure, close to the equilibrium surface pressure (which is indicated in Figure 1 by means of arrows). The acidic sub-phase causes an increase in the pressure of the transition between structures 1 and 2. This transition was observed at surface pressures of 15.4 and 10.5 mN m^{-1} for pH 5 and 7, respectively. However, at pH 9 the transition was not observed. A tail–train structure for β-casein is inferred for surface pressures lower than that of the transition (structure 1). At higher surface pressures, and up to the equilibrium surface pressure, amino-acid segments are assumed to be extended into the underlying aqueous solution and to adopt the form of loops and tails (structure 2). The elasticity derived from the π–A isotherm, $E = -A(\partial\pi/\partial A)_T$, corroborates the existence of two structures for the β-casein monolayer. The elasticity at each surface pressure was higher at pH 5 than at pH 7.

Different structures can be deduced for the DPPC monolayer as a function of pH and surface pressure (see Figure 1). For instance, at pH 7 the π–A isotherm presents three distinct regions: a liquid-expanded (LE) phase (at π < 5.5–9.5 mN m^{-1}), a first-order phase transition—the intermediate region of

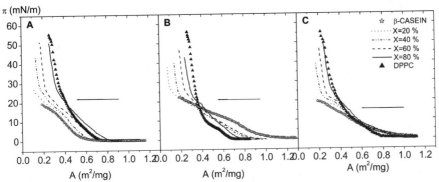

Figure 1 *Surface pressure versus area (π–A) isotherms at 20 °C for DPPC+β-casein monolayers at (A) pH=5, (B) pH=7, and (C) pH=9. The equilibrium surface pressure of β-casein at each pH is indicate by a horizontal line.*

lower slope—between liquid-condensed (LC) and liquid-expanded structures $(9.5 < \pi < 30 \text{ mN m}^{-1})$, a liquid-condensed structure $(\pi > 30 \text{ mN m}^{-1})$, and, finally, monolayer collapse at a surface pressure of about 55 mN m^{-1}. The collapse pressure (of poor reproducibility) is higher than the equilibrium surface pressure $(\pi_e \approx 46\text{–}49 \text{ mN m}^{-1})$. The pH of the aqueous phase also has an effect on the structural characteristics of DPPC. For instance, the surface pressure at the beginning of the first-order phase transition between the LE and LC phases was higher at pH 9 that at pH 5 or 7.

The β-casein + DPPC mixed films at surface pressures lower than that for β-casein collapse (see the lines in Figure 1) adopt a structural polymorphism, as for the pure DPPC monolayer. But, from the E values (data not shown), we have also observed the typical transition between the structures 1 and 2 of β-casein. At pH 9, no transition was observed, as for the pure β-casein monolayer. Moreover, a monolayer expansion was observed at pH 5 (Figure 1A) as the DPPC content in the mixture was increased. However, at pH 7 the behaviour of the mixed monolayers is more complex (Figure 1B). In fact, for $\pi > \pi_e(\beta\text{-casein})$, a monolayer expansion was observed as the DPPC content in the mixture was increased, but the opposite was observed for $\pi < \pi_e(\beta\text{-casein})$. On basic aqueous solutions (pH=9), and at $\pi < \pi_e(\beta\text{-casein})$, the π–A isotherms of pure DPPC and β-casein + DPPC films are practically the same (Figure 1C). At surface pressures higher than that for β-casein collapse, the π–A isotherms for the mixed monolayers appear parallel to that for pure DPPC, irrespective of the pH, which demonstrates that the arrangement of the DPPC hydrocarbon chain in the mixed monolayers is practically the same over the entire interfacial composition range.[15]

The interactions between the components in mixed monolayers can be studied from the point of view of miscibility of the components.[25] In Figure 2 we show the excess area A_{exc} as a function of the surface pressure and the monolayer composition. The excess area is calculated according to the additivity rule

$$A_{exc} = A - (A_1 X_1 + A_2 X_2), \qquad (1)$$

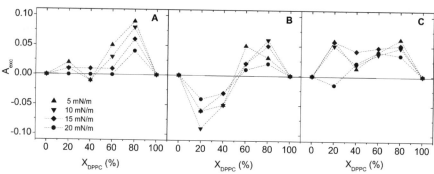

Figure 2 *Excess area A_{exc} as a function of the percentage of DPPC in the mixture and surface pressure for DPPC+β-casein monolayers at 20 °C at (A) pH=5, (B) pH=7, and (C) pH=9.*

where A is the molecular area at a given surface pressure for the mixed monolayer, A_1 and A_2 are the molecular areas of the pure components, and X_1 and X_2 are their mass fractions in the mixed monolayer. The horizontal lines in Figure 2 correspond to an immiscible monolayer or an ideal mixture between film forming components. For β-casein (1)+DPPC (2) monolayers at pH 5, the excess area is positive for $X_2 > 60\%$, especially at the lower surface pressures (Figure 2A). Around pH=5, close to the isoelectric point, β-casein has no net charge, and so the monolayer expansion ($A_{exc} > 0$) may have a steric and/or entropic character, due to the accommodation of DPPC molecules in the residues of β-casein with a tail–train structure at lower π. At pH=7, however, the excess area is negative at $X_2 < 50$ % and positive at $X_2 > 50$ %, with the maximum A_{exc} in absolute terms at the lower surface pressures. The condensation effect, which can be seen at lower contents of DPPC in the mixture (Figure 2B), should be interpreted along the lines of either molecular packing or molecular interactions. The expansion of the monolayer at a higher content of DPPC in the mixture could be due to repulsive interactions between film-forming components, which could overcome the condensation of the monolayer having a steric/entropic character. Finally, at pH=9, the excess area is positive for all mixed monolayers. At this pH both DPPC and β-casein have a negative net charge, and thus the positive values of A_{exc} may be due to electrostatic repulsion between film-forming components. That is, from a macroscopic point of view, the data in figure 2 confirm the existence of electrostatic repulsions and/or steric/entropic interactions between film-forming components, depending on the pH of the aqueous phase. However, the low values of A_{exc} indicate that the magnitude of attractive or repulsive interactions between film-forming components is somewhat weak.

For β-casein+DPPC mixed films, the π–A isotherms for spread monolayers at surface pressures higher than the equilibrium surface pressure of β-casein are practically parallel (Figure 1). This may be attributed to protein displacement by DPPC from the interface. At surface pressures lower than the equilibrium surface pressure of β-casein, both β-casein and DPPC coexist. The

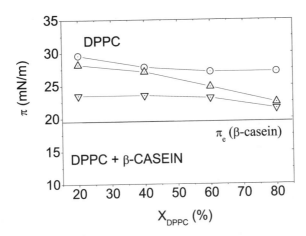

Figure 3 *The displacement surface pressure π_d of β-casein by DPPC from spread mixed monolayers at the air–water interface as a function of the percentage of DPPC in the mixture. The horizontal line represents the equilibrium surface pressure of a pure β-casein monolayers at 20 °C. The surface pressure at the point at which the mixed film behaves like a pure DPPC film is indicated at different values of pH: (O) 5, (Δ) 7, and (∇) 9.*

tendency towards displacement of β-casein by DPPC from the air–water interface depends on the β-casein/DPPC ratio and on the pH. From the π–A isotherms of the mixed monolayers (Figure 1), we define a displacement surface pressure (π_d) as the minimum surface pressure above which there is coincidence of the π–A isotherms for pure DPPC and β-casein + DPPC monolayers (on the basis that only DPPC is present at the interface). Thus, for low π_d, β-casein displacement by DPPC is facilitated. In Figure 3 we show π_d as a function of the mixture composition for different pH values. The π_d values follow the order: π_d (pH 5) > π_d (pH 7) > π_d (pH 9). Thus, the β-casein displacement by DPPC is easier on basic aqueous solutions. In addition, competitive displacement is facilitated at higher DPPC concentrations in the mixed monolayer. That is, β-casein displacement by DPPC is facilitated as the electrostatic repulsion between film-forming components increases, at high pH and at the high DPPC concentrations in the mixture.

Finally, we note that the collapse pressure (π_c) for mixed monolayers (Figure 1) tends to be similar to that for pure DPPC, a phenomenon also observed for protein + monoglyceride monolayers.[15] Unfortunately, we were unable to observe the true collapse pressure of our β-casein + DPPC monolayers because the monolayer overflowed the edges of the trough at the high surface pressures.

In summary, from a systematic study centred on the π–A isotherm of β-casein + DPPC monolayers—including the application of the additivity rule for miscibility—it is concluded that, at a macroscopic level and at π values lower than that for protein collapse, a one-phase mixed monolayer may exist at the air–water interface, but with weak interactions between film-forming components. At higher π values, the collapsed protein is displaced from the interface by DPPC. The existence of weak interactions between β-casein and DPPC, and/or repulsion between these components on a basic aqueous subphase, facilitates the displacement of β-casein by DPPC from the air–water interface.

Penetration of β-Casein into DPPC Monolayers

The penetration of β-casein into the DPPC monolayer was followed by measuring the surface pressure increase at a constant surface area. β-Casein was injected into the aqueous bulk phase, underneath the DPPC monolayer, at different initial surface pressures. Results at pH 7 are shown in Figure 4A.

The evolution of surface pressure with time reflects an adsorption mechanism with three kinetic steps: (i) the pressure increases at low adsorption time, (ii) a plateau is reached, and (iii) the pressure increases again progressively up to a maximum value. However, some differences were observed during the time evolution of surface pressure at different pH values (data not shown). A minimum instead of a plateau was observed in the intermediate region at pH = 5, whereas the central plateau was absent at some interfacial surface pressures at pH = 9. These data may be evidence for β-casein insertion into the DPPC monolayer, because if the protein interacts with the phospholipid head group through electrostatic interactions without penetration into the monolayer this would not affect the surface pressure.[26] The initial increases in π could be associated with a fast penetration of β-casein from underneath the

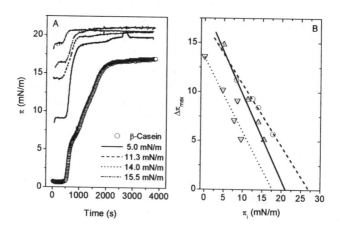

Figure 4 *Penetration of β-casein into a DPPC monolayer at 20 °C. (A) The time dependence of surface pressure after injection of β-casein into a DPPC monolayer at different initial surface pressures at pH 7. (B) Maximum surface pressure increase (Δπ$_{max}$) after injection of β-casein into a DPPC monolayer as a function of the initial surface pressure at different pH values: (O) 5, (Δ) 7, and (∇) 9.*

DPPC monolayer followed by an accumulation of the protein (in the plateau region), which finally destabilizes the DPPC monolayer and promotes a progressive penetration of β-casein into the layer. The differences observed in the π–time plot as a function of the initial surface pressure (Figure 4A) suggest that the penetration is sensitive to the structure of the DPPC monolayer.

Interestingly, the maximum surface pressure attained at longer times (19.8–21.1 mN m^{-1}) is close to the equilibrium surface pressure for a pure β-casein monolayer ($\pi_e \approx 21$ mN m^{-1}). For the same amount of β-casein injected into the sub-phase, adsorption to a clean interface induces an increase in surface pressure up to 16.8 mN m^{-1}. Thus, the DPPC monolayer acts as a promoter of access of β-casein to the air–water interface. However, we do not reject the possibility that a change in the DPPC monolayer structure may be produced as a consequence of the penetration of β-casein into the interface.[27,28]

The maximum surface pressure increase ($\Delta\pi_{max}$) attained after the β-casein injection into a DPPC monolayer as a function of the initial surface pressure (π_i) is shown in Figure 4B. This type of plot generally yields a straight line of negative slope.[1] At low initial surface pressures, larger $\Delta\pi_{max}$ values are observed. The initial surface pressure of the DPPC monolayer for which the addition of β-casein does not induce a surface pressure increase is called the exclusion surface pressure π_{ex}.[1] This is the maximum value of π_i at which interactions (or bonding) affect the insertion of β-casein into the DPPC monolayer. The π_{ex} values obtained by extrapolation of the linear $\Delta\pi_{max}$ *versus* π_i curves are 26.6, 21.1, and 17.5 at pH 5, 7, and 9, respectively. These π_i values are higher than the π_e values for β-casein at the same pH. At these surface pressures, no significant surface pressure changes after addition of β-casein should be observed. As the π_{ex} value is higher at pH 5, the role of electrostatic interactions may be ruled out. The lower π_{ex} value at pH 9 is consistent with the existence of electrostatic repulsion between β-casein and DPPC, both of them carrying a net negative charge at this pH. This hypothesis is corroborated by the values of $\Delta\pi_{max}$ for a diluted DPPC monolayer with a gaseous structure[24] (as $\pi_i \to 0$). In fact, the $\Delta\pi_{max}$ values are 17.3, 16.8, and 13.6 mN m^{-1} at pH 5, 7, and 9, respectively, which compares with $\Delta\pi_{max} = 16.8$ mN m^{-1} for the adsorption of β-casein at a clean interface (Figure 4A), suggesting that the penetration of β-casein is facilitated at acidic pH, but is hindered at basic pH.

In summary, these results indicate that electrostatic interactions have a role in the incorporation of β-casein into DPPC monolayers (Figure 4), as observed for spread monolayers at the air–water interface (Figure 1).

Surface Dilatational Characteristics of β-Casein + DPPC Films

The surface viscoelastic properties of β-casein + DPPC films spread on the air–water interface are shown in Figure 5. The E–π plot shows an irregular shape. It can be seen that E increases with increasing π up to a maximum value at $\pi \approx 10$ mN m^{-1}. Upon further increase of the surface pressure, E decreases to

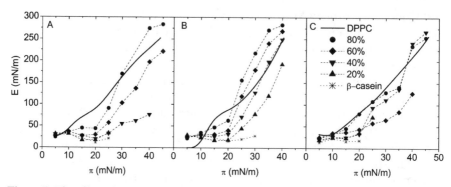

Figure 5 *The effect of surface pressure and monolayer composition on surface dilatational modulus E at 20 °C for spread DPPC+β-casein monolayers at (A) pH=5, (B) pH=7, and (C) pH=9.*

a minimum at $\pi \approx 20$ mN m^{-1}, close to the collapse point of β-casein. After-wards, E increases again with the surface pressure. At these points of inflection we have observed for β-casein the transition between structures 1 and 2 and between structure 2 and monolayer collapse (Figure 1). For the pure DPPC monolayer, the surface dilatational modulus also depends on the monolayer structure. The more condensed the structure is (at higher π), the higher the value of E becomes until collapse is reached. That is, more condensed mono-layer structures may lead to an increase in the forces of interaction between DPPC molecules at the interface, which is consistent with the observed increase in E. The E–π plots (Figures 5A and B) also reflect the structural poly-morphism of DPPC monolayers deduced from the π–A isotherms (Figure 1), involving the LE expanded phase, the transition between the LE and LC phases, and, finally, the LC phase itself, as the surface pressure increases. These results also show that the first-order phase transition between the LE and LC phases takes place at higher surface pressures at pH 9, as compared to pH 5 and pH 7.

Figure 5 shows that the surface pressure dependence of E for the films of β-casein (1)+DPPC (2) depends on the surface pressure, the interfacial composition, and the pH. The same irregular shape in the surface pressure dependence of the surface dilatational modulus occurs for β-casein+monoglyceride spread monolayers[29] At surface pressures lower than that for β-casein collapse, the same irregular shape in the π dependence of E, as for pure β-casein, was observed for the mixed films. In fact, we have observed for the mixed films the transition between structures 1 and 2 and between structure 2 and the monolayer collapse typical of the pure β-casein monolayer (β-casein at pH 9 and β-casein+DPPC monolayers at pH 7 and 9, and at $X_2 > 80\%$, are some exceptions). At these surface pressures β-casein and DPPC may coexist at the air–water interface, with E values between those for the pure components. At surface pressures higher than that for β-casein collapse, the E–π plots for the mixed films tend towards that of DPPC, especially as the content of DPPC in the mixture increases. This demonstrates again that, at higher surface pressures, the behaviour of the mixed films is dominated by the DPPC molecules. However, the data in Figure 5 also demonstrate that some interactions exist between β-casein and DPPC at surface pressures higher than that for β-casein collapse. In fact, the values of E for mixed films are higher than that for a pure DPPC monolayer, even at the collapse point of the mixed films. Thus, the residues of collapsed β-casein might retain the net charge in the remainder of the molecule, thereby producing a physical gel and reducing the electrostatic interactions with the negative net charge of DPPC at pH 7 and 9.

The surface viscoelastic properties of a DPPC monolayer penetrated by β-casein are shown in Figure 6. The surface pressure dependence of E for β-casein + DPPC films depends on the initial surface pressure of the DPPC monolayer and on the pH. The E values are higher for higher values of π in the mixed film. At surface pressures lower than that for β-casein collapse, the same irregular shape in $E(\pi)$ as for pure β-casein was found. This is associated with the transition between structures 1 and 2 and with monolayer collapse. In fact, at pH 5 the E values of the mixed films are similar to those of β-casein, except at higher π values after the β-casein collapse, where the E values tend to those of the DPPC. At pH 7 the maximum E value for structure 2 is displaced towards higher π values due to the existence of attractive interactions between the film-forming components, as deduced from the excess area of the mixed film (Figure 2). However, at pH 9 the mixed film behaves like pure DPPC, especially at higher π, due to the hindrance in the penetration of β-casein into the DPPC monolayer, as deduced from the $\Delta\pi_{max}$ *versus* π_i plots (Figure 4).

In summary, the results of the surface dilatational properties of the spread and penetrated β-casein + DPPC monolayers are complementary. Even at the highest surface pressure, a DPPC monolayer is unable to displace all the β-casein molecules from the air–water interface, unlike the typical behaviour of protein + lipid mixed film.[15] It is clear that the behaviour of ionizable phospholipid monolayers is more complex than that of 'normal' polar lipids.[15]

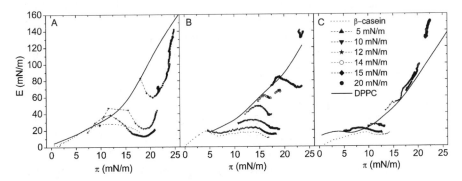

Figure 6 *The effect of surface pressure and monolayer composition on surface dilatational modulus E at 20 °C for spread DPPC monolayers penetrated by β-casein at (A) pH=5, (B) pH=7, and (C) pH=9.*

Topographical Characteristics of β-Casein + DPPC Films

The preceding analysis of spread and adsorbed (penetrated) β-casein + DPPC mixed films was performed by simple application of a combination of surface pressure and surface dilatational measurements. These macroscopic techniques provide information that is averaged over an area that is large relative to the size of individual phospholipid or protein molecules. Combining surface pressure measurements with direct microscopic visualization of the monolayer by BAM, or by forming a monolayer on a solid support and visualizing by AFM, the range of structural domains can be characterized.[30]

The mixed films were imaged by Brewster angle microscopy (BAM) and atomic force microscopy (AFM) as shown in Figures 7, 8 and 9 for different pH values and surfaces pressures. The BAM images were directly obtained from β-casein + DPPC spread monolayers at a DPPC composition of 40% w/w. The AFM images were obtained from mixed films formed from a DPPC monolayer at 5 mN m^{-1} penetrated by β-casein and transferred onto mica slides.

Figure 7 shows the monolayer morphology at a surface pressure lower than the equilibrium surface pressure of pure β-casein. The BAM images of β-casein + DPPC spread monolayers show two different regions—a homogeneous bright region which corresponds to the protein, and a region characterized by the presence of DPPC liquid-condensed domains of irregular shape. This is observed at each pH, although the domain morphology is variable, as also occurs with the pure DPPC monolayer. At pH 9 (image 7C$_1$), both regions are clearly separated due to the existence of repulsive electrostatic forces between the phospholipid and the protein. On the other hand, at the pH values of 5 and 7 (images 7A$_1$ and 7B$_1$), some miscibility between both regions is observed, and more at pH 7 than at pH 5. The different morphology observed with BAM for the different pH values is also observed with AFM for the DPPC monolayers penetrated by β-casein. At pH 9 (images 7C$_2$ and 7C$_3$), the DPPC and β-casein regions have approximately the same thickness, and so only a homogeneous phase is observed in topography (image 7C$_2$). However, the frictional image (image 7C$_3$) shows a completely immiscible monolayer because the DPPC liquid-condensed domains have a higher friction. The DPPC region has a higher friction than the protein region because of the closely packed DPPC chains in the LC state. Multilayers can be distinguished at pH 5 and 7 (images 7A$_2$, 7A$_3$, 7B$_2$, and 7B$_3$). It appears that enhanced interactions at these pH values leads to the formation of a β-casein–DPPC complex.

At a surface pressure closer to the equilibrium value for β-casein, the DPPC begins to squeeze out β-casein from the interface into the region underneath (Figure 8). The BAM images show bright stripes of collapsed β-casein at pH 5 and 7 with zones of different thickness (images 8A$_1$ and 8B$_1$). At these pH values, attractive interactions may exist between the phospholipid and the protein. The protein displaced to the underneath region interacts with DPPC forming a β-casein–DPPC complex that is also observed with the AFM

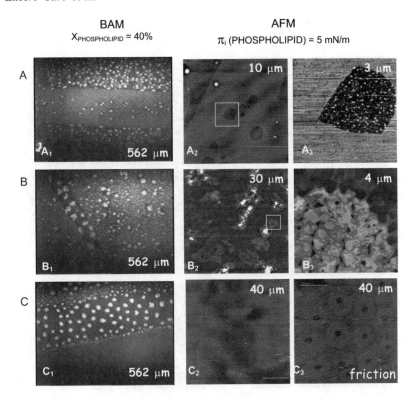

Figure 7 *Visualization of DPPC+β-casein monolayers by Brewster angle microscopy (BAM) and atomic force microscopy (AFM) at 20 °C at surface pressure lower than that for β-casein collapse (π<20 mN m⁻¹): (A) pH=5, (B) pH=7, and (C) pH=9.*

(images $8A_2$, $8A_3$, $8B_2$, and $8B_3$). Therefore, at pH 9, this complex is not formed at any DPPC composition as a consequence of repulsive electrostatic inter-actions between both components (images $8C_1$ and $8C_2$). Although there is no difference in the topography, the frictional images show two different regions corresponding to β-casein and DPPC forming an immiscible monolayer even at this surface pressure (image $8C_3$). As has been already stated, the higher friction corresponds to the DPPC region and the low friction to the β-casein region.

As the mixed monolayer is compressed to surface pressures above the equi-librium surface pressure of β-casein, the protein is displaced from the interface by DPPC. The AFM images were taken in this region after compression of the monolayer once the equilibrium surface pressure was reached. At pH 5, the β-casein is at its isoelectric point, and it tends to form aggregates attached to the phospholipid from below. Big stripes of protein-rich domains could be seen with BAM (image $9A_1$), and a closer look at these stripes could be obtained with the AFM (images $9A_2$ and $9A_3$). At pH 7, the increased

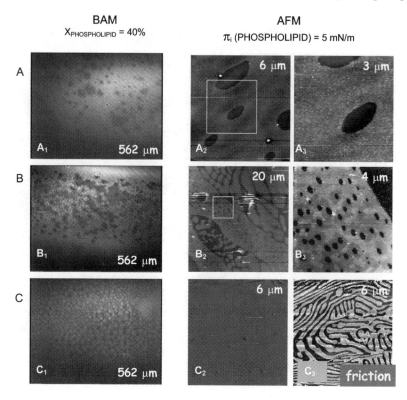

Figure 8 *Visualization of DPPC + β-casein monolayers by Brewster angle microscopy (BAM) and atomic force microscopy (AFM) at 20 °C at surface pressure close to β-casein collapse ($\pi \approx 20\ mN\ m^{-1}$): (A) pH = 5, (B) pH = 7, and (C) pH = 9.*

interactions between DPPC and β-casein favour multilayer formation, and crystals of thickness *ca.* 70 nm are formed (images $9B_1$, $9B_2$ and $9B_3$). The repulsive electrostatic interactions that exist between the components at pH 9 makes complex formation difficult and leads to the early squeezing out of β-casein from the interface (images $9C_1$, $9C_2$ and $9C_3$). As no differences appear in the friction images (data not shown), it is inferred that only DPPC is present at the interface.

The combined experimental observations made using AFM and BAM appear to corroborate the postulates of the orogenic mechanism.[21] They support the importance of the rheology of the protein network during the displacement process, and also the fact that repulsive forces between the protein and phospholipid promote phase separation (*i.e.*, there is a role of specific intermolecular interactions).[31] In fact, when the repulsive electrostatic forces between film-forming components are rather significant (at pH 9), the surface dilatational moduli of the adsorbed β-casein film (Figure 6) and the β-casein + DPPC mixed films (Figure 5) are lower.

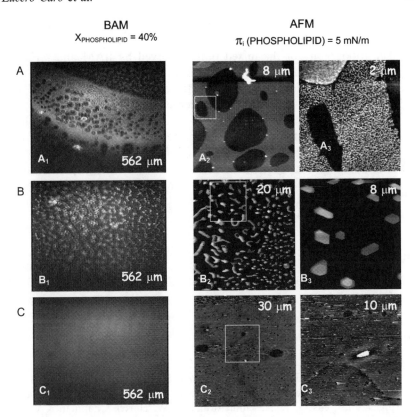

Figure 9 *Visualization of DPPC + β-casein monolayers by Brewster angle microscopy (BAM) and atomic force microscopy (AFM) at 20 °C at surface pressure higher than that for β-casein collapse ($\pi > 20$ mN m^{-1}): (A) pH = 5, (B) pH = 7, and (C) pH = 9.*

4 Conclusions

The main conclusions derived from this study are as follows.

(i) The structural characteristics, miscibility, topography, and surface dilatational properties of the mixed films (either spread or adsorbed) are very dependent on pH, surface pressure, and monolayer composition.

(ii) DPPC and β-casein form a mixed monolayer at the air–water interface on a neutral or acidic aqueous sub-phase. However, significant interactions do exist between the phospholipids and the β-casein in the mixed film, and these electrostatic interactions are more pronounced at pH 9 when the phospholipid molecules become negatively charged.

(iii) Over the entire range of existence of the mixed film, the monolayer presents some heterogeneity with domains of different topography. After β-casein collapse, a characteristic 'squeezing out' phenomenon was observed. At the DPPC monolayer collapse and, especially, at higher

DPPC/β-casein ratios, the mixed film is practically dominated by the presence of DPPC.

(iv) The topography of the mixed films observed by BAM and AFM gives complementary information at different levels of magnification.

(v) The surface dilatational characteristics of spread and adsorbed (penetrated) DPPC + β-casein films are essentially similar.

(vi) At pH 7, when protein is adsorbed into the liquid expanded phase of the lipid, the interactions hold the protein in the layer such that further compression leads to the formation of characteristic multilayers.

(vii) The combined results derived from π–A isotherms, BAM, AFM, and surface dilatational rheology confirm that the competitive displacement of β-casein from the air–water interface by phospholipids follows an orogenic mechanism.

Acknowledgement

The authors acknowledge the support of CICYT through grant AGL2001-3843-C02-01.

References

1. H. Brockman, *Curr. Opin. Struct. Biol.*, 1999, **9**, 438.
2. G. Brezesinski and H. Möwald, *Adv. Colloid Interface Sci.*, 2003, **100/102**, 563.
3. M. Bos, T. Nylander, T. Arnebrant, and D. C. Clark, in 'Food Emulsions and their Applications', eds G. L. Hasenhuettl and R. W. Hartel, Chapman & Hall, New York, 1997, p. 95.
4. S. Damodaran and A. Paraf (eds), 'Food Proteins and their Applications', Marcel Dekker, New York, 1997.
5. E. Dickinson, 'An introduction to Food Colloids', Oxford University Press, Oxford, 1992.
6. G. Cevec, 'Phospholipids Handbook', Marcel Dekker, New York, 1993.
7. I. Golberg (ed.), 'Functional Foods: Designer Foods, Pharmafoods, Nutraceuticals', Chapman & Hall, New York, 1994.
8. D. S. Horne and J. M. Rodríguez Patino, in 'Biopolymers at Interfaces', 2nd edn, ed. M. Malmsten, Marcel Dekker, New York, 2003, p. 857.
9. D. G. Dalgleish, in 'Food Proteins and their Applications', eds S. Damodaran and A. Paraf, Marcel Dekker, New York, 1997, p. 199.
10. J. Benjamins, 'Static and dynamic properties of proteins adsorbed at liquid interfaces', Ph.D. thesis, Wageningen University, Netherlands, 2000.
11. F. MacRitchie and D. Lafiandra, in 'Food Proteins and their Applications', eds S. Damodaran and A. Paraf, Marcel Dekker, New York, 1997, p. 293.
12. J. Örnebro, T. Nylander, and A.-C. Eliasson, *J. Cereal Sci.*, 2000, **31**, 195.
13. J. E. Kinsella, *J. Amer. Oil Chem. Soc.*, 1979, **56**, 242.
14. S. Utsumi, Y. Matsumura, and T. Mori, in 'Food Proteins and their Applications', eds S. Damodaran and A. Paraf, Marcel Dekker, New York, 1997, p. 257.
15. J. M. Rodríguez Patino, M. R. Rodríguez Niño, and C. Carrera, *Curr. Opin. Colloid Interface Sci.*, 2003, **8**, 387.
16. E. Dickinson, *Colloids Surf. B*, 2001, **20**, 197.

17. P. J. Wilde, *Curr. Opin. Colloid Interface Sci.*, 2000, **5**, 176.

18. J. M. Rodríguez Patino, C. Carrera, and M. R. Rodríguez Niño, *Langmuir*, 1999, **15**, 2484.

19. J. M. Rodríguez Patino, C. Carrera, and M. R. Rodríguez Niño, *Food Hydrocoll.*, 1999, **13**, 401.

20. J. M. Rodríguez Patino, M. R. Rodríguez Niño, C. Carrera, and M. Cejudo, *Langmuir*, 2001, **17**, 4003.

21. A. R. Mackie, A. P. Gunning. P. J. Wilde, and V. J. Morris, *J. Colloid Interface Sci.*, 1999, **210**, 157.

22. J. J. Kokelaar, A. Prins, and M. de Geen, *J. Colloid Interface Sci.*, 1991, **146**, 507.

23. M. R. Rodríguez Niño, P. J. Wilde, D. C. Clark, and J. M. Rodríguez Patino, *Ind. Eng. Chem. Res.*, 1996, **35**, 4449.

24. J. Miñones, J. M. Rodríguez Patino, O. Conde, C. Carrera, and R. Seoane, *Colloids Surf. B*, 2002, **203**, 273.

25. G. L. Gaines, 'Insoluble Monolayers at Liquid–Gas Interfaces', Interscience, New York, 1966.

26. F. Ronzon, B. Desbat, J.-P. Chauvet, and B. Roux, *Colloids Surf. B*, 2002, **23**, 365.

27. D. Vollhardt and V. B. Fainerman, *Adv. Colloid Interface Sci.*, 2000, **86**, 103.

28. X. Wang, Y. Zhang, J. Wu, M. Wang, G. Cui, J. Li, and G. Brezesinski, *Colloids Surf. B*, 2002, **23**, 339.

29. J. M. Rodríguez Patino, M. R. Rodríguez Niño, and C. Carrera, *J. Agric. Food Chem.*, 2003, **51**, 112.

30. A. R. Mackie, A. P. Gunning, M. J. Ridout, P. J. Wilde, and J. M. Rodríguez Patino, *Biomacromolecules*, 2001, **2**, 1001.

31. A. R. Mackie, P. A. Gunning, L. A. Pugnaloni, E. Dickinson, P. J. Wilde, and V. J. Morris, *Langmuir,* 2003, **19**, 6032.

Protein Functionality and Aggregation

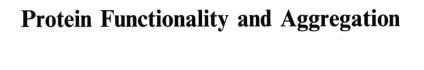

Milk Protein Functionality in Food Colloids

By Harjinder Singh

RIDDET CENTRE, MASSEY UNIVERSITY, PALMERSTON NORTH,
NEW ZEALAND

1 Introduction

Milk proteins, because of their high nutritional value and unique physico-chemical properties, are key functional components in many processed foods. As illustrated in Figure 1, many different grades and types of protein-enriched products are manufactured from milk by the dairy industry: caseins and caseinates, whey protein concentrate (WPC) and whey protein isolate (WPI), milk protein concentrate (MPC) powders, and specific protein blends specifically designed for particular applications. These ingredients perform a wide range of key functions in prepared foods, including emulsification, thickening, gelling and foaming (see Table 1). Through these functions, the proteins also contribute to the sensory characteristics and the stability of the manufactured foods. In practice, milk proteins usually exert several interdependent functional properties simultaneously in each food application.

The caseinates are one of the principal types of functional proteins used by the food industry. This is because they have exceptional water-binding capacity, fat emulsification properties, whipping ability, and a bland flavour. Important applications of caseinates in the food industry are various emulsion-type products, e.g., coffee whiteners, whipped toppings, cream liqueurs and low-fat spreads. In recent years there has been an increase in the use of caseins and caseinates in dietary preparations, in the pharmaceutical industry, and in medical applications; some of these preparations are also oil-in-water emulsions containing relatively small amounts of fat. The caseinates are produced from skim milk by adding acid (hydrochloric or lactic) or microbial cultures to precipitate the casein from the whey at pH 4.6. The acid-precipitated casein can then be resolubilized with alkali or an alkaline salt (using calcium, sodium, potassium or magnesium hydroxide) to about pH 6.7, and spray dried to form caseinate.[1]

Acid and alkaline treatments disrupt the native structure of the casein particles, and so caseinates have no structural similarity to the native casein

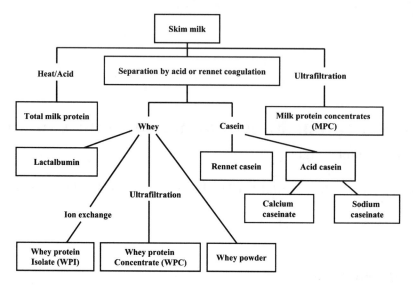

Figure 1 *Functional protein products from milk (reproduced with permission from Elsevier).*

Table 1 *Functional properties of milk proteins in a range of food systems.*

Functional Property	Food System
Solubility	Beverages
Emulsification	Coffee whitener, cream liqueurs, salad dressings, desserts
Foaming	Whipped toppings, shakes, mousses, cakes, meringues
Water-binding	Bread, meats, custard, soups, sauces, cultured foods
Heat stability	UHT and retort-processed beverages, soups and sauces, custard
Gelation	Meats, curds, cheese, surimi, yoghurt
Acid stability	Acid beverages, fermented drinks

micelles in milk. The resulting protein system consists of polymerized casein sub-units that are probably arranged in an ordered structure, which allows maximum interaction through hydrophobic bonding and also retains the polar, acidic groups in an exposed position, where they can be readily influenced by the pH and the ionic composition of the medium.[2] In general, commercial sodium and potassium caseinates are more soluble and possess better functional properties than the calcium derivatives. The larger size and the more strongly interacting character of the calcium caseinate aggregates can be attributed to ionic cross-linking by calcium. Overall, caseinate aggregates are smaller and more sensitive to pH and ionic strength than the colloidal-phosphate-containing casein micelles in milk.[3]

WPC and WPI are concentrated forms of the whey protein components. Ultrafiltration, diafiltration and ion-exchange technology are used to concentrate and separate the protein from other components. The whey protein is then dried to obtain WPC or WPI, both of which are highly soluble, with protein levels ranging from 80 to 95%.[1] The proteins in these products are not denatured, and their functional properties are largely retained. Both WPC and WPI have a wide range of food industrial applications. Due to their high protein content, they function as water-binding, gelling, emulsifying, and foaming agents.[1]

Whey proteins are unique amongst the milk proteins because, in their native conformations, they are soluble at low ionic strength over the entire pH range required for food applications. However, being globular proteins, their solubility decreases at high salt concentrations because of salting out, and they are susceptible to thermal denaturation at temperatures exceeding around 70 °C. Processing treatments used in the manufacture of WPC and WPI may sometimes cause small amounts of denaturation, which tends to affect their functionality.

The milk protein concentrates are processed directly from skim milk by ultrafiltration/diafiltration.[1] Samples of MPC can have a range of protein contents from 56 to 82%. The casein in MPC is in a similar micellar form to that found in milk, and the whey proteins are also in their native form.

The functionality of milk protein products may be considered to be the consequence of the molecular structures of the proteins and their interactions with other food components—such as fats, sugars, polysaccharides, salts, flavours, and aroma compounds. These interactions may take place at two levels. Firstly, the isolation processes alter the native structures of the milk proteins, which can exert a positive or negative effect on the functionality. Secondly, the protein ingredients will tend to interact with other food components during the manufacture of prepared food products. The types and strengths of the various interactions determine the structure, texture, rheology, sensory properties and shelf-life of manufactured food products. Much knowledge on the structure and properties of individual milk protein components has been gained, but less is known about interactions between different components that occur in a food system as a result of processing and formulation. Controlling these interactions is of key significance for the development of novel products and processes, as well as for the improvement of conventional ones.

This paper focuses on the emulsifying properties of milk protein products, as this functional property is very important in all the food applications of milk protein products described above. The adsorption behaviour of different milk protein products is considered, together with the stability of the resulting emulsions, focusing on recent work carried out in our laboratory at Massey University.

2 Formation of Oil-in-Water Emulsions using Milk Proteins

The emulsifying ability of a protein can be determined from the particle size of the emulsion droplets generated at a given protein concentration under defined homogenization conditions. The smaller the droplet size (*i.e.*, the larger the surface area), the better is the protein as an emulsifying agent. Both sodium caseinate and whey protein products (WPC and WPI) show excellent emulsifying ability, and it is possible to make stable emulsions at a relatively low protein/oil ratio (about 1:60). The emulsifying ability of 'aggregated' milk protein products, such as MPC and calcium caseinate, is much lower than that of whey protein and sodium caseinate. In other words, much higher concentrations of protein are required to make stable emulsions and larger droplets are formed under similar homogenization conditions. When the ratio of protein to oil is low in these emulsions, the protein aggregates are shared by adjacent droplets, resulting in bridging flocculation and consequently a marked increase in the effective droplet size.

The relationship between the surface protein concentration and the aggregation state of the protein is well established,[4-6] with more aggregated protein giving rise to a higher surface coverage. In emulsions formed with sodium caseinate or WPC, the surface protein concentration increases with increase in overall protein concentration until it reaches a plateau value of about 2–3 mg m^{-2} (Figure 2). The surface protein concentration of emulsions formed with MPC is in the range 5–20 mg m^{-2} depending on the protein concentration.[5]

It is interesting to note that the surface coverage of sodium caseinate is slightly lower than that of whey proteins at low protein concentrations, which may be attributed to greater flexibility of the caseins than the globular whey proteins, *i.e.*, the caseins can spread more extensively to cover a greater surface area than the whey protein molecules. This means that the adsorbed whey protein molecules on the droplet surface are likely to be closer to each other than the casein molecules, although it should be noted that the whey proteins also unfold to some extent. However, the surface concentration is greater in sodium caseinate emulsions than in WPC emulsions at high protein concentrations (Figure 2), which may be because sodium caseinate molecules in contact with the interface may adopt a more compact conformation.[7] As the caseins exist in

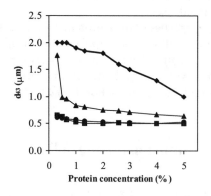
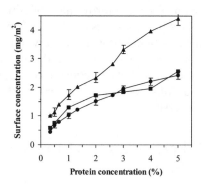

Figure 2 *Influence of protein concentration on average droplet size d_{43} and protein surface coverage in emulsions (30% soya oil) made with sodium caseinate (●), calcium caseinate (▲), WPC (■) or MPC (◆).*

aqueous solution as small aggregates at neutral pH, with the extent of aggregation increasing with increasing concentration of the protein,[8-9] it is also
possible that the caseins are adsorbed to the droplet surface as aggregated
structures (sub-micelles) at high concentrations.[10,11]

In the case of MPC-based and calcium-caseinate-based emulsions, the
spreading of protein at the interface is limited, because the aggregates are held
together by calcium bonds and/or colloidal calcium phosphate, and these
bonds are unlikely to be affected during the emulsification process. The higher
conformational stability of these aggregates also contributes to their reduced
emulsifying ability.[5,11]

3 Competition between Proteins during Emulsion Formation

As the commonly used milk protein ingredients are mixtures of different proteins with various surface activities, there is likely to be competition during
adsorption at oil–water interfaces. Over the last few years, there have been
important advances[5,6,10,11–17] in the area of competitive adsorption. Because of
its greater surface activity, β-casein was found[15] to adsorb in preference to
α_{s1}-casein in emulsions stabilized by mixtures of these proteins, and to displace
α_{s1}-casein rapidly from the droplet surface. No such preference was observed,
however, in sodium-caseinate-stabilized emulsions.[14] More recent work has
shown[10,11] that the competitive adsorption between β- and α_{s1}-caseins from
sodium caseinate is dependent on the ratio of protein to oil in the emulsions
(see Figure 3). For example, in sodium-caseinate-stabilized emulsions, when
the ratio of protein to oil is below 1:10, β-casein is preferentially adsorbed at
the droplet surface; but, when the total amount of protein is in excess of that

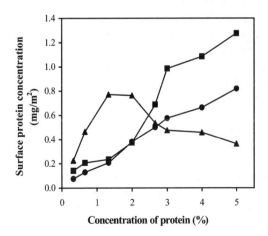

Figure 3 *Surface concentrations of a_{s1}-casein (■), β-casein (▲) and κ-casein (●) in
sodium caseinate emulsions (30% oil).*
(Data from ref. 11.)

needed for full surface coverage, α_{s1}-casein is adsorbed in preference to the other caseins. At all concentrations, κ-casein from sodium caseinate appears to be less readily adsorbed. In contrast, no such concentration dependence of competition is observed in emulsions stabilized with 'aggregated' caseins, such as calcium caseinate and MPC.[5,11] In these systems, the average surface composition is probably determined by the adsorption of protein aggregates of fixed composition. For instance, calcium caseinate solution consists of large α_{s1}-casein-rich aggregates, which appear to dominate the droplet surface after emulsification.[11] Whey proteins also show no significant preferential adsorption between β-lactoglobulin and α-lactalbumin regardless of the protein/oil ratio in the emulsion.[14,16,18,19]

The concentration dependence of the competitive adsorption behaviour of α_{s1}-casein and β-casein in sodium caseinate emulsions is curious. The preferential adsorption of β-casein, because of its high surface activity, appears to exist only at low concentrations where casein exists as monomers. With increasing protein concentration, the caseins aggregate to form various complexes of different compositions and sizes.[8,9] β-Casein may lose its competitive ability because of its self-aggregation to form micelles or through the formation of complexes with other caseins. Therefore, the surface composition of emulsions formed using a relatively high sodium caseinate concentration is likely to be determined by the surface activities of the casein aggregates and complexes. Although extensive information on the surface activity and hydrophobicity of individual caseins is available, little is known about how these characteristics are modified when casein molecules undergo self-association under different conditions.

The aggregation state and the flexibility of protein molecules can be altered by changes in pH, by addition of divalent cations, and by various processing treatments prior to emulsification. These changes will inevitably influence the adsorption behaviour of milk proteins at the oil–water interface. For example, addition of $CaCl_2$ at levels above a certain critical concentration to a sodium caseinate or whey protein solution before homogenization increases the mean droplet size, increases the surface protein coverage, and, in sodium caseinate emulsions, also affects the competition between different proteins.[19,20] The proportions of β-lactoglobulin and α-lactalbumin at the droplet surface remain unaffected by the addition of $CaCl_2$ to a whey protein solution prior to emulsification. In contrast, addition of $CaCl_2$ to a sodium caseinate solution markedly enhances the adsorption of α_{s1}-casein at the droplet surface, but it has much less effect on β-casein adsorption. The effects of calcium on surface coverage and composition can be explained by the binding of the ions to the negatively charged amino-acid residues on the protein. This reduces electrostatic repulsion between the protein molecules and increases the potential for intermolecular associations. Because of the presence of clusters of phosphoserine residues, the caseins have stronger affinity for calcium ions than the whey proteins. Consequently, the caseins (except κ-casein) are precipitated by calcium, with α_{s1}-casein being the most sensitive to aggregation and to precipitation by calcium. In sodium caseinate emulsions, the increased surface coverage upon addition of ionic calcium prior to emulsification is probably due to the adsorption of casein aggregates onto the droplet surface.[20] Greater adsorption of α_{s1}-casein reflects this protein's stronger tendency to be aggregated by calcium ions in solution or at the interface.

The native whey proteins do not bind much calcium and are not precipitated in the presence of ionic calcium.[21] However, the heat-denatured whey proteins are able to bind considerable amounts of calcium and undergo aggregation.[22] The increase in surface protein coverage suggests[19] the formation of aggregates of whey proteins in the presence of ionic calcium, which subsequently become adsorbed during emulsification. This has been attributed to a decrease in the denaturation temperature of the whey proteins in the presence of calcium. Barbut and Foegeding[23] observed that the temperature required for whey protein aggregation decreases from 72 to 45 °C in a 4% protein solution containing 10 mM $CaCl_2$. This would mean that calcium-mediated protein–protein interactions, resulting in the formation of whey protein aggregates, can take place during the emulsification process, which is usually carried out at above 40 °C.

All these results confirm that, under a given set of homogenization conditions, the surface composition is largely dependent on the protein/oil ratio and the aggregation state of the proteins in solution. It seems to be possible, therefore, to produce a range of emulsion droplets with different surface protein concentrations and compositions by manipulating the protein concentration, the protein type, and the ionic environment. Because of their different surface structures, these droplets would be expected to exhibit different reactivities and physico-chemical properties. Further studies are required for a better understanding of the relationship between droplet surface structures and the sensitivity of droplets to different environments.

4 Stability of Emulsions

Besides their emulsifying behaviour, the ability of milk proteins to *stabilize* emulsions is the most important functional property in most food colloids. As long as sufficient protein is present during homogenization to cover the oil droplets, emulsions stabilized by milk proteins are generally very stable to coalescence over prolonged storage. However, these emulsions are susceptible to different types of flocculation, which in turn leads to enhanced creaming or serum separation. At low protein/oil ratios, there is insufficient protein to cover fully the oil–water interface during homogenization. As a result, protein molecules/particles are shared by two or more droplets, causing bridging flocculation. Another consequence of insufficient protein is coalescence of droplets during or immediately after emulsion formation. Bridging flocculation is commonly observed in emulsions formed with aggregated milk protein products, such as calcium caseinate or MPC, in which the droplets are bridged by casein aggregates or micelles. Optimum stability can generally be attained at protein concentrations high enough to allow full saturation coverage at the oil–water

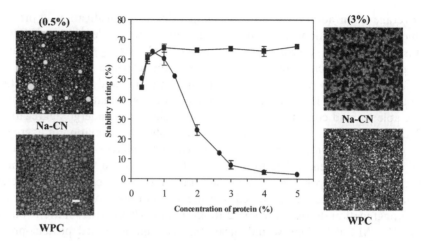

Figure 4 *Creaming stability and microstructure of emulsions made with sodium caseinate (●) or WPC (■) emulsions (30% oil). Scale bar represents 10 μm.*

interface. However, at very high protein to oil ratios, the presence of excess, unadsorbed protein may lead to depletion flocculation in some emulsions. Both depletion flocculation and bridging flocculation cause an emulsion to cream more rapidly.

Depletion flocculation has been observed in sodium-caseinate-based emulsions,[24] but not in emulsions formed with calcium caseinate, MPC or whey proteins[5,6,10] (see Figure 4). In sodium-caseinate-based emulsions, the critical concentration of protein in the aqueous phase (unadsorbed) required to initiate flocculation is above 1%. Because of this depletion flocculation, the effective diameter of the droplets increases, resulting in a marked decrease in creaming stability with increase in the caseinate concentration.

The differences in the creaming stability of emulsions made with different kinds of milk protein products can be explained by the depletion flocculation theory.[25,26] Depletion interactions between particles in a suspension containing polymer molecules arise when the particles approach close enough for the polymer to be excluded from the gap between the particles. The depletion interaction free energy, ΔG_{DEP}, of the order of a few kT, can be estimated from

$$\Delta G_{\text{DEP}} = -2\pi r_m^2 \Pi(r_d - 2r_m/3), \tag{1}$$

where Π is the osmotic pressure of the polymer solution, represented as a fluid of hard spheres of radius r_m, and r_d is the mean droplet radius. The osmotic pressure under ideal conditions is given by

$$\Pi = cRT/M, \tag{2}$$

where R is the molar gas constant, T is the temperature, M is the molecular mass of the polymer, and c is the number concentration of the polymer. For depletion flocculation to occur, the polymer has to have a fairly high value of M, so that r_m is relatively large. However, at a given c, M is inversely proportional to Π. Therefore, an increase in the polymer molecular mass will reduce the osmotic pressure driving the depletion interaction. Hence, at a given concentration, the depletion interaction free energy is low for a polymer of low molecular mass; it increases with an increase in molecular mass until it reaches a maximum; and then it decreases again with a further increase in molecular mass. Similarly, a reduction in the polymer number concentration will reduce the osmotic pressure.

Although the exact state of the casein molecules in concentrated sodium caseinate solutions is unknown, it appears that sodium caseinate exists as small aggregates, called sub-micelles (diameter 10–20 nm), in equilibrium with monomeric caseins.[8] A commercial sodium caseinate solution has been reported[9] to have a radius of gyration of about 20–30 nm, as determined by static light scattering. It is likely[24] that depletion flocculation in sodium caseinate emulsions is caused by the presence of these sub-micelles formed from self-assembly of sodium caseinate in the aqueous phase of the emulsion at concentrations above 1 wt%.

Emulsions formed with whey proteins, MPC or calcium caseinate do not show depletion flocculation, probably because there are no suitably sized protein particles at the required concentration in the aqueous phase. The molecular size of whey proteins is less than the optimum, whereas the casein micelles in MPC are too large to induce depletion flocculation.[5] Calcium caseinate consists of mixtures of casein aggregates of different sizes,[6] but the concentration of aggregates capable of inducing depletion flocculation is probably too low. The extent of creaming in these emulsions is largely determined by the particle size of the droplets. Generally, in these emulsion systems, the creaming stability increases with increasing protein concentration up to a certain concentration and then remains almost constant.[5,6] However, the creaming stability of emulsions formed with calcium caseinate or MPC at relatively high protein concentration tends to be higher than that of whey-protein-stabilized emulsions. This can be attributed to an increase in the droplet density as a result of the presence of a much thicker and denser adsorbed protein layer at the droplet surface.

Depletion flocculation in sodium-caseinate-stabilized emulsions is sensitive to the concentration of electrolytes in the aqueous phase. The addition of moderate amounts of $CaCl_2$ to emulsions containing excess sodium caseinate has been shown[20,27] to eliminate depletion flocculation and to improve the creaming stability, as illustrated in Figure 5. This appears to be due to a growth in the average size of the casein aggregates in the aqueous phase, resulting in a large increase in the molecular mass of the caseins.[28] In addition, there is a reduction in the concentration of unadsorbed caseinate. Both these effects are expected to cause a substantial reduction in the concentration of small particles, which are assumed to be the main depleting species responsible for inducing reversible flocculation in the calcium-free systems. Presumably,

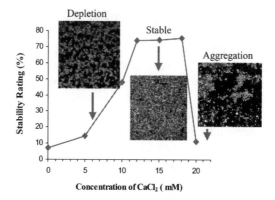

Figure 5 *Influence of CaCl₂ (added before emulsification) on the creaming stability and microstructure of 3% sodium caseinate emulsions (30% soya oil).*
(Data taken with permission from ref. 20.)

the substantial reduction in osmotic pressure makes the magnitude of ΔG_{DEP} predicted from equation (1) too small to cause depletion flocculation. Similarly, the addition of NaCl at levels above a certain concentration reduces the extent of depletion flocculation of sodium caseinate emulsions and improves the creaming stability.[29] This effect is due to increased adsorption of protein at the droplet surface and hence a lower concentration of unadsorbed protein remaining in the solution. Reducing the pH of emulsions formed with excess sodium caseinate also gradually eliminates depletion flocculation, through aggregation of adsorbed protein and a transfer of more protein to the droplet surface (Figure 6). Therefore, it seems to be possible to switch depletion flocculation off and on by controlling the concentration and the aggregation state of the casein molecules in the aqueous phase.

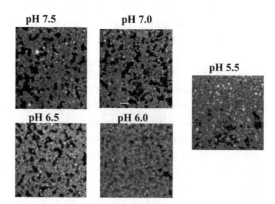

Figure 6 *Effects of pH on the microstructures of sodium caseinate emulsions containing 3% protein and 30% soya oil. The pH was adjusted after emulsion formation.*

5 Effects of Heat Treatment on Emulsion Properties

Many food emulsions are heat treated at relatively high temperatures to give a long shelf-life to the product *via* microbial sterility. Common problems encountered during the heating of emulsions are aggregation or coalescence of droplets and gel formation. Despite this, little work on the effects of heating on the behaviour of milk-protein-based emulsions has been published. Emulsions formed with sodium caseinate (2 wt% protein, 20% soya oil) are stable to heating at 90 °C for 30 min or 121 °C for 15 min, as determined by droplet size analysis.[30] However, the amount of adsorbed protein increases and the relative proportions of adsorbed caseins change upon heat treatment.[31] The increase in surface protein concentration on heating indicates that interactions between unadsorbed caseinate molecules and caseinate at the droplet surface may occur during heating.

Analysis of adsorbed caseins isolated from emulsions heated at 121 °C for 15 min shows[31] that a substantial proportion of the adsorbed caseinate is polymerized to form high molecular weight aggregates. These aggregates are held together through covalent bonds other than disulfide bonds; α_{s1}- and α_{s2}-caseins appear to be particularly sensitive to covalent bond formation. These covalent bonds may be formed between caseinate molecules at the surface of the same droplet because of the higher local concentrations of casein molecules at the droplet surface, or between two molecules at the surface of two different droplets, or between molecules on the surface and those in the bulk aqueous phase. The relatively small changes in droplet-size distributions after heating suggest that there is little or no covalent bond formation between the molecules adsorbed on different droplets. Simultaneously, adsorbed caseins appear to undergo thermal degradation, resulting in the formation of low-molecular-weight species. Relatively high proportions of casein degradation products present at the droplet surface indicate that the adsorbed caseinate molecules are more susceptible to fragmentation during heating than those in solution, and that these peptides remain adsorbed. This is probably due to the different structures and conformations of the caseins at the droplet surface than in bulk solution.

The creaming stability of sodium caseinate emulsions has been found[31] to improve upon heating, with the onset of depletion flocculation occurring at higher protein concentration than in unheated emulsions. This can be attributed to a reduction in the number of unadsorbed caseinate molecules (submicelles) in the aqueous phase as a result of increased surface coverage and heat-induced polymerization and degradation of the casein molecules. The improvement in the creaming stability in heated emulsions at low protein concentrations may be attributed to an increase in droplet density, because of the presence of higher amounts of polymerized protein at the droplet surface.

Emulsions formed with whey proteins at neutral pH are stable against heating when the ionic strength and/or the concentration of protein in the emulsions are low. Addition of KCl at 100 mM or above decreases the heat stability, leading to gel formation.[30] Aggregation of emulsion droplets is more extensive and it proceeds more rapidly as the concentration of protein in the emulsion is increased, whereas removal of unadsorbed protein from the emulsion lowers the rate of droplet aggregation.[32] When emulsions are made with relatively high protein concentrations, the amount of adsorbed protein increases with increasing intensity of the heat treatment; for instance, in emulsions made with 3.0% WPI and 25% soya oil, the amount of adsorbed protein was found to increase from 2.9 to 3.7 mg m^{-2} within the first 10 min of heating at 75 °C, but further heating had no effect.[33] At 90 °C, the plateau value of about 4 mg m^{-2} was reached within 5 minutes.

The effect of varying the heating temperature in the range 50–90 °C at pH 7.0 on WPI emulsions is interesting.[34,35] Droplet aggregation occurs on heating in the range 75–80 °C, which causes an increase in viscosity and loss of creaming stability, but the degree of aggregation and the susceptibility to creaming both decrease on heating above 80 °C.[33,34] It has been suggested[33] that, in the temperature range 75–80 °C, the whey protein molecules at the droplet surface are only partly unfolded and that not all of the hydrophobic amino-acid residues are directed towards the oil phase. Consequently, the surface of the droplet is more hydrophobic, making it susceptible to droplet aggregation.[33,34] At higher temperatures, the proteins become fully unfolded with all of the hydrophobic residues being directed into the oil phase, which makes the droplets less prone to aggregation. The role of sulfhydryl–disulfide interchange reactions in droplet aggregation is not clear. It has been suggested that disulfide-mediated interactions during heat treatment are not critical during the initial stages of aggregation, but that they later tend to strengthen the aggregates.[34]

It is also interesting to note that adsorbed whey proteins are incapable of forming disulfide bonds with κ-casein after heating.[36] The sulfhydryl group of β-lactoglobulin could be inaccessible to other proteins approaching the interface from solution. Both the non-adsorbed protein and the adsorbed protein are necessary for the heat-induced aggregation of whey-protein-stabilized emulsions.[32] The protein-covered droplet appears to interact more readily with the non-adsorbed protein during heat treatment than with another emulsion droplet. This has been explained[32] by assuming that the relative surface hydrophobicities of the emulsion droplet and the unadsorbed denatured whey proteins are different. Interaction of two emulsion droplets through their respective adsorbed protein layers will have a relatively low probability because the surface hydrophobicity is likely to be relatively low. When an emulsion droplet and a non-adsorbed protein molecule aggregate, at least one of them (*i.e.*, the denatured protein unadsorbed molecule) has a relatively high surface hydrophobicity, and this will tend to increase the probability of interaction and aggregation.[32]

6 Emulsions Stabilized by Hydrolysed Whey Proteins

It is generally accepted that the flexibility, and thus the availability of hydrophobic and hydrophilic segments within the protein chain, can be improved by moderate enzymatic hydrolysis of globular proteins (*e.g.*, whey proteins). These changes in protein structure improve the emulsifying properties of protein. However, extensive hydrolysis (above 20% degree of hydrolysis), leading to the production of many short peptides, has been found to be detrimental to the emulsifying and stabilizing properties of whey proteins.[37] The main form of instability in emulsions formed with highly hydrolysed whey proteins is the coalescence that arises because of the inability of the predominantly short peptides to adequately stabilize the large oil surface generated during homogenization.[37,38] Nevertheless, it seems to be possible[38] to make a fairly stable emulsion using highly hydrolysed whey proteins as the sole emulsifier at high peptide concentrations (protein/oil ratio ~1:1) and at low homogenization pressures. Under these conditions, there is a sufficient concentration of high-molecular-weight peptides (>5000 Da) in the emulsion to cover and stabilize the emulsion droplets.

These hydrolysed whey protein emulsions are sensitive to the presence of divalent cations. For example, addition of Ca^{2+} or Mg^{2+} at levels of >20 mM has been found to reduce the stability, as indicated by the formation of large particles.[39] This instability arises from interactions of calcium with the peptides in solution or at the droplet surface, leading to flocculation. The binding of ionic calcium to the adsorbed peptides could cause a reduction in the charge density at the droplet surface, which would reduce the interdroplet repulsion and enhance the likelihood of droplet flocculation. The formation of calcium bridges between peptides present on two different emulsion droplets would enhance flocculation.

In these emulsions, it has been observed that some very large droplets, apparently formed by coalescence, are also produced in the presence of calcium. This is likely to be due to the binding of calcium ions to the negatively charged peptides, causing aggregation of the larger, more surface-active peptides, a situation that would reduce the effective concentration of emulsifying peptides.

Heat treatment of emulsions stabilized by highly hydrolysed whey proteins at 121 °C for 16 min results in destabilization of the emulsions, which appears to occur mainly *via* a coalescence mechanism.[40] As the adsorbed peptide layers in these emulsions lack the cohesiveness of the parent proteins, and have a poor ability to provide steric or charge stabilization, increased collisions between the droplets during heating would cause droplet aggregation, leading to coalescence. It is also possible that desorption of some loosely adsorbed peptides occurs during heating, as indicated by the decrease in the amount of peptides associated with the oil surface after heating, which would also enhance coalescence.

It is well known that the addition of a polysaccharide such as xanthan or guar gum to a protein-stabilized emulsion, in the concentration range 0.01–0.4%, promotes droplet flocculation through the depletion mechanism, which consequently enhances the creaming rate.[41] Interestingly, this depletion flocculation also promotes the coalescence of droplets during the storage of emulsions formed with hydrolysed whey proteins,[42] as shown in Figure 7. The rate of coalescence is enhanced considerably with increasing concentration of

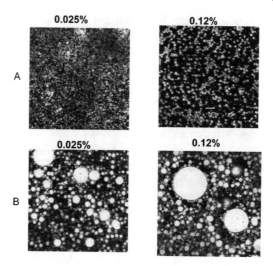

Figure 7 *Microstructure of emulsions formed with whey protein hydrolysate (4%) and corn oil (4%) containing different amounts of xanthan (0.025% or 0.12%): A, fresh emulsions; B, after storage for 24 h at 20 °C.*

polysaccharide in the emulsion up to a certain critical concentration. Whey protein peptides adsorbed at the droplet surface in these emulsions would almost certainly have a reduced surface viscosity compared with intact proteins, and this could lead to reduced stability to drainage and film rupture. Overall, it can be concluded that the flocculation of droplets, through various mechanisms (*e.g.*, depletion flocculation, calcium-induced aggregation, heat treatment) in these types of emulsions, where the interface is rather weak, would tend to lead to coalescence during storage.

References

1. D. M. Mulvihill and M. P. Ennis, in 'Advanced Dairy Chemistry—1. Proteins', 3rd edn, eds P. F. Fox and P. L. H. McSweeney, Kluwer/Plenum, New York, 2003, part B, p. 1175.
2. C. V. Morr, 'Developments in Dairy Chemistry—1', ed. P. F. Fox, Applied Science, London, 1982, p. 375.
3. J. E. Kinsella, *CRC Crit. Rev. Food Sci. Nutr.*, 1984, **21**, 197.
4. D. M. Mulvihill and P. C. Murphy, *Int. Dairy J.*, 1991, **1**, 13.
5. S. R. Euston and R. L. Hirst, *Int. Dairy J.*, 1999, **9**, 693.
6. M. Srinivasan, H. Singh, and P. A. Munro, *J. Food Sci.*, 2001, **66**, 441.
7. D. G. Dalgleish, *Food Res. Int.*, 1996, **29**, 541.
8. L. Pepper and H. M. Farrell, *J Dairy Sci.*, 1982, **65**, 2259.
9. J. A. Lucey, M. Srinivasan, H. Singh, and P. A. Munro, *J. Agric. Food Chem.*, 2000, **48**, 1610.
10. S. E. Euston, H. Singh, P. A. Munro, and D. G. Dalgleish, *J. Food Sci.*, 1995, **60**, 1124.
11. M. Srinivasan, H. Singh, and P. A. Munro, *Int. Diary J.*, 1999, **9**, 337.

12. E. Dickinson, *Food Hydrocoll.*, 1986, **1**, 3.
13. E. Dickinson, *J. Dairy Sci.*, 1997, **80**, 2607.
14. J. A. Hunt and D. G. Dalgliesh, *Food Hydrocoll.*, 1994, **8**, 175.
15. E. Dickinson, S. E. Rolfe, and D. G. Dalgliesh, *Food Hydrocoll.*, 1988, **2**, 193.
16. E. Dickinson, S. E. Rolfe, and D. G. Dalgliesh, *Food Hydrocoll.*, 1989, **4**, 183.
17. D. G. Dalgliesh, in 'Food Macromolecules and Colloids', eds E. Dickinson and D. Lorient, Royal Society of Chemistry, Cambridge, p. 23.
18. S. E. Euston, H. Singh, P. A. Munro, and D. G. Dalgliesh, *J. Food Sci.*, 1996, **61**, 916.
19. A. Ye and H. Singh, *Food Hydrocoll.*, 2000, **14**, 337.
20. A. Ye and H. Singh, *Food Hydrocoll.*, 2001, **165**, 195.
21. J. J. Baumy and G. Brule, *Le Lait*, 1988, **68**, 33.
22. C. P. Pappas and J. Rothwell, *Food Chem.*, 1991, **42**, 183.
23. S. Barbut and E. A. Foegeding, *J. Food Sci.*, 1993, **58**, 867.
24. E. Dickinson and M. Golding, *Food Hydrocoll.*, 1997, **11**, 13.
25. P. Walstra, in 'Food Colloids and Polymers: Stability and Mechanical Properties', eds E. Dickinson and P. Walstra, Royal Society of Chemistry, Cambridge, 1993, p. 3.
26. D. J. McClements, *Collids Surf. A.*, 1994, **90**, 25.
27. E. Dickinson and M. Golding, *Colloids Surf. B.*, 1998, **144**, 167.
28. E. Dickinson, M. G. Semenova, L. E. Belyakova, A. S. Antipova, M. M. Il'in, E. N. Tsapkina, and C. Ritzoulis, *J. Colloid Interface Sci.*, 2001, **239**, 87.
29. M. Srinivasan, H. Singh, and P. A. Munro, *Food Hydrocoll.*, 2000, **14**, 497.
30. J. A. Hunt and D. G. Dalgliesh, *J. Food Sci.*, 1995, **60**, 1120.
31. M. Srinivasan, H. Singh, and P. A. Munro, *Food Hydrocoll.*, 2002, **16**, 153.
32. S. R. Euston, S. R. Finnigan, and R. L. Hirst, *Food Hydrocoll.*, 2000, **14**, 155.
33. E. L. Sliwinski, P. J. Roubos, F. D. Zoet, M. A. J. A. van Boeble, and J. T. M. Wouters, *Colloids Surf. B.*, 2003, **31**, 231.
34. F. J. Monohan, D. J. McClements, and J. B. German, *J. Food Sci.*, 1996, **61**, 504.
35. K. Demetriades and D. J. McClements, *J. Agric. Food Chem.*, 1998, **46**, 3936.
36. D. G. Dalgliesh, in 'Food Emulsions and Foams: Interfaces, Interactions and Stability', eds E. Dickinson and J. M. Rodriguez Patino, Royal Society of Chemistry, Cambridge, 1999, p. 1.
37. A. M. Singh and D. G. Dalgliesh, *J. Dairy Sci.*, 1998, **81**, 918.
38. S. O. Agboola, H. Singh, P. A. Munro, D. G. Dalgleish, and A. M. Singh, *J. Agric. Food Chem.*, 1998, **46**, 84.
39. C. Ramkumar, H. Singh, P. A. Munro, and A. M. Singh, *J. Agric. Food Chem.*, 2000, **48**, 1598.
40. S. O. Agboola, J. Singh, P. A. Munro, D. G. Dalgliesh, and A. M. Singh, *J. Agric., Food Chem.*, 1998, **46**, 1814.
41. H. Singh, M. Tamehana, Y. Hemar, and P. A. Munro, *Food Hydrocoll.*, 2003, **17**, 549.
42. A. Ye, Y. Hemar, and H. Singh, *Food Hydrocoll.*, 2004, **18**, 737.

Aggregation and Gelation of Casein Sub-Micelles

By M. Panouillé, T. Nicolai, L. Benyahia, and D. Durand

POLYMÈRES, COLLOÏDES, INTERFACES, UMR-CNRS,
UNIVERSITÉ DU MAINE, 72085 LE MANS CEDEX 9, FRANCE

1 Introduction

Casein is the major protein component of milk. There are mainly four types of casein molecules called α_{s1}-casein, α_{s2}-casein, β-casein and κ-casein, present in the ratio 3:1:3:1.[1] These molecules differ in their hydrophobicity, global net charge, phosphate content, and calcium sensitivity. All four types are known to self-associate *via* hydrophobic and electrostatic interactions.[2-5] Monomer α_{s1}-casein can be considered as a triblock copolymer, with two hydrophobic extremities and a flexible hydrophilic centre containing the majority of phosphate groups. In solution, α_{s1}-casein associates into long worm-like polymers.[6] Monomer α_{s2}-casein is the most hydrophilic casein, but its association properties are not well known. Monomer β-casein has hydrophobic and hydrophilic extremities, and it associates into temperature-sensitive spherical micelles of diameter ~ 20 nm.[7,8] All these three caseins (α_{s1}, α_{s2}, β) are sensitive to calcium ions and can be precipitated by addition of calcium salts.[9-12] The fourth monomer, κ-casein, also forms spherical micelles of ~ 20 nm, but its aggregation behaviour is insensitive to temperature and ionic calcium.[7,8,13]

In milk, casein exists in the form of colloidal complexes called casein micelles. They are polydisperse, roughly spherical particles with a mean diameter of 160 nm.[14] Casein micelles are composed of the different casein molecules (93%) together with minerals (7%) called colloidal calcium phosphate (CCP). The κ-casein is located mainly on the surface and it stabilizes the casein micelle by steric repulsion.[15,16] The internal structure of the casein micelle has not yet been completely elucidated, and different models have been proposed for the micellar sub-structure. The sub-unit model[1,17,18] is based on the proposed existence of casein 'sub-micelles' linked together by CCP. Another model[4,19] suggests that the casein micelle is a mineralized protein gel with calcium phosphate nanoclusters cross-linking the proteins. Despite their great stability against high temperatures and addition of ethanol or salt, casein micelles can be dissociated into smaller sub-units after removal of the CCP. Procedures to remove or dissolve the CCP and dissociate casein micelles include the use of chelating agents such as EDTA,[20–22] dialysis against Ca^{2+}-free buffers,[23,24] and acidification.[25]

The properties of casein sub-units are strongly dependent on the aqueous solution environment.[26–28] Various studies have been made on casein sub-micelles by different experimental methods such as electron microscopy,[13.29] light scattering,[30] X-ray scattering,[31] small-angle neutron scattering,[32] and chromatography.[17,23,33] They show that the sub-micelles have a diameter ranging from 10 to 20 nm and a molecular weight around 3×10^5 g mol^{-1}.

Many studies have been made[34–37] on the heat stability of casein micelles or their modifications at high temperatures. Heat treatment of casein micelles may induce various changes such as acidification, precipitation of the CCP, or dissociation of individual caseins from the micelle. However, very few studies have dealt with the heat-induced aggregation of casein sub-micelles. These entities have been reported[30] to withstand heating at 140 °C for short periods of time (~ 10 s), but to aggregate progressively into 200 nm particles on prolonged heating (4–10 min). A pre-condition for the heat stability of sub-micelles is the presence of κ-casein.

The casein micelles in UHT milk can also dissociate and aggregate after long times of storage—a process known as 'age-gelation'.[38,39] The milk viscosity decreases slightly, after which it does not change any more over a long period, but then sharply increases just before gelation. The origin of this weak gelation has been attributed to proteolysis during storage, and also separately to physico-chemical changes occurring to the casein micelles. Gelation occurs through a kind of 'bridging flocculation' of the casein micelles, which occurs faster with increasing temperature. When phosphate or citrate is added, the gelation rate increases, and small casein particles—perhaps sub-micelles—are observed in the gel.[40] In this work we study the dissociation of casein micelles by use of a chelating agent, polyphosphate, and the subsequent heat-induced aggregation of the casein sub-units. The aggregation has been monitored by light scattering and rheology. In addition, we aim to compare this behaviour with the heat-induced aggregation of sodium caseinate.

2 Materials and Methods

Materials

Casein sub-micelles were obtained from a native phosphocaseinate powder (NPCP) dissolved in a sodium polyphosphate solution (Joha, BK Giulini Chemie). The NPCP (supplied by INRA-LRTL, Rennes) was prepared as described elsewhere.[41,42] The chemical composition of the NPCP used in this study is reported in Table 1. Sodium caseinate was obtained from DMV and used as received.

Table 1 *Chemical composition of the native phosphocaseinate powder (NPCP).*

Content	$g\ kg^{-1}$		% TS
Total solids (TS)	920	Ca	3.2
Total nitrogen matter	831	P	1.7
Non-casein nitrogen	43	Na	0.0
Non-protein nitrogen	3	Cl	0.2
Ash	80		

Preparation of Solutions

The NPCP or sodium caseinate were dissolved in sodium polyphosphate solutions, with 200 ppm sodium azide added as a bacteriostatic agent, while stirring for several hours. The salt solutions consisted of Millipore water (18 MΩ) containing different concentrations of sodium polyphosphate. The pH of the solution was adjusted to pH = 6 with 0.5 M HCl or NaOH. For light-scattering studies, the casein dispersions were centrifuged (Beckman Coulter, Allegra 64R Centrifuge) at $5 \times 10^4\ g$ and 20 °C for 90 minutes. A small opalescent top layer containing residual fat was formed. The clear supernatant, without the opalescent layer, was carefully removed and filtered through a 0.45 μm pore-size filter. The amount of protein lost by this method was shown to be less than 5%. The protein concentration was determined by UV absorption at 280 nm using an extinction coefficient of 0.81 mL mg^{-1} cm^{-1}. The use of this extinction coefficient was found to give protein concentrations consistent with values obtained by the Kjeldahl method. The value of the coefficient is also consistent with that found by Kamizake *et al.*[43]

Turbidity Measurements

The turbidity T of the casein solutions was determined from the absorbance A at 685 nm using the equation

$$T = (A \times \ln 10)/\Delta l, \tag{1}$$

where Δl is the path length of the light beam. In order to cover a large range of turbidity values, quartz cells were used with path lengths of both 1 mm and 10 mm.

Cryo-Electron Microscopy

Cryo-electron microscopy (cryo-EM) was carried out on a JEOL JEM-2100F instrument in the laboratory LMCP at the University of Paris 6.

Rheology

Viscosity measurements of Newtonian solutions/dispersions (up to 80 g L^{-1} casein) were made using a low-shear rheometer (Contraves Low Shear 40) with a Couette geometry. The concentrated casein suspensions were studied by shear measurements using a stress-imposed rheometer (AR1000, TA Instruments) with a cone-and-plate geometry (20 mm, 0.5°). The zero-shear viscosity was extrapolated from flow curves.

In situ gelation kinetics were studied at 80 °C and 0.1 Hz using a strain-imposed rheometer (ARES, Rheometrics) with a Couette geometry. Paraffin oil was layered on the sample to avoid water evaporation at 80 °C. Gels were studied at 20 °C by frequency sweeps using a plate-plate geometry (20 mm) with stress chosen to lie in the linear regime.

Light Scattering

Static and dynamic light-scattering measurements were made using an ALV-5000 multi-bit multi-τ correlator in combination with a Malvern goniometer and a Spectra Physics laser operating with vertically polarized light of wavelength $\lambda = 532$ nm. Data were collected at scattering angles θ between 10° and 140°, corresponding to a scattering wave-vector q in the range 2.7–30.0 $\times$ 10^{-3} nm^{-1}, where

$$q = (4\pi n/\lambda)\sin(\theta/2), \qquad (2)$$

and n is the refractive index. Temperature was controlled by a thermostat bath set at 20 °C.

The relative scattering intensity I_r was determined by subtracting the solvent scattering from the total scattering intensity and dividing by the scattering intensity of toluene. In general, I_r can be written as[44,45]

$$I_r = KCM_wS(q), \qquad (3)$$

where M_w is the weight-average molar mass, K is a contrast factor that depends on the refractive index increment and the experimental set-up, and $S(q)$ is the q-dependent static structure factor. We have used the value 0.189 for the refractive index increment of casein, and a toluene standard with Rayleigh ratio 2.79×10^{-5} cm^{-1}.

3 Results and Discussion

When casein is dissolved in polyphosphate, the casein micelles are dissociated into smaller sub-units. On heating, these sub-micelles may aggregate in different ways depending on the experimental conditions, especially the concentrations of casein and polyphosphate, but also the pH and ionic strength. In this study we focus on the effects of the casein and polyphosphate concentrations at pH = 6. Solutions of different casein content (c_c) and polyphosphate content (c_p) were heated for a long time (>24 h) at 80 °C. The final states of the heated solutions as a function of system composition are summarized in the state diagram in Figure 1.

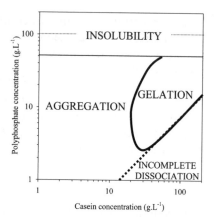

Figure 1 *State diagram of casein solutions heated at 80 °C for a long time. The dotted line represents the dissociation boundary. The solid line represents the gelation limit.*

The state diagram can be characterized in terms of four main regions (aggregation, gelation, insolubility, and incomplete dissociation) as described below.

(1) A minimal quantity of polyphosphate is needed to dissociate the casein micelles. Below this critical concentration, casein micelles and sub-micelles coexist in solution.

(2) In the presence of sufficient polyphosphate, the sub-micelles aggregate at low casein concentrations and precipitate after a heating time that increases with decreasing casein concentration.

(3) Once the casein concentration is high enough, the casein sub-micelles aggregate into a gel.

(4) When too much polyphosphate is added, the casein becomes insoluble. For $c_p > 100$ g L^{-1}, the casein does not dissolve at room temperature. For c_p in the range 50–100 g L^{-1}, dissolution is possible at 20 °C, but the casein precipitates immediately on heating.

Dissociation of Casein Micelles

The polyphosphate is used to dissociate the casein micelles into smaller sub-micelles through removal of calcium ions. This dissociation can be studied by measuring the turbidity of the casein solutions at different polyphosphate concentrations (see Figure 2). For a given casein concentration, the turbidity was found to decrease with increasing polyphosphate concentration before reaching a plateau. The residual turbidity at large values of c_p is caused by the presence of a small weight fraction of large particles, probably fat globules.

When the solutions were centrifuged at a speed at which all the casein micelles were pelleted, the concentration of casein in the supernatant was found to increase with increasing polyphosphate concentration. The decrease in the turbidity can thus be explained by the increasing dissociation of casein micelles into sub-micelles with increasing polyphosphate concentration. The

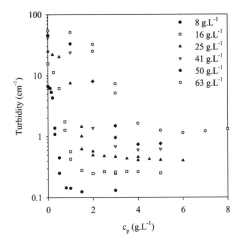

Figure 2 *Turbidity of casein solutions as a function of the polyphosphate concentration c_p for different casein concentrations.*

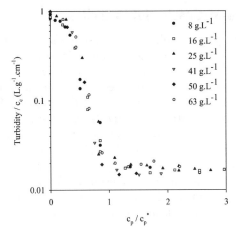

Figure 3 *The same data as in Figure 2 after normalizing the turbidity with the casein content c_c and c_p by c_p*.*

profile of the curves in Figure 2 is similar at each casein concentration. A minimal polyphosphate concentration, c_p*, can be defined as the concentration required to dissociate all the casein micelles and reach the turbidity plateau. Figure 3 shows that all the data can be superimposed if the turbidity is normalized by c_c and the polyphosphate concentration by c_p*. The quantity c_p* is directly proportional to the casein concentration, as shown in Figure 4. We estimate that 0.07 g of polyphosphate is required to dissociate 1 g of casein, which is equivalent to 0.93 mM calcium. As the polyphosphate salt is a mixture of phosphates of different chain lengths, it is difficult to know the exact number of phosphate groups and their average valency. However, a rough estimate shows that complete dissociation of casein micelles requires about 0.6 phosphate groups per calcium ion.

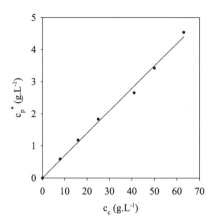

Figure 4 *The minimum polyphosphate concentration required to dissociate all the casein micelles, c_p^*, as a function of the casein concentration c_c.*

Characterization of Casein Sub-Micelles

When dissolved in polyphosphate, the casein micelles are dissociated into smaller sub-units. These have been studied by static and dynamic light-scattering as described previously.[46] Interpretation of the light-scattering measurements is complicated[8,13,46] by the presence of large particles of radius ~ 100 nm. These particles, which could be partially removed by centrifugation and filtration, are assumed to be residual fat globules.

Results extrapolated to infinite dilution show that the casein sub-units have a z-averaged hydrodynamic radius of $R_{hz} \approx 10$ nm and a weight-average molar mass of $M_w = 4 \times 10^5$ g mol⁻¹, *i.e.*, the aggregation number is around 20. The specific volume of the sub-micelles determined from combined viscosity and light-scattering measurements lies in the range 4–6 mL g⁻¹, which is close to the literature values for casein micelles,[3,47,48] sub-micelles,[32] and caseinate aggregates.[49]

Electron microscopy can give information that is complementary to light scattering. A cryo-EM micrograph of the casein sub-micelles ($c_c = 16$ g L⁻¹, $c_p = 20$ g L⁻¹) is shown in Figure 5. The sub-micelles appear as polydisperse spherical particles. They are distributed at a preferred distance apart, which means that electrostatic repulsion between the spheres is important. The size distribution of the sub-micelles (assumed spherical) is shown in Figure 6. The mean diameter is found to be ~ 30 nm, which is somewhat larger than the hydrodynamic size obtained from light scattering. This could be due to an artefact of the electron microscopy, because the film of the casein solution on the grid was very thin, meaning that the sub-micelles could have become adsorbed at the air-liquid surface, resulting in a deformation of their shape.

The zero-shear viscosity of suspensions of casein sub-micelles was measured as a function of the casein concentration (see Figure 7). We see that the apparent viscosity slowly increases with casein concentration up to ~ 50 g L⁻¹ and then sharply increases at higher concentrations. The first part of this curve (up

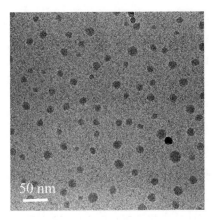

Figure 5 *Cryo-EM micrograph of casein sub-micelles.*

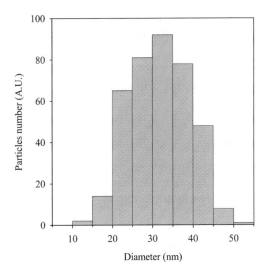

Figure 6 *Size distribution of casein sub-micelles as determined from the cryo-EM micrograph (Figure 5).*

to 100 g L^{-1}) is similar to that observed for hard spheres. It can be well fitted by the semi-empirical equation proposed by Quemada,[50]

$$\eta = \eta_s(1-\phi/\phi_m)^{-2} = \eta_s(1-c/c_m)^{-2}, \tag{4}$$

where η_s is the solvent viscosity, and ϕ_m and c_m are the volume fraction and concentration at which the spheres become jammed. The concentration c_m is found to be ~ 100 g L^{-1} at 20 °C, with the actual value depending weakly on temperature and salt concentration. For high concentrations ($c_c > c_m$) the experimental viscosity does not actually diverge, but it continues to increase

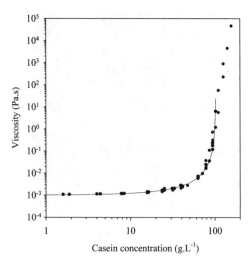

Figure 7 *The zero-shear viscosity of suspensions of different casein content (20 g L⁻¹ polyphosphate, pH 6, 20 °C). The solid line represents the equation* $\eta = 10^{-3} (1 - c_c/100)^{-2}$.

strongly. Casein sub-micelles should therefore be considered to behave as soft deformable spheres rather than as hard spheres.

Aggregation Behaviour

While casein sub-micelles aggregate very slowly at room temperature, the rate of aggregation does increase with increasing temperature. Aggregation causes an increase in the scattering intensity (see, for example, Figure 8). Three different stages can be distinguished: (i) initially, the scattering intensity increases very little; (ii) after a lag time, the intensity increases suddenly as the sub-micelles start to aggregate; and (iii) the aggregation finally leads to gelation or precipitation, depending on the casein and polyphosphate concentrations. At low casein concentrations, the aggregated casein sub-micelles form a precipitate which occupies a volume fraction that increases with increasing concentration, until for $c_c > 20$ g L⁻¹ a self-supporting gel is formed (Figure 9). Gelation causes the development of large slow fluctuations in the intensity at long times, as shown in Figure 8. We may define the gel time t_g as the time of the onset of these large intensity fluctuations.

The aggregation kinetics was found[46] to be influenced by the casein concentration and the temperature. The gel time decreases with increasing casein concentration or temperature. The concentration dependence can be described by a simple power-law relation, $t_g \propto c_c^{-1.6}$, and the temperature dependence can be described in terms of an apparent activation energy of *ca.* 100 kJ mol⁻¹. (A smaller value erroneously appeared in ref. 46).

The structure of the casein aggregates was studied by measuring the q-dependence of the scattered light intensity after dilution in water (to remove

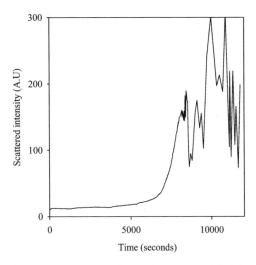

Figure 8 *Example of scattering intensity during heating at 80 °C for a casein solution in 20 g L⁻¹ polyphosphate. The scattering angle is 30°.*

Figure 9 *Aggregation type as a function of casein concentration (polyphosphate 20 g L⁻¹, pH 6).*

the effect of interactions). Figure 10 shows the data for aggregates formed at $c_c = 16$ g L⁻¹ and $c_p = 20$ g L⁻¹ after heating for different times at 80 °C. The light intensity before heating, essentially due to residual fat globules, has been subtracted. The intensity increases with the heating time, indicating a higher molar mass and mean aggregate size. For the largest aggregates we can observe a power-law q-dependence of the intensity which indicates that the aggregates have a fractal structure. At the highest q values there is a deviation from this power-law behaviour because the scattering is probing the local structure of the aggregates.

Casein aggregates were also studied by electron microscopy (see Figure 11). The size of the spherical elementary units appears to be close to that of the casein sub-micelles. The small spheres are linked together to form a ramified structure.

The proportion of the casein involved in gelation or precipitation was determined by measuring the concentration of protein in the supernatant after centrifugation. It was found that the quantity of aggregated casein increases with time up to about 80% of the total casein for long heating times (one or two

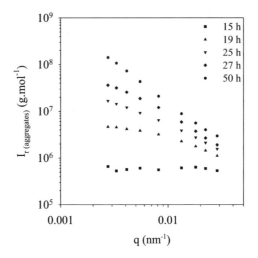

Figure 10 *The q-dependence of the relative excess scattering intensity of aggregates as a function of the heating time at 80 °C.*

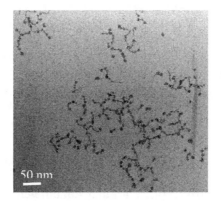

Figure 11 *Cryo-EM micrograph of casein aggregates.*

days). The aggregates were stable following dilution in pure water or in 0.1 M NaCl, but they could be dissociated by dilution in 20 g L^{-1} polyphosphate, which suggests the presence of calcium bridges between the casein molecules. After a very extensive heating period (several days) another type of irreversible bond appears to be produced, with the result that the aggregates could then no longer be dissociated in polyphosphate.

When the casein concentration is high enough, the aggregation leads to gelation of the casein sub-micelles. This gelation was studied by rheology, and an example of the behaviour observed is shown in Figure 12 for $c_c = 78$ g L^{-1} and $c_p = 10$ g L^{-1} at 75 °C. The storage and loss moduli, G' and G'', were found to increase very rapidly as the gel was being formed, after which the increase flattened off. However, prolonged heating led to extensive syneresis and the

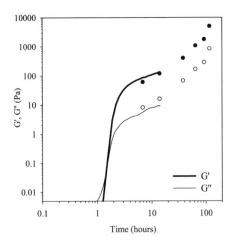

Figure 12 *Gelation kinetics for a system with $c_c = 78$ g L^{-1}, $c_p = 10$ g L^{-1}, pH=6, heated at 75 °C. The lines represent in situ experiments and the points represent separately prepared samples.*

associated slippage rendered *in situ* rheological measurements impossible. In order to study the gels after long heating times, they were prepared in moulds at 80 °C. A thin slice of gel was cut and placed in the rheometer cell with a plate-plate geometry. For short heating times—before any syneresis—the results were consistent with the *in situ* experiments. Figure 12 shows that the gels continue to evolve over a very long period, during which time they can expel up to 70% of their water content. This means that the casein concentration of the gel increases, which explains at least partially the increase in the elastic modulus at long heating times.

As mentioned above, a minimum concentration of polyphosphate is necessary in order to dissociate the casein micelles; but the casein precipitates if too much polyphosphate is added. The polyphosphate concentration also has an influence on the gel time t_g defined as the time when G' and G'' cross over at 0.1 Hz. The gel times for $c_c = 50$ g L^{-1} at 75 °C determined by rheology and light scattering are compared in Figure 13 as a function of the polyphosphate concentration. The values are roughly comparable, except at the highest polyphosphate concentrations where the gels are very heterogeneous. At low polyphosphate concentrations, the dissociation of the casein micelles is incomplete and gelation does not occur. For concentrations between 7 and 30 g L^{-1}, the light-scattering data show a weak increase of t_g with c_p, but the rheology data are not precise enough to reveal this tendency. For concentrations above ~ 40 g L^{-1}, the gel time decreases and the casein solutions are destabilized by excess of polyphosphate.

Comparison with Sodium Caseinate Aggregation

It seems instructive to compare the aggregation of the casein sub-micelles with the state of aggregation in commercial sodium caseinate. We have therefore

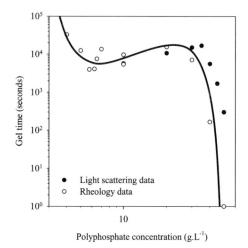

Figure 13 *Polyphosphate concentration dependence of the gel time from light scattering and rheology (*$C_c = 50$* g *L^{-1}*, pH 6). The solid line is a guide to the eye.*

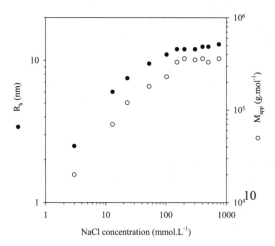

Figure 14 *Apparent hydrodynamic radius *R_h* (left axis, •) and apparent molar mass *M_{app}* (right axis, o) of caseinate as a function the NaCl concentration (*$Cc = 9$* g *L^{-1}*, pH = 6, 20 °C).*

studied the micellization of sodium caseinate at 9 g L^{-1} as a function of the ionic strength using light scattering (see Figure 14). The apparent hydrodynamic radius R_h and the apparent molar mass M_{app} increase with ionic strength up to 200 mM NaCl, and then reach plateau values of $R_h \approx 12$ nm and $M_{app} \approx 4 \times 10^5$ g mol^{-1}. Additional preliminary experiments seem to suggest that the apparent hydrodynamic radius does not depend on the caseinate concentration. Aggregates formed by sodium caseinate thus appear to be similar to casein sub-micelles formed from the native casein micelles in polyphosphate.

On heating, sodium caseinate becomes aggregated and precipitated, both in 0.1 M NaCl and 20 g L^{-1} polyphosphate, but centrifugation shows that only about 10% of the caseinate is implicated in the aggregation. Moreover, gelation is not observed at any caseinate concentration up to at least 80 g L^{-1}. If ionic calcium is added to a sodium caseinate solution, heating results in precipitation of a larger fraction of the caseinate. If both calcium and polyphosphate are added, gelation occurs after heating at 80 °C. This confirms our supposition of calcium bonding in the aggregation, but it shows also the importance of the presence of polyphosphate in order to form a stable gel.

4 Conclusions

The chelation of native casein micelles leads to the formation of spherical casein particles with a radius of around 10 nm and a mean molar mass of 4×10^5 g mol^{-1}. In the presence of polyphosphate, these so-called sub-micelles form aggregates that grow until they precipitate or—at higher concentrations—form a gel. The aggregation rate has been found to increase with the casein concentration and the temperature. The aggregates consist of spherical particles of size close to that of sub-micelles that are connected into a ramified self-similar structure. Comparison with caseinate shows that both Ca^{2+} and polyphosphate are necessary in order to form stable gels.

Acknowledgements

We thank Nicolas Boisset and Eric Larquet for performing the cryo-EM experiments.

References

1. D. G. Schmidt, *Neth. Milk Dairy J.*, 1980, **34**, 42.
2. H. S. Rollema, in 'Advanced Dairy Chemistry—1. Proteins', ed. P. F. Fox, Elsevier Applied Science, London, 1992, p. 111.
3. C. Holt, *Adv. Protein Chem.*, 1992, **43**, 63.
4. D. S. Horne, *Int. Dairy J.*, 1998, **8**, 171.
5. G. A. Morris, *Biotechnol. Genetic Eng. Rev.*, 2002, **19**, 357.
6. A. Thurn, W. Burchard, and R. Niki, *Colloid Polym. Sci.*, 1987, **265**, 897.
7. A. Thurn, W. Burchard, and R. Niki, *Colloid Polym. Sci.*, 1987, **256**, 653.
8. C. G. de Kruif and V. Y. Grinberg, *Colloids Surf. A*, 2002, **210**, 183.
9. D. S. Horne and D. G. Dalgleish, *Int. J. Biol. Macromol.*, 1980, **2**, 154.
10. D. G. Dalgleish, E. Paterson, and D. S. Horne, *Biophys. Chem.*, 1981, **13**, 307.
11. T. G. Parker and D. G. Dalgleish, *J.Dairy Res.*, 1981, **48**, 71.
12. C. Guo, B. E. Campbell, K. Chen, A. M. Lenhoff, and O. D. Velev, *Colloids Surfaces. B*, 2003, **29**, 297.
13. H. M. Farrell, P. H. Cooke, G. King, P. D. Hoagland, M. L. Groves, T. F. Kumosinski, and B. Chu, *ACS Symp. Ser.*, 1996, **650**, 61.
14. D. G. Schmidt, in 'Developments in Dairy Chemistry—1', ed. P. F. Fox, Applied Science, London, 1982, p. 61.

15. C. G. de Kruif, *Colloids Surf. A*, 1996, **117**, 151.
16. C. Holt and D. S. Horne, *Neth. Milk Dairy J.*, 1996, **50**, 85.
17. T. Ono and T. Obata, *J. Dairy Res.*, 1989, **56**, 453.
18. P. Walstra, *Int. Dairy J.*, 1999, **9**, 189.
19. C. Holt, *J. Dairy Res.*, 1982, **49**, 29.
20. M. C. A. Griffin, R. L. J. Lyster, and J. C. Price, *Eur. J. Biochem.*, 1988, **174**, 339.
21. S. H. C. Lin, S. L. Leong, R. K. Dewan, V. A. Bloomfield, and C. V. Morr, *Biochemistry*, 1972, **11**, 1818.
22. L. Pepper and H. M. Farrell, *J. Dairy Sci.*, 1982, **65**, 2259.
23. C. Holt, D. T. Davies, and A. J. R. Law, *J. Dairy Res.*, 1986, **53**, 557.
24. T. Aoki, N. Yamada, Y. Kako, and T. Imamura, *J. Dairy Res.*, 1988, **55**, 180.
25. J. A. Lucey, C. Dick, H. Singh, and P. A. Munro, *Milchwissenschaft*, 1997, **52**, 603.
26. C. V. Morr, R. V. Josephson, R. Jenness, and P. B. Manning, *J. Dairy Sci.*, 1971, **54**, 1555.
27. L. K. Creamer and G. P. Berry, *J. Dairy Res.*, 1975, **42**, 169.
28. C. Holt, *J. Dairy Sci.*, 1998, **81**, 2994.
29. D. G. Schmidt, P. Walstra, and W. Buchheim, *Neth. Milk Dairy J.*, 1973, **27**, 128.
30. B. Chu, Z. Zhou, G. Wu, and H. M. Farrell, Jr, *J. Colloid Interface Sci.*, 1995, **170**, 102.
31. T. F. Kumosinski, H. Pessen, H. M. Farrell, Jr, and H. Brumberger, *Arch. Biochem. Biophys.*, 1988, **266**, 548.
32. P. H. Stothart and D. J. Cebula, *J. Mol. Biol.*, 1982, **160**, 391.
33. T. Ono, S. Odagiri, and T. Takagi, *J. Dairy Res.*, 1983, **50**, 37.
34. K. S. Mohammad and P. F. Fox, *J. Dairy Res.*, 1987, **54**, 377.
35 K. S. Mohammad and P. F. Fox, *NZ J. Dairy Sci. Technol.*, 1987, **22**, 191.
36. H. Singh, *Int. Dairy Fed. Spec. Issue*, 1995, **9501**, 86.
37. M. A. J. S. van Boekel, *Int. Dairy Fed. Spec. Issue*, 1993, **9303**, 205.
38. V. R. Harwalkar, in 'Advanced Dairy Chemistry—1. Proteins', ed. P. F. Fox, Elsevier Applied Science, London, 1992, p. 691.
39 J. A. Nieuwenhuijse and M. A. J. S. van Boekel, personal communication.
40 V. R. Harwalkar and H. J. Vreeman, *Neth. Milk Dairy J.*, 1978, **32**, 204.
41 A. Pierre, J. Fauquant, Y. Le Graet, M. Piot, and J. L. Maubois, *Le Lait*, 1992, **72**, 461.
42 P. Schuck, M. Piot, S. Mejean, Y. Le Graet, J. Fauquant, G. Brulé, and J. L. Maubois, *Le Lait*, 1994, **74**, 375.
43 N. K. K. Kamizake, M. M. Gonçalves, C. T. B. V. Zaia, and D. A. M. Zaia, *J. Food Comp. Anal.*, 2003, **16**, 507.
44 B. Berne and R. Pecora, 'Dynamic Light Scattering', Wiley, New York, 1976.
45 W. Brown, 'Light Scattering: Principles and Developments', Clarendon Press, Oxford, 1996.
46 M. Panouillé, T. Nicolai, and D. Durand, *Int. Dairy J.*, 2004, **14**, 297.
47 P. Walstra, *J. Dairy Res.*, 1979, **46**, 317.
48 G. A. Morris, T. J. Foster, and S. E. Harding, *Biomacromolecules*, 2000, **1**, 764.
49 T. H. M. Snoeren, B. van Markwijk, and R. van Montfort, *Biochim. Biophys. Acta*, 1980, **622**, 268.
50 D. Quemada, *J. Theor. Applied Mech. (Paris)*, Special Issue, 1985, p. 267.

Caseinate Interactions in Solution and in Emulsions: Effects of Temperature, pH and Calcium Ions

By Maria G. Semenova, Larisa E. Belyakova, Eric Dickinson,[1] Caroline Eliot,[1] and Yurii N. Polikarpov

INSTITUTE OF BIOCHEMICAL PHYSICS OF RUSSIAN ACADEMY OF SCIENCES, VAVILOV STR. 28, 119991 MOSCOW, RUSSIA
[1]PROCTER DEPARTMENT OF FOOD SCIENCE, UNIVERSITY OF LEEDS, LEEDS LS2 9JT, UK

1 Introduction

In a recent study[1,2] of the temperature-dependent rheology of sodium caseinate-stabilized oil-in-water emulsions, three categories of behaviour were identified, depending on the pH and the added ionic calcium content. Emulsions at around neutral pH with a rather low content of Ca^{2+} fall into category A. These are temperature-insensitive liquid-like emulsions that remain liquid when heated up to 50 °C, as indicated by the temperature dependence of the complex shear modulus G^*. A category B emulsion, with a higher content of calcium ions (> 7 mM) and a pH in the range 5.5–6.4, is also liquid-like at room temperature, but it is converted to an emulsion gel on heating, as indicated by a high value of the complex shear modulus G^* at around body temperature. This type of emulsion is 'thermoreversible' under the action of a hydrodynamic flow field during back-cooling. Emulsions of pH below 5.5, and with a widely variable range of calcium ion concentration (up to 24 mM), are assigned to be in category C. They are solid-like emulsions below 35 °C. The gelation state for such emulsions is non-reversible on back-cooling.

To gain greater insight into the key molecular factors determining the heat-induced gelation of these caseinate-based emulsions in the presence of calcium ions and pH lowering, we attempt here to correlate temperature-sensitive changes in the rheological behaviour of the emulsion systems with temperaturesensitive changes in the state of aggregation of the caseinate in aqueous solution. The molecular parameters studied are: the weight-average molar mass (M_w), the radius of gyration (R_G), the hydrodynamic radius (R_h), and the structure-sensitive parameter $\rho = R_G/R_h$. The molecular pair interaction strength is defined by the value of the second virial coefficient $A_{\text{pr-pr}}$. This quantity is the coefficient in the expansion of the chemical potential of solvent (or protein) in terms of the protein concentration. It characterizes the thermodynamics of the protein–solvent and protein–protein interactions.[3–5]

These various parameters are determined here in aqueous solution from a combination of static and dynamic light-scattering experiments[3,6–8] at specific experimental conditions corresponding to the three different categories of emulsion composition mentioned above.[1,2] In deciding on the specific experimental conditions for our study, we have chosen to explore two sections of the state diagram:[3] one refers to a reduction in pH from 6.5 to 5.1 at fixed calcium ion content (15 mM), and the other to an increase in the concentration of Ca^{2+} from 5 to 15 mM at fixed pH = 5.5.

2 Protein Aggregation and Interactions in Relation to Emulsion Gelation

From the comparison of the light-scattering and rheological data in Figure 1, it is clear that the most pronounced increase in the extent of caseinate aggregation

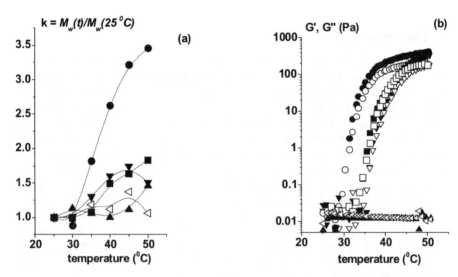

Figure 1 *Comparison of the effects of temperature on caseinate aggregation and emulsion rheology. (a) The relative extent of aggregation of caseinate nanoparticles in aqueous solution (ionic strength 0.01 M), $k = M_w(t)/M_w(25°C)$, is plotted against temperature t. (b) The storage modulus G′ (filled symbols) and loss modulus G″ (open symbols) of caseinate-stabilized emulsions (30 vol% vegetable oil, 4 wt% protein) at experimental conditions corresponding to the three different categories of emulsion composition identified in refs. 1 and 2. Category A systems: △, pH = 6.5, 15 mM Ca^{2+}; ▲, pH = 5.5, 5 mM Ca^{2+}. Category B systems: ▼, pH = 5.5, 15 mM Ca^{2+}, ■, pH = 6.0, 15 mM Ca^{2+}. Category C system: ●, pH = 5.1, 15 mM Ca^{2+}.*

with increasing temperature, in the range from 35 to 45 °C, was observed for experimental conditions (pH 5.1, 15 mM Ca^{2+}) conforming to solid-like emulsions (category C) forming gels below 35 °C. In contrast, the least temperature-sensitive aggregation behaviour was found for experimental conditions (pH 6.5, 15 mM Ca^{2+}) corresponding to liquid-like emulsions remaining liquid when heated (category A). The intermediate extent of temperature sensitivity of caseinate aggregation was found at pH 6.0 and 5.5 (15 mM Ca^{2+}), corresponding to liquid-like emulsions becoming converted to emulsion gels on heating to 35–43 °C (category B). Furthermore, there is a lowering of the aggregation temperature with both decreasing pH at fixed Ca^{2+} content and with increasing Ca^{2+} content at fixed pH, which correlates well with the systematic lowering of the gelation temperature[1,2,9] through the set of emulsion categories A→B→C.

Relying on the light-scattering data, it may be safely suggested that a sharp increase in the emulsion viscoelasticity on heating over an interval of just a few degrees in the range 35–45 °C (see Figure 1b) is attributable to a strengthening of the attractive protein–protein primary interactions and to the formation of a large number of new structurally important bonds between the caseinate nanoparticles. This behaviour is much more marked for the case of the category C emulsion (pH 5.1, 15 mM Ca^{2+}). The light-scattering data also reinforce the tentative explanation[1,2,9] for the heat-induced gelation phenomenon in terms of temperature-dependent changes in the interactions of adsorbed and unadsorbed caseins, *i.e.*, to an increase in the strength of hydrophobic protein–protein attractions in the presence of calcium ions, both between casein-coated emulsion droplets and within casein aggregates in the bulk aqueous phase.

In support of this hypothesis, at intermediate states of protein aggregation, Figure 2 shows a decrease in the positive value of the second virial coefficient, as well as a change in the sign from positive to negative. The increasingly negative values of A_{pr-pr} with heating in the presence of calcium ions is a reflection of an increase in the thermodynamic affinity between caseinate nanoparticles in the aqueous medium.[3,10,11] This is most likely attributable to an increase in the total level of nanoparticle surface hydrophobicity.

Closer inspection of the combined results on the temperature-dependent aggregation and protein–protein attraction suggests that the addition of calcium ions lowers the electrostatic repulsion between the negatively charged groups of the caseinate nanoparticles. Simultaneously there is enhanced casein self-association due to the general strengthening of hydrophobic interactions on heating.[12] The high level of intrinsic hydrophobicity of the caseins, together with their particular charge distributions, implies a strong tendency towards hydrophobically driven self-association in aqueous solution, as is well known from the literature.[13-17] A further contribution of interparticle Ca^{2+} bridging to the aggregation mechanism may also be expected.[18-22]

Figure 3 shows a comparison of the pH dependencies of M_w for sodium caseinate in the presence of 15 mM Ca^{2+} before heating (25 °C) and the calcium ion activity in the sodium caseinate aqueous solutions. We can infer the decisive role of calcium ion binding in the enhancement of the attractive

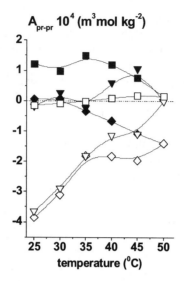

Figure 2 *The effect of heating (filled symbols) and subsequent back-cooling (open symbols) on the character of the protein–protein pair interaction in aqueous solution, as reflected in the second virial coefficient A_{pr-pr} determined from static light scattering:* ■, *pH=6.0, 15 mM Ca^{2+};* ▼, *pH=5.5, 15 mM Ca^{2+};* ◆, *pH=5.5, 10 mM Ca^{2+}.*

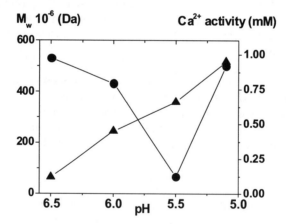

Figure 3 *Effect of pH on calcium ion activity in caseinate solutions and weight-average molar mass M_w of caseinate nanoparticles in the presence of 15 mM Ca^{2+} at 25 °C:* ●, *Ca^{2+} activity;* ▲, *M_w.*

protein–protein interactions as the pH is lowered from 6.5 to 5.5. This shows itself in the higher values of M_w for the protein aggregates and lower values of the Ca^{2+} activity in the aqueous medium, *i.e.*, increased binding of calcium ions by the sodium caseinate. However, Figure 3 also shows that the Ca^{2+} binding seems no longer to play such a decisive role on lowering the pH towards the protein's isoelectric point (p$I \sim 4.8$). That is, there is a rather high value of M_w

Table 1 *The effect of temperature on the properties of sodium caseinate nano-*
particles at pH 5.5 in the absence of calcium ions (ionic strength
0.01 M).

Molecular parameters	Heating				Back-cooling		
	Temperature (°C)						
	20	35	45	50	45	35	20
$M_w \times 10^{-6}$ (Da)	15	14	17	17	17	15	15
R_G (nm)	200	182	237	235	234	190	197
R_h (nm)	92	88	65	–	64	86	93
$A_{pr-pr} \times 10^4$ (m^3 mol kg^{-2})	0.2	0.7	0.9	1.1	1.0	0.6	0.3

at pH = 5.1 despite this corresponding to a low extent of Ca^{2+} binding, *i.e.*, a high value of measured calcium ion activity. An explanation[23,24] for this may be that the protonation of glutamic acid residues adjacent to phosphorylated serine residues may reduce the binding of calcium ions to sodium caseinate near p*I*.

Finally, we should emphasize the governing role of calcium ions in the behaviour of the solutions and emulsions followed in the pH range from neutrality to pH = 5.5. In the absence of added ionic calcium, the molecular parameters of sodium caseinate remain practically invariable over the temperature cycle of heating and subsequent back-cooling, particularly at pH = 5.5 (Table 1). Correspondingly, also in the absence of ionic calcium, the rheology of the sodium caseinate emulsions exhibits no pronounced dependence on temperature.[9]

3 Thermoreversibility of Protein Aggregation and Emulsion Gelation

Further examination of the light-scattering data gives insight into the molecular basis of the key features of temperature reversibility (category A and B systems) and irreversibility (category C behaviour) of the viscoelasticity of acid-induced caseinate emulsion gels in the presence of calcium ions.[1,2,9] Figure 4 shows the effects of heating and back-cooling on the states of caseinate aggregation. The results correlate well with the corresponding degrees of thermoreversibility found for the heat-induced emulsion gels.

It was reported previously[1,4] that the flocculated network of protein-coated droplets formed on heating a category B system is not spontaneously disrupted on cooling under the influence of Brownian motion alone. This can be attributed to the strengthening of the protein–protein attractive interactions under back-cooling, as shown clearly in Figure 2 by the change in second virial coefficient. It is possible that there is some change in interactions of the surface of caseinate sub-micelles as a result of continual realignment of their constituent elements, involving both intra- and intermolecular Ca^{2+} cross-linking during heating and subsequent back-cooling.

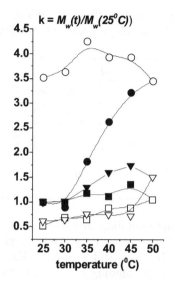

Figure 4 *The effect of heating (filled symbols) and subsequent back-cooling (open symbols) on the reversibility of the state of aggregation of caseinate nanoparticles in aqueous solution under experimental conditions corresponding to the three categories of emulsion compositions identified by the state diagram of refs 1 and 2: category A (■), pH=6.5, 15 mM Ca²⁺; category B (▼), pH=5.5, 15 mM Ca²⁺; category C (●), pH=5.1, 15 mM Ca²⁺.*

4 Structural Features of the Caseinate Aggregates

Figure 5 shows the temperature sensitivity of the ratio of the radius of gyration to the hydrodynamic radius, $\rho = R_G/R_h$. This parameter provides specific information about the architecture of the protein aggregates that can be considered as the initial structural elements in the protein gel network, both in bulk solution and in the adsorbed protein layer, and which in turn determine the overall microstructure and rheology of the caseinate gels and emulsion gels.

A reduction on heating was found in the value of the structure-sensitive parameter ρ for two specific cases. Firstly, it occurs at pH=5.5 and with a concentration of calcium ions that is lower than 15 mM. It seems that under these conditions the binding of Ca^{2+} occurs principally in the 'interior' of the same protein molecule rather than between different protein molecules.[21] This sort of behaviour also occurs at a pH close to p*I*, *i.e.*, when the net protein charge is negligible. The observed decrease in the structure-sensitive parameter indicates a transformation in the conformation of the protein aggregates from a rather rigid and open architecture to one characteristic of a hard sphere. This transition is most likely caused by a strengthening of hydrophobic attraction in the interior of the caseinate nanoparticles, under conditions where much of the protein charge is neutralized by bound calcium ions. This is shown in Figure 5a.

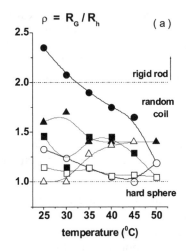 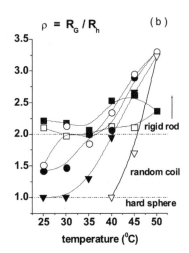

Figure 5 *The effect of heating (filled symbols) and subsequent back-cooling (open symbols) on the reversibility of the conformation of caseinate nanoparticles reflected in the values of the structure-sensitive parameter ρ: (a) ●, pH =5.5, 5 mM Ca^{2+}; ▲, pH =5.5, 10 mM Ca^{2+}; ■, pH =5.1, 15 mM Ca^{2+}; (b) ●, pH =6.5, 15 mM Ca^{2+}; ■, pH =6.0, 15 mM Ca^{2+}; ▼, pH =5.5, 15 mM Ca^{2+}.*

The implied difference in the size and architecture of the caseinate aggregates determined under heating and subsequent back-cooling suggests the occurrence of a continual change and realignment of the constituent elements of the caseinate nanoparticles with respect to one another. This involves both intramolecular and intermolecular cross-linking of calcium ions throughout the whole temperature cycle.

In line with our data, there is evidence elsewhere in the literature[25] of a decrease in the average hydrodynamic diameter of aggregating primary particles of caseinate with increasing temperature. The higher particle voluminosity at the lower temperatures could be due to weaker hydrophobic interactions in the interior of the caseinate nanoparticles.[26]

In contrast, at higher calcium ion concentrations and at less acidic pH values, the inferred structure of the caseinate aggregates becomes more open on heating, *i.e.*, the structure-sensitive parameter ρ increases with temperature (Figure 5b). It is as if the concentration of calcium ions is sufficient to allow cross-linking both in the interior of caseinate nanoparticles and amongst them. In addition, we would expect a contribution to the more open architecture of caseinate aggregates from both the total protein charge and the higher binding extent of calcium ions at less acidic pH values (see Figure 3). Again, a significant difference between the size and architecture of caseinate aggregates determined under heating and subsequent back-cooling (Figure 5) suggests a continual reorganization of the constituent elements of the caseinate nanoparticles with respect to one another, involving both intramolecular and intermolecular cross-linking of calcium ions, throughout the whole temperature cycle.

Hence, from these light-scattering results, we can postulate that heating alters not only the nature and strength of the physical (hydrophobic) interactions between the emulsion droplets covered by caseinate. It most likely transforms the structural characteristics of the protein network in the bulk and at the interface, thereby affecting the viscoelastic properties of the emulsions.

5 Conclusions

On the basis of new light-scattering measurements, it is here inferred that the following molecular processes are central to the mechanism of heat-induced gelation of caseinate emulsions and its thermoreversibility:

(i)   an enhanced development of the aggregation of caseinate nanoparticles with heating, and its essential thermoreversibility;
(ii) an increase in the thermodynamic affinity between the caseinate nanoparticles in the aqueous medium over the whole temperature cycle of heating and subsequent cooling; and
(iii) a marked transformation in the size and architecture of the caseinate aggregates under heating and subsequent back-cooling.

Like the temperature-dependent emulsion rheology itself, each of these molecular processes is very sensitive to pH and calcium ion concentration.

Acknowledgements

We are grateful to Dr Lara Matia Merino for the data on the pH dependence of calcium ion activity of the caseinate solutions and to Mr. V. Yu. Polikarpov for the development of a computer programme for light-scattering data treatment. We thank DMV International (Veghel, Netherlands) for donating the sample of spray-dried sodium caseinate.

References

1. C. Eliot, S. J. Radford, and E. Dickinson, in 'Food Colloids, Biopolymers and Materials', eds E. Dickinson and T. van Vliet, Royal Society of Chemistry, Cambridge, 2003, p. 234.
2. C. Eliot and E. Dickinson, *Int. Dairy J.*, 2003, **13**, 679.
3. C. Tanford, 'Physical Chemistry of Macromolecules', Wiley, New York, 1961.
4. E. Edmond and A. G. Ogston, *Biochem. J.*, 1968, **109**, 569.
5. M. Nagasawa and A. Takahashi, in 'Light Scattering from Polymer solutions', ed. M. B. Huglin, Academic Press, London, 1972, p. 671.
6. J. M. Evans, in 'Light Scattering from Polymer Solutions", ed. M. B. Huglin, Academic Press, London, 1972, p. 89.
7. W. Burchard, in 'Physical Techniques for the Study of Food Biopolymers', ed. S. B. Ross-Murphy, Blackie, Glasgow, 1994, p. 151.
8. W. Burchard, *Chimia*, 1985, **39**, 10.
9. E. Dickinson and H. Casanova, *Food Hydrocoll.*, 1999, **13**, 285.
10. M. G. Semenova and L. B. Savilova, *Food Hydrocoll.*, 1998, **12**, 65.

11. M. G. Semenova, *ACS Symp. Ser.*, 1996, **650**, 37.
12. W. P. Jencks, 'Catalysis in Chemistry and Enzymology. Part II', McGraw-Hill, New York, 1969, chap. 8.
13. P. Walstra and R. Jenness, 'Dairy Chemistry and Physics', Wiley, New York, 1984.
14. J. Leman and J. E. Kinsella, *Crit. Rev. Food Sci. Nutr.*, 1989, **28**, 115.
15. H. D. Belitz and W. Grosch, 'Food Chemistry', Springer-Verlag, Berlin, 1987, chap. 10.
16. P. F. Fox and D. M. Mulvihill, 'Physico-Chemical Aspects of Dehydrated Protein-Rich Milk Products', Proceedings of International Dairy Federation Symposium, Denmark, 1983, p. 188.
17. H. S. Rollema, in 'Advanced Dairy Chemistry—1. Proteins', ed. P. F. Fox, Elsevier Applied Science, London, 1992, p. 111.
18. E. Dickinson, J. A. Hunt, and D. S. Horne, *Food Hydrocoll.*, 1992, **6**, 339.
19. D. S. Horne and J. Leaver, *Food Hydrocoll.*, 1995, **9**, 91.
20. A. Ye and H. Singh, *Food Hydrocoll.*, 2000, **15**, 195.
21. E. Dickinson, M. G. Semenova, L. E. Belyakova, A. S. Antipova, M. M. Il'in, E. N. Tsapkina, and C. Ritzoulis, *J. Colloid Interface Sci.*, 2001, **239**, 87.
22. A. S. Antipova, E. Dickinson, B. S. Murray, and M. G. Semenova, *Colloids Surf. B*, 2002, **27**, 123.
23. H. J. M. van Dijk, *Neth. Milk Dairy J.*, 1990, **44**, 111.
24. L. F. Mallee, *Industrial Proteins*, 1999, **7**, 5.
25. N. Vetier, S. Desobry-Banon, M. M. Ould Eleya, and J. Hardy, *J. Dairy Sci.*, 1997, **80**, 3161.
26. J. Chen and E. Dickinson, *Int. Dairy J.*, 2000, **10**, 541.

Nanorheological Properties of Casein

By Kristina M. Helstad, Alan D. Bream,[1] Jana Trckova,[2]
Marie Paulsson, and Petr Dejmek

DEPARTMENT OF FOOD TECHNOLOGY, ENGINEERING AND
NUTRITION, FOOD ENGINEERING, LUND UNIVERSITY,
P.O BOX 124, SE-221 00 LUND, SWEDEN
[1]DEPARTMENT OF CHEMISTRY, PHYSICAL CHEMISTRY 1,
LUND UNIVERSITY, P.O. BOX 124, SE-221 00 LUND, SWEDEN
[2]DEPARTMENT OF DAIRY AND FAT TECHNOLOGY,
INSTITUTE OF CHEMICAL TECHNOLOGY, TECHNICKA 5,
166 28 PRAGUE 6, CZECH REPUBLIC

1 Introduction

Knowledge of the nanorheological properties of casein layers and casein micelles may be important for understanding dairy processes and improving the quality of casein-based products. The atomic force microscope (AFM) has become a useful tool to investigate the mechanical properties of biological samples, due to its capability for measuring forces as small as nanonewtons, and the ability to make measurements in hydrated environments *via* the liquid cell technique.[1] Burnham and Colton[2] reported early measurements of nano-mechanical properties. The AFM has recently been used to measure the force exerted by end-grafted polymer layers[3,4] and adsorbed polymer layers[5] on a solid surface. The apparent modulus of bacterial cells was investigated by Touhami,[6] thin polymer films by Domke and Radmacher,[7] and the displacement of β-casein Langmuir–Blodgett films from a graphite surface by Gunning *et al.*[8] The local structure and the elasticity of soft gelatin gels have also been studied by AFM, and the properties compared with the macroscopic elastic modulus measured by rheometry.[9] And Elofsson *et al.*[10] have used AFM to study dried whey protein concentrate and cold gelling whey protein dispersions on mica surfaces.

The subject of adsorbed protein layers at fluid interfaces, including their interactions, structure and surface rheology, has been reviewed by Dickinson.[11] The adsorption of β-casein on hydrophobic and hydrophilic surfaces has been studied by Nylander and Tiberg,[12] and the forces between adsorbed layers of β-casein on mica surfaces have been measured[13] as a function of surface separation.

In bovine milk the major part of the protein fraction—the casein part—exists as roughly spherical aggregates called casein micelles. The properties of these casein micelles influence to a large extent the technological use of milk, especially in cheese making. Casein micelles are formed by association of the individual casein components, α_{S1}-, α_{S2}-, β- and κ- casein,[14] the proteins being also bound together with calcium phosphate and small amounts of citrate. The casein micelles are voluminous, containing on average about 4 mL per gram of casein. Each micelle consists of 2–15×10^4 individual casein molecules. The size range has been established by electron microscopy and dynamic light scattering as 40–300 nm, with an average diameter of ~150 nm. At the surface of the casein micelle, a hairy layer with a thickness of at least 5 nm, which is usually taken to consist of the hydrophilic parts of κ-casein, accounts for the predominantly steric stabilization of the casein dispersion.[14] However, it has been suggested[15] that β-casein and α_s-casein may share the surface with κ-casein. Two basic models for the structure of casein micelles have been put forward. One type of model proposes[14,16] that the micelle core is built up from discrete sub-units (sub-micelles) with distinctly different properties from the outside 'hairy' layer. Another family of models suggests[17–19] that the inside of the casein micelle resembles a mineralized network with no regular internal structure. (For the most recent views, see Horne[20] and Holt *et al.*[21]) The rheological properties of casein networks have been thoroughly investigated[22,23] as they are important for the quality of milk products.

In this study casein micelles are adsorbed on a hydrophobic graphite surface and studied by AFM in a liquid cell. For the first time, as far as we know, the nanorheological properties of casein micelles have been measured by the technique of nano-indentation. The aim is to relate the estimated casein micelle modulus to that for cheese.

2 Materials and Methods

Sample Preparation

The 0.02% casein solution was prepared from a frozen casein concentrate, produced from microfiltered, diafiltered and ultrafiltered pasteurized cow's skim milk (22% w/w casein, 0.1% w/w fat, 2.6% w/w lactose, 0.85% w/w calcium, 0.51% w/w phosphorus, Arla Foods, Sweden). The pH of the concentrate was 6.7, and azide was added to prevent bacterial growth. The casein concentrate was diluted in an imidazole buffer[24] (50 mM NaCl, 5 mM $CaCl_2 \cdot 6H_2O$ (Prolabo), 20 mM imidazole, $C_3H_4N_2$ (Merk, Darmstadt). The pH of the buffer solution was adjusted to pH = 7.0 using HCl diluted in MilliQ-water.

The casein solution was prepared two days before the AFM imaging and stored at 10 °C. A freshly cleaved graphite surface (HOPG Advanced Ceramics Brand Grade ZYPTM /SPI Supplies, West Chester, PA) was mounted in a fluid cell. Imidazole buffer was injected into the cell for calibration of the photodiode. The casein solution was injected into the cell and held for 2 h to allow the protein to adsorb onto the graphite surface. Then the casein solution was replaced by the buffer, and the AFM measurements were performed. The temperature of the sample in the liquid cell was 25 °C.

Instrumentation

The AFM studies were performed with a NanoScope III instrument (Veeco Instruments, Santa Barbara, CA) equipped with a scanner (J-type, maximum scan size 120 μm) and a fluid cell for exchange of buffer and protein solution. NanoScope software version 5.12r3 was used to control the instrument and for data acquisition. The V-shaped cantilevers used for the measurements were made of silicon nitride with a spring constant of 0.06 N m^{-1}. The tip shape was square pyramidal, with a tip half-angle of 35°. The tips were oxide sharpened (Veeco Model # MSCT-AUNM, P/N 00-103-0970, wafer#170-71#19 10/01). Imaging the tip using a TGT01 substrate and MIDAS computer program gave a radius of curvature of 20 nm.

Force Measurement by AFM

In addition to imaging in contact mode, some measurements were made in force–volume mode, which allows force curves to be acquired as a function of the lateral position. The scan size was 3 μm × 3 μm and the ramp size was 600 nm. Force–volume images consisting of arrays of 16 × 16 force curves were recorded with 512 data points each. The data were recorded in the relative trigger mode, which ensures that the tip is only pressed against the sample with a preset maximum force. The trigger threshold was set to 240 nm and the deflection sensitivity was varied in the range 70–90 nm V^{-1}. To obtain a reference for the sensitivity of the photo diode, the first measurements were performed on bare graphite under buffer. Single penetration measurements were made at the same point of a sample to investigate the elastic behaviour of an individual casein micelle. The tip velocity for single penetration was 6700 nm s^{-1}. Force–volume measurements were performed at tip velocities that varied from 120 nm s^{-1} to 6700 nm s^{-1}.

Data Analysis

After conversion from the raw voltage data, the instrument provides a series of force curves as the tip is raster scanned laterally over the sample. Each force curve consists of 512 value pairs (d, z), where d is a measure of the deflection of the cantilever and z is a measure of the distance between the sample holder and the cantilever holder (see Figure 1). With knowledge of the cantilever spring constant, each deflection value can be converted to a force. The zero force position for each force curve was determined as the average of the deflection values for 200 z positions furthest from the sample surface, none closer than 25 nm to the position where a force value above the noise (standard deviation 0.004 nN) could be detected.

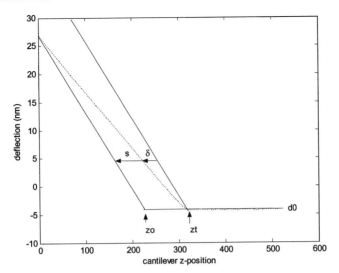

Figure 1 *Definitions of symbols used in data analysis of plots of deflection d versus cantilever position z. See text for further details.*

For a substrate covered with a layer of adsorbed material, the AFM does not allow an unambiguous assignment of the position of the substrate surface. It is customary to consider that the tip has reached the hard surface of the substrate when the slope of the *d versus z* plot is indistinguishable from unity, *i.e.*, when the apparent stiffness of the adsorbed layer is very large compared to the stiffness of the cantilever spring. To compensate for small errors of calibration, we have individually fitted the first 100 readings in the linear part of the force curves, and have used the local slope k, which was found not to deviate more than 1% from the calibrated value. Then, for any (d, z) pair on the linear part of the curve, if the deflection for zero force is d_0, the position z_0 of the hard surface is given by

$$k = (d - d_0)/(z - z_0). \tag{1}$$

For the force curves which did not contain 100 linear points, *i.e.*, for the areas with large casein aggregates, the k and z_0 values were obtained by interpolation from the neighbouring positions. At any value of d, the separation s between the hard surface and the cantilever tip is given by

$$s = z - z_0 + d/k. \tag{2}$$

The indentation depth δ is given by

$$\delta = z_t - z_0 - s, \tag{3}$$

where z_t is coordinate of the contact point, *i.e.*, the point on a spline-smoothed force curve where the force significantly deviates from zero.

For a stiff sphere indenting a planar elastic sample, the Hertz model[25] describes the force F on the cantilever as a function of the indentation depth δ, i.e.,

$$F = \frac{4E\sqrt{R}}{3(1-v^2)}\delta^{3/2}, \tag{4}$$

where v is Poisson's ratio of the sample, E is Young's modulus, and R is the curvature radius of the tip. The scaling relationships of Alexander and de Gennes have been shown[26] to describe the force opposing the compression of a layer of polymer adsorbed to a hard surface. For adsorbed polymer at a low level of surface coverage, in the so-called mushroom regime, the force scales as

$$F \sim (s/s_0)^{-8/3}. \tag{5}$$

For adsorbed polymer at a moderate level of surface coverage, in the so-called brush regime, we have

$$F \sim (s/s_0)^{-9/4} - (s/s_0)^{3/4}, \tag{6}$$

where s_0 is the extension of the polymer layer, here taken as $s_0 = z_t - z_0$. These relationships hold strictly for forces between infinite planes. So we calculate numerically the integral over the surface of an indenting sphere, corresponding to our tip of radius $R = 20$ nm, as a function of the separation distance and the extension of the polymer layer.

3 Results

Casein on a Graphite Surface

Under our experimental conditions, casein aggregates corresponding to casein micelles could adsorb and remain stable on the hydrophobic graphite surface. No stable absorbed casein aggregates could be found on hydrophilic mica surfaces. As seen in the topographic image in Figure 2, the casein micelles were affected by the presence of the graphite surface, having more the shape of a thick 'pancake' than a sphere. In between the casein micelles, the surface was not bare. This can be seen in Figure 3, which contrasts the force curves of a freshly prepared hydrophobic surface under buffer and the surface exposed to the dilute casein solution. Individual casein molecules, which become attached to the graphite with their hydrophobic part, may form the thin layer observed.

Thin Layer of Casein Compared to the Scaling Law Model

In Figure 4 some randomly chosen force *versus* separation curves from the thin layer of casein are fitted with the scaling theory of Alexander and de Gennes. The brush-scaling model gives a marginally better fit to the experimental data. A good fit occurs assuming that the carbon surface is 3 nm below the experimentally observed hard surface, and that the total reach of the polymer

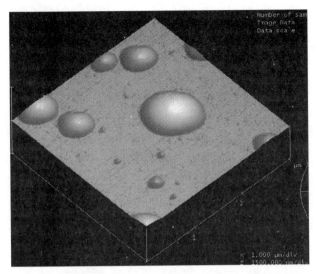

Figure 2 *Topographic NanoScope image of casein on graphite surface in a liquid cell. The scan size is 3 μm × 3 μm. The largest micelle in the picture is 970 nm in diameter and 160 nm high.*

layer is 15 nm. This is compatible with existing evidence, since whole casein gives adsorbed layers in emulsions which have similar thickness to those produced by β-casein.[27] Nylander and Wahlgren observed[13] the onset of an attractive force between layers of β-casein adsorbed on hydrophobic mica surfaces at 25 nm separation, and they hypothesized that the chains overlap at this distance. Caldwell reported[28] a 15 nm thickness of β-casein layers on polystyrene latex by light scattering. A 3 nm thickness of a fully compressed β-casein layer is conceivable, since the ellipsometry data of Nylander *et al.* were interpreted[29] as a 6.6 nm thickness for a protein volume fraction of 0.3.

Casein Micelle Force Curves

The shape of the force curves over the surface with adsorbed casein micelles is illustrated in Figure 5. Almost identical curves were obtained on approach and retraction, with the changes brought about by the tip being reversible on a timescale of 0.1 s. None of the curves showed any regular discontinuities, and thus there is no evidence of sub-structure on the tens of nanometers scale. Curve (2) in Figure 5 shows fracture-like behaviour. This phenomenon appeared only in some curves, usually close to the edge of the casein micelle at surface separations of 15–20 nm. The more common pattern was that observed in curves (3) and (4). Surprisingly, the curves show a more-or-less pronounced inflexion point, *i.e.*, the apparent stiffness goes through a maximum. For force curves recorded closer to the edge of a micelle, a significantly lower stiffness was observed. This could be caused by the fact that the tip engages a sloping

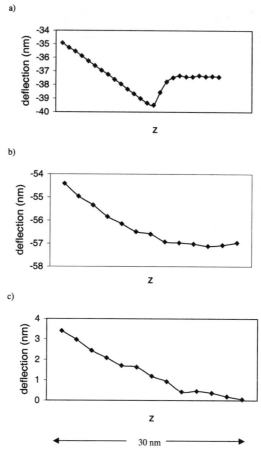

Figure 3 *Typical deflection versus position plots for (a) bare graphite, (b) a thin casein layer, and (c) a casein micelle.*

surface, or it may be due to temporary escape of the polymer from under the tip.

Two single penetrations at the same point are shown in Figure 6. The tip velocity was the same (6700 nm s^{-1}) for both penetrations, and the time between the indentations was ~10 s. The deflection curves show the same slope and the same contact point for both curves. No changes in structure could be detected in second or subsequent penetrations.

Determination of Young's Modulus

An attempt was made to fit the data to the Hertz model for the initial indentation of 20 nm, corresponding to the tip radius. The model predictions with two different moduli and two different contact points are plotted with the initial part of the force curve (see Figure 7) to illustrate the effect of changing these parameters. This part of the curve could be made to fit reasonably well,

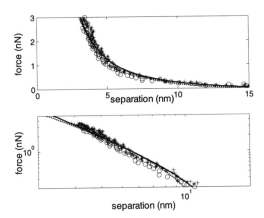

Figure 4 *Force curves on linear and logarithmic scales for three different points on the thin casein layer, and the fits to the scaling law for an adsorbed polymer: the mushroom regime (dotted line); the brush regime (solid line).*

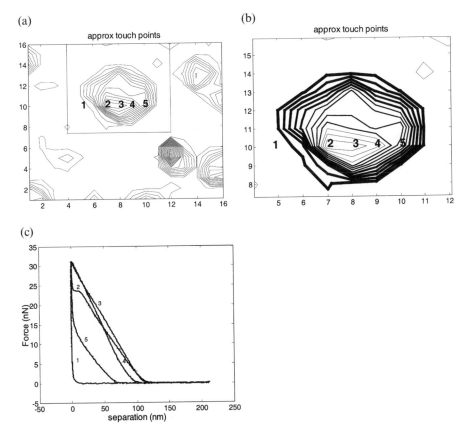

Figure 5 *Force curves at various points on an imaged surface: (a) touch point map of the scanned surface; (b) touch point map of one micelle; (c) force curves for the five indicated measuring points.*

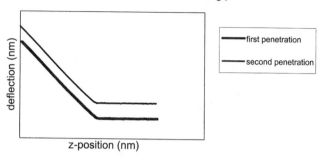

Figure 6 *Deflection versus position plot for two single penetrations at the same measuring point.*

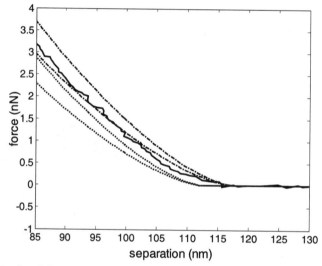

Figure 7 *The fit of the Hertz model with two different moduli and two different contact points.*

with just some minor indications of lack of fit. Assuming a constant Poisson ratio of 0.5, as expected for soft biological materials,[6] and a curvature of the casein micelle surface much larger than the radius of the tip, a Young's modulus value of 0.06–0.08 MPa was determined. This can be compared to cheese, which has a modulus of 0.01–1.0 MPa.[14]

The softer initial part of the casein micelle force curve resembles the behaviour of the thin layer on carbon (Figure 3). One could reasonably assume that the surface of the casein micelle is very similar to the surface of the adsorbed layer. If we attribute to the micelle surface layer the same stiffness as the thin layer, we can evaluate the underlying force *versus* separation behaviour of the bulk of the casein micelle using a simple 'springs in series' calculation. Two representative results are presented in Figure 8. The initial softer part was then found to disappear completely and a linear behaviour was

(a)

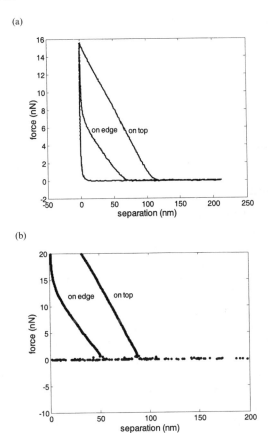

(b)

Figure 8 *(a) Force curves from three measuring points: the thin casein layer, the edge of a casein micelle, and the top of a casein micelle. (b) The same force curves, but where the 'thin-layer force curve' has been subtracted.*

observed in all cases, at variance with the expected power of 3/2 in equation (4) for the Hertz-like behaviour of a homogenous elastic material.

Viscoelasticity

Further force–volume measurements were made at three different tip velocities of 120, 1200 and 6700 nm s^{-1}. In Table 1, the results of repeated measurements, nominally at the same point at the same micelle, are shown as the stiffness at an indentation corresponding to a 2.5 nN force. The values in Table 1 are almost constant and the minor variations do not follow any pattern. On the probed timescale, there is therefore no evidence of viscous or viscoelastic behaviour. The deformation rate $(\Delta h/h_0)/\Delta t$ for these samples varied between 0.8 and 41.9 s^{-1}. This deformation rate is in the same range as applied in typical rheological measurements on casein gels and cheese products. In casein curd, a change of some 50% in the modulus would have been expected over this range of deformation rate.

Table 1 *Variation in measured stiffness for six different tip velocities.*

Experiment	Tip Velocity (nm/s)	Stiffness (au)
1	1200	76
2	120	78
3	6700	74
4	1200	66
5	6700	71
6	6700	67

4 Conclusions

Casein micelles and individual casein molecules have been adsorbed onto a graphite surface from dilute solution. The mechanical behaviour of the thin adsorbed casein layer has been found to be well described by the scaling law of Alexander and de Gennes. The total thickness of the adsorbed casein molecules was found to be approximately 15 nm, which is comparable to the expected thickness of a mixed layer dominated by β-casein. The adsorbed casein micelles show deformation-rate independent elastic behaviour under nano-indentation and no sign of internal structure on the tip scale. These measurements could not detect any permanent fractures in the casein micelle structure on repeated penetrations. From the initial stages of indentation, Young's modulus could be estimated as 0.06–0.08 MPa, but the deeper indentation behaviour could not be made to agree with the Hertz model. Alternatively, the AFM results could be interpreted as a 'hairy' layer surrounding the casein micelle, with properties similar to a layer of casein adsorbed onto a hard hydrophobic surface, and with the inside of the micelle behaving as a linear spring.

Acknowledgements

Prof. Karsten Qvist at the Royal Agricultural University, Copenhagen, Denmark, and Dr Tommy Nylander at Physical Chemistry 1, Lund University, Sweden, contributed with valuable comments. The Danish Dairy Research Foundation supported this study. J.T. acknowledges support from the EU Erasmus program and the Swedish Institute.

References

1. G. Binning and C. F. Quate, *Phys.Rev. Lett.*, 1986, **56**, 930.
2. N. A. Burnham and R. J. Colton, *J. Vac. Sci. Technol.*, 1989, **A7**, 2906.
3. J. Jimenez and R. Rajagopalan, *Eur. Phys. J. B*, 1998, **5**, 237.
4. D. Goodman, J. N. Kizhakkedathu, and D. E. Brooks, *Langmuir*, 2004, **20**, 3297.
5. A. D. Braem, B. Biggs, C. D. Prieve, and R. D Tilton, *Langmuir*, 2003, **19**, 2736.
6. A. Touhami, B. Nysten, and Y. F. Dufrene, *Langmuir*, 2003, **19**, 4539.
7. J. Domke and M. Radmacher, *Langmuir*, 1998, **14**, 3320.

8. A. P. Gunning, A. R. Mackie, P. J. Wilde, and V. J. Morris, *Langmiur*, 1999, **15**, 4636.
9. V. I. Uricanu, M. H. G. Duits, R. M. F. Nelissen, M. L. Bennink, and J. Mellema, *Langmuir*, 2003, **19**, 8182.
10. C. Elofsson, P. Dejmek, M. Paulsson, and H. Burling, *Int. Dairy J.*, 1997, **7**, 813.
11. E. Dickinson, *Colloids Surf. B*, 1999, **15**, 161.
12. T. Nylander, F. Tiberg, and N. M. Wahlgren, *Int Dairy J.*, 1999, **9**, 313.
13. T. Nylander and N. M. Wahlgren, *Langmuir*, 1997, **13**, 6219.
14. P. Walstra (*et al.*), 'Dairy Technology: Principles of Milk Properties and Processes', Marcel Dekker, New York, 1999.
15. D. G. Dalgleish, *J. Dairy Sci.*, 1998, **81**, 3013.
16. D. G. Schmidt, P. Walstra, and W. Buchheim, *Neth. Milk Dairy J.*, 1973, **27**, 128.
17. H. Visser, 'Protein Interactions', ed. H. Visser, VCH, Weinheim, 1992, p. 135.
18. C. Holt and D. S Horne, *Neth. Milk Dairy J.*, 1996, **50**, 85.
19. D. S. Horne, *Int. Dairy J.*, 1998, **8**, 171.
20. D. S. Horne, *Curr. Opin. Colloid Interface Sci.*, 2002, **7**, 456.
21. C. Holt, C. G. de Kruif, R. Tuinier, and P. A. Timmins, *Colloids Surf. A*, 2003, **213**, 275.
22. J. A. Lucey, M. E. Johnson, and D. S. Horne, *J. Dairy Sci.*, 2003, **86**, 2725.
23. M. Mellema, P. Walstra, J. H. J van Opheusden, and T. van Vliet, *Adv. Colloid Interface Sci.*, 2002, **98**, 25.
24. D. S. Horne and C. M. Davidson, *Colloid Polym. Sci.*, 1986, **264**, 727.
25. H. Hertz, *J. Reine Angew. Math.*, 1881, **92**, 156.
26. T. D. Cloud and R. Rajagopalan, *J. Colloid Interface Sci.*, 2003, **266**, 304.
27. Y. Fang and D. G. Dalgleish, *J. Colloid Interface Sci.*, 1993, **156**, 329.
28. K. D. Caldwell, J. Li, J. T. Li, and D. G. Dalgleish, *J. Chromatogr.*, 1992, **604**, 63.
29. T. Nylander, F. Tiberg, and N. M. Wahlgren, *Int. Dairy J.*, 1999, **9**, 313.

Whey Protein Aggregation Studies by Ultrasonic Spectroscopy

By Milena Corredig and Douglas G. Dalgleish

DEPARTMENT OF FOOD SCIENCE, UNIVERSITY OF GUELPH,
GUELPH, ONTARIO N1G 2W1, CANADA

1 Introduction

Ultrasound spectroscopy is a valuable addition to the more established techniques such as light scattering, other spectroscopic methods, and rheology. It is a non-destructive technique that can be applied to optically opaque samples. It can be used, for example, to determine molecular and structural changes at concentrations high enough for proteins to interact and form gels. High-frequency sound waves have been employed to probe intermolecular forces in protein solutions.[1] High-resolution ultrasound spectroscopy differs from the other ultrasound techniques so far employed in food systems, as it does not determine the attenuation of sound over a large range of frequencies,[2] but rather it measures, with high precision, both the attenuation and the velocity at fixed frequencies. By measuring ultrasonic velocity, it is possible to derive information on the changes in density and compressibility of the medium. In protein solutions, changes in velocity are affected by the hydration and apparent compressibility of the molecules.[3,4]

Whey proteins are often employed as ingredients to improve the texture of food products. For this reason, the aggregation and gelation of these proteins are of great interest for the food industry. The ability to control these processes more effectively should result in better food products. To be able to control the formation of heat-induced aggregates, it is important to understand the molecular interactions that occur during heating. By understanding the dynamics of gel formation, especially at the early stages, it should become easier to manipulate and control the texture of foods.

While most studies on heat-induced conformational changes and aggregation have been conducted on β-lactoglobulin alone, the combined behaviour of α-lactalbumin and β-lactoglobulin in whey protein isolate (WPI) is less well understood. The most abundant individual whey protein, β-lactoglobulin (β-lg), is thought to dominate the overall aggregation behaviour. In the present study, both commercial WPI and pure β-lg were chosen as model systems, as many studies have been reported[5,6] with other techniques on the aggregation of pure β-lg at neutral pH. Different macromolecular whey protein complexes can arise depending on system composition, protein concentration, and heating conditions. It has also been shown[6,7] that the type of aggregates formed during heating of whey proteins is affected by the ionic strength of the medium. Most of these previous studies have been conducted using either rheological measurements or light scattering under dilute conditions. Ultrasound spectroscopy is employed here in an attempt to make the connection between known macroscopic events (gelation and structure formation of whey proteins) and known molecular changes occurring in dilute solution. Whey protein solutions were heated *in situ*, and ultrasound spectroscopy was employed to determine if differences exist, especially in the pre-gelation stages, between solutions of WPI and those of β-lg. The preparations were tested after equilibration with NaCl (0.01 or 0.1 M).

2 Experimental

Whey protein isolate (WPI) (Alacen 895) was donated by New Zealand Milk Proteins (Mississagua, ON). Purified β-lactoglobulin (β-lg) was prepared from the same isolate using ion-exchange chromatography. Samples were dialysed overnight against solutions containing 0.01 or 0.1 M NaCl, centrifuged at $7 \times 10^3 g$ for 20 min, and filtered through 0.22 μm filters. Solutions were adjusted to 10% w/v concentration and thoroughly degassed under vacuum.

Ultrasound measurements were carried out using an HR-US102 instrument (Ultrasonic-Scientific, Dublin, Ireland). This instrument uses a differential measurement method, with two identical cells placed in a temperature-controlled block. The frequency and bandwidth of the sound waves at fixed frequencies were measured continuously in both the sample cell and the reference cell. The former was filled with the whey protein solution, while the latter was filled with the filtered dialysis solution. The attenuation of the sound and the difference in ultrasonic velocities were calculated from the frequency and bandwidth of the sound wave using the Ultrasonic Scientific software (version 4-50-25-0). The instrument was tuned to measure at two frequencies corresponding to 5099 and 7835 kHz for water at 25 °C. Samples were subjected to isothermal runs in the temperature range of 50–75 °C, or linear temperature ramps of 20–85 °C at a rate of 0.4 °C min^{-1}. All runs were analysed and plotted using routines written in Microsoft Excel. To improve clarity, the data shown in the figures represent a running average of the collected data points.

3 Results

In order to derive information on the molecular structure and interactions of the whey proteins, changes were studied *in situ* at a gelling concentration

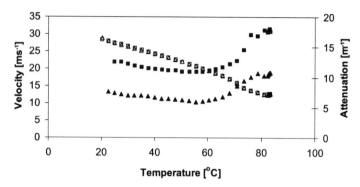

Figure 1 *Ultrasonic relative velocity (△, □) and attenuation (▲, ■) of 10% WPI in 0.01 M NaCl during heating from 20 to 85 °C (0.4 °C min⁻¹) as measured at 5 MHz (▲, △) and 7 MHz (■, □). Data are plotted as a function of temperature. The ultrasonic relative velocity is the difference between the velocity in the WPI solution and the velocity measured in the reference cell.*

(10% w/v). Figure 1 illustrates the variation in ultrasonic velocity and attenuation during heating of a solution of WPI equilibrated with 0.01 M NaCl. The results are reported as relative velocity, and defined as the difference between the velocity measured in the sample and that measured in the reference cell. The relative ultrasonic velocity decreased linearly up to 60 °C followed by a drop at 70 °C. The steady decrease in velocity was attributed to the differences in the density and the compressibility of the protein hydration shell—both are temperature-dependent quantities—compared to the same properties of the bulk solution. The compressibility is affected by the molecular organization and intermolecular interactions occurring in the aqueous solution. The steady decrease in the relative ultrasonic velocity of the protein solution with heating indicates a general increase of the compressibility of the protein with temperature. This suggests an overall loosening of the protein structure with increase in temperature, possibly caused by the exposure of hydrophobic sites to the aqueous environment. Between 45 and 50 °C the attenuation was found to decrease with temperature, and a change in direction was noted at 60 °C. This reduction in attenuation was more evident in β-lg solutions. At 70 °C, which is the temperature marking 'classical' protein denaturation and the formation of a gel network, there was found to be a sharp decrease in the relative velocity and an increase in the attenuation.

The different stages in the heat-induced aggregation of the whey proteins could be better distinguished by following the rates of change of the velocity and attenuation as functions of temperature. Figure 2 shows the first derivatives of the relative velocity and attenuation with respect to heating temperature for WPI and β-lg at different ionic strengths. The plot of the first temperature derivative of the relative velocity *versus* temperature shows a higher transition temperature for a pure β-lg solution containing 0.1 M NaCl,

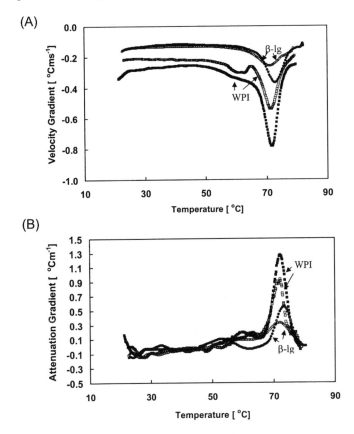

Figure 2 *Temperature differentials (first derivative with respect to temperature) of the relative ultrasonic velocity (A) and ultrasonic attenuation (B) at 7 MHz for WPI (■, □) and β-lg (●, ○) as a function of heating temperature, for solutions containing 0.01 M NaCl (□, ○) or 0.1 M NaCl (■,●). Heating was carried out from 20 to 85 °C at a rate of 0.4 °C min⁻¹.*

as compared to a β-lg solution containing 0.01 M NaCl, or a WPI solution at either of the two ionic strengths studied. In addition, the protein heated in the presence of NaCl indicated sharper transitions than protein heated at low ionic strength. This was noted for both WPI and β-lg solutions. At low temperatures (< 50 °C) there was a decrease in ultrasonic attenuation that could reflect conformational changes of the whey proteins. In particular, this may have been caused by a change in the quaternary structure of β-lg, which at neutral pH dissociates from dimers to the monomeric form.

The change in behaviour appeared to occur in a cooperative fashion, as the molecular compressibility of the protein showed a continuous increase with heating temperature. At *ca.* 50 °C, there was a marked decrease in velocity at the lower temperatures for the WPI solutions as compared to the β-lg solutions, perhaps indicating conformational changes occurring at lower

temperatures in solutions containing α-lactalbumin. These transition temperatures are much lower than those commonly reported for whey protein denaturation, *e.g.*, as measured by differential scanning calorimetry.[8] At a temperature
of *ca.* 60 °C, a shoulder can be observed in the gradient of ultrasonic attenuation
for both WPI and β-lg samples. This may indicate the formation of intermediate
aggregates, but with very little change in the compressibility of the sample.
Conversely, however, there was no apparent shoulder at 60 °C in the velocity
gradient, especially for the β-lg samples. A sharp rise in the gradient of the attenuation and a decrease in the velocity accompanies the most important thermal
transition of the protein at 72 °C, corresponding to the overall denaturation
of the protein and the onset of irreversible aggregation.

To confirm the results obtained with temperature ramps, heating studies
were also carried out at constant temperature. The attenuation and velocity
of the sound wave propagating through the protein solution were measured
isothermally over time in the range 50–75 °C (Figure 3). These results confirm
what was observed in the temperature ramp experiments. The initial values
of relative velocity decreased with heating temperature, and at <65 °C there
were found to be no changes in the measured velocity with time. This indicates

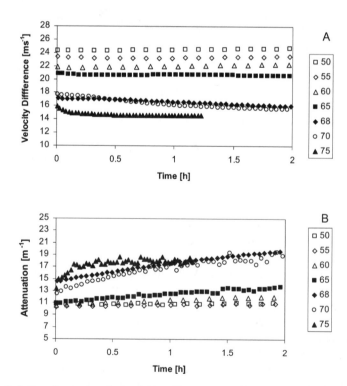

Figure 3 *Relative ultrasonic velocity (A) and attenuation (B) at 7 MHz for WPI heated
at various temperatures under isothermal conditions for solutions containing 0.01
M NaCl.*

that the changes affecting the bulk compressibility occur rapidly. Above 65 °C, the value of the set temperature affected the relative velocity of sound propagating through the solution, and the changes in compressibility became time dependent. These isothermal experiments have indicated that the onset of the main transition occurred at 70 °C, confirming the results obtained in the temperature ramp experiments. At this temperature (70 °C) the aggregation took place at a much faster rate. Ultrasonic attenuation also was also found to be affected by temperature. While there were no changes of attenuation with time below 60 °C, at higher temperatures the attenuation increased with time, up to a plateau value, indicating the existence of an effect of protein–protein interactions.

4 Conclusions

High-resolution ultrasound spectroscopy allows the observation of the main thermal transitions in whey protein solutions: denaturation and gelation are detected at *ca.* 70 °C. Using this technique it has been possible, for the first time, to detect transitions and changes in molecular conformations and the state of aggregation at temperatures lower than the main transition temperature (70 °C). These results confirm previously reported work[9,10] on the existence of reversible structural changes of β-lg below 60 °C. The results also demonstrate that sharper transitions occur at higher ionic strength, and that there are marked differences in the aggregation behaviour between pure β-lg and the mixed protein ingredient WPI. However, the attenuation or velocity changes could not be quantitatively related to gel strength. Both velocity and attenuation measurements indicate that the presence of other proteins in WPI solutions influences the mechanism of denaturation and the nature of the protein–protein interactions.

Using ultrasound spectroscopy, therefore, we have found that it is possible to observe the aggregation behaviour of whey proteins *in situ* at concentrations higher than what is normally feasible with other spectroscopic techniques. However, the parameters measured with ultrasound are related to conformational and hydration changes, and to associated denaturation and aggregation, rather than to the actual process of gelation itself. In conclusion, this study carried out using a relatively well-characterized system indicates the potential of this novel technique for probing molecular interactions under more 'realistic' conditions in food systems that are concentrated and optically opaque.

References

1. V. Buckin and C. Smyth, *Sem. Food Anal.*, 1999, **4**(2), 113.
2. M. J. W. Povey, *J. Food Eng.*, 1989, **9**, 1.
3. V. A. Buckin, *Biophys. Chem.*, 1988, **29**, 283.
4. K. Gekko and H. Noguchi, *J. Phys. Chem.*, 1979, **83**, 2706.
5. C. Le Bon, T. Nicolai, and D. Durand, *Int. J. Food Sci. Technol.*, 1999, **31**, 451.

6. S. Ikeda and V. J. Morris, *Biomacromolecules,* 2002, **3**, 382.
7. P. Puyol, M. D. Perez, and D. S. Horne, *Food Hydrocoll.*, 2001, **15**, 233.
8. P. Relkin, *Thermochim. Acta,* 1994, **246**, 371.
9. S. Iametti, B. De Gregori, G. Vecchio, and F. Bonomi, *Eur. J. Biochem.*, 1996, **237**, 106.
10. H. Li, C. C. Harding, and E. A. Foegeding, *J. Agric. Food Chem.*, 1994, **42**, 241.

Critical Concentration for Fibrillar Aggregation of Bovine β-Lactoglobulin

By Luben N. Arnaudov[1,2] and Renko de Vries[1,2]

[1]LABORATORY OF PHYSICAL CHEMISTRY AND COLLOID SCIENCE, WAGENINGEN UNIVERSITY, DREIJENPLEIN 6, 6700 EK WAGENINGEN, THE NETHERLANDS
[2]FOOD PHYSICS GROUP, WAGENINGEN UNIVERSITY, BOMENWEG 2, 6703 HD WAGENINGEN, THE NETHERLANDS

1 Introduction

Fibrillar aggregation of proteins is a generic phenomenon. It can be observed in the fine stranded heat-set gels formed from globular proteins such as ovalbumin,[1] bovine serum albumin[2] and β-lactoglobulin (β-lg).[3,4] It is also the cause of a number of debilitating diseases such as the Alzheimer's disease, Creutzfeldt–Jacob disease, systemic amyloidoses, and spongiform encephalopathies.[5] Almost all of the proteins that cause diseases by aggregation *in vivo* are also found to form fibrillar aggregates also *in vitro*.[6] Many globular proteins not associated with disease are also found to form fibrillar aggregates under specific conditions. Recently, Gosal *et al.* have shown[7] that bovine β-lg forms amyloid aggregates in solutions containing trifluorethanol (TFE). Bovine β-lg is of interest for the food industry because it is a major whey protein.[7] The formation of fine-stranded β-lg gels upon heating the protein at acidic pH has been extensively studied,[3,4,9–12] as has the process of fibril formation by β-lg.[13–20]

Aymard *et al.*[16] investigated the kinetics of heat-induced aggregation of β-lg at pH 2.0 by static and dynamic light scattering and small-angle neutron and X-ray scattering. They found that not all the protein was converted to fibrils by using a precipitation technique. Separately, Veerman *et al.*[17] obtained similar results—between 40 to 70% of the protein in the solution was converted into fibrils independent of the ionic strength (up to 0.08 M). The morphology of β-lg aggregates was studied by Kavanagh *et al.*[3] and Hamada and Dobson[18] using transmission electron microscopy. Atomic force microscopy (AFM) was used by Ikeda and Morris[19] to visualize fine-stranded aggregates (at pH 2.0) and particulate aggregates (at pH 7.0) of β-lg and whey protein isolate (WPI) aggregates. AFM was also used by Gosal *et al.*[7] to investigate fibrillar networks derived from β-lg in aqueous solution or in TFE–water solutions. In the latter case, a 'beaded' appearance of the fibrils was observed. Aggregates formed by β-lg and whey protein isolates at pH 2 were studied by Ikeda[20] using AFM and Raman scattering spectroscopy. The fine-stranded aggregates of β-lg appeared to be composed of strings of monomers, whereas the equivalent WPI aggregates were granular in appearance.

In the present work we use AFM to make a detailed study of the morphology of the fibrillar aggregates of β-lg formed by heating the protein at 80 °C in aqueous solution at pH 2.0. We also study the kinetics of fibril formation using static and dynamic light scattering and proton nuclear magnetic resonance spectroscopy.

2 Materials and Methods

Materials

The bovine β-lactoglobulin (β-lg), a mixture of genetic variants A and B, was obtained from Sigma (3 × crystallized and lyophilized, ref. L0130). All solutions were prepared with deionized water (Barnstead) and contained 200 ppm NaN_3 to prevent bacterial growth. The pH was adjusted by addition of small amounts of 1 M HCl (Merck). Prior to use, a concentrated solution of β-lg at pH = 2.0 was extensively dialysed against the solvent. After dialysis, the β-lg solutions were centrifuged for 3 h at 45,000 g using a Beckman Avanti J–25I high performance centrifuge, and subsequently filtered through 0.45 μm low-protein adsorbing syringe filters (Sterile Acrodisk, Gelman Sciences) into the glass tubes in which the experiments were subsequently carried out. The solutions used in all the experiments were prepared by dilution from the dialysed sample. The protein concentrations were determined by spectrophotometry at $\lambda = 278$ nm, using an extinction coefficient of 0.83 L g^{-1} cm^{-1}. Care was taken to minimize contamination by dust. Glassware was cleaned with chromic acid, rinsed with plenty of deionized water, and dried in a clean environment. The solutions for the light-scattering experiments were filtered through 0.1 μm syringe filters (Sterile Acrodisk) directly into the clean glass tubes prior to the experiments.

Tapping-Mode Atomic Force Microscopy

Tapping-mode AFM was carried out with a Nanoscope III, Multimode Scanning Force Microscope (Digital Instruments, USA). The samples for observation were prepared as protein solutions of different concentrations, and were put into tightly closed glass tubes in a water bath preheated at 80 ± 0.1 °C. At time intervals ranging from 1 to 48 h, the tubes were taken out of the bath, aliquots were taken, and the tubes were returned into the bath. The aliquots were quenched in ice-cold water and diluted on the basis of the initial β-lg monomer concentration to a final protein concentration of 0.1 wt%. The protein dilution was done to facilitate quantitative comparison of the results obtained from the experiments performed at different initial protein concentrations. Protein monomers and/or protein aggregates were adsorbed onto clean silicon substrates.

The observations were performed on dry samples in air prepared as follows. The plates were cleaned first in pure ethanol by ultrasound, then rinsed with pure ethanol, and dried with pure dry nitrogen. The dry silicon plates were subsequently subjected to plasma cleaning. After cleaning, the silicon plates were dipped into the test protein solutions for 1 h, and then taken out and dried immediately using pure dry nitrogen.

Light Scattering

Static and dynamic light-scattering data were obtained at a scattering angle of 90° using a ALV/SLS/DLS–5000 light-scattering apparatus (Langen, Germany) equipped with an argon ion laser (Lexel, Palo Alto, CA) emitting vertically polarized light at a wavelength of 514.5 nm. The intensity of scattered light was calibrated from the intensity of the light scattered from pure toluene measured before each series of experiments at 25 °C. The scattering from the solvent was accounted for by subtracting the intensity of scattered light by the solvent from the scattered intensity of the protein sample. Before starting the heating, the presence of only monomeric protein in the sample tubes was established by dynamic light scattering at 25 °C for each sample. The aggregation process was then followed *in situ* by directly inserting the sample into the preheated sample holder of the LS set-up, which was at 80 °C, keeping it there for a period of time ranging from 1 to 48 h, while collecting scattering data, and subsequently quenching the sample in ice-cold water.

Nuclear Magnetic Resonance

NMR experiments were carried out using a Bruker AMX 500 spectrometer operating at a ^{1}H frequency of 500 MHz. Solutions of β-lg (500 μl of 8.2, 4.1, 3.1, 2, and 1 wt% β-lg, respectively) were prepared from stock solutions including 10% v/v D_2O, adjusted to pH = 1.92 with HCl, and were transferred to Wilmad (Buena, NJ, USA) PP-528 5 mm NMR tubes. The NMR tubes containing the protein solutions were quickly introduced into a preheated probe at the desired temperature of 80 °C in the spectrometer. The data acquisition started after the sample temperature had equilibrated for approximately 3 to 5 minutes. To follow aggregation of β-lg in real time, many one-dimensional proton NMR spectra were sequentially acquired over an extended period of time ranging from 19 to 64 h in total and stored into a 2D serial file. Depending on the protein concentration of the sample, 96 to 256 scans were collected per sequential proton NMR spectrum. The residual water resonance in the middle of the spectrum was suppressed by presaturation of the solvent signal during the relaxation delay period of 2 seconds. After data acquisition, the sequential FIDs were multiplied by a cosine bell window function and Fourier transformed. The transformed spectra were baseline corrected by a third-order polynomial function. All spectral processing was done with the Bruker (Rheinstetten, Germany) Xwinnmr Linux 3.1 software package. The same software was also used to integrate the remaining soluble proton NMR signals in the spectra of β-lg that were acquired as a function of incubation time. Signals in different spectral regions were integrated, including the total spectral proton resonance range (–1 to 11 ppm), the region containing the combined amide and aromatic protons (5.5 to 11 ppm), and the region containing the non-exchangeable protons (–1 to 4 ppm).

3 Results and Discussion

Atomic Force Microscopy

Figure 1 shows tapping-mode AFM height images of samples of β-lg at pH 2.0 subjected to different heating periods at 80 °C. In the upper two rows are the images of samples of different concentrations heated for 2 h: (a) 1.0 wt%, (b) 3.0 wt%, (c) 4.0 wt%, and (d) 4.9 wt%. In the lower two rows are the images of samples of 2 wt% heated for different periods of time: (e) 2 h, (f) 5 h, (g) 24 h, and (h) 48 h. Based on images (a) to (e), we can infer that the fibril formation is concentration dependent—the higher the protein concentration, the more fibrils that can be observed. Images (e) to (h) show that the process of fibril formation takes about 24 h to complete.

Figure 2 presents 3D projections of tapping-mode AFM images of the detailed structure of β-lg fibrils obtained upon heating at 80 °C. We can see the periodic structure of the β-lg fibrils. As these images are the result of a convolution between the real sample topography and the AFM tip geometry, the representative dimension is the height of the observed objects. The mean thickness of a fibril is 3 ± 1 nm obtained by averaging the heights of more than 100 separate fibrils. The average periodicity calculated from the samples that we could observe in great detail is 28 ± 3 nm. The average peak-to-valley height difference is *ca.* 1 nm. Taken together these data suggest that the fibrils have structure that is twisted and helical, but not completely regular. The reason for our not observing this structure in all the AFM pictures is the convolution of the sample topography with the tip geometry. The standard tips for tapping-mode AFM imaging have a radius of curvature at the tip of about 10 nm, which means that details finer than 20 nm are practically impossible to distinguish.

Light Scattering

The intensities of scattered light from 10 different β-lg samples at pH 2.0, inserted into the preheated apparatus at 80 °C, are plotted in Figure 3. The initial part of the curves, where the intensity is almost constant, is not due to a lag time in the process of aggregation, but rather to the process of temperature

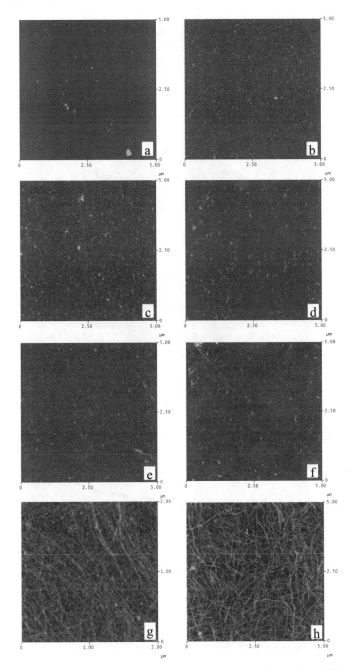

Figure 1 *Tapping-mode AFM height images (5 μm × 5 μm) of samples of β-lg at pH 2.0 subjected to different heating periods at 80 °C. The upper set of four images refers to samples of different concentration heated for 2 h: (a) 1.0 wt%, (b) 3.0 wt%, (c) 4.0 wt%, and (d) 4.9 wt%. The lower set of four images refers to samples of 2 wt% heated for different times: (e) 2 h, (f) 5 h, (g) 24 h, and (h) 48 h.*

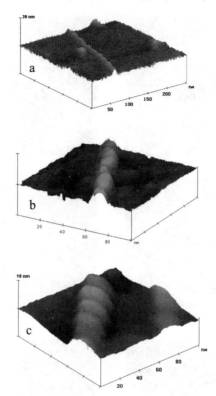

Figure 2 *3D projection of tapping-mode AFM images of the detailed structure of β-lg fibrils obtained upon heating at 80 °C: (a) 0.5 wt% β-lg, pH=2.0, 17 h; (b) higher magnification of the (a) sample; (c) 1.0 wt.% β-lg, pH=2.0, 48 h.*

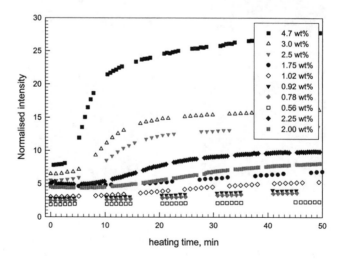

Figure 3 *Normalized light-scattering intensity of β-lg solutions at 10 different concentrations at pH=2.0 and ionic strength I=0.013 M, as a function of heating time at 80 °C.*

equilibration in the sample. After the initial equilibration, there is a region of steep linear increase in the scattered intensity due to the aggregation process. This region is best seen for concentrations higher than 2 wt%. Immediately after the linear region, the rate of increase of the scattered intensity becomes lower due to the strong repulsive interaction between the formed fibrils, since each individual β-lg molecule at pH 2.0 carries 20 positive charges.

The linear region of the plot of normalized intensity *versus* heating time can be used to extract more information about the kinetics of β-lg fibrillar aggregation at pH 2.0. Figure 4 shows a plot of the rate of increase of the normalized scattered intensity with heating time *versus* the protein concentration. The change in the slope of the data plotted in Figure 4 indicates the presence of a critical aggregation concentration for β-lg fibril formation at pH 2.0 upon heating at 80 °C. The value of the critical concentration obtained from the point of intersection of the linear regressions carried out on the two distinct regions of the data is 2.2 wt%. This value is in disagreement with the data of other researchers, who either give a lower value of 0.5 wt%[16] or a conversion that is dependent on the concentration.[17] However, the value obtained by us is the result of a very sensitive *in situ* experiment, whereas the data of the other authors[16,17] arise from a complicated procedure involving a pH quench, with subsequent precipitation and centrifugation, and determination of the monomer concentration in the supernatant by spectrophotometry. In our opinion the pH quench can lead, not only to the precipitation of the fibrils formed, but also to the precipitation of the denatured protein, as has previously been reported by Harwalkar and Kalab.[21] (See also the NMR data below.)

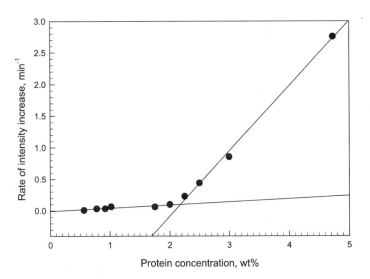

Figure 4 *Rate of light-scattering intensity increase versus protein concentration for β-lg solutions at 10 different concentrations at pH = 2.0 and ionic strength I = 0.013 M, heated at 80 °C.*

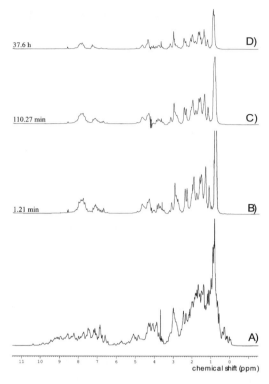

Figure 5 *Set of 500 MHz proton NMR spectra of 8.2 wt% β-lg at pH=2.0 and ionic strength I=0.013 M: A, native β-lg at 35.2 °C; B, C, and D, β-lg heated at 80 °C for three different times.*

Nuclear Magnetic Resonance

In Figure 5 the proton NMR spectra of a 8.2 wt% β-lg solution at pH 2.0 are plotted. Figure 5A is the native β-lg spectrum, and the spectra in Figure 5B–D correspond to the largely unfolded protein. As can be seen, the protein unfolds very quickly upon heating at 80 °C. Another important observation is that the overall intensity of the NMR spectrum decreases with the heating time. This is due to the formation of large aggregates: the rotational diffusion coefficient of an aggregate larger than 200 kDa is so small that the protein molecules incorporated in such an aggregate practically do not contribute to the NMR spectrum anymore. Therefore, the overall proton NMR spectrum intensity can be used as a measure of the protein monomer concentration in the studied solution.[22]

Figure 6(a) shows the extrapolated fibril concentration as a function of the initial protein concentration. The fibril concentration is calculated by using the overall intensity of the protein NMR spectrum normalized with respect to the intensity of the spectrum for 8.2 wt% β-lg. The fibril concentration is assumed to be equal to the difference between the initial protein concentration and the extrapolated final monomer concentration. The data for the change of the overall intensity of the proton NMR spectrum for 8.2, 4.1, and 3.1 wt%

(a)

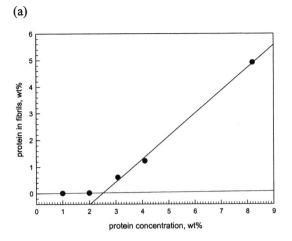

(b)

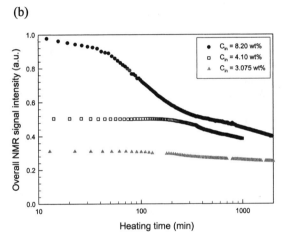

Figure 6 *Estimation of the critical concentration for β-lg fibril formation. (a) Conversion of β-lg into fibrils as extrapolated from the change of the integral intensity of the proton NMR spectra for five protein concentrations. (b) Change in the integral of the whole 500 MHz proton NMR spectrum of 8.2 wt% (●), 4.1 wt% (□), and 3.1 wt% (Δ) β-lg solutions at pH 2.0 and ionic strength I=0.013 M as a function of heating time at 80 °C.*

β-lg solutions are given in Figure 6(b). In the NMR experiments carried out with 1 and 2 wt% β-lg, no change in the overall intensity of the NMR spectra for both concentrations could be observed for more than 24 h (data not shown). As one can see, the plot in Figure 6(a) is very similar to the one in Figure 4, and it strongly supports the existence of a critical concentration for fibril formation from β-lg at pH 2.0 upon heating at 80 °C. The value for the critical concentration obtained as the point of intersection between the lines drawn through the two distinct regions of the data in Figure 6(a) is 2.5 wt%. This is in good agreement with the value obtained by light scattering.

4 Conclusions

The linear aggregation of β-lg at pH 2.0 and at low ionic strength following prolonged heating at 80 °C appears to be a multi-step process. Competing reactions lead to two basic outcomes—long linear aggregates and unfolded protein that can no longer form fibrils.

There is a 'critical' concentration for fibril formation of 2.2 wt% as found by NMR and confirmed by static light scattering.

AFM suggests that the final fibrils have an irregular helical structure with a thickness of one protein monomer and a periodicity of *ca.* 28 nm.

References

1. A. Koike, N. Nemoto, and E. Doi, *Polymer*, 1996, **37**, 587.
2. J. Lefebvre, D. Renard, and A. C. Sanches-Gimeno, *Rheol. Acta*, 1998, **37**, 345.
3. M. Langton and A.-M. Hermansson, *Food Hydrocoll.*, 1992, **5**, 523.
4. G. M. Kavanagh, A. H. Clark, and S. B. Ross-Murphy, *Int. J. Biol. Macromol.*, 2000, **28**, 41.
5. C. M. Dobson, *Phil. Trans. Roy. Soc. Lond.*, 2001, **B356**, 133.
6. A. K. Chamberlain, E. M. Cait, J. Zurdo, L. A. Morozova-Roche, H. A. O. Hill, C. M. Dobson, and J. J. Davis, *Biophys. J.*, 2000, **79**, 3282.
7. W. S. Gosal, A. H. Clark, P. D. A. Pudney, and S. B. Ross-Murphy, *Langmuir*, 2002, **18**, 7174.
8. D. W. S. Wong, W. M. Camirand, and A. E. Pavalath, *Crit. Rev. Food Sci. Nutr.*, 1996, **36**, 807.
9. A. Tobitani and S. B. Ross-Murphy, *Macromolecules*, 1997, **30**, 4845.
10. A. Tobitani and S. B. Ross-Murphy, *Macromolecules*, 1997, **30**, 4855.
11. G. M. Kavanagh, A. H. Clark, and S. B. Ross-Murphy, *Langmuir*, 2000, **16**, 9584.
12. D. Renard and J. Lefebvre, *Int. J. Biol. Macromol.*, 1992, **14**, 287.
13. E. P. Schokker, H. Singh, D. N. Pinder, and L. K. Creamer, *Int. Dairy J.*, 2000, **10**, 233.
14. Ch. Le Bon, T. Nicolai, and D. Durand, *Macromolecules*, 1999, **32**, 6120.
15. T. Lefevre and M. Subirade, *Biopolymers*, 2000, **54**, 578.
16. P. Aymard, T. Nicolai, D. Durand, and A. Clark, *Macromolecules*, 1999, **32**, 2542.
17. C. Veerman, H. Ruis, L. M. C. Sagis, and E. van der Linden, *Biomacromolecules*, 2002, **3**, 869.
18. D. Hamada and C. M. Dobson, *Protein Sci.*, 2002, **11**, 2417.
19. S. Ikeda and V. J. Morris, *Biomacromolecules*, 2002, **3**, 382.
20. S. Ikeda, *Spectroscopy—An International Journal*, 2003, **17**, 195.
21. V. R. Harwalkar and M. Kalab, *Milchwissenschaft*, 1985, **40**, 665.
22. J. Belloque and M. Ramos, *Trends Food Sci. Technol.*, 1999, **10**, 313.

Properties of Fibrillar Food Protein Assemblies and their Percolating Networks

By Cecile Veerman, Leonard M. C. Sagis, and
Erik van der Linden

FOOD PHYSICS GROUP, DEPARTMENT OF AGROTECHNOLOGY
AND FOOD SCIENCES, WAGENINGEN UNIVERSITY,
P.O. BOX 8129, 6700 EV WAGENINGEN, THE NETHERLANDS

1 Introduction

The relationship between the macroscopic properties of complex systems and the molecular properties of their ingredients has received considerable attention in recent years. To bridge the gap between properties at these two length scales, it is important to focus on the intermediate mesoscopic level (10–1000 nm). The description of complex systems at this intermediate length scale offers challenges from both the fundamental and the applied points of view.

The objective of this paper is to explore the assembly of food proteins into fibrils, and to describe the resulting percolating systems at rest and under shear flow, in terms of mesoscopic fibril properties. Protein fibrils are interesting structures, which can serve as model systems to study the behaviour of semi-flexible polymers; and they can be used for structuring at minimal weight fractions of material—a major issue in foods. For example, the food protein β-lactoglobulin (β-lg) can form a network consisting of long semi-flexible fibrils. Neither the formation of this type of mesostructure, nor the resulting mesoscopic properties, have been fully investigated. For industrial applications the effect of shear flow of fibrillar structures is especially important. A better understanding of this behaviour will allow more accurate prediction of the behaviour of fibrillar structures during the processing of materials. Hence, a detailed exploration of the making and properties of these fibril-based meso-structures should yield improved fundamental understanding as well as novel product concepts.

The outline of this paper is as follows. In section 2, we discuss the properties of fibrillar protein assemblies in the steady state. We show that the dependence of the critical percolation concentration on ionic strength can be explained in terms of an adjusted random contact model, in which the critical percolation concentration is related to the average number of contacts per particle and the excluded volume per rod. In section 3, the influence of shear flow on the critical percolation concentration of fibrillar assemblies of ovalbumin is discussed. Experimental viscosity measurements are analysed using percolation theory. These calculations are based on a random contact model for rod-like particles, making use of a shear-dependent excluded volume per fibril. Lastly, in section 4, we discuss a new multi-step process to make β-lg gels at extremely low weight fractions.

2 Fibrillar Protein Assemblies in the Steady State

Fibrillar protein assemblies have been obtained from three different globular food proteins: β-lactoglobulin (β-lg), bovine serum albumin (BSA) and ovalbumin. Both β-lg and BSA are whey proteins, which are by-products of cheese manufacturing.[1,2] About 60% of the whey protein fraction consists of β-lg, and 8% consists of BSA.[1-3] Ovalbumin is the major globular protein component in egg white.[4,5] All three of these proteins are often used as ingredients in foods because of their unique functional properties. An important general property is their contribution to the consistency of foods through the formation of heat-induced gels.[3-8] The gel properties depend on protein concentration, ionic strength, pH, and the heating procedure.[3,9] The formation of heat-induced gels from globular proteins takes place in three steps: denaturation, aggregation, and the formation of a macroscopic network.

Fibrillar protein assemblies of β-lg, BSA and ovalbumin were obtained by heating these three globular proteins around or above their denaturation temperatures, at pH = 2 and at low ionic strength.[10-12] The resulting TEM micrographs in Figure 1 show long semi-flexible structures with lengths *ca.* 4.5 ± 2.5 μm for β-lg and *ca.* 250 ± 50 nm for BSA and ovalbumin. The average contour length of the fibrils showed only minor variations with ionic strength within the measured ionic strength regime.[10-12]

To gain more insight in the assembly mechanism for BSA, we have determined the amount of BSA monomers converted into fibrils as a function of protein concentration.[10] Firstly the conversion after heating, but without dilution, was measured. After that, we determined the conversion of this sample with increasing dilution, *i.e.*, with decreasing protein concentration. A sharp decrease in the amount of converted monomers was found at a concentration of 0.1%, which indicates that below this concentration the fibrils fall apart.[10] This implies that the aggregation process of BSA is reversible, and that there is a critical concentration below which no self-assembly occurs. We determined the average contour length of ovalbumin fibrils at various times after dilution (0–24 h). This was found to be independent of the time following dilution,[12] which shows that the self-assembly process of ovalbumin is irreversible. Also for β-lg we found the assembly process to be irreversible.

To investigate the minimum concentration necessary to form a network structure, we performed rheological measurements as a function of protein

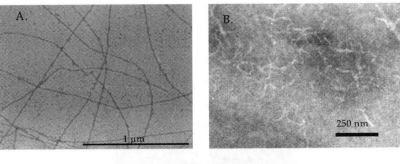

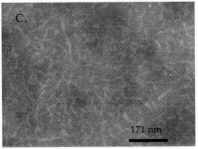

Figure 1 *TEM micrographs of fibrillar assemblies produced by heating pure globular proteins at pH=2: (A) β-lg, (B) BSA, and (C) ovalbumin.*

concentration for all three proteins.[10–12] We determined the critical percolation concentration c_p within the framework of percolation theory, which assumes the scaling relation $G' \sim (c - c_p)^t$, where G' is the elasticity modulus, c is the monomer concentration, and t is a universal scaling exponent.[13,14] The c_p values varied between 0.5 and 7.6%.[10–12] For all three proteins, a decreasing value of c_p with increasing ionic strength was found. This decrease can be explained quantitatively in terms of a random contact model. The 'mechanical equation of state' is given by[15]

$$\phi_{p,m} = 2\langle\alpha\rangle V_0 / V_{ex}^{iso}, \qquad (1)$$

where $\phi_{p,m}$ is the percolation mass fraction of the fibrils, and α is the average number of contacts per particle (assumed to be of the order of 1 at c_p). The quantity V_0 is the volume of a semi-flexible charged fibril, $(L_c/L_p)\frac{1}{4}\pi L_p D_{eff}^2$, and V_{ex}^{iso} the excluded volume of the semi-flexible charged fibril at rest, $(L_c/L_p)\frac{1}{2}\pi L_p^2 D_{eff}$. Here, L_c is the contour length, L_p is the persistence length of the fibrils, and D_{eff} is an effective diameter of the fibrils.[16] We find the following relation for the percolation mass fraction of a semi-flexible charged system:[11,17]

$$\phi_{p,m} = \alpha(D_{eff}/L_p). \qquad (2)$$

Figure 2 shows $\phi_{p,m}$ *versus* D_{eff} for the three proteins. The value of $\phi_{p,m}$ depends linearly on D_{eff} as predicted by equation (2). So the decrease of the critical

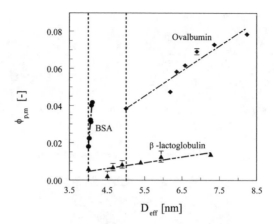

Figure 2 *Percolation mass fraction of fibrils, $\phi_{p,m}$, as a function of the effective diameter of the fibrils, D_{eff}, for β-lg, BSA, and ovalbumin at pH 2.*

percolation concentration, or mass fraction, with increasing ionic strength can be attributed solely to a decrease in the electrostatic repulsion between the fibrils. Equation (2) can be used to estimate a persistence length for each of the proteins, if we assume that α is of order 1. We find $L_p = 1.6 \pm 0.4$ μm for β-lg, $L_p = 16 \pm 4$ nm for BSA, and $L_p = 300 \pm 75$ nm for ovalbumin.[17]

Summarizing the above, we note that the dependence of c_p on ionic strength for the three different food proteins has been determined from results of elasticity measurements. The results can be explained with an adjusted random contact model, which is related to the excluded volume per fibril (at zero shear).

3 Fibrillar Protein Assemblies under Shear Flow

In the previous section, we described the determination of c_p using G' data in the limit of zero shear.[18] Now we consider the effect of an external shear field on c_p for solutions of the fibrillar protein assemblies. The effect of shear flow on the arrangement of fibrillar structures in solution is of interest in many applications.[19–21] A better understanding of the dynamics of sheared solutions of fibrils would allow, for example, a more accurate prediction of the behaviour of fibrillar structures during processing in industrial applications.[19–21]

We have investigated the effect of a shear field on c_p for solutions of fibrillar ovalbumin assemblies. The following relation holds for viscosity near the percolation threshold:[18]

$$\eta = A(c_p(\dot{\gamma}) - c)^{-k}. \tag{3}$$

The quantity A in equation (3) is a proportionality constant and k a universal scaling exponent. Expanding $c_p(\dot{\gamma})$ as a Taylor series, we can use equation (3) obtain a relationship for c_p in terms of the shear-rate $\dot{\gamma}$.[18]

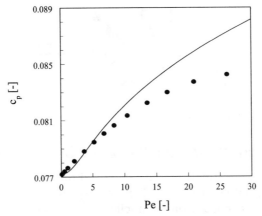

Figure 3 *Critical percolation concentration c_p versus Peclet number Pe. The points represent the experimental data and the line indicates the results from the theoretical analysis.*

From rheo-optical measurements, we can obtain values for c_p as a function of the Peclet number (Pe). For various ovalbumin concentrations, the birefringence data *versus* shear-rate could be fitted using the Doi–Edwards model for monodisperse, rigid hard rods in the semidilute regime.[22] From this fit could be found a rotational diffusion coefficient D_r for each concentration. By extrapolation, a value of $D_r \sim 1$ s^{-1} at $c = c_p$ was obtained,[18] which was used to calculate the Peclet number (Pe $= \dot{\gamma}/D_r$). This then allows us to rewrite the experimentally obtained curve of c_p *versus* shear-rate as a relationship between c_p and the Peclet number (see Figure 3).

We can compare c_p(Pe) as obtained from the experimental data with c_p(Pe) as obtained from theory. The theoretical analysis is based on a random contact model for rod-like particles, making use of an excluded volume per fibril as a function of shear, instead of the usual zero-shear limit of this volume. The excluded volume per fibril as a function of shear-rate, V_{ex}(Pe), can be calculated from the Doi–Edwards model[22] using

$$V_{ex}(\text{Pe}) = V_{ex}^{iso} Q, \tag{4}$$

where Q describes the orientation of the rods as as function of Peclet number. This excluded volume under shear is related[18] to the critical percolation concentration *via* a random contact model for rigid hard rods under shear:[15]

$$c_p(\text{Pe}) = 2\alpha \frac{V_0}{V_{ex}(\text{Pe})}. \tag{5}$$

Using equations (4) and (5) we can theoretically estimate c_p as a function of the Peclet number. Figure 3 compares the experimental results with the theory. Good agreement is found, considering the fact that polydispersity of the ovalbumin fibrils has been neglected and that the fibrils are semi-flexible.[18]

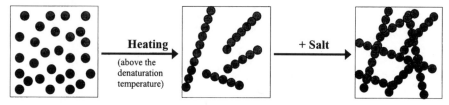

Figure 4 *Schematic representation of the cold-set gelation method.*

4 An Application of Fibrillar Protein Assemblies

In the previous two sections, we have described three fibrillar protein systems from a fundamental point of view. Now we explain how we can make use of fibrillar protein assemblies from an applied point of view.

The traditional heat-induced gelation method consists of heating a protein solution and then cooling it, during which time gelation takes place.[23-29] It is also possible, however, to make gels *via* a cold-set gelation method. To obtain gels in this alternative way, it is first necessary to prepare a heat-denatured *solution*, with the protein concentration below the critical gelation concentration. The *gelation* is subsequently induced at low temperatures by the addition of monovalent or polyvalent cations,[3,30-36] as illustrated in Figure 4. This process of cold-set gelation can lead to potential new uses for whey proteins in a variety of foods.[37]

In the cold-set gelation method, as typically referred to in the existing literature[3,30-33] (here called the 'conventional cold-gelation method'), salt-free whey protein isolate or β-lg is heated at pH 7, cooled, and then cross-linked by the addition of NaCl or $CaCl_2$. The cold-set gels formed using this conventional cold-gelation method show a fine-stranded structure, better water-holding capacity, and higher gel strength than gels formed by normal heat-induced gelation.[33,37,38]

The objective of using a new multi-step cold gelation procedure is to obtain β-lg gels at a very low protein concentrations.[39] In the new procedure, long linear β-lg fibrils are formed at pH 2 and low ionic strength, after heating at 80°C for 10 h.[11,40] Solutions of these fibrils are cooled, and subsequently the pH is adjusted to 7 or 8. TEM micrographs show (Figure 5) that the long linear fibrils are stable against changes of pH. In the final step of the procedure, the fibrils are cross-linked using $CaCl_2$.

To compare the fibrils formed by the new method with those formed by the conventional cold-gelation method, we have also made TEM micrographs for 3% β-lg heated at pH 7 or 8 at 80°C for 30 min, and subsequently cooled on ice to 0°C. The fibrillar structures formed were found[39] to be more than a factor of 10 shorter in length, as compared with the fibrils formed after heating at pH 2.

Using rheological measurements, the critical percolation concentration was determined using the method described by van der Linden and Sagis.[14] The lowest value found for c_p, 0.07%, was observed[39] when a 2% β-lg solution was heated at pH 2, and when, after cooling, the pH was adjusted to 7 and 0.01 M $CaCl_2$ was added. For this set of conditions, there is an optimal interplay

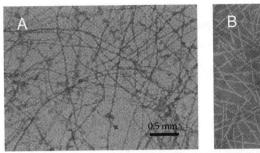
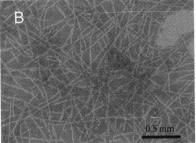
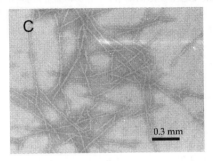

Figure 5 *TEM micrographs of 2% β-lg at pH=2 heated at 80°C for 10 h, and subsequently cooled on ice to 0°C: (A) before pH adjustment; (B) after pH adjustment to pH=7; (C) after pH adjusted to pH=8.*

between the screening of the electrostatic interactions between the fibrils and the formation of intermolecular ion-bridges between charged or carboxylic groups of β-lg by the Ca^{2+} ions.[39] On comparing our new multi-step cold-gelation method with the conventional cold-gelation method, we have found[39] for the latter method a value for c_p which is an order of magnitude higher than that found with the new method.

Using this multi-step cold-gelation process, a gel network can be formed at much lower concentrations than possible with the conventional cold-gelation method. This novel processing route opens up possibilities for more efficient use of protein ingredients.

5 Conclusions

We have discussed the conditions leading to gel formation, and the critical percolation concentration, in terms of mesoscopic fibril properties under static (*i.e.*, non-flow) conditions. The critical percolation concentration has been determined from the results of elasticity measurements, and explained in terms of an adjusted random contact model, which is related to the excluded volume per fibril (at zero shear). The influence of shear flow on the critical percolation concentration has been described both experimentally and theoretically. The critical percolation concentration *versus* shear flow could also be expressed in terms of an excluded volume per fibril, in this case as a function of shear, again with good agreement between experiment and theory. With the use of a new multi-step cold-gelation process, a gel network can be formed at extremely low protein concentrations (0.07%), offering possibilities for more efficient use of food ingredients.

References

1. J. E. Kinsella and D. M. Whitehead, in 'Advances in Food and Nutrition Research', ed. J. E. Kinsella, Academic Press, San Diego, 1989, p. 343.
2. D. M. Mulvihill and M. Donovan, *Irish J. Food Sci. Technol.*, 1987, **11**, 43.
3. C. M. Bryant and D. J. McClements, *Trends Food Sci. Technol.*, 1998, **9**, 143.
4. N. Nemoto, A. Koike, K. Osaki, T. Koseki, and E. Doi, *Biopolymers*, 1993, **33**, 551.
5. Y. Mine, *J. Agric. Food Chem.*, 1996, **44**, 2086.
6. M. Paulsson, P. Dejmek, and T. van Vliet, *J. Dairy Sci.*, 1990, **73**, 45.
7. J. I. Boye, I. Alli, and A. A. Ismail, *J. Agric. Food Chem.*, 1996, **44**, 996.
8. N. Hagolle, P. Relkin, D. G. Dalgleish, and B. Launay, *Food Hydrocoll.*, 1997, **11**, 311.
9. E. Doi, *Trends Food Sci. Technol.*, 1993, **4**, 1.
10. C. Veerman, L. M. C. Sagis, J. Heck, and E. van der Linden, *Int. J. Biol. Macromol.*, 2003, **31**, 139.
11. C. Veerman, H. Ruis, L. M. C. Sagis, and E. van der Linden, *Biomacromolecules*, 2002, **3**, 869.
12. C. Veerman, G. de Schiffart, L. M. C. Sagis, and E. van der Linden, *Int. J. Biol. Macromol.*, 2003, **33**, 121.
13. D. Stauffer, A. Coniglio, and M. Adam, *Adv. Polym. Sci.*, 1982, **44**, 103.
14. E. van der Linden and L. M. C. Sagis, *Langmuir*, 2001, **17**, 5821.
15. A. P. Philipse, *Langmuir*, 1996, **12**, 1127.
16. A. Stroobants, H. N. W. Lekkerkerker, and T. Odijk, *Macromolecules*, 1986, **19**, 2232.
17. L. M. C. Sagis, C. Veerman, and E. van der Linden, *Langmuir*, 2004, **20**, 924.
18. C. Veerman, L. M. C. Sagis, P. Venema, and E. van der Linden, submitted for publication.
19. E. L. Meyer and G. G. Fuller, *Macromolecules*, 1993, **26**, 504.
20. A. W. Chow and G. G. Fuller, *Macromolecules*, 1985, **18**, 786.
21. L. Hilliou, D. Vlassopoulos, and M. Rehahn, *Macromolecules*, 2001, **34**, 1742.
22. M. Doi and S. F. Edwards, *J. Chem. Soc. Faraday Trans. 2*, 1978, **74**, 918.
23. T. Koseki, N. Kitabatake, and E. Doi, *Food Hydrocoll.*, 1989, **3**, 123.
24. E. Doi and N. Kitabatake, *Food Hydrocoll.*, 1989, **3**, 327.
25. T. Koseki, T. Fukuda, N. Kitabatake, and E. Doi, *Food Hydrocoll.*, 1989, **3**, 135.
26. N. Matsudomi, D. Rector, and J. E. Kinsella, *Food Chem.*, 1991, **40**, 55.
27. M. Murata, F. Tani, T. Higasa, N. Kitabatake, and E. Doi, *Biosci. Biotechnol. Biochem.*, 1993, **57**, 43.
28. M. Stading and A.-M. Hermansson, *Food Hydrocoll.*, 1990, **4**, 121.
29. M. Langton and A.-M. Hermansson, *Food Hydrocoll.*, 1992, **5**, 523.
30. S. Barbut, *Lebensm. Wiss. Technol.*, 1997, **29**, 590.
31. P. Hongsprabhas, *Lebensm. Wiss. Technol.*, 1999, **32**, 196.
32. Z. Y. Ju and A. Kilara, *J. Agric. Food Chem.*, 1998, **46**, 3604.
33. P. Hongsprabhas and S. Barbut, *J. Food Sci.*, 1997, **62**, 382.

34. C. M. Bryant and D. J. McClements, *J. Food Sci.*, 2000, **65**, 259.
35. C. M. Bryant and D. J. McClements, *J. Food Sci.*, 2000, **65**, 801.
36. P. Hongsprabhas and S. Barbut, *Int. Dairy J.*, 1997, **7**, 827.
37. P. Hongsprabhas and S. Barbut, *Food Res. Int.*, 1996, **29**, 135.
38. P. Hongsprabhas and S. Barbut, *Food Res. Int.*, 1998, **30**, 523.
39. C. Veerman, H. Baptist, L. M. C. Sagis, and E. van der Linden, *J. Agric. Food Chem.*, 2003, **51**, 3880.
40. P. Aymard, T. Nicolai, and D. Durand, *Macromolecules*, 1999, **32**, 2542.

Foams and Emulsions

Disproportionation Kinetics of Air Bubbles Stabilized by Food Proteins and Nanoparticles

By Brent S. Murray, Eric Dickinson, Zhiping Du,
Rammile Ettelaie, Thomas Kostakis, and Julien Vallet

PROCTER DEPARTMENT OF FOOD SCIENCE, UNIVERSITY OF
LEEDS, LEEDS LS2 9JT, UK

1 Introduction

Disproportionation is the migration of gas from regions of high concentration to low concentration, *via* diffusion, where the higher concentration is provided by a higher local curvature and hence a higher Laplace pressure. Adsorbed films of either low-molecular-weight surfactants or polymers at fluid–fluid interfaces offer little resistance to the diffusion of gases through them. Thus, even if the bubbles are stable to coalescence and aggregation, in a foam with a fluid continuous phase, the gas will diffuse between bubbles of different sizes, with the result that the smaller bubbles tend to shrink and the larger bubbles tend to grow. Such foams would therefore seem to be inevitably unstable. Of course, if the continuous phase has become solidified in some way, in practice usually by cooking, or by freezing, then the bubble-size distribution can be effectively arrested in time. However, this immobilization may only occur after significant disproportionation has already taken place, thereby changing the product microstructure compared with the original, freshly formed foam. For example, ageing due to disproportionation is still evident during the hardening of highly viscous ice-cream mixes.[1]

The coarsening of the bubble-size distribution *via* disproportionation can itself accelerate other instability mechanisms such as aggregation and coalescence. Changes in bubble-size distribution can also have a profound effect on the texture and appearance of a foamed food product. Nevertheless, theory does indicate[2] that, if the adsorbed film surrounding the bubbles has significant *interfacial* viscoelasticity, then this film by itself can provide sufficient resistance to stop disproportionation. The crucial question is whether or not this stabilization mechanism actually occurs in practice. In this paper we report experiments designed to answer this question.

259

2 Materials and Methods

Materials

Pure bovine β-lactoglobulin (β-lg) (crystallized, lyophilized, lot no. 114H7055) was from Sigma Chemicals (Poole, Dorset). Cationic gelatin (220 Bloom PS, 18 Mesh, lot no. 09198) was obtained from Sanofi Bio-Industrie (Brussels, Belgium). Commercial whey protein isolate (WPI) (PSDI 2400) with $>95\%$ β-lg content was obtained from MD Food Ingredients (Vidabaek, Denmark). Commercial spray-dried sodium caseinate ($>82\%$ dry protein, $<6\%$ moisture, $<6\%$ fat and ash) was supplied by DMV International (Veghel, Netherlands). Soy glycinin was a gift from the Wageningen Centre for Food Sciences, prepared as described elsewhere.[3] All other reagents were *AnalR* grade from Sigma, as was the dye Rhodamine B. The fumed silica particles were a gift from Wacker-Chemie GmbH (Burghausen, Germany); they had been treated with dichlorodimethylsilane, such that 33% of the surface Si–OH groups were grafted with dimethylsilane in order to make the silica partially hydrophobic.[4] Unless otherwise stated, the protein concentration was 0.05 wt% in 0.05 mol dm^{-3} imidazole buffer (pH 7) containing 0.05 mol dm^{-3} NaCl.

Methods

Detailed descriptions of the techniques and apparatus have been given elsewhere.[4-7] Figure 1 is a schematic illustration of the essence of the experiment. Basically, an air–water (A–W) interface is created by introducing the aqueous test solution into a chamber, and after a certain time air bubbles are injected beneath the A–W interface contained within a hole in a thin mica sheet floating

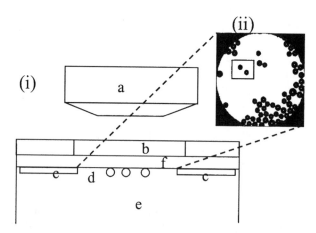

Figure 1 *Investigation of the stability of individual bubbles at a planar air–water interface. (i) Schematic diagram of experimental arrangement: (a) microscope; (b) upper glass window; (c) mica float with hole; (d) bubbles that have risen to air–water interface after injection (size grossly exaggerated in proportion to rest of diagram); (e) aqueous phase; (f) air gap. (ii) A typical image of (d) contained within (c). In this particular case, only the two bubbles surrounded by the box are considered 'isolated' and therefore suitable for analysis.*

at the interface. On arrival at the interface, depending on the solution conditions, a certain fraction of the bubbles is stable to coalescence, and these remain beneath the interface. A certain amount of ageing of the A–W interface is required to allow significant protein adsorption to take place, and so to give a satisfactory number of stable bubbles for subsequent observation. For all the measurements reported here, a 10 min ageing time was used unless otherwise stated. Similarly, in experiments with silica particles, sufficient time had to be allowed for a layer of silica particles to be adsorbed at the interface. For the silica particle systems, the bubbles were generated by sonicating the dispersion for 10 min in a Grant model XB14 ultrasound bath (Grant Instruments, Shepreth, UK), adding the dispersion to the cell, pressurizing the dispersion to 5 bar overnight, and then abruptly lowering the pressure in the chamber.[4] Stable bubbles were observed from above *via* a microscope and video camera, and their sizes were monitored with time *via* digital video recording and appropriate image analysis.

All the protein-stabilized systems showed at least some bubble shrinkage with time (see later), which meant that the surface concentration of protein on the bubble surface probably increased (assuming limited protein desorption), while the surface tension γ decreased. To measure the change in γ under these conditions for some of the systems, a similar adsorbed protein film was established in a Langmuir trough, and this was compressed at a rate to match the relative area change of the bubble surface. The change in γ was measured with a Wilhelmy plate. Full details of the Langmuir trough apparatus have been given elsewhere.[8]

Images of some systems of particle-stabilized bubbles were obtained using a confocal laser scanning microscope (CLSM). The CLSM was a Leica TCS SP2 instrument combined with a Leica Model DM RXE microscope and a 20 × dry objective lens of 0.7 numerical aperture. A 0.1 vol% solution of Rhodamine B was used to stain the silica particles. Samples were excited with the 543 nm line of the He–Ne laser.

3 Theory of Bubble Dissolution

The theory developed to explain the experimental results has been described in detail elsewhere.[5,6] Basically, the approach consists of solving the equations for the diffusive mass transport of gas between the bubble, the A–W interface, and the surrounding solution. At the same time, as disproportionation proceeds, a dilatational elasticity (ε) can be built into the model to account for the reduction in γ as the bubble shrinks. This interfacial elasticity therefore modifies the Laplace pressure driving force and provides additional mechanical resistance to further shrinkage. The diffusion equations for an isolated bubble are relatively easy to solve (by the method of 'images' due to Jeans[9]); but for two (or more) bubbles *close together* at the interface, this becomes a multi-body problem, and the solution is not straightforward. As for the definition of 'close' in this context, it can be shown that bubbles behave as if they are effectively isolated when their surfaces remain approximately twice the mean bubble diameter away from each other.

As will be described later, the microscopic details of bubble juxtaposition and relative bubble size have a large influence on dissolution kinetics. This type of information cannot be obtained from classical theories of disproportionation (or Ostwald ripening) involving global particle-size distributions such as the Lifshitz, Slyozov and Wagner (LSW) theory.[10] The LSW theory uses a mean-field approach in a closed system, such that the concentration of gas is assumed to rise uniformly throughout the aqueous phase as smaller bubbles dissolve. Kloek *et al.*[2] recently updated this sort of approach to include the detailed effects of dilatational or bulk elasticity and viscosity on the rates of bubble dissolution. The theory indicates[2] that, when the elastic moduli are large enough, they can arrest dissolution completely. Viscous moduli can actually only slow down dissolution, although very high values can lead to negligible bubble shrinkage over significant periods of time. However, our experimental system is open, as are most real systems, and quite clearly some deviations in the local gas concentration do occur. It is these local changes in gas distribution between the bubbles and the planar interface that our theoretical analysis seeks to address. One of the key questions to answer is whether or not adsorbed proteins, which are known to be capable of forming films with a high interfacial viscoelastic modulus, can cause a significant reduction in the rate of disproportionation by virtue of such interfacial behaviour.

Most of the theoretical parameters required to describe the diffusive mass transport of air are known to reasonable accuracy: the solubility S and diffusion constant D of air in water are taken here as 6.9×10^{-6} mol m^{-1} N^{-1} and 1.99×10^{-9} m^2 s^{-1}, respectively, and the surface tension γ of the A–W interface in the presence of protein is taken from previous measurements.[11,16] The separation L of the bubble surface from the planar interface is by far the least well-known quantity. From a consideration of our experimental conditions, an initial estimate of L was taken as 10 nm, which corresponds to approximately twice the thickness of the adsorbed protein layer. In fact, our theoretical fitting of the experimental data involves floating all the other parameters as a group variable τ, defined as

$$\tau = \frac{P_0}{2 D \gamma S R T}, \tag{1}$$

where T is the temperature, R is the gas constant, and P_0 is the pressure above the interface, taken as 1×10^5 N m^{-2}. Fixing the value of L at 10 nm seems to be reasonably justified, in that the numerical values of τ required to fit the data (see below) are found to lie within the bounds expected, given the uncertainty resulting from the combination of errors in the individual terms contributing to τ.

4 Results and Discussion

Protein-Stabilized Bubbles

Figure 2 shows typical results obtained for several different proteins. The bubble radius R is plotted as a function of time t for isolated bubbles only.

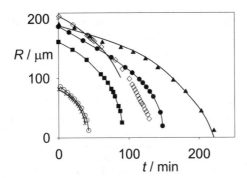

Figure 2 *Typical plots of radius R versus time t for isolated bubbles with different proteins at pH 7: caseinate (X), WPI (○), gelatin (■), ovalbumin (◇), β-lg (●), and soy glycinin (▲). The lines drawn through the points indicate the best fits to the data, i.e., requiring the following ε values: 0 mN m⁻¹ for caseinate and WPI; 2.3 mN m⁻¹ for gelatin; 7 mN m⁻¹ for β-lg; 8.5 mN m⁻¹ for soy glycinin; 12.5 mN m⁻¹ for ovalbumin.*

As it is generally not possible to control very precisely the size of the stable isolated bubbles obtained following injection, the results presented here have been selected so that the starting sizes of bubbles are close enough to allow proper comparison of the effects of the different proteins. However, many replicate experiments have shown[5,6] that the results chosen are entirely representative of each protein. On the whole, it may seem surprising that the results for the various proteins are not so very different, given that the interfacial rheology amongst the group of proteins varies widely. For example, caseinate forms films of low interfacial viscoelasticity compared to the globular proteins, β-lg and soy glycinin. Bubbles of the same size do appear to shrink more slowly with a globular protein as stabilizer, but not markedly so. The overall impression is that in none of systems are the bubbles particularly stable: bubble shrinkage accelerates as the bubble radius increases, as predicted from the inverse dependence of the Laplace pressure on R. The one possible exception is ovalbumin, where some tendency for levelling off in the plot of R *versus* t was sometimes observed. In fact, ovalbumin presents a rather special case, and so it is discussed at greater length below.

The theoretical fits of the experimental data agree with the general impression indicated above. Figure 2 gives examples of such fits, including finite values of the dilatational elasticity ε. Table 1 lists the values of ε required to give good fits to the data. For caseinate and WPI a constant surface tension (ε = 0) gives an almost perfect fit to the data. Although the WPI sample used consisted largely of β-lg, the likely presence within it of small amounts of low-molecular-weight contaminants, such as fatty acids, is probably the reason why the adsorbed WPI appeared to behave more like a reversibly adsorbed film, such as that formed from a small-molecule surfactant. For gelatin, β-lg and soy glycinin, increasing values of ε are needed to give good fits to the data; but the numerical values required are not particularly high, and it is notable

Table 1 *Values of the measured surface tension (γ) and dilatational elasticity (ε_{fit}) obtained from theoretical fits to bubble shrinkage data. Also shown are approximate values of apparent surface shear viscosity (η_{app}), surface shear elasticity (G_{app}), surface storage dilatational modulus (ε') and surface loss dilatational modulus (ε'') for the various protein solutions under the conditions described in text, measured at a shear rate of 10^{-3} s^{-1}, a strain of <5 %, and a frequency of 10^{-3} Hz, respectively.*

Protein	γ (mN m^{-1})	ε_{fit} (mN m^{-1})	η_{app} (mN s m^{-1})	G_{app} (mN m^{-1})	ε' (mN m^{-1})	ε'' (mN m^{-1})
sodium caseinate	48	0	7	0.5	4.5	0.5
WPI	49	0	–	–	–	–
gelatin	53	2.3	80	0.5	–	–
β-lactoglobulin	50	7.0	450	–	5.5	2
soy glycinin	50	8.5	1000[a]	–	–	–
ovalbumin	48	12.5	300[a]	–	18	5

[a] Estimated from ref. 3.

that in no cases are they sufficiently large to arrest bubble shrinkage completely. It should be stressed that the fitted value of ε was reproducibly found to be identical for the same protein system, within the errors shown, independent of the initial bubble size chosen.

It can reasonably be concluded from these results that bubble dissolution proceeds slowly enough to allow rearrangement of the protein film as the bubble surface area decreases, thereby maintaining an essentially fixed protein load and therefore an almost constant surface tension. In other words, the interfacial film is *viscoelastic*, and the presence of the viscous rheological element allows the tension in the film to relax sufficiently quickly that little stress can build up in the film to retard appreciably the rate of shrinkage. The dilatational moduli measured for such proteins are typically obtained under conditions of oscillatory area variation of a few per cent strain at ~1 Hz.[12] These conditions have little resemblance to the conditions applicable to the fitted values of ε during bubble shrinkage, and they cannot explain the relative differences in shrinkage rates, even though these are quite small. This is hardly surprising, however, given the low rate of deformation of the bubbles, except in the very late stages of their collapse. In this respect the fitted values of ε are closer to those measured at very low frequencies, *e.g.*, *via* Fourier transform of the relaxation of γ after expansion.[13]

Further credence is given to the reliability of our fitted values of ε in the results of carefully controlled experiments on a Langmuir trough with β-lg or caseinate. Having observed the rate of bubble shrinkage, the barrier motion of the trough was programmed to match the decrease in relative area A of the film contained by the barriers to that of the bubble surface area change. The gradient of the plot of γ *versus* ln A in the trough containing β-lg was found to be ε $= 7 \pm 0.5$ mN m^{-1} (see Figure 3). This value perfectly matches the

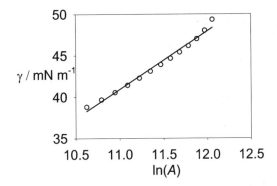

Figure 3 *Measured γ versus ln A in the Langmuir trough (where A is the trough area) for an adsorbed β-lg film compressed at the same rate as that of corresponding bubble collapse. The straight line fit to the data has gradient ε=7 mN m⁻¹, matching the theoretical value of ε obtained from the bubble shrinkage data.*

theoretical value obtained from fitting the bubble shrinkage kinetics. Similarly, when a caseinate film was compressed on the Langmuir trough at a rate of bubble area decrease corresponding to that of the bubble surface area decrease, the value of γ was found to be practically constant (*i.e.*, implying $\varepsilon=0$), which again is consistent with the theoretical fitted value.

Table 1 also gives values of interfacial *shear* moduli for the proteins measured under similar conditions.[12,13] Stability towards disproportionation seems to be correlated at least as well with the interfacial shear modulus as it is with the dilatational modulus. This correlation may be connected to the fact that, in practice, the distortion of the interfacial film during extended shrinkage may not take the form of a simple areal deformation. If the protein film has sufficient coherence, then during bubble shrinkage the film may behave more like the skin of a deflating balloon—wrinkling, bending, and buckling in on itself, rather than simply remaining uniform and flat with simultaneous protein desorption into the bulk. Of course, the appearance of such a wrinkled protein 'skin' has been reported on the surface of large oil droplets for well over a century.[14] However, in monitoring this behaviour in the context of our collapsing bubbles, some further interesting observations were made with the globular proteins.

We refer first to bubbles stabilized by β-lg or soy glycinin. When such a bubble was seen to disappear under the microscope, there was visible an optically dense, irregularly shaped particle located where the bubble centre once was, but with nothing resembling the spherical bubble remaining afterwards. (The resolution of the microscope meant that bubbles with radii below ~ 10 μm could not be monitored accurately: in any case they shrank so rapidly that their images could not be recorded quickly enough). Some examples of the images obtained are given in Figure 4. The sizes of the irregular particles formed seem to match the idea that they consist entirely of the collapsed protein film surrounding the original bubble on injection. For example, if the protein surface load on a 350 μm diameter bubble is taken as 2 mg m⁻²,

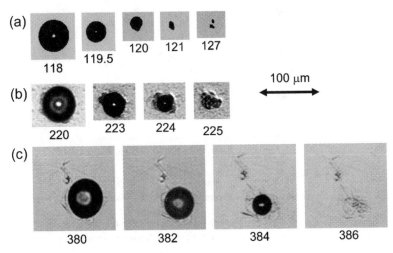

Figure 4 *Typical sequences of images of bubbles in their final stages of shrinkage for (a) β-lg, (b) soy glycinin, and (c) ovalbumin. The numbers beneath the images indicate the bubble ages (in minutes).*

and the mean density of the protein material is assumed to be 1.2 ± 0.2 g cm^{-3}, it is easy to calculate that this protein could form a solid spherical particle of diameter 24–27 μm. With time (tens of minutes or hours), these particles appeared to become increasingly transparent or to decrease in size, suggesting that they might be gradually redissolving and/or breaking up into smaller fragments.

It was already hinted at above that ovalbumin may represent a special case. This relates to the nature of the R *versus* t plots, their fitted ε values, and the appearance of the 'protein particles', just mentioned above. In general, the $R(t)$ plots with ovalbumin were found to be much less reproducible—sometimes with a distinctly flatter shape, and occasionally with a sudden decrease in the steepness of the curve, as illustrated by the examples in Figure 5. At pH 7, the ovalbumin solutions were not completely clear, and protein aggregates appeared to be present at the A–W interface at the start of the experiment. Also, thin striations were sometimes visible at the interface, suggesting the presence of an insoluble skin of protein. Of course, ovalbumin is well known for its sensitivity to interfacial coagulation,[15] and so it might be presumed that the presence of such aggregates is responsible in some way for the greater variability in the results. However, filtering the solutions to remove aggregates in the bulk before starting the experiment, and/or carefully dissolving the ovalbumin with minimum stirring and agitation, did not seem to improve the reproducibility significantly. There were also problems in fitting the theory to the experimental data. In general, it was not possible to obtain a good fit to the whole $R(t)$ curve. A consistent value of $\varepsilon = 12.5$ mN m^{-1} fitted the first hour of shrinkage fairly well, but this same value did not give a good fit also to the final stages of shrinkage. Again this is perhaps not surprising, because the

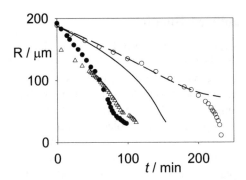

Figure 5 *Examples of time-dependent radius R(t) plots for ovalbumin at pH=7 (●, △) and at pH=3 (○) which are not well fitted by the diffusion model. For the pH 3 example, the best fit to the early part of the curve gives ε=24 mN m⁻¹ (dashed line); the best fit for ε=12.5 mN m⁻¹ (full line) is also shown for comparison.*

process of bubble collapse also appeared to be rather unusual. Ovalbumin-stabilized bubbles often became ellipsoidal in shape in their final stages of shrinkage, surrounded by a fuzzy layer of material connected to the surrounding interface. Finally, the bubbles were seen to collapse completely to form particles similar to those obtained with β-lg and soy glycinin, although the outer edges of the particles still appeared to consist of the less dense, fuzzy material originally present. An example of such images is given in Figure 4. The surface of the ovalbumin-stabilized bubbles appeared to be connected to a film of insoluble material in the surrounding interface right from the start of the experiment. As the bubbles shrank they appeared to drag some of this material towards them, which then became incorporated into the protein particle formed from the collapsed protein film around the bubble.

As a result of the complications of ovalbumin aggregation in the bulk under neutral pH conditions, some additional experiments were conducted in citric acid buffer at pH 3, where clear solutions were obtained, and far fewer aggregates or striated material were visible at the interface at the start. An example of the type of *R versus t* data obtained is included in Figure 5. The bubble shrinkage now was found to be much slower than at pH 7, and the shape of the curve even flatter, such that even less satisfactory fits could be obtained to the whole curve. Fitting of the first hour of shrinkage gives even higher values of ε, *i.e.*, 24±2 mN m⁻¹. In calculating the fitted values of ε at pH 3 and pH 7, the initial surface tension was taken as 48 mN m⁻¹, from the data of Relkin *et al.*[16] It should be pointed out that, although the surface activity is apparently similar at both pH values, the net charge on the ovalbumin molecule is quite different: approximately +42e at pH 3 and −10e at pH 7. Increased repulsion between adsorbed protein molecules is expected to inhibit film compression. However, protein particles with entrained interfacial material were still seen to be formed at pH 3, in the same way as at pH 7, which suggests that surface coagulation also takes place under these more acidic conditions.

Nanoparticle-Stabilized Bubbles

Our results with protein-stabilized bubbles suggest that no food protein film has the capacity to inhibit disproportionation completely. Such protein films are always likely to be viscoelastic, with associated mechanisms for film desorption or distortion that allow slow but relentless bubble shrinkage. The question therefore remains whether any interfacial film is capable of arresting disproportionation completely. Low-molecular-weight surfactants, which are reversibly adsorbed, seem even less likely candidates. On the other hand, adsorbed insoluble particles, which can apparently aid the formation of highly stable emulsion droplets,[17,18] are worthy of further study. Fat crystals and other insoluble particles are frequently observed in films surrounding bubbles and emulsions in foods, although their contribution to stability (or instability) is often far from clear. In the experiments described here and elsewhere,[4,7] we have followed the example of Binks *et al.*[17,18] in using partially hydrophobic silica nanoparticles as model surface-active particles. Figure 6 shows an example of the bubble stability obtained using a 1 wt% suspension of such silica particles. In order to obtain stable bubbles, the choice of the type and concentration of the nanoparticles appears to be crucial. It is especially important to optimize the degree of hydrophobicity, in order to achieve an appropriate balance between maximizing the extent of particle adsorption and minimizing the tendency for particle aggregation in the bulk before adsorption.[4] Addition of 3 mol dm^{-3} NaCl to the aqueous phase was used to modify this balance.

For comparison with the protein results, Figure 6 shows R *versus* t plots, not only for particle-stabilized bubbles, but also for β-lg-stabilized bubbles of similar initial size. Clearly the particle-stabilized bubbles are far more stable, even for quite small initial bubble sizes; and for larger initial bubble sizes they appear to be almost indefinitely stable. Microscopic observations suggest[7] that the formation of a rigid shell of nanoparticles around the bubbles is what prevents their collapse. The origin of this resistance might be purely mechanical,

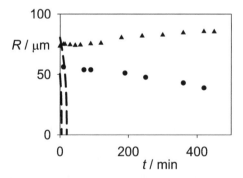

Figure 6 *Typical plots of bubble radius R versus time t for two bubbles formed with 1 wt% partially hydrophobic silica particles in 3 mol dm^{-3} NaCl (▲, ●). For comparison with proteins, the dashed lines show typical results for β-lg-stabilized bubbles of similar initial size.*

(a)   (b)

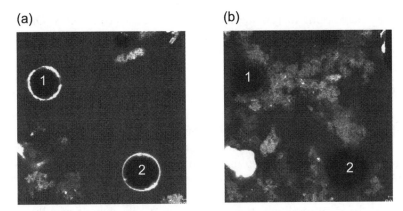

Figure 7 *CLSM images of two bubbles (labelled 1 and 2) stabilized by aggregated silica nanoparticles (the particles appear bright against a dark background). In picture (a) the bubbles appear separate from the surrounding silica particle network. Picture (b) shows the same two bubbles (which appear as dark shadows because they are no longer in focus) at an optical section 38 μm below image (a); the solid material on the bubble surfaces is clearly connected to a surrounding network of aggregated silica.*

or it could be due to the fact that even the occasional particle desorption event is very unlikely, due to the extremely high energies of adsorption of such particles (up to several thousand kT).[17,18] On the other hand, recent CLSM images of these systems,[4] such as the one shown in Figure 7, suggest that a more subtle mechanism may also be involved. Although the primary particle size is 20 nm, clearly much of the solid adsorbed material is highly aggregated, and it is therefore unlikely to be present as a perfect, close-packed monolayer on the bubble surface. Furthermore, it appears that, whenever significant bubble stabilization by particles takes place, the nanoparticles themselves interact to form a weak gel network throughout the aqueous phase,[4] and the particles adsorbed around the bubbles appear to be continuous with this particle network. This is illustrated in Figure 7. What this suggests is that it could be the resistance to collapse of the whole network, as much as the resistance to the collapse of the individual particle films around the bubbles, that determines whether or not the resulting foam remains stable.

From the point of view of the food technologist, it is perhaps interesting to compare our model nanoparticle-stabilized system with that stabilized by ovalbumin, which is the food protein that most readily appears to form particles in the bulk and at the interface. Although our fits to the R *versus* t data are not perfect, they do still suggest the highest elastic modulus for this protein film, and the greatest resistance to bubble shrinkage. The ovalbumin particles are obviously incapable of forming a perfectly rigid shell like spherical silica nanoparticles. But it could be that the well-known functionality of ovalbumin as an excellent whipping agent in foods is somehow connected with its susceptibility for forming coagulated protein aggregates and surface-active particles. Indeed, this is the explanation often given in textbooks,[19] though the work presented here seems to provide the first direct evidence for this. One might speculate that other protein particles could be directly engineered in some way to provide even more effective food foam stabilization properties.

Effects of Bubble Clustering

It should be noted that the consideration of bubble shrinkage kinetics presented above has been entirely for isolated bubbles at the planar interface. When bubbles are in close proximity, it can be shown,[6] both experimentally and theoretically, that there are two main consequences. Firstly, when equal-sized bubbles are close together, they have the effect of screening one another from the surrounding 'sink' of lower gas concentration in the bulk. As a result, each bubble shrinks more slowly than when it is on its own. Secondly, if two neighbouring bubbles are just slightly different in size, then this has the effect of accelerating the shrinkage of the smaller bubble, whilst the larger bubble shrinks more slowly, or even grows, as it 'feeds off' the gas from the adjacent smaller bubble(s). However, these two effects are moderated by the variable tendency for some groups of bubbles to become aggregated, and so shrink whilst sticking together, whereas other bubbles shrink away from one another, as if they are stuck to the planar A–W interface.

The tendency for bubble aggregation varies according to the type of protein in the system. For example, caseinate-stabilized or gelatin-stabilized bubbles frequently aggregate at the planar A–W interface, and then stay aggregated as

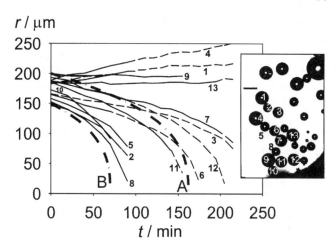

Figure 8 *An example of the complex shrinkage kinetics of an aggregated bubble system, in this case for 0.05 wt% gelatin. The inset shows the configuration of the bubbles at the interface after 40 min (bar = 400 μm). The set of identification numbers on the bubbles corresponds to the set of numbers on the R versus t plots. A full line indicates the period when a bubble is apparently in contact with at least one other bubble. A dashed line indicates when a bubble is apparently not in contact with another bubble. The two thick dash–dot lines labelled A and B show the theoretical behaviour for completely isolated bubbles of starting radii 200 and 150 μm, respectively.*

the bubbles shrink. The subsequent variation in the size of its near neighbours means that a bubble within a cluster of bubbles can exhibit both shrinkage or growth at any time throughout the ageing of the cluster.[6] On the other hand, β-lg-stabilized bubbles often stick to the planar A–W interface, but shrink away from each other, so that with time they gradually begin to behave like isolated bubbles. An example of the complex shrinkage kinetics for a clustered system is seen in Figure 8, together with some theoretical plots for non-clustered systems. The larger bubbles on the periphery of the cluster, such as the bubbles labelled 1 and 4, initially grow rather than shrink, but the bubbles initially sandwiched between larger bubbles, such as bubbles 2, 5 and 8, are observed to shrink quickly from the start.

The net effect of bubble clustering is therefore a much more complex evolution in the local bubble-size distribution than the mean-field theories can account for. This behaviour is likely to be of significance in relation to the microstructural properties of real food foams.

Acknowledgements

B.S.M. and E.D. thank the Wageningen Centre for Food Sciences, and especially Ton van Vliet and Martin Bos, for generously supporting this research with a studentship to Z.D. We also acknowledge support from BBSRC (UK) and Unilever Research (Colworth Laboratory) for the equipment used in this project.

References

1. S. Turan, M. Kirkland, and R. Bee, in 'Food Emulsions and Foams: Interfaces, Interactions and Stability', eds E. Dickinson and J. M. Rodriguez Patino, Royal Society of Chemistry, Cambridge, 1999, p. 151.
2. W. Kloek, T. van Vliet, and M. Meinders, *J. Colloid Interface Sci.*, 2001, **237**, 158.
3. A. Martin, in 'Mechanical and conformational aspects of protein layers on water', Ph.D. thesis, Wageningen University, Netherlands, 2003.
4. E. Dickinson, R. Ettelaie, T. Kostakis, and B. S. Murray, *Langmuir*, 2004, **20**, 8517.
5. E. Dickinson, B. S. Murray, R. Ettelaie, and Z. Du, *J. Colloid Interface Sci.*, 2002, **252**, 202.
6. R. Ettelaie, E. Dickinson, Z. Du, and B. S. Murray, *J. Colloid Interface Sci.*, 2003, **263**, 47.
7. Z. Du, M. P. Bilbao-Montoya, B. P. Binks, E. Dickinson, R. Ettelaie, and B. S. Murray, *Langmuir*, 2003, **19**, 3106.
8. B. S. Murray and P. V. Nelson, *Langmuir*, 1996, **12**, 5973.
9. J. Jeans, 'The Mathematical Theory of Electricity and Magnetism', Cambridge University Press, Cambridge, 1948.
10. A. S. Kabalnov and E. D. Schukin, *Adv. Colloid Interface Sci.*, 1992, **38**, 69.
11. B. S. Murray, B. Cattin, E. Schüler, and Z. O. Sonmez, *Langmuir*, 2002, **18**, 9476.
12. B. S. Murray, in 'Proteins at Liquid Interfaces', eds R. Miller and D. Möbius, Elsevier, Amsterdam, 1998, p. 179.
13. B. S. Murray, C. Lallemant, and A. Ventura, *Colloids Surf. A.*, 1998, **143**, 211.

14. F. M. Ascherson, *Arch. Anat. Physiol.*, 1840, 44.
15. F. MacRitchie, *Adv. Colloid Interface Sci.*, 1986, **25**, 341.
16. P. Relkin, N. Hagolle, D. G. Dalgleish, and B. Launay, *Colloids Surf. B*, 1999, **12**, 409.
17. B. P. Binks and S. O. Lumsdon, *Langmuir*, 2000, **16**, 8622.
18. B. P. Binks and J. H. Clint, *Langmuir*, 2002, **18**, 1270.
19. P. Walstra, 'Physical Chemistry of Foods', Marcel Dekker, New York, 2003, p. 539.

Coarsening and Rheology of Casein and Surfactant Foams

By A. Saint-Jalmes, S. Marze, and D. Langevin

LABORATOIRE DE PHYSIQUE DES SOLIDES,
UNIVERSITE PARIS-SUD, 91405 ORSAY, FRANCE

1 Introduction

For aqueous foams, as for many dispersed systems, the macroscopic behaviour is the result of microscopic processes occurring on smaller length scales. In order to understand foam macroscopic properties, one thus has to determine the crucial and relevant ingredients needed to describe foam on its different scales. From a physical point of view,[1] the most important parameters are the bubble diameter d (or radius r), the foam liquid fraction ε, and the surface tension σ. The parameter σ quantifies the surface energy of the dispersion, and it appears in the Laplace capillary pressure σ/r. An important issue is to find out if foam properties can be described by universal functions of d, ε and σ, or if other parameters are needed to explain the behaviour.

In fact, it is already known that the adsorbed surfactant layer plays an important role, and that not all surfactants are equivalent. The viscoelasticity, the net charge and the thickness of the interface—all are crucial for foamability and thin film stability. Regarding foam drainage, which depends on the mobility of the surface layer, as described by microscopic parameters like surface shear viscosity[2,3] or elasticity,[4] one can find different macroscopic regimes. The macroscopic mechanical properties, and other processes such as coarsening or film rupture, are less well understood, and it is not yet known if surface viscoelasticity or other microscopic parameters must be included. In this spirit, recent experiments have been performed on coarsening,[5] as well as theoretical works conjecturing that surface properties can actually play a role in coarsening.[6,7] One can also wonder about possible different effects arising on the scale of the thin films. Beside bubble surfaces and thin films, the gas type itself also has its role, for instance in relation to coarsening.[8] However, quantitative experimental tests are scarce, especially because it is difficult to separate coarsening from drainage. Indeed, the two processes are strongly coupled,[2,8] resulting in a non-direct but important role of the gas in the drainage process. The gas may also affect foam rheology. Finally, the bulk viscoelasticity of the surfactant liquid solution may also have to be taken into account.[7,9]

Here we focus on the interfacial and gas contributions to coarsening and rheology, by using two different solutions (providing very different interfacial layers) and two different gases. We have used various techniques to collect results at different length scales: measurement of surface tension, video microscopy of the thin films, static and dynamic properties on the bubble scale using multiple light-scattering techniques, and bulk rheology. We have also developed a new experimental procedure to study coarsening, keeping the liquid fraction constant.

2 Materials and Foam Making

We have studied foams made from solutions of micellar casein and sodium dodecyl sulfate (SDS). The casein and SDS were obtained from Sigma (St. Louis, MO). For the pure solutions, we have used a casein concentration of 4.5 g L^{-1} in a 10 mM phosphate buffer (pH = 5.7), and an SDS concentration of 6 g L^{-1}. We have also studied mixtures, with a small amount of SDS (0.3–1.5 g L^{-1}) added to the casein solution. Casein adsorbed layers are very different from SDS layers, the former being thicker and more viscous and rigid, with strong differences especially in terms of shear viscosities. The thin films are also very different, as discussed in this paper. However, the surface tensions are roughly similar, *i.e.*, 42 mN m^{-1} for the casein solution and 36 mN m^{-1} for the SDS solution. The bulk viscosities are also similar, and rather close to that of pure water. Nitrogen (N$_2$) and hexafluoroethane (C$_2$F$_6$) have been used as the two gases. The solubility and bulk diffusivity of C$_2$F$_6$ in water are, respectively, 11 and 2.8 times smaller than those for N$_2$, which means that coarsening should be faster with N$_2$.

The foams were produced by a turbulent mixing method,[11] which provided foams with bubble sizes around 150 μm, independent of the type of solution used. With this setup, the initial liquid fraction ε can be tuned from 3% to 35%. The experiments presented here have been performed with relatively wet foams (ε = 15%).

3 Coarsening Experiments

In order to study the coarsening process, the foam liquid fraction ε should ideally remain constant over long periods of time, which is difficult to achieve because of drainage. Accordingly, we have developed a rotating cell to suppress drainage. This involves allowing drainage to proceed freely for a short time period *T*, and then at the end of the time period rotating the cell upside down, thereby inverting the drainage direction. Optimizing the cell size and the period *T* allows us to keep a constant value of ε at the cell centre for an extended period of time. The cell is 40 cm high, 12.5 cm wide and 2.5 cm thick; it is made of transparent Plexiglas. The period *T* was set to lie between 20 s to

200 s depending on the type of foam being investigated. With time, the value of T has to be reduced, as bubbles get larger and the drainage becomes faster. This procedure appears well adapted to foams made of small bubbles, for which coarsening can be rather fast, while drainage is relatively slow. Two different light-scattering experiments have been performed on the foam samples during and after rotation.

Bubble Size Growth: Gas Type and Thin Film Effects

We have measured the time dependence of the intensity transmitted by the foams. In the limit of multiple scattering,[12] the transmitted intensity I_t is simply related to d and ε by

$$I_t \sim d / \sqrt{\varepsilon}.$$ (1)

This means that, if one can keep ε constant, the increase in the transmitted intensity is attributable to bubble growth. This remains true as long as the light is multiply scattered in the foam, meaning that the sample thickness H must be large compared to the bubble size.[12] (In practice, for $\varepsilon = 15\%$, a minimum thickness of around 25 bubbles is needed). The setup is very simple, consisting of a light source on one side of the sample, and a CCD camera on the other side. Here we have used a laser, needed for the inelastic scattering measurements (see below), but white light illumination could also be used (as in earlier studies[2] of foam drainage).

Figure 1 shows transmission images of foam sections with typical dimensions of 3 cm at different ages of the foam. One can see that the overall intensity increases over time. The intensity profiles along a vertical axis in these sections are reported in Figure 2. These profiles can be fitted by gaussian curves, from which the maximum intensity at the centre can be obtained. Typical data are presented in Figure 3. The initial part ($t < t^*$) corresponds to pure coarsening; at $t = t^*$ (vertical dotted line) the rotation is stopped; and, for $t > t^*$, a steeper increase is observed due to combined drainage and coarsening. Theoretically, the coarsening is predicted to be a self-similar process,[13] *i.e.*,

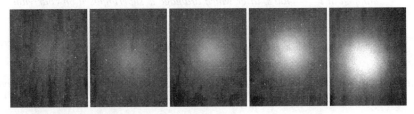

Figure 1 *Light transmission images of foam sections observed at different times during coarsening. With time, the intensity of light transmitted increases (from left to right). The size of the sections is typically 3 cm.*

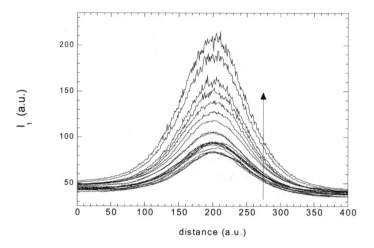

Figure 2 *Light transmission profiles along a vertical axis at different times. The arrow indicates increasing times.*

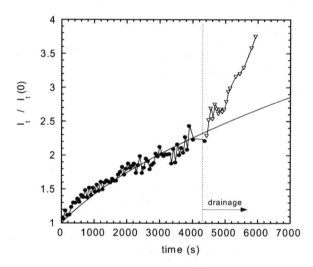

Figure 3 *Normalized transmitted intensity I_t/I_t $(t=0)$ during cell rotation. Rotation is stopped at $t^* \sim 4400$ s, after which drainage is superimposed on coarsening.*

$$\langle d \rangle^2 - d_0^2 = d_0^2 (t - t_0)/t_c, \qquad (2)$$

which asymptotically gives $d \sim t^{1/2}$. The characteristic coarsening time is given by

$$t_c = \frac{d_0^2 h}{2 K_{geo} K_{gas} f(\varepsilon)}, \qquad (3)$$

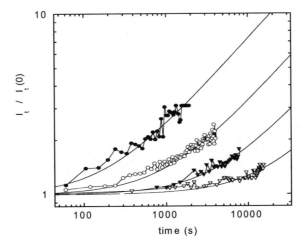

Figure 4 *Normalized transmitted intensity $I_t/I_t(t=0)$ for foams of SDS (solid symbols) and casein (open symbols), made either with N_2 (O) or C_2F_6 (▽). The lines are fits with the self-similar model, from which the coarsening time t_c is found.*

where K_{geo} is a constant reflecting the bubble geometry, K_{gas} is a constant including the gas diffusivity and gas solubility, $f(\varepsilon)$ a function of the liquid content ε, and h is the thin film thickness.[8]

The results for the normalized transmitted intensity, $I_t/I_t(t=0)$, obtained for the different solutions and gases for $t<t^*$ are shown in Figure 4. It appears that the data can be correctly fitted by the self-similar coarsening model. In this way we find $t_c=190$ s, 980 s, 4.2×10^3 s and 1.91×10^4 s for the SDS/N_2, casein/N_2, SDS/C_2F_6 and casein/C_2F_6 foams. The ratio $t_c(C_2F_6)/t_c(N_2)$ is around 20 for both the SDS and the casein systems. According to the theoretical expression for t_c, this ratio must be equal to the ratio of the gas constants, $K_{gas}(N_2)/K_{gas}(C_2F_6)$, which has a value of 30 (with gas solubility and diffusivity included in the model[8]). Our results are thus in rough agreement with the predictions. Despite the discrepancy that remains still to be understood, the results effectively show that both diffusivity and solubility are important—otherwise the effect of the gas on the coarsening rate would have been considerably smaller. The values of t_c found for the SDS foams with N_2 or C_2F_6 lead to a film thickness of 40 nm for an initial bubble size of 150 µm. This value of film thickness is comparable to the equilibrium film thickness predicted from a balance of van der Waals and electrostatic forces between film surfaces, and it is also close to the measured value (see below). Besides the gas effect, we have also found a clear difference between the foaming agents used: t_c for casein is always around 5 times larger than t_c for SDS. Our results also show that, on adding SDS (≥0.5 g L^{-1}) to a pure casein solution, the coarsening time decreases towards the t_c value for pure SDS.

In order to find the origin of the above differences, we have studied pure SDS and casein thin films using the thin-film balance apparatus. In this technique,[14] a single film is formed on a porous-glass holder, the film is

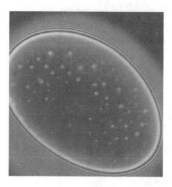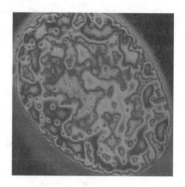

Figure 5 *Pictures of thin films obtained with the thin-film balance apparatus: left, SDS film; right, casein film.*

visualized from above by videomicroscopy in white light, and the thickness is determined by measuring the reflected light intensity. The picture on the left in Figure 5 represents a typical SDS thin film after drainage due to the capillary suction exerted by its thicker edges. The film is homogeneous and of thickness $h \approx 50$ nm, corresponding to a common black film. The casein film on the right in Figure 5 looks quite different: it is very heterogeneous with a non-uniform thickness, and the image is coloured, meaning that its average thickness is of the order of a few hundreds of nanometres. The casein film does not drain over time; it looks gel-like. Similar films have also been observed[15] for surfactant + polyelectrolytes mixtures close to their precipitation boundaries. It is thus quite tempting to interpret the differences between the casein and SDS foam coarsening rates in terms of simple differences in film thickness, and indeed these are compatible with a t_c ratio of *ca.* 5. The thin film properties can also explain the coarsening behaviour of the mixtures: indeed, if 0.5 g L^{-1} of SDS is added to the casein solution, the film begins to resemble that of pure SDS. So, the effect of casein on coarsening would seem to be mainly a film thickness effect, and not due to a surface gas barrier built by the casein surface layers. We note finally that, despite the important role of the film thickness in the coarsening process, these thin films do not contribute appreciably to the overall liquid fraction, as the liquid volume contained in the films remains always negligible compared to the liquid volume in the Plateau borders and in the nodes connecting them.

Coarsening-Induced Bubble Dynamics

As coarsening proceeds, stresses and strains are built up locally within the foam, inducing bubble rearrangements, above a local critical strain γ_c, typically involving a few bubbles.[16] Such dynamic processes can be monitored by diffusing wave spectroscopy (DWS), the rearrangements leading to time fluctuations of the light scattered by the foam.[17,18] A DWS experiment produces an intensity auto-correlation function $g_2(\tau)$, which is related to the electric field correlation function $g_1(\tau)$ by the Siegert relation,

$$g_2(\tau) \sim 1 + \beta g_1^2(\tau). \tag{4}$$

In the back-scattering setup, and for localized intermittent dynamics, the DWS formalism gives

$$g_1(\tau) \sim \exp(-2\sqrt{6\tau/\tau_m}), \tag{5}$$

where τ_m is the average of the time intervals between coarsening-induced rearrangements in the sample.[16–18] Experimentally, a small part of the back-scattered light is collected by an optical fibre, which is connected to a photon detector and fed to a correlator.[17–19] Since the signal has no angular dependence, the setup is quite simple. An important requirement is that the laser must have a long coherence length, so that the decorrelation between light paths can be assigned only to sample dynamics.[17,18]

We report in Figure 6 some typical intensity auto-correlation functions measured for the casein and SDS foams ($\varepsilon = 15\%$) made with N_2 and C_2F_6. The fit to the predicted form for intermittent dynamics is good, allowing us to determine τ_m for each curve. We find that τ_m does not vary much during the rotation of the cell. Indeed, during our experiments the variation of the bubble size is too small to affect τ_m significantly.[19] We note, however, that the accuracy is limited by the rather short integration time (typically 1–2 min). We have found that τ_m ranges between 0.5 s for the fastest coarsening foam (SDS/N_2) to around 2 s for the slowest one (casein/C_2F_6). So it turns out that τ_m and

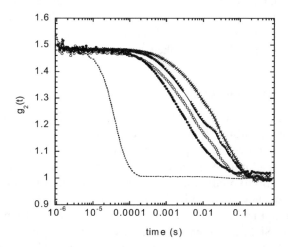

Figure 6 *DWS back-scattering intensity auto-correlation function $g_2(t)$ for the different coarsening foams: •, SDS/N_2; O, casein/N_2; ▼, SDS/C_2F_6; ×, casein/C_2F_6. The dotted line is a curve obtained for a sheared foam at constant shear-rate 1 s^{-1}.*

t_c are correlated, but τ_m is much smaller than t_c, and the maximum ratio of the values of τ_m is also much smaller than for t_c. The fact that τ_m is small (~ 1 s) may seem surprising, but this is a general finding: for shaving cream foam (similar bubble sizes, but a lower liquid fraction) the value $\tau_m \sim 5$ s has been reported.[19]

We have also performed DWS experiments with a continuous shear applied to the foam. The applied shear-rate was 1 s^{-1}. Figure 6 shows that the resulting value of τ_m is much smaller than for the non-sheared coarsening foam. We note also the different shapes of the curves, reflecting different types of dynamics.

The results obtained for $t > t^*$ (the foam at rest) are also interesting. The τ_m values of the different systems, and also the maximum ratio between them, increase as the foams get drier. The value of τ_m increases up to ~ 200 s for the casein/C_2F_6 foams, while it increases up to only 6 s for the SDS/N_2 foams, although the latter have larger bubbles, which implies a higher τ_m. All together, the results show that τ_m increases not only with the bubble size,[19] but it also depends strongly on the liquid fraction.

We believe that the small ratio between the values of τ_m (when compared to t_c) for the wet foams, and its increase with the liquid fraction, is the result of two combined effects. First, it reflects a dependence of γ_c on ε: a very small value of γ must be needed to produce rearrangement in a wet foam. Secondly, the coarsening-induced strains may not increase at the same rate everywhere in the foam.[20] The measured value of τ_m arises from an average over a distribution of time intervals between rearrangements in different regions of the foam. If one considers that the shortest time intervals are close, and almost independent of the gas diffusion rate, and that γ_c is low, then the observed dynamics can be mainly controlled by these fastest rates, leading to small differences between the foams. Nevertheless, further theoretical and experimental work is needed to understand properly these local coarsening-induced dynamics, in order finally to make the link between the local rheological properties scanned by such coarsening-controlled rheometry, and those obtained when a controlled macroscopic shear is applied.

4 Rheology experiments

Rheological experiments were performed using an MCR 300 rheometer from Paar Physica fitted with a home-made cone-and-plate geometry (diameter 12 cm, angle 10°). To reduce slippage, the surfaces were roughened by sand blasting and sand gluing. We have first investigated the linear rheological response of the foams to oscillation. In such an experiment, the strain was varied at a constant frequency, $\omega = 1$ rad s^{-1}, and the storage and loss moduli, G' and G'', were measured. Figure 7 shows G' and G'' for different C_2F_6 foams made with pure casein, pure SDS, and their mixtures. For all these systems, the classical curves for G' and G'' usually found for concentrated dispersed systems are recovered.[21,22] Although there are some small differences in behaviour between the foams, these differences are definitely not as large as seen with emulsions.[23,24]

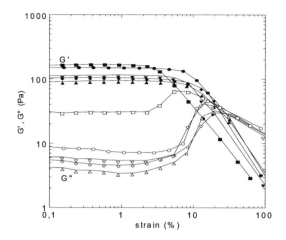

Figure 7 *Storage modulus G' (solid symbols) and loss modulus G" (open symbols) versus applied strain at a frequency 1 rad s^{-1} for C_2F_6 foams with casein (■), SDS (●) or casein/SDS mixtures (▼, 0.3 g L^{-1} of SDS added; ♦, 0.5 g L^{-1} of SDS added; ▲, 1.5 g L^{-1} of SDS added).*

To analyse these results further, one has to normalize the data with the Laplace pressure σ/r. Taking a mean bubble size of 150 μm and a surface tension of 40 mN m^{-1}, and a mean value of G' at zero strain of 130 Pa, we get $(\sigma/r)^{-1}G' = 0.24$, in good agreement with values for other surfactant foams and emulsions with $\varepsilon = 15\%$.[21,22] So, at first sight, once normalized by the Laplace pressure, the elastic moduli of these systems follow rather universal behaviour, independent of the species adsorbed at the surfaces. However, one can note some small differences in rheology between the foams, which may be associated with variations in surface properties. The viscous modulus of the pure casein foam is around 4–5 times higher than that of the SDS foam or the SDS/casein foam. As for the coarsening kinetics, this may be related to the thin-film thickness. Though the casein foam has the highest elastic modulus, the yield strain appears slightly smaller, reflecting a more brittle character. One can also note that the lower values of G' for the mixtures cannot be completely explained by the surface tension scaling: the values of $(\sigma/r)^{-1}G'$ are lower for these foams than for the pure systems. These interesting differences need to be confirmed by further investigation. Let us record here also that, although we did not see any large viscosity differences in the continuous shear experiments, visual observations revealed pronounced shear-banding effects in casein foams, occurring already at small applied deformation. The impact on the macroscopic rheology of these non-uniform strains remains to be understood.

The effect of the gas type on the rheology is related to foam ageing. At short times, the rheological response of the foams was found to be independent of the gas used. However, over time, one can see that G' and G'' evolve differently (see Figure 8). With C_2F_6, the coarsening is slow, and the foam mainly gets drier by drainage, and so G' increases.[21,22] With N_2, even if the foam drains,

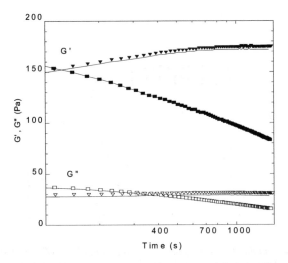

Figure 8 *Effect of the type of gas on the time evolution of the viscoelastic properties (G', solid symbols; G'', open symbols) for a casein foam:* ■, N_2; ▼, C_2F_6.

the bubble size also strongly increases by coarsening, and this appears to dominate, since G' decreases.

5 Conclusions

We have shown that both the coarsening time t_c and the coarsening-induced dynamics can be measured at constant liquid fraction by combining a rotating cell setup with multiple light-scattering techniques. The coarsening time depends on the gas properties; it is also dependent on the foaming agent used. Thin-film studies reveal significant thickness differences that can account for the different coarsening rates. The bubble rearrangement time τ_m associated with the local yield strains is much smaller than t_c, and it depends on the bubble size and the liquid fraction.

The rheological properties of the surfactant foam and the casein foam are similar in the linear response regime. When SDS is added to the casein solution, the coarsening and rheological behaviour of the pure SDS foam is recovered, as soon as the film thickness becomes identical, thereby demonstrating its crucial role.

References

1. D. Weaire and S. Hutzler, 'The Physics of Foam', Oxford University Press, Oxford, 1999.
2. A. Saint-Jalmes and D. Langevin, *J. Phys. Cond. Mat.*, 2002, **14**, 9397.
3. S. A. Koehler, S. Hilgenfeldt, E. R. Weeks, and H. A. Stone, *Phys. Rev. E*, 2002, **66**, 040601(R).
4. M. Durand and D. Langevin, *Eur. Phys. J. E*, 2002, **7**, 35.

5. E. Dickinson, R. Ettelaie, B. S. Murray, and Z. Du, *J. Colloid Interface Sci.*, 2002, **252**, 202.

6. M. Meinders, W. Kloek, and T. van Vliet, *Langmuir*, 2001, **17**, 3923.

7. W. Kloek, T. van Vliet, and M. Meinders, *J. Colloid Interface Sci.*, 2001, **237**, 158.

8. S. H. Hilgenfeldt, S. A. Koehler, and H. A. Stone, *Phys. Rev. Lett.*, 2001, **86**, 4704.

9. M. Safouane, M. Durand, A. Saint-Jalmes, D. Langevin, and V. Bergeron, *J. Physique IV*, 2001, **11** (PR6), 275.

10. W. G. Mallard and P. J. Lindstrom (eds), 'NIST Chemistry WebBook', NIST Standard Reference Database, 2000.

11. A. Saint-Jalmes, M. U. Vera, and D. J. Durian, *Eur. Phys. J. B*, 1999, **12**, 67.

12. M. U. Vera, A. Saint-Jalmes, and D. J. Durian, *Appl. Optics*, 2001, **40**, 4210.

13. W. W. Mullins, *J. Appl. Phys.*, 1986, **59**, 1341; J. A. Glazier, *Phys. Rev. Lett.,* 1993, **70**, 2170.

14. V. Bergeron, *J. Phys. Cond. Mat.*, 1999, **11**, R215.

15. V. Bergeron, D. Langevin, and A. Asnacios, *Langmuir*, 1996, **12**, 1550.

16. D. J. Durian, D. A. Weitz, and D. J. Pine, *Science*, 1991, **252**, 686; A. D. Gopal and D. J. Durian, *Phys. Rev. Lett.*, 1995, **75**, 2610.

17. D. J. Pine, D. A. Weitz, P. M. Chaikin, and E. Herbolzheimer, *Phys. Rev. Lett.*, 1988, **60**, 1134.

18. D. J. Pine, D. A. Weitz, J. X. Zhu, and E. Herbolzheimer, *J. Phys. (France)*, 1990, **51**, 2101.

19. D. J. Durian, D. A. Weitz, and D. J. Pine, *Phys. Rev. A*, 1991, **44**, R7902.

20. S. Cohen-Addad and R. Hohler, *Phys. Rev. Lett.*, 2001, **86**, 4700.

21. T. G. Mason, J. Bibette, and D.A. Weitz, *Phys. Rev. Lett.*, 1995, **75**, 2051; T. G. Mason, M. D. Lacasse, G. S. Grest, D. Levine, J. Bibette, and D. A. Weitz, *Phys. Rev. E*, 1997, **56**, 3150.

22. A. Saint-Jalmes and D. J. Durian, *J. Rheol.*, 1999, **43**, 1411.

23. L. Bressy, P. Hebraud, V. Schmitt, and J. Bibette, *Langmuir*, 2003, **19**, 598.

24. T. A. Dimitrova and F. Leal-Calderon, *Langmuir*, 2001, **17**, 3235.

Interfacial and Foam Stabilization Properties of β-Lactoglobulin–Acacia Gum Electrostatic Complexes

By Christophe Schmitt, Eric Kolodziejczyk, and Martin E. Leser

NESTLE RESEARCH CENTER, DEPARTMENT OF FOOD SCIENCE, VERS-CHEZ-LES-BLANC, CH-1000 LAUSANNE 26, SWITZERLAND

1 Introduction

Foams occur in a number of daily applications—both food (coffee, whipped desserts, ice cream, beer) and non-food (sponges, shaving cream, cleaning agents). Less well-known applications of foams are in insulation, fire fighting, mineral aero-flotation, and construction (polymeric, metallic or ceramic foams).[1] Foams are basically made of a dispersion of gas cells within a continuous solid or a liquid matrix.

Food foams are generally wet, *i.e.*, the gas cells remain more or less round. The matrix is often viscous and contains a high fraction of water. Drainage and coalescence of the bubbles are reduced and the main destabilizing mechanism is disproportionation or coarsening.[2] Disproportionation is due to the pressure difference between the inside and the outside of the bubble, leading to a gradient of gas solubility.[3] As a consequence, the gas diffuses out of the bubble and the bubble shrinks. A common way of circumventing this coarsening process is to increase the viscosity of the continuous phase. A more elegant way is to increase the viscoelasticity of the interface so that it becomes less permeable to the gas.[2] In order to produce and stabilize foams, especially in the food business, proteins and surfactants are often used alone or in mixtures.[4,5] These surface-active molecules are generally structured into hydrophilic and hydrophobic groups that allow interfacial adsorption.[6] In the case of globular proteins, as compared to random coil or micellar proteins, the amphiphilic character combined with the reduced size favours foam formation. A subsequent interfacial unfolding favours foam stabilization.[7]

Few studies have investigated the foam stabilization properties of protein–polysaccharide complexes.[8,9] Interestingly, it has been reported[10–13] that protein–polysaccharide electrostatic complexes can significantly improve foaming properties compared to protein alone. A likely explanation is that the complexes form a kind of 'giant surfactant', the hydrophilic part being formed by the polysaccharide and the hydrophobic part being formed by the protein. The relatively limited interest expressed so far in the interfacial and foaming properties of electrostatic protein–polysaccharide complexes might be explained[14] by the non-equilibrium nature of the system. As soon as the protein and the polysaccharide are mixed, phase separation occurs through complex coacervation.[15] Structures formed within such a system are highly time-dependent, leading to various functional properties.[16]

In this study, we have investigated the interfacial and foaming properties of a mixture of β-lactoglobulin (β-lg) and acacia gum (AG) forming electrostatic complexes and coacervates at acidic pH.[17] We aim to draw a parallel between the structure formed during bulk phase separation and that involved in the two-dimensional structure/stabilization of the air–water interface.

2 Materials and Methods

Materials

The β-lactoglobulin (β-lg) (lot JE 001-1-922) was obtained from Davisco (Le Sueur, MN). It was purified from sweet whey using filtration and ion-exchange chromatography. The powder composition (g per 100 g of powder) was: 93.2% protein, 4.3% moisture, 0.2% fat, 2.3% ash (0.019% Ca^{2+}, 0.002% Mg^{2+}, 0.009% K^+, 0.848% Na^+, < 0.04% Cl^-). Acacia gum (AG) (Instant Gum IRX 40693, lot OS-780) was obtained from Colloïdes Naturels (Rouen, France). The polysaccharide was solubilized from the exuded pellets of *Acacia senegal* trees and filtered before spray drying. The powder composition (g per 100 g of powder) was: 4% protein, 80.56% polysaccharide, 12.23% moisture, 3.21% ash (0.552% Ca^{2+}, 0.191% Mg^{2+}, 0.698% K^+, 0.041% Na^+, 0.05% Cl^-). From the mineral contents of the powders, and taking the tested protein to polysaccharide ratio as 2:1, and the highest total biopolymer concentration used as 5 wt%, the ionic strength of our samples was estimated to be in the range 7–11 mM equivalent NaCl. Such an ionic strength has been shown[18] to have a negligible effect on complex coacervation. However, ionic strength has to be taken into account when it is higher than 100 mM, as could be the case when using commercial whey protein concentrates or isolates.[19] All reagents used were of analytical grade (Merck, Darmstadt, Germany) unless otherwise stated.

Biopolymer dispersions were prepared (on a protein or polysaccharide wt% basis) by drop-wise addition of the required amount of powder into Milli-Q water (18.6 MΩ cm). The pH was then adjusted to 4.2 by addition of 0.1 or 1 M NaOH or HCl. The protein or polysaccharide concentration was adjusted by addition of Milli-Q water. Sodium azide (0.03 %) was added in order to prevent bacterial growth. Dispersions were centrifuged at 10^4 g for 1 hour (Sorvall RC5C centrifuge, DuPont, Newtown, CT) to discard insoluble matter and air bubbles before storage at 4 °C.

For most of the following experiments, we investigated the protein alone and mixtures at ratio 2:1 (w/w). All experiments were duplicated, unless otherwise stated.

Phase Composition Determination

Phase separation kinetics of a 1 wt% β-lg + AG mixture of ratio 2:1 and pH 4.2 was followed at 24 °C by determination of the protein and polysaccharide contents in the upper phase using total nitrogen (Kjeldahl, N × 6.38) and total sugars,[20] respectively. Biopolymer compositions in the lower phase were obtained from the difference between the initial composition of the mixture and the upper phase composition.

Rheological Characterization of the Coacervate

After mixing 5 wt% β-lg + AG dispersions at pH = 4.2, small-deformation oscillatory measurements were performed on the coacervate phase. Mechanical spectra were recorded 24 and 48 hours after mixing by mean of a controlled-strain rheometer (ARES 100 FRT, Rheometric Scientific, Piscataway, NJ). The strain amplitude was set to 1.5%, which was in the linear region of the stress *versus* strain response. Grooved parallel plates (40 mm diameter) were used and the gap was set to 1 mm. Samples were poured into the measuring geometry at 24 ± 1 °C and submitted to frequency sweeps over the range 10^{-2}–10 Hz). A film of paraffin oil was used to cover the sample to reduce evaporation.

Confocal Scanning Laser Microscopy

Bulk phase separation was followed in the supernatant and in the coacervate phase using confocal scanning laser microscopy (CSLM) in a 1 wt% mixture. In order to visualise β-lg using the fluorescence mode of the CSLM, the latter was labelled with fluorescein isothiocyanate (FITC) following the procedure described elsewhere.[21] A 100 mL β-lg dispersion at 5 wt% was incubated for 90 min at room temperature and at pH = 8.5 in presence of 125 μL of a 2 wt% FITC solution in DMSO. The system was then adjusted to pH = 7.0 and the dispersion was extensively dialysed (molecular cut-off 10^4 g mol^{-1}) against Milli-Q water to remove non-bound FITC. The resulting spectroscopic determination gave an FITC to protein molar ratio of 3.7%. This low labelling ratio was nevertheless sufficient to achieve efficient imaging of the phase separation due to the sensitivity of the equipment. In addition, mass spectroscopy showed that the binding was not covalent, but hydrophobic in nature. Determination of equilibrium surface tension and surface elastic modulus on labelled and non-labelled β-lg gave no significant differences. From these results, we expect that the labelled β-lg behaves similarly to the native protein. For imaging, we used a Carl-Zeiss LSM 510 system, equipped with an Ar–Ne laser emitting at 533 nm, coupled to an Axioplan 2 imaging confocal microscope (Carl-Zeiss,

Darmstadt, Germany). A volume of 100 μL of sample was dropped onto a glass slide sealed with an additional cover slide to avoid sample dehydration. The data were processed using the Zeiss LSM Image Browser software.

Diminishing Bubble Experiments

The diminishing bubble technique was used to follow the stability of a single air bubble immersed within a liquid drop composed of a 0.1 wt% dispersion of β-lg or β-lg+AG mixture at pH=4.2. The principle of the technique[22] is to blow an air bubble into a drop of dispersion containing the surface-active species. A microscopic film is formed at the top of the bubble floating at the surface of the drop. Because of the pressure difference between both sides of the film, the bubble spontaneously shrinks because of gas diffusion (Ostwald ripening). The bubble diameter and consequently the film diameter decrease with time. The air bubble diameter was in the range 400–600 μm. Bubbles were formed under the air–water interface with an age of 5 min to allow the formation of stable bubbles.[3] The teflon cell containing the bubble was thermostated at 24 ± 1 °C using a Peltier cooling device. The film was imaged from the top using a bright-field Polyvar microscope (Reichert-Jung, CH) and digitized with a Kappa CF 15/3 video camera (Gloor Instruments AG, Uster, CH). The images were then compiled using image analysis software developed in-house to plot the normalized radius of the bubble *versus* time. In addition, the variation of the contact angle θ between the film and the bulk was calculated from

$$2\theta = \arcsin(r/R - 2\rho g R^3/3\sigma r) - \arcsin(2\rho g R^3/3\sigma r). \tag{1}$$

The gas permeability K of the film was thereafter calculated from

$$K = \frac{P_{atm}/2\sigma(R_0^4 - R_t^4) + 8/9(R_0^3 - R_t^3)}{\int_0^t r^2 dt}, \tag{2}$$

where R_0 is the initial bubble radius ($t=0$), R_t is its radius at time t, r is the film radius, P_{atm} is atmospheric pressure, σ is the surface tension of the dispersion, and ρ is the density of the solution. Experiments were continued until the bubble size remained constant.

Interfacial Rheology

Dynamic surface tension σ was measured for β-lg, AG and β-lg+AG dispersions using a dynamic pendant-drop Tracker tensiometer (IT Concept, Longessaigne, France) as described earlier.[23] Basically, the principle is to form an axisymmetric air bubble at the tip of a needle of a syringe. A computer drives the plunger position of the syringe *via* a motor drive into a thermostated optical glass cuvette containing 7 mL of biopolymer dispersion. The image of the bubble is recorded with a CCD camera and digitized. The interfacial tension γ is calculated by analysing the profile of the bubble according to the Laplace equation. When an equilibrium surface tension was obtained, we determined the dynamic elastic modulus,

$$|E| = A(\Delta\gamma/\Delta A) = E' + iE'', \tag{3}$$

where $\eta_d = E''/\omega$, by fluctuating sinusoidally the area of the bubble at a frequency ω in the range 0.01–0.1 Hz and a relative amplitude of $\Delta A/A = 0.1$. Experiments were performed at $24 \pm 0.5\,°C$. The biopolymer concentration was 0.1 wt% and the surface area of the bubble was set to 16 mm^2 (volume $\approx$ 48 mm^3).

Foaming Properties

The foaming properties of 1 wt% β-lg and β-lg + AG mixtures were determined by the method of Guillerme and co-workers.[24] The principle is to foam a defined quantity of dispersion by gas sparging through a glass frit with porosity and gas flow controlled. The generated foam rises along a glass column where its volume is followed by image analysis using a CCD camera. The amount of liquid incorporated in the foam, and the extent of foam homogeneity, was followed by measuring the conductance in the cuvette containing the liquid at different heights in the column by means of electrodes.[25] In addition, a second CCD camera equipped with a large objective was used to image the variation of the air bubble size within the foam every 15 s at a height of around 10 cm (corresponding to half of the foam height). To perform these experiments, a modified version of the Foamscan apparatus (IT Concept, Longessaigne, France) was used. A square glass column replaced the standard cylindrical glass column in order to avoid optical artifacts. It was obvious that the effect of the glass surface on the drainage of the first layer of bubbles was not negligible, but it was anyway the same for all samples. In addition to bubble imaging, the top of the column was sealed and connected to a pressure transducer in order to follow foam destabilization.[26] The final pressure ΔP_∞ outside the foam when all the foam interface had disappeared could be measured, and therefore the interfacial area A_t could be calculated from

$$A_t = 3V(\Delta P_\infty - \Delta P_t)/2\sigma, \tag{4}$$

where V is the initial foam volume and σ is the equilibrium surface tension. The temperature within the cuvette and the column was set to $24 \pm 1\,°C$.

The foaming properties of the β-lg and the mixtures were measured by pouring 50 mL of the dispersion into the cuvette and sparging N$_2$ at 80 mL min^{-1}. (Pure AG was not able to form a self-supporting foam at 1 wt%.) This flowrate was found to be suitable to allow efficient foam formation before strong gravitational drainage occurred. The porosity of the glass frit used for testing these foaming properties allowed formation of air bubbles of diameter in the

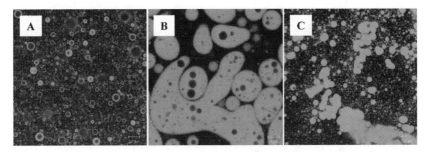

Figure 1 *Confocal micrographs of a mixture of 2.5 wt% FITC-β-lg and 1.25 wt% of AG at pH=4.2 and 20 °C. A, 10 s after mixing; B, bottom phase after 1.44 × 10⁴ s; C, bottom phase after 8.64 × 10⁴ s. Scales bars represent 20, 10 and 20 μm, respectively.*

range 10–16 μm. Bubbling was stopped after a foam volume of 180 cm³ was obtained. The extent of drainage, the change in total foam volume, the pressure above the foam, and the gas bubble size were followed with time.

3 Results and Discussion

Bulk Phase Separation

Figure 1 shows evidence of phase separation in the β-lg + AG mixture at pH = 4.2. Just 90 s after mixing, some liquid droplets have already formed (see Figure 1A). These structures are the so-called coacervates[15] made mainly of biopolymers with a low amount of solvent. Coacervates originate from aggregation of soluble β-lg–AG complexes that progressively lose solubility upon reaching charge neutrality. When a critical size is reached, macroscopic liquid–liquid phase separation occurs.[19,27] With time the coacervates coalesce and lose their vacuoles (Figure 1B). This phenomenon can be explained in terms of an internal structural reorganization because of the different surface properties of the β-lg and the AG.[14,19] Such a specific structure has already been described[28] for mixture of charged surfactants. After one day, the bottom phase had become very rich in the biopolymers, since the fluorescence intensity—the overall 'whiteness' of the coacervates—had increased compared to the initial mixture (Figure 1C). It is interesting to note that the higher fluorescence intensity also gives an indication of a local increase of viscosity in the bottom phase. In addition, Figure 1C clearly shows the loss of vacuoles within the coacervates, contributing to increased biopolymer concentration.

The most salient feature of the phase separation phenomenon is its high speed of demixing. Regarding the composition of the supernatant after mixing of β-lg and AG at 1 wt%, almost 80% of the biopolymers have reacted and begun to sediment after 1 minute (Figure 2). This fast reaction time was already described[29] using diffusive wave spectroscopy, and it can be explained by the fact that the reaction conditions chosen for this study correspond exactly to total charge neutralization of the two biopolymers. Upon ageing,

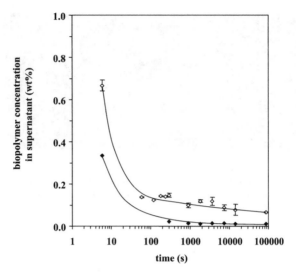

Figure 2 *Time-evolution of protein and polysaccharide compositions in the supernatant after mixing β-lg and AG at 1 wt%, pH = 4.2, ratio 2:1, and 24 °C: ◇, β-lg; ◆, AG.*

the composition of the supernatant shows a decrease in biopolymer concentration. After 24 hours, the β-lg and AG remaining in the supernatant could be in the form of the free biopolymers or soluble complexes, as a result of incomplete electrostatic neutralization.

The small-deformation rheological characterization of the coacervate phase reveals interesting features of the behaviour of the concentrated phase upon ageing (Figure 3). Until 24 hours, the loss modulus G'' is higher than the storage modulus G'. In addition, these two moduli remain parallel over the frequency range tested, showing a gradual increase with increasing frequency. The coacervate thus typically exhibits the mechanical spectrum of a dilute biopolymer solution,[30,31] *i.e.*, liquid-like behaviour for all frequencies tested (Figure 3A). After 48 hours of ageing, the coacervate shows a crossover of G' and G'' (Figure 3B), revealing that the behaviour shifts from liquid-like to gel-like $(G' > G'')$.[31] Hence, the mechanical spectrum is now comparable to that of a concentrated biopolymer solution, with liquid-like behaviour at low frequencies and a solid like-behaviour at high frequencies. The complex viscosity of the coacervate increases by a factor of 100 between the two experiments. Such a viscosity increase has already been reported[32] for gelatin/carboxymethylcellulose coacervation. Obviously, this important increase in viscosity is due to biopolymer concentration arising from water release from the coacervate phase, as a consequence of the tendency for the overall entropy of the system to increase upon maturation.[18]

Summarizing the results obtained on the bulk coacervation involving β-lg and AG, we have found that phase separation leads to a fast concentration of the biopolymers into a concentrated phase. This latter phase results from

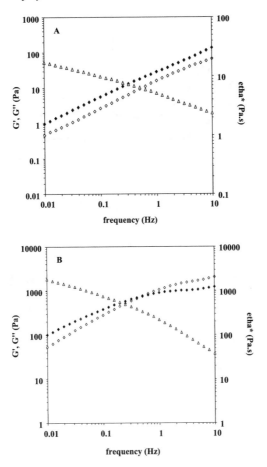

Figure 3 *Frequency spectra obtained by small-deformation oscillation for the storage modulus G′ (◇), the loss modulus G″ (◆) and the complex viscosity η* (Δ) of the bottom phase of a 5 wt% mixture of β-lg + AG at ratio 2:1, pH=4.2 and 24 °C: A, after 24 hours; B, after 48 hours.*

coalescence of liquid coacervate droplets. Interestingly, the coacervate phase is characterized by a maturation process leading to viscous-like properties.

Interfacial Properties

The surface tensions of the β-lg, AG and β-lg + AG dispersions were measured as a function of time until equilibrium was reached (Figure 4A). The β-lg and β-lg + AG dispersions showed similar adsorption kinetics with a rapid decrease of the surface tension within 10^3 s to reach around 50 mN m^{-1}. By contrast, acacia gum alone was far less efficient in decreasing the surface tension on the same timescale (inset to Figure 4A). Equilibrium surface tensions obtained for β-lg and β-lg + AG were found to be similar (45 mN m^{-1}), but the equilibrium tension for AG only reached 53 mN m^{-1}. These differences in behaviour can be

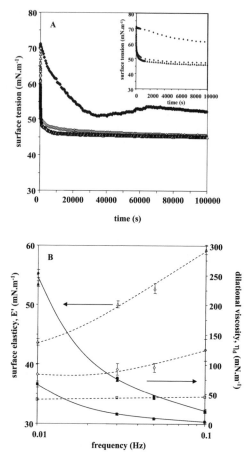

Figure 4 *Surface tension and viscoelastic moduli for 0.1 wt% β-lg, AG and β-lg + AG dispersions at pH = 4.2, ratio 2:1, and 20 °C. (A) Time evolution of surface tension (inset shows initial decrease):* ◇, *β-lg;* ◆, *AG;* △, *β-lg + AG. (B) Frequency spectrum for surface elasticity E' (dashed line) and dilational viscosity* η_d *(full line):* ◇, *E', β-lg;* □, *E', AG;* △, *E', β-lg + AG;* ◆, η_d *β-lg;* ■, η_d *AG;* ▲, η_d *β-lg + AG.*

due to the differences in hydrodynamic radius for β-lg (R_h = 2.5 nm) and AG (R_h = 14.2 nm), which in turn lowers the diffusion coefficient of the latter by a factor of almost six.[17,33] In addition, AG is a complex polysaccharide composed of 3 different fractions exhibiting various surface properties. The most surface-active fraction (with the smallest R_h) only represents 2–4% of total gum, whereas the second most surface-active (with the highest R_h) represents 80% of total gum.[34,35] This product heterogeneity would seem to fit in with the 'unusual' adsorption curve of AG which exhibits three different regions. In addition, the multiple-step behaviour points to structural modifications within the interface. It is worth noting that the characteristic adsorption profile obtained for AG alone could not be detected in its mixture with β-lg. As coacervation may lead to the formation of complexes having a larger size than the protein alone, this result is questionable, since one would have expected a different surface tension profile than for the β-lg alone. A putative explanation of this fact is that non-bound β-lg goes first to the interface and reacts with AG, or that the initially formed complexes are as surface-active as the β-lg (with a preferential conformation of the protein?). To resolve this question would require more investigation, involving, for example, variation of the biopolymer concentration or fractionation of the gum.

On measurement of the dilational viscoelastic modulus E obtained at equilibrium, β-lg and AG gave similar profiles, reaching steady-state values of ~ 40 mN m^{-1} over the frequency range tested (data not shown). In contrast, the β-lg + AG mixture was characterized by a significantly higher modulus ($E = 60$ mN m^{-1}) indicating the higher viscoelasticity of the interface.[36] Dickinson and Galazka[37] reported a similar increase in surface shear viscosity for the case of electrostatic complexes formed between bovine serum albumin and dextran sulfate at pH = 7.0. Our data for the elastic modulus E' and dilational viscosity η_d are plotted in Figure 4B. As far as surface viscosity is concerned, the curves obtained for β-lg alone or β-lg + AG appear superimposable, *i.e.*, there is a decrease in η_d with frequency increase. A similar trend was also obtained for AG, but with lower values. This frequency dependence of η_d can be explained in terms of the rate of molecular exchange from the interface to the bulk.[36] That is, once a critical frequency has been reached, the molecules do not have time to leave the interface anymore. Considering the lower diffusion coefficient for AG as compared to β-lg, the low-frequency dependence of η_d can be explained. However, it seems that the interface is mainly composed of β-lg that is able to leave the interface in the β-lg + AG mixture. Nevertheless, the surface elasticity is significantly higher for the mixture, supporting the assumption that a two-dimensional gel-like structure is formed, giving cohesiveness to the interface, which is then more difficult to deform.

To obtain insight into the structure of the interfacial film formed, some pictures of the diminishing bubble experiments were taken (see Figure 5). A few seconds after the bubble was formed in the β-lg dispersion, a black film could be observed (Figure 5A). Based on this observation, we infer that the interfacial film is mainly composed of protein molecules and some water molecules.[38] The presence of a white ring formed around the film—the interference fringes—shows that the film still contains some water. After a longer time, the film thickness and area were both reduced (Figure 5B). The interference fringes now have disappeared, but some white spots are visible within the film. They might be due to the presence of protein aggregates, as already observed[39] for β-lg films close to isoelectric pH.

Observations of the film structure obtained for the β-lg + AG mixture (Figures 5C and 5D) reveal a different pattern compared to that for β-lg alone. The film is initially very heterogeneous, with a dark background and some white spots 'floating' in it. It is worth mentioning here that the white spots appear iridescent in colour pictures, indicating a greater thickness than the

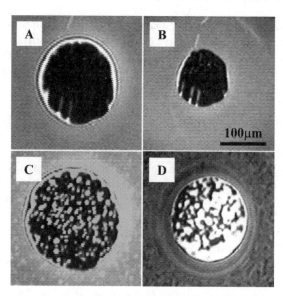

Figure 5 *Micrographs of the thin film formed at the top of an air bubble immersed either in 0.1 wt% β-lg or 0.1 wt% β-lg + AG at pH = 4.2, ratio 2:1, and 25 °C. A, β-lg film after 25 s; B, β-lg film after 1.28 × 10⁴ s; D, β-lg + AG film after 32 s; D, β-lg + AG film after 1.25 × 10⁴ s.*

dark region (Figure 5C). The thin film also has a non-circular structure, demonstrating its intrinsic viscoelasticity. Upon ageing, the film decreases in area because of gas loss from the bubble (Figure 5D). Interestingly, the black/white area ratio decreases to the benefit of the white regions, *i.e.*, to the thick and viscous regions. The likely explanation of the structural change is that, upon drainage of the film, the surface-active β-lg–AG complexes are concentrated at the interface. This allows further interaction, as already shown in the bulk phase separation experiments, leading to a more continuous interfacial film. Several authors have reported similar structures[40,41] in thin films stabilized by complexes of cationic surfactants and anionic polyelectrolytes. Recently, Monteux and co-workers have described[42] the formation of interfacial microgels within the dodecyltrimethylammonium bromide + polystyrene sulfonate system based on similar structural observations and ellipsometry measurements.

Bubble and Foam Stability

Figure 6 presents the variation of the normalized bubble ratio as a function of time. Use of normalized values allows direct investigation of the diffusive nature of the gas loss from the single bubble as already described.[43,44] The most salient feature when comparing bubbles stabilized by β-lg and β-lg + AG is the presence of an initial lag time in the latter case. This reveals that little gas diffuses out of the bubble for around 4 × 10³ seconds. In contrast, the bubble

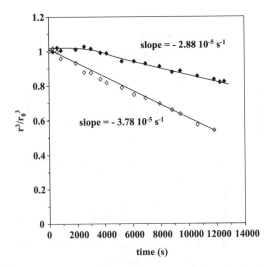

Figure 6 *Time evolution of the normalized size of a single air bubble, r/r_0, stabilized by 0.1 wt% β-lg or 0.1 wt% β-lg+AG at pH=4.2, ratio 2:1, and 25 °C: ◇, β-lg; ◆, β-lg+AG.*

stabilized by protein alone exhibited a linear decrease of bubble radius from the beginning of the experiment. In addition, after gas diffusion had started from the β-lg–AG-stabilized bubble, the rate of gas loss was still slower than for the β-lg-stabilized bubble, proving that interfacial permeability and structure were definitely different. Dickinson and co-workers[3] have described a similar decrease of bubble diameter with time for 0.05 wt% protein solutions of whey protein isolate, β-lg, casein or gelatin at pH = 7.0. However, their bubbles had an initial diameter of 50–150 μm, and they never could be fully stabilized against disproportionation. In addition, the same authors[3] did not find substantial differences in the rate of disproportionation depending on the protein used. In our case, the relative rate of gas flow was around 30% greater for the protein alone. Taking into account the relative effects of the interfacial tension and the interfacial elasticity, these results should not be surprising, since $\sigma/E' = 1.39$ for the β-lg+AG mixture as against 0.93 for β-lg alone.[2] In addition to the previous results, these data support the hypothesis of the formation of an interfacial film having strong viscoelastic properties able to counteract gas diffusion. From the plot of the variation of the film contact angle with time, the magnitude of film permeability could be approximated. It was around 0.521 m s⁻¹ for β-lg alone as against 0.021 cm s⁻¹ for β-lg+AG. The former value is very high and suggests a very coarse packing of the β-lg molecules in the interfacial film. By contrast, the lower value obtained for the β-lg+AG mixture suggests a dense packing of the molecules, as was recently described for β-lg alone at its isoelectric point.[39]

The stability of foams obtained by sparging nitrogen through β-lg and β-lg+AG dispersions was studied. The volume stability of the two foams exhibited different profiles (Figure 7). Interestingly, the time required to form

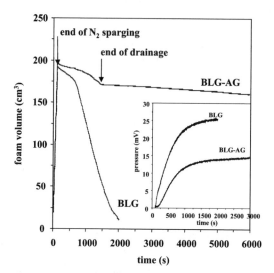

Figure 7 *Foam volume stability obtained after N_2 sparging in a 1 wt% dispersion of β-lg or β-lg + AG at pH = 4.2, ratio 2:1, and 25 °C. Inset shows variation of pressure above the foams upon coarsening.*

the 180 cm³ foam volume was similar (100 s for β-lg and 106 s for β-lg + AG). The small difference could be attributed to the higher protein concentration in the former case, 1 wt%, as compared with 0.67 wt% in the latter. However, both dispersions were found to exhibit similar adsorption properties, as already shown from surface tension measurements. After bubbling had stopped, both foam volumes exhibited a two-step decrease. Keeping in mind that foam destabilization results from the superimposition of several mechanisms (drainage, coalescence, disproportionation), we tentatively attribute the first step to gravitational drainage within the foam and the second to subsequent gas loss from coarsening and bubble collapse.[44,45]

The rate of drainage of the β-lg-stabilized foam was similar to that of the β-lg + AG mixture (data not shown). This result supports the fact that the interfacial film viscosity was similar, thereby preventing to a similar extent any water motion within the thin films and Plateau borders.[1] However, it can be seen from Figure 7 that after the drainage stage, when there was a steady liquid conductance below the foam, the foam volume decreases drastically for pure β-lg, whereas it remains almost constant for β-lg + AG. The likely explanation is that the higher interfacial viscoelasticity and lower gas permeability of the bubbles in the β-lg + AG foam prevents gas loss, contributing to volume stability. This assumption is clearly supported by the measurements of the outside pressure in the column during foam coarsening (inset to Figure 7). The increase in pressure above the foam is due to gas release and is directly proportional to the loss of interfacial area.[26] The initial rate of gas loss is much higher for the β-lg foam compared to the β-lg + AG foam, which is in agreement with results obtained with the diminishing bubble experiments.

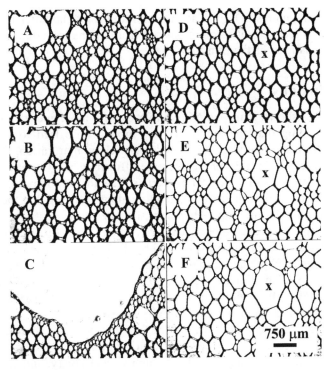

Figure 8 *Micrographs of gas bubbles obtained after drainage of 1 wt% β-lg or β-lg + AG foams obtained at pH = 4.2, ratio 2:1, and 25 °C. Pictures were taken at a height of 10 cm from the bottom of the foam: A, β-lg foam after 810 s; B, β-lg foam after 915 s; C, β-lg foam after 1110 s; D, β-lg + AG foam after 810 s; E, β-lg + AG foam after 3509 s; β-lg + AG foam after 5953 s.*

Assuming that the initial interfacial area, A_0, was almost equal for both foams (a reasonable assumption, as the final times of foaming were comparable), it turns out that the interfacial area A_t after 10^3 s in the β-lg foam was about half of that in the β-lg + AG foam.[13,26] This implies that the average air bubble size must be higher in the β-lg foam compared to that in the β-lg + AG one in the coarsening stages. In fact, the air bubble sizes were qualitatively not different at the foam height investigated in the β-lg or mixed biopolymer foams (Figures 8A and 8D). The β-lg foam initially has smaller bubbles that rapidly coarsen (Figure 8B), leading to a complete foam destabilization from the top after 2×10^3 s (Figure 8C). Hence, β-lg foam destabilization is likely to be due to a loss of gas from the top bubble layer upon contact with the air within the column (bubble collapse). A continuous 'creaming' of the big bubbles is visible, and this is the reason for the large change in the pressure measured above the foam. In contrast, the foam stabilized by β-lg + AG initially has larger bubbles compared to the β-lg foam, maybe because of its lower protein content. However, the bubbles remain very stable with time, exhibiting a very slow coarsening rate, but no creaming (see, for example, the evolution of the bubble labelled 'x' in Figures 8D–F). The greater stability of the latter foam is thus mainly due to reduced diffusion of gas out from the bubbles due to the higher interfacial elasticity of the thin film.

4 Conclusions

This study has shown that β-lg–AG complexes form a polymer-rich phase very rapidly providing that the initial pH and mixing ratio lead to a maximum charge neutralization between the biopolymers. Upon maturation, the biopolymer concentration in this phase increases, forming a 3D network, which exhibits highly viscous rheological properties. The adsorption properties of the β-lg–AG complex are similar to those of β-lg alone, proving that the protein is mainly covering the interface. However, the interfacial elastic modulus was 30% higher for the protein–polysaccharide complex than for the pure protein. The interface of a single bubble stabilized by the β-lg–AG complex is stabilized by a viscoelastic film that concentrates upon drainage leading to a low gas permeability. Similarly, foam stabilization properties of β-lg–AG complexes are considered to be mainly achieved through a reduced gas loss from the bubbles rather than to an increase in bulk viscosity.

Put all together, these data support the fact that β-lg–AG complexes form a 2D interfacial network upon complex coacervation, as is the case for bulk phase separation. The interface exhibits high viscoelasticity, and consequently low gas permeability, providing that the adsorption of the complexes is achieved soon after mixing of the two biopolymers in order to allow structure development in the interfacial film.

Several questions still remain, relating to the exact structure and thickness of the interfacial film, and the effect of biopolymer concentration on film/foam stability. The adsorption kinetics of the complexes is not fully understood. In addition, further experiments would be required to understand if specific AG fractions are preferentially involved in film stabilization.

Acknowledgements

The authors gratefully acknowledge Claudine Bovay, Tania Palma Da Silva, Philippe Frossard and Martin Beaulieu (Nestlé Research Center) for their technical assistance. Alain Cagna from the IT Concept company is thanked for the development of the imaging system for the Foamscan apparatus. Monica Fischer and Dinakar Panyam from the Nestlé Research Center are thanked for their critical reading of the manuscript.

References

1. H. A. Stone, S. A. Koehler, S. Hilgenfeldt, and M. Durand, *J. Phys. Cond. Mat.*, 2003, **15**, 283.
2. W. Kloek, T. van Vliet, and M. Meinders, *J. Colloid Interface Sci.*, 2001, **237**, 158.

3. E. Dickinson, R. Ettelaie, B. S. Murray, and Z. Du, *J. Colloid Interface Sci.*, 2002, **252**, 202.
4. D. T. Wasan, in 'Emulsions, Foams, and Thin Films', eds K. Mittal and P. Kumar, Marcel Dekker, New York, 2000, p. 1.
5. P. J. Wilde and D. C. Clark, in 'Methods of Testing Protein Functionality', ed. G. M. Hall, Chapman & Hall, London, 1996, p. 110.
6. P. J. Wilde, *Curr. Opin. Colloid Interface Sci.*, 2000, **5**, 176.
7. L. G. Phillips, D. M. Whitehead, and J. Kinsella, 'Structure–Function Properties of Food Proteins', Academic Press, San Diego, 1994.
8. C. Schmitt, C. Sanchez, S. Desobry-Banon, and J. Hardy, *CRC Crit. Rev. Food Sci. Nutr.*, 1998, **38**, 689.
9. E. Dickinson, *Food Hydrocoll.*, 2003, **17**, 25.
10. P. M. T. Hansen and D. H. Black, *J. Food Sci.*, 1972, **37**, 452.
11. M. Ahmed and E. Dickinson, *Food Hydrocoll.*, 1991, **4**, 395.
12. B. Mann and R. C. Malik, *J. Food Sci. Technol.*, 1996, **33**, 202.
13. E. Dickinson and E. Izgi, *Colloids Surf. A*, 1996, **113**, 191.
14. C. Sanchez, G. Mekhloufi, C. Schmitt, D. Renard, P Robert, C.-M. Lehr, A. Lamprecht, and J. Hardy, *Langmuir,* 2002, **18**, 10323.
15. H. G. Bungenberg de Jong, in 'Colloid Science', ed. H. G. Kruyt, Elsevier, Amsterdam, 1949, vol. 2, p. 232.
16. S. L. Turgeon, M. Beaulieu, C. Schmitt, and C. Sanchez, *Curr. Opin. Colloid Interface Sci.*, 2003, **8**, 401.
17. C. Schmitt, C. Sanchez, F. Thomas, and J. Hardy, *Food Hydrocoll.*, 1999, **13**, 483.
18. D. J. Burgess, *J. Colloid Interface Sci.*, 1990, **140**, 227.
19. F. Weinbreck, R. de Vries, P. Schrooyen, and C. G. de Kruif, *Biomacromolecules*, 2003, **4**, 293.
20. M. Dubois, K. A. Gilles, J. K. Hamilton, P. A. Rebers, and F. Smith, *Anal. Chem.*, 1956, **28**, 350.
21. A. Lamprecht, U. F. Schäfer, and C.-M. Lehr, *Eur. J. Pharm. Biopharm.*, 2000, **49**, 1.
22. D. Platikanov, M. Nedyalkov, and V. Nasteva, *J. Colloid Interface Sci.*, 1980, **75**, 620.
23. J. Benjamins, A. Cagna, and E. H. Lucassen-Reynders, *Colloids Surf. A*, 1996, **114**, 245.
24. C. Guillerme, W. Loisel, D. Bertrand, and Y. Popineau, *J. Texture Stud.*, 1993, **24**, 287.
25. A. Kato, A. Takahashi, N. Matsudomi, and K. Kobayashi, *J. Food Sci.*, 1983, **48**, 62.
26. M.-A. Yu and S. Damodaran, *J. Agric. Food Chem.*, 1991, **39**, 1555.
27. D. Leisner and T. Imae, *J. Phys. Chem. B*, 2003, **106**, 12170.
28. F. M. Menger, A. V. Peresypkin, K. L. Caran, and R. P. Apkarian, *Langmuir*, 2000, **16**, 9113.
29. C. Schmitt, C. Sanchez, A. Lamprecht, D. Renard, C.-M. Lehr, C. G. de Kruif, and J. Hardy, *Colloids Surf. B*, 2001, **20**, 267.
30. A. H. Clark and S. B. Ross-Murphy, *Adv. Polym. Sci.*, 1987, **83**, 57.
31. J. F. Steffe, 'Rheological Methods in Food Process Engineering', 2nd edn, Freeman, East Lansing, MI, 1996.
32. G. L. Koh and I. G. Tucker, *J. Pharm. Pharmacol.*, 1988, **40**, 233.
33. C. Sanchez, D. Renard, P. Robert, C. Schmitt, and J. Lefebvre, *Food Hydrocoll.*, 2002, **16**, 257.

34. S. Damodaran and L. Razumovsky, *Food Hydrocoll.*, 2003, **17**, 355.
35. R. C. Randall, G. O. Phillips, and P. A. Williams, *Food Hydrocoll.*, 1989, **3**, 65.
36. K.-D. Wantke and H. Fruhner, in 'Drops and Bubbles in Interfacial Research', eds D. Möbius and R. Miller, Elsevier, Amsterdam, 1998, p. 327.
37. E. Dickinson and V. B. Galazka, in 'Gums and Stabilisers for the Food Industry', eds G. O. Phillips, P. A. Williams, and D. J. Wedlock, IRL Press, Oxford, 1992, vol. 6, p. 351.
38. J. J. Benattar, F. Millet, M. Nedyalkov, and D. Sentenac, in 'Emulsions, Foams, and Thin Films', eds K. Mittal and P. Kumar, Marcel Dekker, New York, 2000, p. 251.
39. V. Petkova, C. Sultanem, M. Nedyalkov, J. J. Benattar, M. E. Leser, and C. Schmitt, *Langmuir*, 2003, **19**, 6942.
40. V. Bergeron, D. Langevin, and A. Asnacios, *Langmuir*, 1996, **12**, 1550.
41. V. Bergeron, J. Hanssen, and F. Shoghl, *Colloids Surf. A*, 1997, **123–124**, 609.
42. C. Monteux, C. Williams, J. Meunier, O. Anthony, and V. Bergeron, *Langmuir*, 2004, **20**, 57.
43. I. M. Lifshitz and V. V. Slyozov, *J. Phys. Chem. Solids*, 1961, **19**, 35.
44. D. Weaire and S. Hutzler, 'The Physics of Foams', Clarendon Press, Oxford, 1999.
45. A. Saint-Jalmes and D. Langevin, *J. Phys. Cond. Mat.*, 2002, **14**, 9397.

Interactions between β-Lactoglobulin and Polysaccharides at the Air–Water Interface and the Influence on Foam Properties

By Rosa I. Baeza, Cecilio Carrera Sánchez,[1] Juan M. Rodríguez Patino,[1] and Ana M. R. Pilosof

DEPARTAMENTO DE INDUSTRIAS, FACULTAD DE CIENCIAS EXACTAS Y NATURALES, UNIVERSIDAD DE BUENOS AIRES, 1428 BUENOS AIRES, ARGENTINA
[1]DEPARTAMENTO DE INGENIERÍA QUÍMICA, FACULTAD DE QUÍMICA, UNIVERSIDAD DE SEVILLA, PROF. GARCÍA GONZÁLEZ S/NÚM, 41012 SEVILLE, SPAIN

1 Introduction

Proteins and polysaccharides are present in many kinds of foamed foods. The main functional role of proteins is to stabilize the air–water interface through their capacity to lower the surface tension of water.[1] Most high-molecular-weight polysaccharides, being hydrophilic, do not have much of a tendency to adsorb at the air–water interface, but they can strongly enhance the stability of protein foams by acting as thickening or gelling agents. Alongside the use of non-surface-active polysaccharides in food foams as thickeners, there is some evidence to support the concept of an additional role for some polysaccharides in the interfacial film.[2,3]

Some of the following questions are expected to be addressed in the present work. (1) Does the hydrophilic character of polysaccharides like xanthan or λ-carrageenan mean that they do not adsorb at air–water interfaces in protein + polysaccharide mixtures? (2) What are the mechanisms by which polysaccharides can influence the interfacial properties of proteins? (3) Is a surface-active polysaccharide more effective than a non-adsorbing polysaccharide in lowering the surface tension in a protein + polysaccharide mixture? (4) What is the relative impact of bulk and interfacial properties on foam stability?

In order to answer the questions above, it seems appropriate to study protein+polysaccharide mixtures at pH values above the protein isoelectric point, where complex formation as a result of net electrostatic interactions is

301

absent. Above the protein isoelectric point, thermodynamic incompatibility between the protein and polysaccharide generally occurs because of repulsive electrostatic interactions and different affinities of the two biopolymers for the solvent.[4] Therefore the protein and polysaccharide species may co-exist in a single phase (miscibility) in domains in which they mutually exclude one another, or, above a critical concentration, they may segregate into different phases. Excluded volume effects can have the following manifestations: enhancement of the association of macromolecules, reduction in the critical concentration for gelation, increase in the rate of gelation,[5,6] and enhancement of protein adsorption at fluid interfaces.[7]

In this work we study the interfacial behaviour of mixed β-lactoglobulin (β-lg) + polysaccharide (PS) systems at the air–water interface and their influence on foam properties at neutral pH. The structural and dynamic properties of β-lg at the air–water interface have been extensively studied in the past years,[8–10] and so this protein is a good model to study interactions with either non-adsorbing or surface-active polysaccharides. We here consider one sort of polysaccharide with interfacial activity, propylene glycol alginate (PGA), and two without interfacial activity, λ-carrageenan and xanthan gum. And to evaluate the effect of the degree of esterification and the viscosity on the PGA surface properties, three different commercial samples were examined.

Adsorption of pure xanthan or λ-carrageenan at the air–water interface seems unlikely. Xanthan does not cause a decrease in surface tension.[11,12] But xanthan does promote soy protein sub-unit aggregation at the air–water interface in native soy proteins foams.[2] Specific effects further influence the foam stability of the mixed systems.[11]

The propylene glycol esters of alginic acid (PGA) are high-molecular-weight linear polysaccharides composed of 1,4 linked-D-mannuronic acid and L-guluronic acid.[13] They are produced commercially with a range of viscosities and degrees of esterification. Increasing the degree of esterification reduces the overall hydrophilic character of the molecules and imparts surface-active character. PGA is used in beer to aid the 'head' retention of the foam.[14] Due to its surface-active character, competitive adsorption of PGA could occur in mixtures of this polysaccharide with proteins. Whereas numerous studies have been published on competitive adsorption of milk proteins and low-molecular-weight surface-active agents such as monoglycerides,[15,16] surfactants,[17,18] and lecithins,[19] little research has been done on competitive adsorption of proteins and polysaccharides. In addition, there is the possibility that the formation of protein–PGA complexes at the interface could also occur.[20]

2 Experimental

The β-lactoglobulin (β-lg) was supplied by Danisco Ingredients (Denmark). The powder composition was: protein 92%, β-lg > 95%, α-lactalbumin < 5%. The polysaccharides λ-carrageenan (λC) and xanthan gum (X) were provided by Biotec (Argentina) and the samples of propylene glycol alginate (PGA) were from ISP Alginates. The PGA samples used were Kelcoloid O

Table 1 *Characteristics of propylene glycol alginate samples.*

PGA Sample	Degree of esterification	Viscosity[a]	Surface tension[b] (mN m⁻¹)
Manucol ester (MAN)	High	High (11.8 cps)	62.5
Kelcoloid LVF (KLVF)	Medium	High (13.9 cps)	63.5
Kelcoloid O (KO)	High	Low (4.7 cps)	61.4

[a] Apparent viscosity (60 s⁻¹) of 0.5% w/w solution.
[b] Steady-state surface tension of 0.1% w/w solution measured by the drop tensiometer method.

(KO), Kelcoloid LVF (KLVF), and Manucol ester (MAN). Their degrees of esterification, viscosities and surface tensions are shown in Table 1.

Time-dependent surface pressure and surface dilatational properties of β-lg, polysaccharide (PS) and β-lg + PS mixed films at the air–water interface were determined in an automatic drop tensiometer IT Concept at 20 ± 0.2 °C as described elsewhere.[21,22] The powder samples were dissolved in Milli-Q ultrapure water at room temperature and the pH was adjusted to pH = 7 with Trizma buffer. The bulk concentrations of the PS and protein solutions were 0.5 wt% and 2 wt%, respectively.

The measurements of the surface pressure π *versus* average area per molecule *A* at 20 °C and the surface dilatational properties of the spread films were performed on a fully automated Wilhelmy-type film balance (KSV 3000, Finland) as described elsewhere.[23] To form the protein surface film, the β-lg was spread in the form of a solution using Milli-Q ultrapure water at pH 7. The sub-phase solutions consisted of 0.05 wt% polysaccharide solutions prepared with Milli-Q water. Aliquots of 400 μl of aqueous solutions of β-lg (1.6×10^{-4} mg μL⁻¹) at pH 7 were spread on the interface.[23] The elasticity, a measure of film resistance to a change in area, is calculated from the slope of the π–*A* isotherm, *i.e.*, $E = -A(\partial \pi / \partial A)_T$.

A volume of 50 mL of 2 wt% β-lg solution or 2 wt% β-lg + 0.5 wt% PS solution was foamed in a graduated cylinder (3 cm diameter) at 20 °C for 3 min with a Griffin and George stirrer operating at 6×10^3 rpm. Kinetics of drainage and disproportionation were determining by a rheological method[24] in a specially designed glass tube with a T-spindle attached to a Brookfield DV-LVT viscometer with helipath stand. Simultaneously, the liquid drained at the bottom of the tube was recorded over time. This method also allowed measurement of foam height decay (collapse) in the graduated tube. The following mathematical model[25] was applied to describe the foam apparent viscosity as a function of time:

$$\mu_{app}(t) = a[\exp(-K_2 t^{0.5}) - \exp(-K_1 t)]. \qquad (1)$$

In equation (1), $\mu_{app}(t)$ is the apparent viscosity of foam at time *t*, *a* is a constant related to the maximum viscosity, and K_1 and K_2 are the rate constants for drainage and disproportionation, respectively. The expression for liquid drainage from the foam is[26]

$$v(t) = V_{\text{Lmax}} t^n / (c + t^n) , \tag{2}$$

where $v(t)$ is the volume of drained liquid at time t, V_{Lmax} is the maximum volume of drained liquid, n is a constant related to the sigmoidal shape of the drainage curve, and c is a parameter related to the drainage half-time as $c^{1/n}$. The rate constant of drainage was calculated as $K_{\text{dr}} = n/Vc^{1/n}$. The time when the foam height started to decrease (the lag time) was taken as the characteristic parameter for describing foam collapse.

The bulk viscosity and the yield stress of the solutions were measured at 20 °C with a cone-and-plate geometry using a DV-LVT Brookfield viscometer.

3 Results and Discussion

Kinetics of Adsorption

The surface pressure evolution over time for the set of β-lg (2 wt%) + PS (0.5 wt%) mixed films adsorbed at the air–water interface is plotted in Figure 1. Except for MAN, the most highly esterified PGA sample, the presence of the PS in the bulk phase was found to lead to an increase of surface pressure when compared to the protein alone. Amongst the PGA samples, KLVF promoted the highest surface pressure increase, but on its own it showed the least surface activity (see Table 1). In the presence of non-surface-active PS, there was observed to be a small increase in surface pressure at low adsorption times (up to approximately 10^3 s). At higher adsorption times, however, a marked increase of π in the mixed systems could be observed, reaching in the presence of xanthan a surface pressure (28 mN m^{-1}) even higher than in the presence of the most surface-active PGA sample.

The results indicate that the interfacial behaviour of the mixed films is mainly determined by the protein, since the shape of the π–time curves were similar to that for the protein alone. As the protein is more surface-active

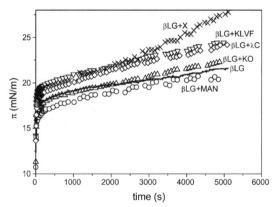

Figure 1 *The transient surface pressure π for β-lg (2%) + surface-active polysaccharides (0.5%) adsorbed at the air–water interface: —, pure β-lg, (○) β-lg + Manucol, (▽) β-lg + Kelcoloid LVF, (△) β-lg + Kelcoloid O, (◊) β-lg + λ-C, and (✕) β-lg + xanthan.*

than the polysaccharides, and as 2 wt% concentration in the bulk phase allows film saturation,[10] the protein is assumed rapidly to form a primary surface monolayer. However, even though the protein is responsible for the high surface pressure observed at short adsorption times, the PS strongly affects the long-term adsorption.

The existence of competitive or cooperative adsorption between β-lg and PS can be deduced by comparing the π–time curves for the single biopolymers and for the mixtures (Figures 2–5). As xanthan did not increase the surface pressure (Figure 2), the significant increase in π of the mixed system indicated a strong synergism between those biopolymers. The existence at pH = 7 of limited thermodynamic incompatibility between protein and polysaccharide[6,27] may increase the amount of adsorbed protein.[3,28] Similarly, λC appears to show a synergistic behaviour with the protein up to approximately 2×10^3 s (Figure 3). During this time the λC produced on its own a non-significant increase in π so that the existence of a limited thermodynamic incompatibility between protein and polysaccharide may account for the observed surface

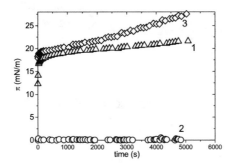

Figure 2 *The transient surface pressure π for (1) pure β-lg (2%), (2) pure xanthan (0.5%) and (3) β-lg (2%) + xanthan (0.5%).*

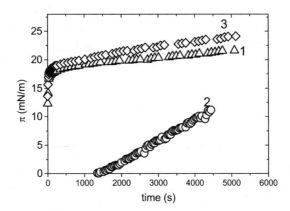

Figure 3 *The transient surface pressure π for (1) pure β-lg (2%), (2) pure λ-carrageenan (0.5%) and (3) β-lg (2%) + λ-carrageenan system (0.5%).*

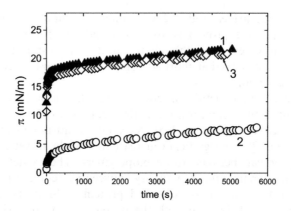

Figure 4 *The transient surface pressure π for (1) pure β-lg (2%), (2) pure Manucol (0.5%) and (3) β-lg (2%) + Manucol (0.5%).*

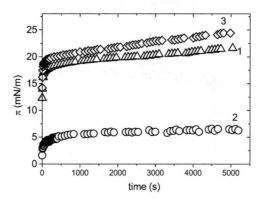

Figure 5 *The transient surface pressure π for (1) pure β-lg (2%), (2) pure Kelcoloid LVF (0.5%) and (3) β-lg (2%) + Kelcoloid LVF (0.5%).*

pressure increase of the mixed system. Nevertheless, during the long-term adsorption the interaction appears to turn to antagonistic, as the surface pressure of the pure λC film started to increase. Possibly, a surface-active contaminant of the λC preparation, which was not removed from the aqueous solution by suction, could displace some of the protein from the interface. Displacement of β-lg from the air–water interface has been reported to occur in the presence of low-molecular-weight surfactants.[16–18]

The π values of the β-lg + KO film are close to that of the protein alone (Figure 1). Therefore, as KO alone shows π ≈ 10 mN m^{-1} after 5×10^3 s, it can be concluded that an antagonistic effect takes place between these biopolymers. This effect was more marked for the Manucol ester sample (Figure 4). This antagonism reflects the competition between the protein and KO or MAN for the interface. On the contrary, the surface pressure of the β-lg + KLVF film shows an almost additive behaviour (Figure 5), reflecting some degree of cooperation between these biopolymers at the air–water interface.

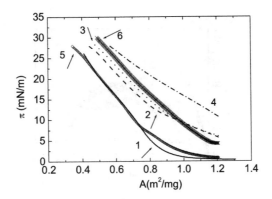

Figure 6 *Pseudo-steady-state π–A isotherms of β-lg monolayers spread on aqueous PS solutions at 20 °C and pH = 7: 1 (——), β-lg spread on pure water; 2 (- - -) with Kelcoloid O; 3 (······) with Kelcoloid KLF; 4 (-··-), with Manucol ester; 5 (◊), with λ-carrageenan; 6 (XXX) with xanthan.*

How Does Polysaccharide Affect Protein Monolayer Structure?

The β-lg monolayer structural characteristics can be obtained through the π–A isotherm for β-lg spread on the aqueous sub-phase at pH = 7 (Figure 6). The protein monolayer has a liquid-expanded-like structure,[10] and at the critical surface pressure of 12 mN m⁻¹ the transition between a monolayer with a more expanded structure (state 1) towards a monolayer in which the protein forms a more condensed structure (state 2) takes place. Nevertheless, according to Graham and Phillips,[29] β-lg retains elements of the native structure not fully unfolded at the interface, even in the more expanded configuration.[10] On adding PS to the sub-phase, the π–A isotherms were found to shift to higher surface pressure values as the time increased, due to the adsorption and penetration of the polysaccharides into the protein monolayer. The pseudo-steady state was reached after 18–22 hours.[23] The π–A isotherms for β-lg + PGA and β-lg + X (Figure 6) are separated from the surface pressure axis, apparently indicating a less condensed monolayer structure because in the molecular area estimation we assume that only β-lg molecules are spread on the interface. Amongst the PGA samples studied, MAN shows the highest surface activity as it induces the highest π increase. In spite of the lack of surface activity on its own, xanthan shows behaviour similar to PGA, and at low molecular areas it induces a higher surface pressure than KLVF and KO. The λC slightly increases the surface pressure at low molecular areas.

By plotting the elasticity *versus* surface pressure for pure β-lg and the mixed systems (Figure 7), more clear evidence of the changes in monolayer structure may be obtained. The elasticity of the pure β-lg monolayer increases with surface pressure, reaches a maximum as the change between states 1 and 2 takes place, and then decreases at the higher surface pressures, as the monolayer collapses and the formation of multilayers takes place.[10] The influence of PS prevails at low surface pressures (< 12 mN m⁻¹) as β-lg adopts the less

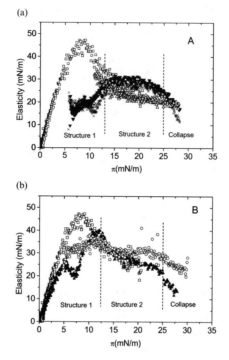

Figure 7 *Surface pressure dependence of the elasticity at 20 °C and pH =7 of β-lg mono-layers spread on water (□) and after the addition of polysaccharides to the aqueous phase: (A) Surface-active PS: ×, Kelcoloid O; ▼, Kelcoloid LVF; △, Manucol. (B) Non-surface-active PS:▲, λ-carrageenan; o, xanthan.*

condensed state (state 1) with the elasticity of the film reduced by up to 50% by the presence of PGA (Figure 7A). This effect may be due to the penetration of PGA into the protein film which hinders the interactions within the protein amino-acid residues at the air–water interface. The non-surface-active X and λC also reduce the elasticity of the pure β-lg monolayer (Figure 7B), but to a lower extent than PGA. This effect may be due to the interactions between protein and polysaccharide that affect the original structure of the protein monolayer. In the region where the protein adopts state 2, smaller differences were observed among the systems, the interfacial elasticity varying within the range 20–30 mN m^{-1}. Above a surface pressure of 25 mN m^{-1}, in the monolayer collapse region, there was found to be a tendency for the elasticity to decrease no matter what kind of polysaccharide was present.

Surface Dilatational Elasticity

The values of the surface dilatational modulus (data not shown) were found to be very similar to those of the dilatational elasticity E_d. The increase of E_d of the β-lg monolayer with adsorption time (Figure 8) is associated with protein adsorption at the interface,[10,29,30] and it reveals the increasing solid-like character

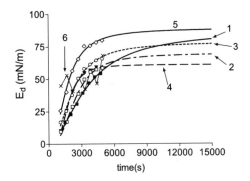

Figure 8 *Time-dependence of the surface dilatational modulus E_d at 20 °C and pH=7 for adsorbed monolayers formed from pure β-lg (2%) and β-lg (2%) +PS (0.5%): 1 (■), pure β-lg; 2 (□), Kelcoloid O; 3 (○), Kelcoloid KLF; 4 (▼), Manucol; 5 (◊), λ-carrageenan; 6 (×) xanthan.*

of the film. The evolution of the dilatational elasticity of the β-lg+PGA and β-lg+λC films shows a similar trend (Figure 8). The dilatational elasticity of the β-lg+xanthan film, however, attained a high value on starting the sinusoidal oscillations (after 10^3 s), showing an oscillating pattern around 45 mN m⁻¹, and increasing up to 70 mN m⁻¹. Over the time of the experimental measurements (5×10^3 s), the dilatational elasticity of the mixed films was higher than that of the pure protein film.

Because of the extensive molecular rearrangement of protein at the interface, the dilatational elasticity did not reach an equilibrium value during the time of measurement. In order to estimate the equilibrium E_d values for the films, the evolution of the dilatational elasticity with time in Figure 8 was fitted by the following empirical equation:

$$E_d(t) = (E_{d\infty}t^n)/(c+t^n). \tag{3}$$

Here, $E_{d\infty}$ is the maximum E_d value, n describes the sigmoidal shape of the curve, and c is related to the time needed to reach $E_{d\infty}/2$ (*i.e.*, $c^{1/n}=t_{1/2}$). Equation (3) fits the experimental data very well (with R^2 lying in the range 0.990–0.998). The data in Figure 8 indicate that, except for λC, the $E_{d\infty}$ values of the mixed films are lower than for the protein alone.

The above behaviour was corroborated by evaluation of the surface dilatational elasticity of β-lg spread monolayers in the presence of PS. The dilatational elasticity of the pure β-lg film was found to increase with surface pressure up to a maximum close to the surface pressure at which the monolayer collapses (Figure 9). The presence of λC in the aqueous sub-phase slightly modifies the dilatational elasticity of the monolayer. A small decrease in E_d was observed at low surface pressure and at the collapse point. The PGA was found to reduce E_d considerably at all surface pressures. Manucol ester is the PGA sample that most reduces the solid character of the film, especially at the highest surface pressures. Xanthan gum was found to give a strong effect,

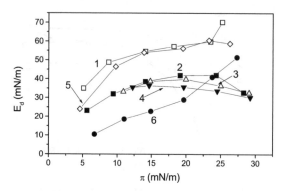

Figure 9 *Surface pressure dependence of the surface dilatational modulus E_d at 20 °C and pH=7 of β-lg monolayers spread on water before and after PS addition to the aqueous phase: 1 (□), pure β-lg; 2 (■), Kelcoloid O; 3 (△), Kelcoloid KLF; 4 (▼), Manucol; 5 (◊), λ-carrageenan; 6 (●), xanthan.*

producing the largest reduction in E_d, mainly at surface concentrations below 20 mN m⁻¹.

Mechanisms of β-Lactoglobulin–Polysaccharide Interaction at the Interface

The results obtained from these studies reveal a significant effect of polysaccharides, even the non-surface-active xanthan and λC, on the properties of the β-lg film. The observed behaviour is related to the protein–polysaccharide interactions in the bulk and at the interface, and to the relative concentrations of these biopolymers. When β-lg adsorbs at the air–water interface in the presence of PS, three phenomena can occur (Figure 10): (i) the PS may adsorb at the interface on its own in competition with the protein (competitive adsorption); (ii) the PS may complex with the adsorbed protein, mainly by electrostatic interactions or hydrogen bonding;[3] or (iii) because of the existence of limited thermodynamic compatibility between the biopolymers, the PS may concentrate the adsorbed protein.

Anchorage of the polysaccharide to the interfacial film may occur by mechanisms (i) or (ii), depending on the chemical structure of the polysaccharide and on the pH. Once the PS is at the interface, or attached to it by complexation, excluded volume effects between the unlike biopolymers at neutral pH could lead to a rise of chemical potential, or, in other words, to a modification of the thermodynamic activity of the protein at the interface.[31] Therefore, the protein at the air–water interface would tend to behave as a more concentrated film, leading to an increase in the surface pressure. It has been demonstrated[2] that xanthan addition to soy protein solutions at neutral pH can have an effect similar to that observed here in increasing the protein concentration due to excluded volume effects. Evidence of the phase separation of macromolecules in mixed films at the air–water interface has been recently given.[32]

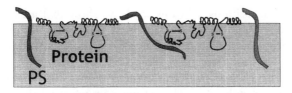

(i)Competitive adsorption

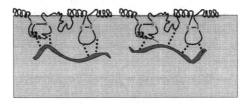

(ii)Complexation

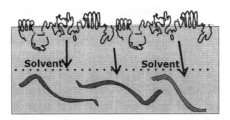

(Iii)Flow out of solvent due to
depletion in the vicinity of
the air-water interface

Figure 10 *Schematic representation of the three main types of phenomena that can occur during the adsorption of β-lg at the air–water interface in the presence of polysaccharide: (i) competition, (ii) complexation, and (iii) incompatibility.*

Even if the PS does not participate directly in the interfacial structure (*i.e.*, does not adsorb on its own, nor complex with adsorbed protein), the existence of a limited thermodynamic compatibility between protein and polysaccharide in the vicinity of the air–water interface could lead to the concentration of adsorbed protein by a depletion mechanism (Figure 10(iii)). That is, there may be an associated osmotic driving force that favours protein aggregation and results in a surface pressure increase.

Interactions in β-Lactoglobulin + Surface-Active Polysaccharide Films

Competitive adsorption has often been reported to occur between proteins and low-molecular-weight surface-active agents such as monoglycerides,[15,16] surfactants,[17,18] and lecithins.[19] In this work we show that a surface-active PS of high molecular weight, such as KO or MAN, can also show competitive behaviour with proteins. Competitive adsorption affects the surface pressure in a direct way, by displacement of the protein by the surface-active polysaccharide, and in an indirect way by thermodynamic incompatibility between the adsorbed macromolecules. The strong competitive behaviour of MAN and KO should be attributed to their high degree of esterification (high hydrophobicity).

Conversely, the less surface-active PGA sample, KLVF, behaves like a non-surface-active PS (*i.e.*, xanthan), showing cooperative behaviour that arises from complexation and incompatibility (mechanisms (ii) and (iii)). Due to its low degree of esterification, KLVF has insufficient hydrophobicity to enable it to compete with the protein for the interface. Furthermore, because of its low degree of esterification, it displays a higher negative charge, which allows it to complex with positive patches exposed on the adsorbed protein. The higher net charge would also tend to increase the incompatibility with the protein at neutral pH.

When PGA is adsorbed onto a previously spread β-lg film, the order of effectiveness in increasing surface pressure is MAN > KLVF > KO. This result cannot only be attributed to the high degree of esterification; it is necessary point out the role of the viscosity (*i.e.*, the molecular weight) of the PGA, which is indicative of the degree of incompatibility with the protein. Increasing the molecular weight of the PGA can be expected to reduce the biopolymer compatibility.[33] Nevertheless, the possibility of interfacial complexing must not be discarded. Even if KO had a degree of esterification similar to MAN, due to its low viscosity it would contribute to a lesser extent in concentrating the protein at the interface.

The analysis of the viscoelastic properties of the mixed films, adsorbed or spread, as compared to β-lg alone shows that PGA (mainly MAN) reduces the long-term solid-like character of the films due to the competitive behaviour described above. Partial displacement of protein from the surface during competitive adsorption or penetration into the spread protein film could hinder the interactions amongst the protein amino-acid residues at the air–water interface which constitute the gel network responsible for the solid-like character of the interface.

Interactions in β-Lactoglobulin + Non-Surface-Active Polysaccharide Films

As pure xanthan and λC are not surface-active, the modification of surface pressure and rheological properties of adsorbed or spread β-lg films necessarily suggests the participation of X and λC at the interface directly, through complexation (mechanism (ii)) or indirectly by excluded volume effects (mechanism (iii)).

In bulk solution, mixtures of whey proteins with X or λC at pH = 7 appear to be governed by segregative or limited thermodynamic compatibility phenomena.[27] However, local net attractive interactions between protein and PS may also occur. Following adsorption of protein at the interface, the character of protein–polysaccharide interactions may be different from in bulk solution because of the altered conformation of the protein at the interface. The strong increase of surface pressure of spread β-lg films in the presence of xanthan (Figure 6), or the synergistic surface pressure increase during adsorption from β-lg + X solutions (Figure 2), may be taken as evidence of the 'concentrating effect' arising from limited thermodynamic compatibility between the biopolymers at or in the vicinity of the interface. Xanthan reduces the elasticity and surface dilatational modulus of spread β-lg films, suggesting that the interaction with the protein structure may weaken the protein network. Because of the presence of residual adsorbing impurities, λC shows a more complicated behaviour. Nevertheless, the small effect that λC has on the surface pressure and rheology of β-lg spread monolayers, as compared to xanthan, should probably be attributed to a lower degree of thermodynamic incompatibility associated with its lower molecular weight.

Impact of Interfacial and Bulk Properties on Foam Stability

The bulk viscosity and the yield stress of the β-lg + PS solutions, as well as the rate of liquid drainage, the rate of disproportionation, and the time before the foam starts to collapse (the lag time for collapse), are shown in Table 2.

The presence of xanthan (0.5 wt%) was found to halt liquid drainage, disproportionation and collapse during the time frame of the experiments (8 h). This is because of the existence of a yield stress[34] that imparts a gel-like character to the continuous phase of the foam. Recently, Kloeck *et al.*[35] demonstrated theoretically that bulk elasticity can stop bubble shrinkage. The presence of PGA or λC strongly increased the stability of protein foams against liquid drainage and collapse as indicated by the decrease of the rates of drainage and the increase of the lag times for foam collapse of the mixed systems as compared to the β-lg foam. Nevertheless, the rate of disproportionation was enhanced by the presence of the PS.

Table 2 *Viscosity μ and yield stress for β-lg and β-lg + PS solutions and characteristic parameters of foam stability (K_{dr} = drainage rate, K_{disp} = disproportionation rate, t_{lag} = collapse lag time) for β-lg and β-lg + PS mixed foams.*

System	μ (at 60 s⁻¹) (cp)	τ_0 (dyn cm⁻²)	$K_{dr} \times 10^3$ (mL min⁻¹)	$K_{disp} \times 10^3$ (min⁻¹)	t_{lag} (min)
β-lg	1.08	–	7.92	6.11	10.7±0.1
β-lg + KO	4.85	–	2.72	6.87	17.3±0.3
β-lg + KLVF	10.9	–	1.76	9.28	43.5±2.5
β-lg + MAN	9.85	–	2.24	11.28	32±4
βLG + λC	31.8	1.0	1.56	2.11	82±5
βLG + X	132.8	27.4	–	–	–

Table 3 *Linear correlation between stability parameters of mixed foams and bulk viscosity μ and surface dilational elasticity E_d.*

Parameter	μ	E_d (after 5×10^3 s) (mN m^{-1})	$E_{d\infty}$ (mN m^{-1})
K_{dr}	–	82.2	–
t_{coll}	97.2	81.6	–
K_{disp}	–	–	65

The majority of previous studies of protein-stabilized foams have focused on either foaming properties or surface rheology, but only a few publications have attempted to relate interfacial rheology to bulk stability of foams. Recently, Murray[36] reviewed a number of theoretical and experimental studies that have tried to relate interfacial rheology to various aspects of stability. We attempt here to relate the surface properties (π, E_d, tan δ) and the bulk rheology (apparent viscosity) to the different mechanisms of foam destabilization (stability parameters in Table 2) by linear regression analysis. The relevant timescale for drainage is 3–6×10^3 s, as during this time most of the liquid was found to be drained off the foams (data not shown). Therefore, for the correlation analysis, the dynamic surface parameters were taken at 5×10^3 s. Table 3 shows that E_d explains on its own 83% of the variation in the rate of drainage. The higher the surface dilatational elasticity, the lower the rate of drainage: this may be attributed to the ability of strong elastic films to enhance local viscosity in the foam lamellae which tends to inhibit liquid drainage. As shown in Figure 8, during the early stages of adsorption (up to 5×10^3 s), the protein + PS films exhibit a more elastic or solid-like character than the β-lg film. Similarly, the foam collapse starting within 600 s (β-lg) and 4920 s (λC) (lag times in Table 2) shows a high correlation ($R^2 = 81.6\%$) with the E_d value after 5×10^3 s. Nevertheless, the bulk viscosity on its own could explain 97% of the variation of the lag time before collapse. It may be concluded that both these parameters are relevant to this mechanism of foam destabilization.

Bubble disproportionation generally occurs after a significant amount of liquid has drained.[37] The foams containing PGA or λC started to coarsen after approximately 5×10^3 s, and in the presence of λC the process continued up to 1.8×10^4 s (data not shown). Rates of disproportionation do not correlate with any of the tested dynamic surface parameters at low adsorption times (5×10^3 s), nor with the bulk viscosity. The best correlation ($R^2 = 65\%$) was found with the long-term surface dilatational elasticity ($E_{d\infty}$), indicating that a high surface elasticity reduces the rate of bubble ripening. As shown in Figure 8, the long-term surface dilatational modulus of the mixed β-lg + PGA films were found to be lower than that for the β-lg film, which tended to enhance the rate of disproportionation (Table 2). Kloeck et al.[35] have examined the critical role of interfacial elasticity in the bubble dissolution process. Their theory provides guiding values of elasticity that can stop shrinkage. More recently, Dickinson et al.[38] pointed out the role of interfacial elasticity of β-lg films in explaining the dynamics of bubble shrinkage.

In conclusion, both bulk and surface rheological properties have been shown to impact upon foam destabilization. The stability of β-lg + xanthan foam is controlled by the high viscosity and the gel-like behaviour of the bulk phase. For lower viscosity foams (β-lg + PGA or λC), the surface rheological behaviour is more important, and stability is determined by a combination of bulk and interfacial properties. However, the relative impact of each property on foam stability depends on the specific destabilization mechanism and on the relevant timescales.

Acknowledgements

This research was supported by CYTED through project XI.17 and CICYT through grant AGL2001-3843-C02-01. The authors also acknowledge the support from Universidad de Buenos Aires and Consejo Nacional de Investigaciones Científicas y Técnicas de la República Argentina.

References

1. S. Damodaran, in 'Food Proteins and their Applications', eds S. Damodaran and A. Paraf, Marcel Dekker, New York, 1997, p. 57.
2. D. J. Carp, G. B. Bartholomi, and A. M. Pilosof, *Colloids Surf. B*, 1999, **12**, 309.
3. E. Dickinson, *Food Hydrocoll.*, 2003, **17**, 25.
4. V. B. Tolstoguzov, in 'Food Proteins and their Applications', eds S. Damodaran and A. Paraf, Marcel Decker, New York, 1997, p. 171.
5. I. Capron, T. Nicolai, and D. Durand, *Food Hydrocoll.*, 1999, **13**, 1.
6. R. Baeza and A. M. R. Pilosof, in 'Food Colloids: Fundamentals of Formulation', eds E. Dickinson and R. Miller, Royal Society of Chemistry, Cambridge, 2001, p. 392.
7. E. N. Tsapkina, M. G. Semenova, G. E. Pavlovskaya, and V. B. Tolstoguzov, *Food Hydrocoll.*, 1992, **6**, 237.
8. D. S. Horne and J. M. Rodríguez Patino, in 'Biopolymers at Interfaces', ed. M. Malmsten, Marcel Dekker, New York, 2003, p. 857.
9. M. R. Rodríguez Niño, C. Carrera, M. Cejudo, and J. M. Rodríguez Patino, *J. Amer. Oil Chem. Soc.*, 2001, **78**, 873.
10. J. M. Rodríguez Patino, C. Carrera, M. R. Rodríguez Niño, and M. Cejudo, *J. Colloid Interface Sci.*, 2001, **242**, 141.
11. D. J. Carp, G. B. Bartholomai, P. Relkin, and A. M. R. Pilosof, *Colloids Surf. B*, 2001, **21**, 163.
12. A. R. Yilmazer, A. R. Carrillo, and J. Kokini, *J. Food Sci.*, 1991, **56**, 513.
13. J. D. Dziezak, *Food Technol.*, 1991, **45**, 117.
14. D. K. Sarker and P. J. Wilde, *Colloids Surf. B*, 1999, **15**, 203.
15. E. Dickinson and S. Tanai, *J. Agric. Food Chem.*, 1992, **40**, 179.
16. J. M. Rodriguez Patino, M. R. Rodríguez Niño, and C. Carrera Sánchez, *Curr. Opin. Colloid Interface Sci.*, 2003, **8**, 387.
17. A. R. Mackie, A. P. Gunning, P. J. Wilde, and V. J. Morris, *J. Colloid Interface Sci.*, 1999, **210**, 157.
18. A. R. Mackie, A. P. Gunning, P. J. Wilde, and V. J. Morris, *Langmuir*, 2000, **16**, 8176.

19. J.-L. Courthaudon, E. Dickinson, and W. W. Christie, *J. Agric. Food Chem.*, 1991, **39**, 1365.
20. M. Ahmed and E. Dickinson, *Food Hydrocoll.*, 1991, **91**, 395.
21. M. R. Rodríguez Niño and J. M. Rodríguez Patino, *Ind. Eng. Chem. Res.*, 2002, **41**,1489.
22. J. M. Rodríguez Patino, C. Carrera, S. E. Molina, M. R. Rodríguez Niño, and C. Añón, *Ind. Eng. Chem. Res.*, 2004, **43**, 1681.
23. R. Baeza, C. Carrera, A. M. R. Pilosof, and J. M. Rodríguez Patino, *Food Hydrocoll.*, 2004, **18**, 959.
24. D. J. Carp, B. E. Elizalde, G. B. Bartholomai, and A. M. R. Pilosof, in 'Engineering and Food', ed. R. Jowitt, Academic Press, London, 1997, p. 69.
25. D. J. Carp, J. Wagner, G. B. Bartholomai, and A. M. R. Pilosof, *J. Food Sci.*, 1997, **62,** 1105.
26. D. J. Carp, G. B. Bartholomai, and A. M. R. Pilosof, *Lebensm. Wiss. Technol.*, 1997, **30**, 253.
27. C. Sanchez, C. Schmitt, V. G. Babak, and J. Hardy, *Nahrung*, 1997, **41**, 336.
28. T. Burova, N. Grinberg, V. Ya. Grinberg, A. L. Leontiev, and V. B. Tolstoguzov, *Carbohydr. Polym.*, 1992, **18**, 101.
29. D. E. Graham and M. C. Phillips, *J. Colloid Interface Sci.*, 1979, **70**, 427.
30. F. McRitchie, *Adv. Protein Chem.*, 1978, **32**, 283.
31. G. E. Pavlovskaya, M. G. Semenova, E. N. Thzapkina, and V. B. Tolstoguzov, *Food Hydrocoll.*, 1993, **7**, 1.
32. T. Sengupta, L. Razumovsky, and S. Damodaran, *Langmuir*, 2000, **16**, 6583.
33. C. Schmitt, C. Sanchez, S. Desobry-Banon, and J. Hardy, *Crit. Rev. Food Sci. Nutr.*, 1998, **38**, 689.
34. P. Walstra, in 'Foams: Physics, Chemistry and Structure', ed. A. Wilson, Springer, Basel, 1989, p. 1.
35. W. Kloeck, T. van Vliet, and M. Meinders, *J. Colloid Interface Sci.*, 2001, **237**, 158.
36. B. S. Murray, *Curr. Opin. Colloid Interface Sci.*, 2002, **7**, 426.
37. M. A. Yu and S. Damodaran, *J. Agric. Food Chem.*, 1991, **39**, 1563.
38. E. Dickinson, R. Ettelaie, B. S. Murray, and Z. Du, *J. Colloid Interface Sci.*, 2002, **252**, 202.

Proposing a Relationship between the Spreading Coefficient and the Whipping Time of Cream

By Natalie E. Hotrum,[1,2] Martien A. Cohen Stuart,[2] Ton van Vliet,[1] and George A. van Aken[1,3]

[1]WAGENINGEN CENTRE FOR FOOD SCIENCES, P.O. BOX 557, 6700 AN WAGENINGEN, THE NETHERLANDS
[2]WAGENINGEN UNIVERSITY, LABORATORY OF PHYSICAL CHEMISTRY AND COLLOID SCIENCE, P.O. BOX 8038, 6700 EK WAGENINGEN, THE NETHERLANDS
[3]NIZO FOOD RESEARCH, P.O. BOX 20, 6710 BA EDE, THE NETHERLANDS

1 Introduction

The whipping of cream has both fascinated and baffled scientists for decades.[1-3] Whipped cream is an example of an aerated emulsion: through the incorporation of air bubbles, a three-phase partially crystallized fat-in-water emulsion (cream) is transformed into a four-phase system (whipped cream) stabilized by a network of partially coalesced fat globules.[4]

Several types of cream can be distinguished, including natural cream, homogenized cream, and recombined cream; all are dispersions of oil (fat) in a continuous aqueous phase. In homogenized cream, the fat droplets are stabilized by adsorbed casein and whey protein instead of the milk fat globule membrane, as is the case for natural cream.[5] In recombined cream, the protein and fat phase may differ considerably from that in natural cream. For example, recombined cream may consist of a butterfat emulsion stabilized by skim milk proteins,[6] or in order to make creams more suitable for use in warm climates, it might be desirable to use a fat with a higher melting point, such as fractionated milk fat,[7] or hydrogenated palm oil.[8]

Irrespective of the type of cream, when whipped, the adhesion and partial wetting of the air bubble surface with liquid fat from the droplets is a critical step in the development of a partially coalesced fat droplet network, which traps and stabilizes air bubbles.[1,9-12] Therefore, in order to understand fully the

317

parameters controlling the stabilization of air bubbles during the whipping of cream, we require knowledge of the parameters controlling the entering and spreading of fat droplets at the air–water interface. However, until recently,[13–17] very few studies[18,19] have addressed this research topic.

In this article, we present a few key aspects of our recent work on the spreading behaviour of emulsion droplets at the air–water interface. We report results for the whipping time of cream as a function of the same parameters that are investigated in emulsion droplet spreading work. We introduce a model which we believe correlates the spreading behaviour of emulsion droplets at the air–water interface with the whipping properties of the model creams used in our work.

2 Emulsion Droplet Spreading Behaviour

The balance of the three interfacial tensions (γ_{AW}, γ_{OW} and γ_{OA}) at the air–water–oil phase boundary determines the tendency for an emulsion droplet to spread at the air–water interface (see Figure 1). This can be expressed in terms of a spreading coefficient:

$$S = \gamma_{AW} - (\gamma_{OW} + \gamma_{OA}) . \tag{1}$$

When S is positive, the oil spreads. Triglyceride oils (sunflower, soybean, butter oil, *etc.*) spread at clean air–water interfaces; but paraffin oil, for which S is negative, does not spread.[14,16,17,19,20]

Surface-active species such as proteins can adsorb at air–water and oil–water interfaces and so lower the surface tension. This can cause S to become negative, thereby inhibiting oil spreading.[16,19,20] However, the whipping of cream is a highly dynamic process during which equilibrium surface tensions cannot be achieved. Therefore, at expanding and compressing air bubble surfaces, the tendency for an oil droplet to spread at the air–water interface will depend on the values of the three dynamic tensions (γ_{AW}, γ_{OW} and γ_{OA}). These values may be considerably higher than the equilibrium surface tensions. For example, the dynamic air–water surface tension γ_{SS} that can be achieved for a protein solution depends on the protein concentration and the expansion rate,[13] as shown in Figure 2. The dynamic steady-state air–water surface pressure, $\Pi_{SS} = (72 \text{ mN m}^{-1}) - \gamma_{SS}$, decreases with increasing expansion rate and with decreasing protein concentration.

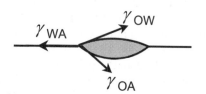

Figure 1 *Schematic representation of the interfacial tensions acting on an oil lens resting on an air–water interface.*

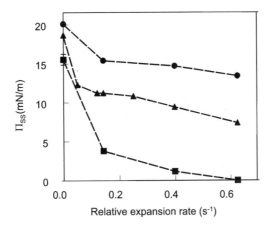

Figure 2 *Steady state surface pressure Π_{SS} versus relative expansion rate for WPI solutions of different concentrations:* ■, *0.01 wt%;* ▲, *0.1 wt%;* ●, *1 wt%.*

The spreading behaviour of protein-stabilized sunflower oil-in-water emulsions has been investigated[13,14] for solutions of whey protein isolate (WPI) at the same concentrations and expansion rates as those reported in Figure 2. In these experiments, oil droplet spreading was observed for all protein concentrations and expansion rates that yielded $\Pi_{SS} \leq 15$ mN m^{-1}. Oil droplet spreading was never observed in these systems for $\Pi_{SS} > 15$ mN m^{-1}. Thus, for the sunflower oil system, the value of 15 mN m^{-1} represents the critical surface pressure, Π_{cr}, above which spreading is inhibited. The relatively large value of Π_{cr} for sunflower oil spreading out of protein-stabilized emulsion droplets indicates that the air–water interface does not need to be completely void of adsorbed protein in order for spreading to occur. This is relevant to the whipping of cream, since the air bubbles formed during whipping are initially stabilized by adsorbed protein, which can potentially inhibit oil spreading.[21]

The value of Π_{cr} is determined by the dynamic oil–water and oil–air surface tensions. When an oil droplet begins to spread at an air–water interface, that droplet will experience a large sudden increase in area. For protein-stabilized emulsion droplets, this results in surface tension values equivalent to those measured for the pure oil–water and oil–air interfaces.[14] However, for low-molecular-weight surfactants, which are more surface-active than proteins, higher Π_{cr} values have been observed, since surfactants can effectively lower the oil–water surface tension under the dynamic conditions encountered during oil droplet spreading.[15] The data in Table 1 for β-lactoglobulin (β-lg) show that Π_{cr} is significantly higher in the presence of sufficiently high concentrations of Tween 20 (22 mN m^{-1}) than it is in the absence of the added surfactant (14 mN m^{-1}). Less rigorous surface expansion conditions would be required to reach Π_{cr} for these systems, demonstrating that the presence of low-molecular-weight surfactants can facilitate oil droplet spreading at the air–water interface. This may be relevant to the aeration of emulsions, where an increase in Π_{cr} may facilitate the adhesion of droplets to the air bubble

Table 1 *Values of Π_{cr} for emulsion droplets stabilized by the surface-active species indicated. For the measurement, a 5 µL aliquot of emulsion (40 wt% sunflower oil) was added under the surface of a solution containing the corresponding surface-active species. The air–water interface was subsequently expanded at a rate of 0.12 s^{-1} (see refs. 14 and 16). For the mixed β-lg+Tween 20 systems, the emulsion was added under the surface of a β-lg solution. For all the data, an error of 1 mN m^{-1} applies.*

Emulsion System	Π_{cr} (mN m^{-1})
WPI	14
β-lactoglobulin	14
β-lactoglobulin + 0.14 mM Tween 20	16
β-lactoglobulin + 2.3 mM Tween 20	17
β-lactoglobulin + 9.1 mM Tween 20	22
Tween 20	22

surface, and subsequent partial wetting of the air–water interface, thereby resulting in a faster build-up of the whipped cream structure.

3 The Whipping Time of Cream

During the whipping of cream, the fat (emulsion) droplets adhere to the air bubble surface. The adsorbed droplets form fat clumps *via* partial coalescence at the interface.[1,21] This process is henceforth referred to here as surface-mediated partial coalescence (Figure 3). During whipping, the process continues until a partially coalesced network of fat droplets is built up, which acts to trap the air bubbles, to retain the serum phase, and to give structure to the whipped cream. The time required to reach the end point of whipping depends on the rapidity with which the partially coalesced network of fat globules can be built up, which in turn is influenced by a number of system parameters.

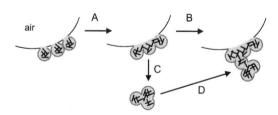

Figure 3 *Schematic depiction of surface-mediated mechanism for partial coalescence. For S>0, fat droplets can attach to the protein-covered air bubble surface and interfacial flocculation takes place; partial coalescence (A) of the interfacial flocs ensues. If, at this point, the bubble bursts, a partially coalesced fat clump (C) remains. If the air bubble remains stable (B), fat clumps from the bulk may partially coalesce with the adsorbed fat clump (D).*

We present results here demonstrating the effect of whisk rotational speed (Figure 4) and Tween 20 concentration (Figure 5) on the whipping time of cream. Our model cream contained a blend of hydrogenated palm oil (14 wt%) and sunflower oil (26 wt%) stabilized by whey protein isolate (1 wt%) and, where applicable, a known concentration of Tween 20. Creams were whipped using the Ledoux whipping apparatus.[22,23] In this equipment, the power input required to maintain a constant whisk rotational speed could be monitored. As the cream increases in viscosity and stiffness, the power input increases, reaching a maximum at the end point of whipping.

Figure 4 shows that the time to reach the end point of whipping (the whipping time) decreases with increasing whisk rotational speed. This implies that the transition from free fat globules to clumped fat proceeds faster when the cream is whipped at a higher rate. Van Aken observed a similar dependence[23] of whipping time on whisk rotational speed for the whipping of natural cream.

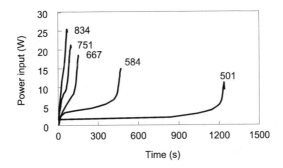

Figure 4 *Power input transferred by the whisks during the whipping of a model cream (14 wt% hydrogenated palm fat, 26 wt% sunflower oil, 1 wt% WPI). Numbers near the curves denote the rotational speed of the whisks in rpm.*

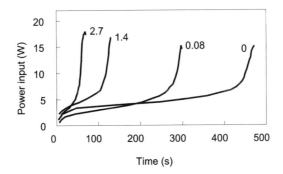

Figure 5 *Power input transferred by the whisks during the whipping of a model cream (14 wt% hydrogenated palm fat, 26 wt% sunflower oil, 1 wt% WPI) containing added Tween 20. Numbers near the curves denote the concentrations (in mM) of Tween 20 present in the emulsion system. A whisk rotational speed of 584 rpm was used.*

In the presence of the surfactant Tween 20, the whipping times of the model creams are drastically reduced (Figure 5). The addition of 0.08 mM Tween 20 causes the whipping time of the model cream to decrease by nearly half. And, with the addition of 2.7 mM Tween 20, the whipping time of the cream is lowered to ~70 s, which, for comparison, is similar to the time required to reach the end point of whipping for natural cream whipped at the same rotational speed.[23] This decrease in whipping time in the presence of added surfactant has also been reported[6] for the addition of glycerol monostearate to recombined cream; additionally, fat clumping has been reported[24,25] to proceed faster for ice-cream mix emulsions containing added surfactant. In the following section we propose a mechanism to describe this aspect of the surfactant functionality.

4 Relationship of Spreading Behaviour to Whipping Time

It was noted above that increasing the whisk rotational speed and adding a surfactant (emulsifier) both lead to a significant decrease in the whipping time of emulsions. In this section, we present a model in which a parallel is drawn between the spreading behaviour of emulsion droplets at the air–water interface and the whipping properties of recombined cream.

It has been established[14] that an adsorbed protein layer can effectively inhibit droplet entering and spreading provided that the surface pressure exerted by the adsorbed layer exceeds a critical value Π_{cr}. Since the air bubbles formed during the whipping of cream are initially stabilized by adsorbed protein,[11,21] it seems reasonable to assume that, during whipping, fat droplet adsorption and spreading can only occur at the portion of the air bubble surface for which the air–water surface pressure is lower than Π_{cr}. Furthermore, the whipping of cream is a highly dynamic process, and the surface pressure at the air bubble–serum interface will not be homogeneous. It is likely that the surface pressure of the interfaces around (different) air bubbles can be better described by a surface pressure distribution, where newly formed air bubble surfaces have low surface pressures and portions of the air bubble surface that are compressed, or to which protein is adsorbed, have a higher surface pressure, which will depend on the adsorbed amount. This distribution is represented schematically in Figure 6a, for a reference case. The total area under the curve represents the sum of the surface areas of the individual air bubbles, or the total air bubble surface present in the whipped cream. The Π_{cr} value required for fat droplet spreading is indicated by a vertical dotted line. The area under the curve to the left of this line is shaded to indicate that, for this fraction of the total air bubble surface, the air–water surface pressure will be low enough for adhesion and (partial) spreading of fat droplets to occur. The remainder of the bubble area is inactive.

Manipulating the fraction of the total bubble surface area for which $S > 0$ is expected to influence the ease and rapidity of fat droplet adhesion and spreading, which will ultimately govern the whipping time of the system. In this way, the spreading behaviour of emulsion droplets at the planar air–water interface

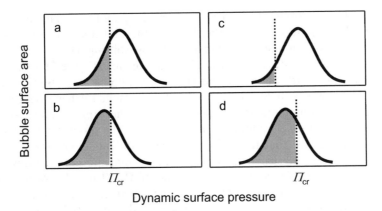

Figure 6 *Schematic representation of the surface pressure distribution at the surface of air bubbles during whipping. The vertical line indicates the value of Π_{cr}. For the reference system (a), the area under the curve to the left of Π_{cr} (shaded) indicates the fraction of the air bubble surface area with $S > 0$. This area can be increased or decreased by shifting the surface pressure distribution to the left (b) or to the right (c), respectively. In addition, if Π_{cr} is shifted to a higher value (d) the fraction of the bubble surface area with $S > 0$ will increase. Increasing the proportion of the bubble surface area with $S > 0$ leads to shorter whipping times, whereas reducing the area leads to longer whipping times.*

is directly related to the whipping behaviour of cream. For example, if the surface pressure distribution is shifted to the left with respect to the reference case, as in Figure 6b, we expect the fraction of the total bubble surface area for which $S > 0$ to increase and the whipping time to decrease. In the experiments at the planar air–water interface (Figure 2), lower steady-state surface pressures could be achieved by increasing the relative expansion rate (*i.e.*, the surface pressure distribution in Figure 6a would be shifted to the left). Similarly, for the model whipped cream, increasing the rotational speed (Figure 4) results in shorter whipping times. Conversely, if the surface pressure distribution is shifted to the right with respect to the reference case, as in Figure 6c, the fraction of the total bubble surface area with $S > 0$ would be expected to decrease and the whipping time to increase. This can be achieved, for example, by lowering the relative expansion rate (rotational speed). The experimental results for both the expansion of the planar air–water interface (Figure 2) and the aeration by whipping (Figure 4) confirm this, supporting the concept that the whipping time is related to the ease with which fat droplets can adhere to and spread at the air–water interface. Moreover, experimental results reported elsewhere[22] for the influence of protein concentration and protein type on the whipping time of cream are consistent with the model presented in Figure 6.

For the variable whisk rotational speed, it was assumed that the value of Π_{cr} is independent of the surface expansion rate. However, if Π_{cr} could be shifted to higher Π_{AW} values, then $S > 0$ could be achieved for a greater proportion of the air bubble surface without having to shift the surface pressure distribution, which would be expected to lead to shorter whipping times. This is depicted schematically in Figure 6d. In earlier work[15] (and in Table 1) it was observed that surfactants facilitate droplet spreading by shifting Π_{cr} to higher Π_{AW} values. Correspondingly, in the whipping experiments (Figure 5) a significant decrease in the length of the second stage of whipping was observed as the concentration of Tween 20 was increased. These results further strengthen the hypothesis that whipping time is dependent on the rapidity and ease with which emulsion droplets adhere to and spread at the air–water interface. It should be mentioned that the whipping time of the model creams (Figure 5) begins to decrease at surfactant concentrations that are considerably lower than those required for a shift in the value of Π_{cr} (Table 1). Although the cause of this requires further study, it may be explained by the ability of low-molecular-weight surfactants to promote shear-induced partial coalescence,[1] in addition to promoting surface-mediated partial coalescence.

5 Summary

The model presented here represents a contribution towards understanding the physical principles underlying structure development during the whipping of cream. It has been demonstrated that the same parameters that promote emulsion droplet spreading also improve the whippability of cream, which leads to the conclusion that partial wetting of the air–water interface by adsorbed emulsion droplets is a key parameter controlling the interaction between emulsion droplets at the air–water interface during whipping. This model is fundamentally different from earlier models that attempted to describe the influence of variables including whipping speed and the presence of low-molecular-weight surfactants on whipping behaviour of cream in terms of their influence on the static properties of the adsorbed layer at the fat droplet surface or the contact angle between fat crystals and the oil–water interface.

This contribution to understanding the interactions between air bubbles and emulsion droplets is expected to enable the food industry to take a more strategic approach to product formulation.

Acknowledgement

The authors thank Franklin Zoet and Serena Avino for performing some of the experiments.

References

1. D. F. Darling, *J. Dairy Res.*, 1982, **49**, 695.
2. H. D. Goff, *J. Dairy Sci.*, 1997, **80**, 2620.
3. H. Mulder and P. Walstra, 'The Milk Fat Globule', Pudoc, Wageningen, 1974.
4. B. E. Brooker, *Food Struct.*, 1993, **12**, 115.
5. D. F. Darling and D. W. Butcher, *J. Dairy Res.*, 1978, **45**, 197.

6. J. G. Zadow and F. G. Kieseker, *Aust. J. Dairy Technol.*, 1975, **30** , 114.

7. A. M. P. Jochems and G. A. van Aken, Proceedings of the 2nd World Congress on Emulsion, Bordeaux, France, September 23–26, 1997, p. 364.

8. K. Shamsi, Y. B. Che Man, M. S. A Yusoff, and S. Jinap, *J. Am. Oil Chem. Soc.*, 2002, **79**, 583.

9. E. Graf and H. R Müller, *Milchwissenschaft*, 1965, **20**, 302.

10. J. van Camp, S. van Calenberg, P. van Oostveldt, and A. Huyghebaert, *Milchwissenschaft*, 1996, **51**, 310.

11. B. E. Brooker, M. Anderson, and A. T. Andrews, *Food Microstruct.*, 1986, **5**, 277.

12. D. G. Schmidt and A. C. M. van Hooydonk, *Scan. Electron Microsc.*, 1980, **III**, 653.

13. N. E. Hotrum, M. A. Cohen Stuart, T. van Vliet, and G. A. van Aken, in 'Food Colloids, Biopolymers and Materials', eds E. Dickinson and T. van Vliet, Royal Society of Chemistry, Cambridge, 2003, p. 192.

14. N. E. Hotrum, T. van Vliet, M. A. Cohen Stuart, and G. A. van Aken, *J. Colloid Interface Sci.*, 2002, **247**, 125.

15. N. E. Hotrum, M. A. Cohen Stuart, T. van Vliet, and G. A. van Aken, to be submitted for publication.

16. E. P. Schokker, M. A. Bos, A. J. Kuijpers, M. E. Wijnen, and P. Walstra, *Colloids Surf. B*, 2003, **26**, 315.

17. N. E. Hotrum, M. A. Cohen Stuart, T. van Vliet, and G. A. van Aken, *Colloids Surf. A*, 2004, **240**, 83.

18. N. King, 'The Milk Fat Globule Membrane and Some Associated Phenomena', Lamport Gilbert, Reading, 1955.

19. H. A. Sirks, Verslag van het Rijks-Landbouwproefstation te Hoorn over 1938, The Hague, 1939, p. 11.

20. C. G. J. Bisperink, 'The influence of spreading particles on the stability of thin liquid films', Ph.D. thesis, Wageningen Agricultural University, Netherlands, 1997.

21. M. Anderson, B. E. Brooker, and E. C. Needs, in 'Food Emulsions and Foams', ed. E. Dickinson, Royal Society of Chemistry, London, 1987, p. 100.

22. N. E. Hotrum, M. A. Cohen Stuart, T. van Vliet, and G. A. van Aken, unpublished results.

23. G. A. van Aken, *Colloids Surf. A*, 2001, **190**, 333.

24. H. D. Goff and W. K. Jordan, *J. Dairy Sci.*, 1989, **72**, 18.

25. B. M. C. Pelan, K. M. Watts, I. J. Campbell, and A. Lips, *J. Dairy Sci.*, 1997, **80**, 2631.

Utilization of a Layer-by-Layer Electrostatic Deposition Technique to Improve Food Emulsion Properties

By D. Julian McClements, Tomoko Aoki, Eric A. Decker, Yeun-Suk Gu, Demet Guzey, Hyun-Jung Kim, Utai Klinkesorn, Lydie Moreau, Satoshi Ogawa, and Parita Tanasukam

DEPARTMENT OF FOOD SCIENCE, UNIVERSITY OF MASSACHUSETTS, AMHERST, MA 01003, USA

1 Introduction

Food emulsions of the oil-in-water (O/W) type are normally created by homogenizing oil and aqueous phases together in the presence of one or more emulsifiers.[1-3] An emulsifier is a surface-active substance that adsorbs at the surfaces of droplets formed during homogenization, where it reduces the interfacial tension and facilitates further droplet disruption.[4,5] In addition, the adsorbed emulsifier forms a protective membrane around the droplets that prevents them from aggregating with one another.[1-3] Many different kinds of emulsifiers are available for utilization in food products,[6-8] the most important being small-molecule surfactants, proteins, polysaccharides, and phospholipids. Each type of emulsifier varies in its effectiveness at producing small droplets during homogenization, and in its ability to prevent droplet aggregation under different environmental stresses, such as pH, ionic strength, heating, freezing and drying.[2,3,9,10] Food emulsifiers also differ in their cost, reliability, ease of use, ingredient compatibility, 'label friendliness', and legal status.[7-8] For these reasons, there is no single emulsifier that is ideal for every type of food product. Instead, the selection of a particular emulsifier (or combination of emulsifiers) for a specific food product depends on the type and concentration of other ingredients, the homogenization conditions used to produce it, and the environmental stresses that it may experience during its manufacture, storage and utilization.

Using conventional food emulsifiers and homogenization techniques, there is only a limited range of functional attributes that can be achieved in emulsion-based food products. This has motivated a number of researchers to examine alternative means of improving emulsion stability by developing novel emulsifier-based strategies. One strategy has been to create emulsifiers with improved or extended functional properties by covalently linking together proteins and polysaccharides.[9-11] An alternative strategy is to create oil-in-water emulsions containing oil droplets surrounded by multi-layered interfacial membranes consisting of emulsifiers and/or biopolymers.[12-22]. In this 'layer-by-layer' (*LbL*) electrostatic deposition approach, an ionic emulsifier that rapidly adsorbs to the surface of lipid droplets during homogenization is used to produce the *primary* emulsion containing small droplets. Then an oppositely charged biopolymer is added to the system to adsorb to the droplet surface and to produce a *secondary* emulsion containing droplets coated with a two-layer emulsifier–biopolymer interfacial membrane (Figure 1). This procedure can be repeated to form oil droplets coated by interfacial membranes consisting of three (or even more) layers, *e.g.,* emulsifier – biopolymer 1 – biopolymer 2. These emulsions containing oil droplets surrounded by multi-layered interfacial membranes have been found[18-22] to have better stability to environmental stresses than conventional oil-in-water emulsions under certain circumstances.

The *LbL*-electrostatic deposition method described above therefore offers a promising way to improve the stability and physico-chemical properties of food emulsions. Nevertheless, the correct choice of the appropriate combination of emulsifier and biopolymers is essential to the success of this approach, as is the determination of the optimum preparation conditions (pH, ionic strength, temperature, biopolymer mixing procedure, droplet separation,

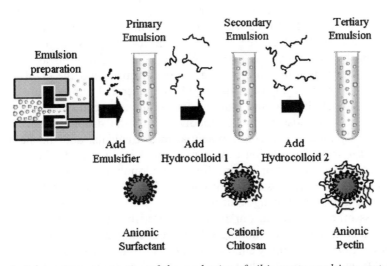

Figure 1 *Schematic representation of the production of oil-in-water emulsions containing droplets stabilized by three layers (emulsifier – biopolymer 1 – biopolymer 2).*

emulsion washing, floc disruption, *etc.*). This article provides an overview of recent research carried out in our laboratory on the development, characterization and application of O/W emulsions containing oil droplets surrounded by multiple layers of emulsifier and biopolymer. In particular, we focus on the use of the *LbL* electrostatic deposition technique to create emulsions with improved resistance to environmental stresses, such as pH, ionic strength, thermal processing, freezing, drying, and lipid oxidation.

2 Production of Multi-Layered Emulsions

Oil-in-water emulsions containing droplets surrounded by multi-layered interfacial membranes have been prepared using a multiple-step process,[19–21] which has been referred to as the *sequential adsorption* approach.[23] For example, the following procedure can be used to create emulsion droplets coated by three layers, *e.g.,* emulsifier – biopolymer 1 – biopolymer 2 (Figure 1). Initially, a *primary* emulsion containing electrically charged droplets surrounded by a layer of emulsifier is prepared by homogenizing oil, aqueous phase and an ionic surfactant (emulsifier). In the second stage, a secondary emulsion containing charged droplets stabilized by emulsifier–biopolymer 1 membranes is formed by incorporating biopolymer 1 into the primary emulsion. For this to work, biopolymer 1 normally has to have an opposite electrical charge from that on the droplets in the primary emulsion—although this is not always necessary if there are patches of opposite charge on the droplet. Mechanical agitation may have to be applied to the secondary emulsion to disrupt any flocs formed by the bridging of droplets by biopolymer 1. In addition, the secondary emulsion may have to be washed (*e.g.,* by filtration or centrifugation) to remove any free biopolymer remaining in the continuous phase. In the third stage, the tertiary emulsion containing droplets stabilized at the interface by emulsifier – biopolymer 1 – biopolymer 2 membranes is formed by incorporating biopolymer 2 into the secondary emulsion. Biopolymer 2 normally has an opposite electrical charge from that on the droplets in the secondary emulsion (but see above). If necessary, mechanical agitation is again applied to the tertiary emulsion to disrupt any flocs formed, and the emulsion may be washed to remove any non-adsorbed biopolymer 2. The procedure could be continued indefinitely to add more layers to the interfacial membrane.

The sequential adsorption of biopolymers to the droplet surface can be conveniently monitored using ζ-potential measurements, and the stability of the emulsions to flocculation can be monitored by light scattering or microscopy.[19–21] Since the major driving force for biopolymer adsorption onto the droplet surface is electrostatic in origin, it is important to control the pH and ionic strength of the mixing solution.

Unless otherwise stated, our results reported below are for 0.2 wt% corn oil-in-water emulsions containing oil droplets stabilized by interfacial membranes of SDS (the primary emulsion), SDS–chitosan (the secondary emulsion) and SDS–chitosan–pectin (the tertiary emulsion) prepared using the sequential adsorption approach at pH = 3.0 in 100 mM acetate buffer.

3 Improved Stability to Environmental Stresses

Food emulsions experience a variety of different environmental stresses during their manufacture, storage, transport and utilization. These include pH extremes, high ionic strengths, thermal processing, freeze–thaw cycling, drying, and mechanical agitation. Many of the emulsifiers currently available for utilization within the food industry provide limited stability to these environmental stresses. In this section, we present some of the recent work we have done on utilizing the *LbL*-electrostatic deposition technique to improve emulsion stability with respect to various environmental stresses.

Sensitivity to pH

The influence of pH on the ζ-potential, mean particle diameter, and creaming stability of diluted primary, secondary and tertiary emulsions was measured after the emulsions had been stored at room temperature for 10 days. The ζ-potential of the SDS-stabilized droplets in the primary emulsions was found to be negative at all pH values due to the presence of adsorbed anionic surfactant (Figure 2). The ζ-potential of the SDS–chitosan-stabilized droplets in the secondary emulsions was positive at relatively low pH values (≤pH 6), but became negative at higher pH. Charge reversal probably occurred because chitosan lost some of its positive charge at high pH values ($pK_a \approx 6.5$), which may also have caused it to desorb from the droplet surfaces. The ζ-potential of the SDS–chitosan–pectin-stabilized droplets in the tertiary emulsion was found to be negative at all pH values, which suggests that the chitosan layer did not desorb from the surface of the emulsion droplets at higher pH values as it did in the secondary emulsions. This may have been because the effective pK_a

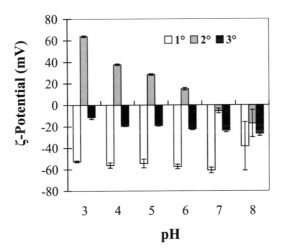

Figure 2 *Dependence of droplet ζ-potential on pH for diluted primary, secondary and tertiary emulsions (originally 0.2 wt% corn oil, pH=3.0, 100 mM acetate buffer).*

value of the positively charged groups on the chitosan molecule is increased appreciably when it is sandwiched between two negatively charged biopolymer layers, as has been reported[24,25] for other polyelectrolytes. There was found to be a substantial increase in the magnitude of the net negative charge on the droplets in the tertiary emulsion with increasing pH. This is probably because of the increase in negative charge on the pectin molecules ($pK_a \sim 4.5$) and the reduction in positive charge on the chitosan molecules ($pK_a \sim 6.5$).

The droplets in the primary, secondary and tertiary emulsions were all stable to droplet aggregation at pH 3–5 when stored for 10 days (Figure 3), as well as to creaming (data not shown). The droplets in the primary and tertiary emulsions were also stable at higher pH values, but extensive droplet aggregation and rapid creaming were observed in the secondary emulsions at $pH \geq 7$. This was probably because the lowering of droplet surface charge (Figure 2) reduced the electrostatic repulsion between the droplets, which led to extensive droplet flocculation. The tertiary emulsion was stable to creaming and droplet aggregation over the whole pH range, presumably because of strong electrostatic and steric repulsion associated with the relatively thick and moderately charged three-layer interfacial membrane.

Sensitivity to Ionic Strength

The influence of NaCl concentration (0–500 mM) on the ζ-potential, the mean particle diameter and the creaming stability of diluted primary, secondary and tertiary emulsions at pH = 3.0 was determined. The ζ-potential of the SDS-stabilized droplets in the primary emulsions was negative at all ionic strengths (Figure 4), but its magnitude decreased as the NaCl concentration increased, which can be attributed to electrostatic screening effects.[26] The

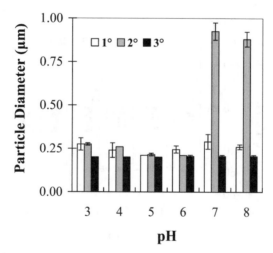

Figure 3 *Dependence of mean particle diameter d_{32} on pH for diluted primary, secondary and tertiary emulsions (originally 0.2 wt% corn oil, pH = 3.0, 100 mM acetate buffer).*

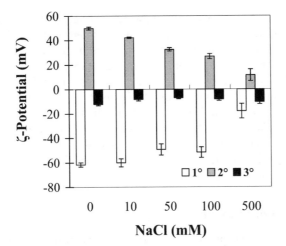

Figure 4 *Effect of NaCl concentration on droplet ζ-potential for primary, secondary and tertiary emulsions at pH=3.*

ζ-potential of the SDS–chitosan-stabilized droplets in the secondary emulsions was positive at all ionic strengths, but its magnitude also decreased as the NaCl concentration increased. This can also be attributed to electrostatic screening effects, as well as to possible desorption of chitosan from the emulsion droplet surface when the ionic strength increases, due to weakening of the electrostatic attraction between the negatively charged droplets and the positively charged chitosan molecules. The ζ-potential of the SDS–chitosan–pectin-stabilized droplets in the tertiary emulsion was negative at all ionic strengths, and its magnitude did not change appreciably as the NaCl concentration was increased (Figure 4). This suggests that neither chitosan nor pectin was desorbed from the emulsion droplet surface with increasing ionic strength.

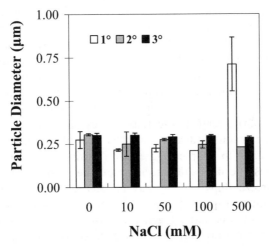

Figure 5 *Effect of NaCl concentration on mean particle diameter d_{32} for primary, secondary and tertiary emulsions at pH=3.*

The secondary and tertiary emulsions were relatively stable to droplet aggregation at all ionic strengths (0–500 mM NaCl), as demonstrated by the particle size data (Figure 5) and creaming stability measurements (data not shown). On the other hand, the primary emulsion was unstable at high ionic strength (500 mM NaCl), as shown by an increase in mean particle diameter (Figure 5) and rapid creaming (data not shown). Droplet aggregation was probably promoted in the primary emulsion due to the reduction of the charge repulsion between the droplets associated with electrostatic screening effects.[26,27] The secondary and tertiary emulsions are stable to droplet aggregation at high ionic strength probably because the relatively thick interfacial membrane provides good steric stabilization.

Sensitivity to Thermal Processing

The influence of thermal processing on the stability of the primary, secondary and tertiary emulsions was studied. The emulsions were held isothermally for 30 minutes at temperatures ranging from 30 to 90 °C, cooled to room temperature, and then stored for 24 hours. The mean particle diameter and the creaming stability of the emulsions were then determined. There was found to be no significant change in the mean particle diameter ($d_{32} = 0.27 \pm 0.03$ μm) of any of the emulsions, and there was no evidence of creaming, which indicated that primary, secondary and tertiary emulsions were all stable to thermal processing in the temperature range considered.

Sensitivity to Freeze–Thaw Cycling

Emulsion samples (10 ml) were transferred into cryogenic test tubes and incubated in a freezer at −20 °C for 22 hours. After incubation the emulsion samples were thawed by placing them in a water bath at 30 °C for 2 hours. This freeze–thaw cycle was repeated six times and its influence on emulsion properties was measured after each cycle (Figure 6). The primary emulsion exhibited extensive oiling-off after only one freeze–thaw cycle, suggesting that extensive droplet coalescence had occurred. The secondary emulsion exhibited extensive droplet flocculation (but no coalescence or oiling-off) after only one freeze-thaw cycle, which led to rapid droplet creaming. On the other hand, the mean diameter of the particles in the tertiary emulsions remained the same ($d_{32} \sim 0.3 \pm 0.05$ μm) after 6 freeze–thaw cycles and there was no evidence of creaming. These results indicate that it is possible to create emulsions that are highly resistant to droplet aggregation during freeze–thaw cycling using the *LbL* electrostatic deposition technique. The high stability of the tertiary emulsions to freeze–thaw cycling may be because the thick interfacial membrane is resistant to rupture by ice or fat crystals. Or it may be because the repulsive colloidal interactions generated by the thick electrically charged interfacial membranes are sufficiently large to overcome any attractive colloid interactions or mechanical forces that tend to push the droplets together during the freezing process.

Before Freezing   **After 1 Freeze-Thaw Cycle**

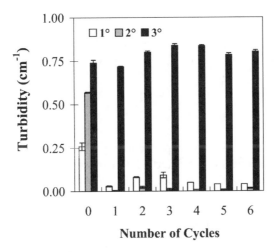

Figure 6 *Dependence of the creaming stability of primary, secondary and tertiary emulsions on the number of freeze–thaw cycles. The stability was determined as turbidity (at 600 nm) measured at 30% of emulsion height after 24 h storage: a high creaming instability is indicated by a low turbidity. The two photographs show the change in visual appearance of the primary, secondary and tertiary emulsions after one freeze–thaw cycle.*

Lipid Oxidation

We have found that the secondary and tertiary emulsions have a much better stability to lipid oxidation than the primary emulsions (Figure 7). The instability of the primary emulsions to lipid oxidation has been attributed to the ability of positively charged Fe^{2+} ions to adsorb to the surface of the negatively charged droplets, where they come into close proximity with the unsaturated lipids within the oil droplets. The greater stability of the secondary emulsions can be attributed to the fact that the SDS–chitosan-stabilized droplets are positively charged, and therefore they repel the Fe^{2+} ions electrostatically. In addition, the greater thickness of the interfacial membrane may help to prevent the iron catalyst from coming into close contact with the lipid substrate. It is interesting to note that the tertiary emulsion was still found to be stable to lipid

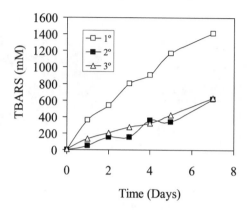

Figure 7 *Progress of lipid oxidation in primary, secondary and tertiary emulsions as monitored by TBARS (Thiobarbituric Acid Reactive Substances), a colorimetric method that provides a measure of the extent of lipid oxidation in a fatty material.*

oxidation even though it contained negatively charged droplets. This again suggests that the thickness of the interfacial membrane plays an important role in retarding iron-catalyzed lipid oxidation in O/W emulsions.

Sensitivity to Freeze-Drying

We have carried out preliminary experiments comparing the stability of primary and secondary emulsions to freeze-drying. We have found that the secondary emulsions exhibit a much better stability than the primary emulsions with respect to droplet aggregation during freeze-drying. This was especially the case when maltodextrin was incorporated into the emulsions (data not shown).

4 Other Multi-Layered Systems

In addition to the SDS–chitosan–pectin system described above, we are examining the suitability of other types of emulsifiers and biopolymers for preparing stable oil-in-water emulsions containing droplets surrounded by multi-layered interfacial membranes. For example, in systems containing β-lactoglobulin (β-lg), we have shown[18,28] that secondary emulsions containing droplets stabilized by β-lg–pectin membranes have better stability to pH and NaCl than do primary emulsions stabilized by β-lg alone under certain conditions. Similarly, we have shown[29] that, under some circumstances, secondary emulsions containing droplets stabilized by β-lg–carrageenan membranes have a better stability to pH, NaCl, and thermal processing than do primary emulsions stabilized by β-lg alone.

5 Conclusions

Our work so far has shown that stable emulsions containing multi-layered lipid droplets can be prepared using a simple cost-effective method and food-grade ingredients. The stability of these multi-layered emulsions to environmental stresses is better than that of conventional emulsions under certain conditions. More research is still needed to establish, at a fundamental level, the factors that influence the preparation of stable multi-layered emulsions with specific functional attributes, including emulsifier characteristics (the sign and magnitude of droplet charge), biopolymer characteristics (molecular weight, charge density, flexibility), and the compositions of the mixing and washing solutions (in terms of ionic strength and pH). In addition, research needs to be carried out to establish where these multi-layered emulsions can be used practically within the food industry.

Acknowledgements

This article is based upon work supported by the National Research Initiative Competitive Grants Program, United States Department of Agriculture, and the CREES IFAFS Program, United States Department of Agriculture (Award Number 2002-35503-12296). We also thank the Tokyo University of Marine Science and Technology (Japan) for financial support to T.A.

References

1. S. Friberg, K. Larsson, and J. Sjöblom (eds), 'Food Emulsions', 4th edn, Marcel Dekker, New York, 2004.
2. D. J. McClements, 'Food Emulsions: Principles, Practice and Techniques', CRC Press, Boca Raton, FL, 1999.
3. E. Dickinson, 'An Introduction to Food Colloids', Oxford University Press, Oxford, 1992.
4. P. Walstra, *Chem. Eng. Sci.*, 1993, **48**, 333.
5. P. Walstra, in 'Food Chemistry', 3rd edn, ed. O. R. Fennema, Marcel Dekker, New York, 1996, chap. 3.
6. G. Charlambous and G. Doxastakis, 'Food Emulsifiers: Chemistry, Technology, Functional Properties and Applications', Elsevier, Amsterdam, 1989.
7. N. J. Krog and F. V. Sparso, in 'Food Emulsions', 4th edn, eds S. Friberg, K. Larsson and J. Sjöblom, Marcel Dekker, New York, 2004, chap. 2.
8. C. E. Stauffer, 'Emulsifiers', Eagen Press, St Paul, MN, 1999.
9. E. Dickinson, *Food Hydrocoll.*, 2003, **17**, 25.
10. N. Garti and D. Reichman, *Food Microstruct.*, 1993, **12**, 411.
11. C. Schmitt, C. Sanchez, S. Desobry-Banon, and J. Hardy, *Crit. Rev. Food Sci. Nutr.*, 1998, **38**, 689.
12. P. Calvo, C. Remunan-Lopez, J. L. Vila-Jato, and M. J. Alonso, *Colloid Polym. Sci.*, 1997, **275**, 46.
13. F. Caruso, *Aust. J. Chem.*, 2001, **54**, 349.
14. F. Caruso and C. Schuler, *Langmuir*, 2000, **16**, 9595.
15. E. Dickinson and J. D. James, *Food Hydrocoll.*, 2000, **14**, 365.

16. P. Fäldt, B. Bergenståhl, and P. M. Claesson, *Colloids Surf. A.*, 1993, **71**, 187.
17, S. Magdassi, U. Bach, and K. Y. Mumcuoglu, *J. Microencaps.*, 1997, **14**, 189.
18. L. Moreau, H. J. Kim, E. A. Decker, and D. J. McClements, *J. Agric. Food Chem.*, 2003, **51**, 6612.
19. S. Ogawa, E. A. Decker, and D. J. McClements, *J. Agric. Food Chem.*, 2003, **51**, 2806.
20. S. Ogawa, E. A. Decker, and D. J. McClements, *J. Agric. Food Chem.*, 2003, **51**, 5522.
21. S. Ogawa, E. A. Decker, and D. J. McClements, *J. Agric. Food Chem.*, 2004, **52**, 3595.
22. X. Y. Shi and F. Caruso, *Langmuir*, 2001, **17**, 2036.
23. A. Voigt, H. Lichtenfeld, G. B. Sukhorukov, H. Zastrow, E. Donath, H. Buamler, and H. Mohwald, *Ind. Eng. Chem. Res.*, 1999, **38**, 4037.
24. S. E. Burke and C. J. Barrett, *Biomacromolecules*, 2003, **4**, 1773.
25. S. E. Burke and C. J. Barrett, *Langmuir*, 2003, **19**, 3297.
26. R. J. Hunter, 'Foundations of Colloid Science', Oxford University Press, Oxford, 1986, vol. 1.
27. J. N. Israelachvili, 'Intermolecular and Surface Forces', Academic Press, London, 1992.
28. D. Guzey, H. J. Kim, and D. J. McClements, *Food Hydrocoll.*, 2004, **18**, 967.
29. Y. S. Gu, E. A. Decker, and D. J. McClements, *J. Agric. Food Chem.*, 2004, **52**, 3626.

Sensory Perception

Perceiving the Texture of a Food: Biomechanical and Cognitive Mechanisms and their Measurement

By David A. Booth

FOOD QUALITY RESEARCH GROUP, SCHOOL OF PSYCHOLOGY, UNIVERSITY OF BIRMINGHAM, EDGBASTON, BIRMINGHAM B15 2TT, UK

1 Introduction

Perception is the objective achievement of acquiring knowledge about the environment through the senses. Therefore, measuring a consumer's perception of the eating quality of a sample of a food requires the scientific use of information produced by the eater to recover sensed information generated by the food material.

This information-transmitting performance by an individual consumer can be indicated by selecting foods for eating, whether at the table, in the kitchen, from the menu at an eating place, or off the shelves in a shop. However, by themselves these choices among foods tell us nothing about what in the foods was perceived that produced the observed selection: in order to study perception, there must also be evidence that distinguishes among the many possibilities of stimulation to the senses from the material. That is, the influential physico-chemical characteristics of the available foods have to be known before consumers' choices among foods and their expectations about food quality can be used to measure their perceptual achievements.[1-3]

Thus, sensory description is not the perception of materials; nor does the mapping of sensory scores onto hedonic scores (or preference votes) model the perception of quality. Neither can sensor readings or other instrumental parameter values serve as quality criteria without having been validated objectively against the perception of sensed quality by representative consumers.

The science of objective perceptual performance is psychology. The sciences of the senses also include neuroscience: this ranges from the molecular neuroscience of the outer membrane of the sensory receptor cell, through the neurophysiology of discharges in the axon from the receptor to the first synapses in the brain, to the functioning of the networks of cells in the cerebral cortex, whose activity is influenced by stimulation of particular types of receptor activity. This can be summed up crudely in, say, event-related electrical waves or nuclear magnetic resonance images of oxygenated blood flow. However, the way in which any part of this neuronal machinery contributes to perception can only specified by analysis of the overall information-processing performance of the individual who has the senses, brain and muscles for stimulation and responding. False-colour brain images may look very pretty but they tell us nothing about the perception of food unless differences in brain activation are related in a mechanistically interpretable manner to differences in the performance of the individuals whose heads have been scanned.

This processing of information in the mind of the life-taught consumer (or of the laboratory-trained sensory panelist), like any other type of causal process invoked by scientists, can only be inferred from observations by use of theory that has been confirmed by long periods of empirical testing of critical hypotheses derived from plausible speculations. The transmission of information from the senses to actions can be regarded as going through communication 'channels' or perceptual processes. All types of mental process, conscious or unconscious, tend to be referred to by psychologists these days as 'cognitive'. Thus calculations to diagnose the sorts of channel or perceptual process that are in use by an assessor's evaluations of a set of food samples are called cognitive models.

2 Preference and Quality as Objective Perceptions

Disconcerting though the implications below may be, there is no fundamental distinction between the perception of material characteristics of a food and the perception of the food's quality with respect to those characteristics. The notion that sensory scores are objective while hedonic scores are subjective is a delusion based on scientifically invalid analysis of the numbers generated by the assessors.

Dispositions to select foods having certain characteristics rather than others, *i.e.*, preferences among foods, can be expressed very precisely and economically in words and numbers. The word used for the attitude—liked, pleasant, just right, preferred, likely to be chosen, good quality—makes little or no difference. The way the numbers are generated matters little either, so long as the test sample (not a magnitude of sensation) is placed at a point between the highest preference possible (ideal, perfect, always choose) and a lack of quality that borders on the unacceptable.

Such quantitative judgements (ratings) can also be used with descriptions of foods in more or less particular terms, *e.g.*, 'runny' or 'thick' on a line from 'not at all' to 'extremely'. These descriptive ratings can include the degree of liking or dislike for the particular level of the named characteristic in the food sample, *e.g.*, "so runny that I'd never choose it" or "exactly as thick as I like it". That is, sensory description and hedonic evaluation do not have to be separated, but can be incorporated into a single score.[2-6]

Traditionally, however, the descriptive assessments in sensory panels are supposed to have been emptied of attitudes. This intention is rooted in the mistaken view that preferences do not measure sensed characteristics. On the contrary, judgements of the preferred of two levels of a sensed food property can discriminate those levels just as well as the traditional profiling scores; the preference-based descriptive judgements have the bonus of specifying the individual assessor's ideal point too, when related to values of a stimulatory physico-chemical parameter.[7]

This error of regarding hedonic, affective preference or overall quality judgements as 'subjective' was corrected 20 years ago.[8] Yet the implications are still not generally recognized. A consumer's scores of liking or choice come to a peak at the personally ideal level of any sensed property of a food. Hence, before averaging such data across the individuals in a panel, a lack of preference below the ideal point must be given the opposite sign to a lack of preference above the ideal point. These unfolded scores are as comparable among assessors as the plain descriptions are.

An even more basic scientific fact about perception has been neglected since the beginnings of sensory evaluation over 40 years ago. This neglect has diverted an enormous amount of effort into developing descriptive vocabulary without any criteria for determining what the words describe. Indeed, the numbers generated in descriptive profiling can never identify what the raters have perceived in the rated samples of food. Even if the panellists have been trained with standard materials to illustrate the meanings of the descriptors, as recommended for sensory analysis of texture,[9] the information transmitted by each descriptor is seldom if ever identified as specific to a source of stimulation to the senses from foods of the sort to be descriptively profiled, let alone calibrated on chemically or physically measured levels of the specific property.

Very recently it has become fashionable to replace the traditional phrases containing the word 'sensory' by the term 'perception'. This makes the scientific mistake even worse, by claiming the one thing for sensory data which is logically impossible for them to provide—evidence on what sensed material properties of a food are reflected in a point on a panel's profile of scores.

In short, numerical scores of descriptions and/or preferences, however sophisticated their statistical modelling, by themselves tell us nothing about the science of what has been sensed in the food. The theoretical nature of a descriptive or 'hedonic' rating is a social communication, not any sort of surrogate chemical assay or physical measurement. Ratings cannot be related successfully to sensed physico-chemical properties until we have evidence as to what information from the food was actually processed through the senses to produce the observed attitude or ascription.

A science of sensory perception has to be an integration of the cognitive science of the perceivers' language with the physical and chemical science of the materials.

3 Food Psychophysics

The traditional name in psychological science for the relationship between a graded assessment of various samples of a particular material and a graded

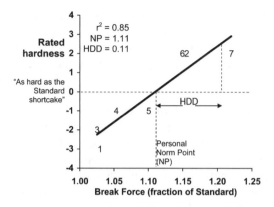

Figure 1 *The psychophysical function—an example of raw data on the texture of a highly heterogeneous food (shortcake biscuit) from one assessor rating seven samples in a single session. (It is not known if breaking force is sensed by audition, touch, kinaesthesis or some combination.) When the samples are not too dissimilar from what is perceived as top quality for that food (the personal Norm Point, NP, in this case 11% higher in 3-point break force than the Standard biscuit, as marketed), the rated intensities of a described characteristic (positive numbers =increasingly harder than standard; negative numbers=less hard) are linear against the amounts in the samples of a stimulus to that response (semi-logarithmic for material stimuli, i.e. ratios of amounts). Data points 1–7 refer to the sequence of presentation of the sample biscuits (r^2=variance accounted for by the regression line; HDD=half-discriminable difference between amounts of the stimulus.)*

physico-chemical property of those samples is the *psychophysical function* (Figure 1). This graph specifies the transformation achieved between input through the senses from patterns in the environment and the output of patterns of action on the environment. This transformation is a causal process in the mind of the assessor, totally dependent both on neural processing in the person's brain and on social processes in the person's culture, but not reducible or deconstructable into either.

Perhaps the most familiar practical example of a psychophysical function is a plot of ratings of how sweet a solution is against various concentrations of sucrose or some other sugar or sweetener in a beverage, *i.e.*, in water also containing some or all of other tastants, aroma volatiles, colourings, macro-molecules, and suspended, emulsified or colloidal particles (solid, liquid or gaseous). Unfortunately, academic psychophysics got caught up in a chase of the 'will of the wisp' of a general equation for intensity scores from stimulation of supposedly single types of sensory receptor from undetectable levels to saturation. This approach abstracted from its context the well-learnt perceptual process (*e.g.*, judging how sweet a food is, relative to some familiar or preferred level), and produced the bizarre enterprise of rating preference for pure solutions of sucrose or saccharin, and unreal claims that there is a universal innate preference for 10% sucrose.

In fact, very precise psychophysical functions can be obtained if the amount of a particular source of stimulation is varied (regardless of which or how many types of receptor are stimulated) around the amount that is familiar to an assessor. The test samples should otherwise be fairly similar to the usual form of the food, and either vary little in other ways or vary substantially in one or more other sensed features, but independently of each other—that is, with no two factors substantially correlated.

The way that psychophysical data are talked about often fails to allow for the logical necessity that data from only one input variable and one output variable are insufficient to provide evidence for or against any one type of mediating process in the mind. The perceptual performance could have been achieved by sensory processes close to the receptors, by linguistic processes close to the descriptive or attitudinal response, or by a phenomenological process such as a sensation or an emotion. That is to say, there is no scientific justification for regarding the construction of a psychophysical function from merely one response and one stimulus as the estimation either of subjective magnitudes (the fallacy of direct scaling) or of firing rates in afferent nerve fibres (the fallacy of neural reductionism).

When there are two or more observed patterns of output, on the other hand, it may be feasible to specify the type of mental process involved. Such covert functions are certainly diagnosable when there are also two or more observed patterns of input. These indirectly evidenced functions represent mental processes, either conscious or unconscious; they are otherwise known as cognitive mechanisms (that account for observed performance) or perceptual channels (of communication through the perceiver). The relationship between an observed stimulus and an observed response can be regarded as the limiting case of a mental process—one that is totally evident in overt performance.

4 Dependence of Precision on Context

Theoretically, a psychophysical function follows the intersection of a vertical plane with the surface of a cone, as shown in Figures 2 and 3. Sensed amounts of a characteristic are perceived as further away from the standard amount (at the cone's peak) as they go further below or above that level, *e.g.*, away from 'best quality' towards 'too little' or 'too much', respectively. When scores for amounts above the usual level are plotted above the score for the highest quality level, the peaked curve becomes the familiar monotonic function (Figure 1).

When the test samples are all of the highest quality (or are maximally preferred by the individual assessor when rating is relative to ideal), except for variation in one characteristic, the intersecting plane goes through the apex of the cone. Then the distances from the apex form an isosceles triangle (Figure 2, dashed line) and the 'unfolded' function is a straight line. However, if the basic preparation is of lower quality (or not personally ideal), the folded function will be a conic section (Figure 2, continuous line) of the form

$$R = m(S^2 + D^2)^{-0.5} + c, \tag{1}$$

where R is the response score to a sample, S is the level of the varied stimulus in that sample, and D is the size of the defect(s) in the preparation. (If the score

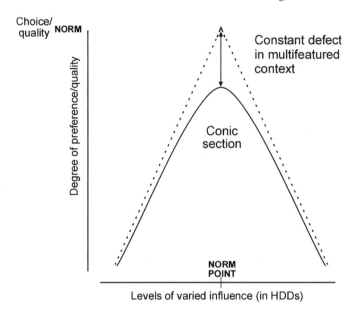

Figure 2 *Cognitive theory of quality recognition: a single influence in its familiar context (ref. 12). The effect of levels of the textural or other influence (scaled in half-discriminable differences) on a consumer's disposition to choose a food or perception of its quality (y-axis) falls on the surface of a cone described by that influence (x-axis) and the familiar context integrated from all other influences (z-axis). When the context is less than perfect in any respect, then no level of the varied influence under test can create perfect preference or quality, and the function outlines a vertical section through the cone away from the peak along the z-axis at a distance in HDD measuring the size of the contextual defect.*

for top quality or ideal is set at zero for analysis, then the constant c disappears, and the slope m becomes the proportionality between response units and stimulus-model units.)

Thus, a rounding of the preference triangle to an inverted U-shape diagnoses an unrealistic set of food samples (characteristic of hedonic scores for plain sugar + water). Even when experimentally varied factors are at the perfect level, a defective food preparation can never reach the highest quality.

It is of fundamental importance that the triangle (Figure 2), the two-feature cone (Figure 3), and the multiple-feature 'hypercone' is each defined by its apex; there is no particular base to the cone. There are limits to the linearity of the function at the extremes of stimulation below and above the level at the apex. As that sort of stimulus becomes hard to detect, the line steadily decreases in slope, changing from a logarithm of physical values to the square root.[10] At high levels (probably beyond anything that can be incorporated in food), the slope will also decrease as receptors approach saturation. More relevantly to the perception of foods, as the level of a sensed property deviates further from usual levels, so the whole formulation becomes less familiar and may even change in character. Hence preference will reach the limit of

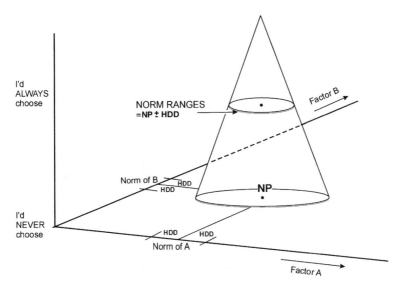

Figure 3 *Cognitive theory of quality recognition: the simplest case of two distinct influences on perceived overall quality or personal preference. (More than two influences are represented by a 'hypercone'.) The cone is drawn through the ideal ranges of the two factors (A and B) under investigation that influence a consumer (when the context is perfect for that person): the ideal range is the ideal point (IP), or more generally the norm point (NP) plus and minus one half-discriminable difference or fraction (HDD, HDF).*

acceptability, and different factors of unacceptability will operate. Thus the folded function is more like an omega (Ω) than a delta (Δ).

These limits on tolerance of deviations from the norm invalidate a key assumption made 30 years ago by the psychologists who developed the multi-dimensional scaling of preferences that are now widely used in sensory statistics (*e.g.*, MD-PREF). In these calculations, all potentially peaked data are fitted to a quadratic function. Yet the set of samples to be assessed is not selected for familiarity to, or tolerance by, the individual assessor. Hence, some of the data are outside the linear region and the theoretical function has a shape something like a Greek omega: therefore the data should be fitted to cubic or quartic functions, not quadratic.

On the other hand, the limits on the unfolded function provide a hard mechanistic basis for the use of fuzzy logic in analysis of psychophysical data collected with regard for individuals' ranges of preference. The non-linear extremes correspond to two-value logic and the linear phase between them carries all the useful information, albeit in far from fuzzy form.

5 'Tasting' and Texture Perception

The descriptors for sensory ratings that are thought to be strongly influenced by the purely physical characteristics of liquids or solids are generally called *textural* terms. 'Texture' is commonly taken to be delimited by mechanical properties sensed by touch, *i.e.*, usually 'mouthfeel', and possibly by kinaesthesis (senses of the muscles, tendons and joints). However, the cracking of hard solids can also be heard; indeed, there are no known mechano-receptors in the mouth that are sensitive to frequency. In addition, the spatial distribution of (near-)surface reflectance by liquids and solids can be sensed by eye (textural appearance). Therefore experiments on perception of the physical properties of foods must not assume that textural information is purely tactile.

Furthermore, the physical properties of foods sensed in the mouth are often related to their sensed chemical composition. In addition, they can greatly affect the release of solutes and volatiles, generating gustatory, irritative (pain) and astringent (texture) stimulation during mastication, as well as retronasal olfactory stimulation during swallowing. The mouthfeel of dairy creams is strongly confounded by the release of lipophilic volatiles, with 'creamy' aroma signalling fat content at least as strongly as apparent viscosity does.[11] The barely describable tastes of sodium chloride and lactose in the aqueous phase may also be important in preserving a genuinely dairy feel at high fat contents against an unnatural 'oily' or 'empty' impression.

Appreciation of the sensed material quality of a food or beverage is commonly termed 'tasting', whether during deliberate sampling of the material or incidentally to eating or drinking it. Thus the 'taste' of a food includes the effects of volatiles as well as solutes, and it may well extend to aspects of stimulation from spatio-temporal patterns of force or displacement generated by the food in the mouth. For example, chocolate may taste smooth as well as sweet, bitter and chocolatey (aroma). The 'flavour' of a food may also extend to textural aspects—such as the feel of cream perhaps. However, except for astringency, it is probably too strong to speak of an illusion of texture as taste in the mouth, in the way that retronasally sensed aroma is normally confused with oral gustation in the combined flavour of a food.

6 Perception of Particular Characteristics

It follows from the preceding discussion that mere differences in wording do not provide evidence for the perception of distinct characteristics of a food. Indeed, when sensory profiling scores are tested by factor analysis or multidimensional scaling, they are often found to be full of redundancies. This shows that, in practice, the panel used was incapable of using the vocabulary to distinguish among most of the chemical or physical properties which investigators tend to assume that the words to refer to.

There is, however, a far more basic reason why descriptive profiling does not work, and cannot ever work, as it has been practised for the last 40 years. The samples presented are never designed to make it logically possible for any method to distinguish the effects of different chemical or physical characteristics of a food or a type of food or beverage. This is a requirement of any sort of investigation on a whole material. The effects of apparent viscosities at low and high rates of shear, on passage through a pipe, pouring into a pot, or setting as a shape, cannot be separately measured unless samples are prepared in which these two characteristics vary independently of each other—that is, the two sets of values are uncorrelated. Otherwise, the effects of the two factors must be confounded and ratings on highly specific descriptors (or different physical measures) cannot give distinct results.

This is a matter of basic logic, not of statistics. It has to be addressed first by a fundamental examination of the hypotheses to be tested. If a breakfast drink is made up at different strengths without dissociating the concentrations of sugar from those of acid, no experimental design or statistical modelling can ever distinguish the effects of sugar and acid, or of sweetness and sourness, on aroma strength (for example) or anything else (such as shelf-life or sales). (The switch here to examples involving taste has no scientific significance: it is merely to clarify the point by use of well-understood examples of stimuli and responses.)

Food scientists are aware of an aspect of this issue as a statistical problem to be solved by blocked experimental designs. However, regression is a much more economical measure of the strength of an effect, such as in the slope of a psychophysical function. With only two levels of each factor whose effect is to be investigated, it needs only eight samples to disconfound seven factors (Table 1). Four samples exhaust the possible combinations of two factors, and eight samples cover the dissociations among three factors.

Another great advantage of regression over ANOVA, in fundamental as well as in applied research, is that the value of a factor's level does not have to be identical in replicate samples. The higher value (in Table 1) can be a range

Table 1 *Formulation requirements for sets of 8 samples of a food in which up to 7 sensed characteristics vary independently in strength and thus their effects can be measured separately. 'Hi' means values in a higher range. 'Lo' means values in a lower range. To avoid serious confounding, the two ranges must not overlap. To study interactions of mental processes in perception, both ranges in the factors studied must all be either above the top-quality point or below it. 'Lo' and 'Hi' ranges may be reversed without altering the disconfounding. For fewer than 7 factors, any set of 2–6 columns may be chosen. This may allow all or at least some of the eight variants to be selected from already existing formulations after their factor levels have been measured.*

| Food variant | Sensed characteristic to be varied | | | | | | |
	A	B	C	D	E	F	G
S	Lo	Lo	Lo	Lo	Hi	Hi	Hi
T	Lo	Lo	Hi	Hi	Lo	Lo	Hi
U	Lo	Hi	Lo	Hi	Hi	Lo	Lo
V	Lo	Hi	Hi	Lo	Lo	Hi	Lo
W	Hi	Lo	Lo	Hi	Lo	Hi	Lo
X	Hi	Lo	Hi	Lo	Hi	Lo	Lo
Y	Hi	Hi	Lo	Lo	Lo	Lo	Hi
Z	Hi	Hi	Hi	Hi	Hi	HI	Hi

of values in the different samples, so long as it does not overlap with the low range of values. The correlation coefficient r will not then be zero; nevertheless, as long as we have the condition $r < 0.3$, then over 90% of the variance in the results will be separately attributable to one or other of the two factors.

7 Sensory Continua and Complex Characteristics

A sensory characteristic that combines two or more simpler features behaves quite differently from a simple characteristic. Even after the contextually appropriate level for a simple factor has been learnt, there is still a continuum from 'too little' through 'just right' until 'too much'. However, two characteristics have to be in balance in order for a complex characteristic composed of them to vary in strength. The quantity may still change if the ratio changes, but the quality certainly changes.[12]

Thus, for example, it is far from certain that there is a genuine continuum of creaminess. Over a range between two familiar milks or creams, creaminess ratings are very closely related to judgement of fat content: indeed, creaminess explains fat content judgements.[13] However, the balance of features in 'double' cream is likely to differ qualitatively from that in high-fat milk. Thus, fat content may be judged by deciding upon the closest two familiar creams, judging the distance of the test product from each of them, and then interpolating the believed fat contents.

One of the greatest difficulties in research on a complex material is to estimate the nature or even the importance of the effect of a factor that has not been tested. For example, Stanley[14] varied only oil fraction, and not droplet sizes, in her study of the physical bases of creaminess. This may explain why she estimated that aroma and taste each contributed over 40% of creaminess, viscosity about 15%, and fat content only about 3%. The average distance between droplets can have major effects on viscosity. At high oil fractions, droplet size is likely also to have a serious impact on viscosity. Furthermore, colloidal and emulsion particles might affect the sense of touch independently of their influences on viscosity (see below).

8 Model Systems

For a model system to be of scientific value, it has to be sufficiently close to reality to provide opportunities for advancing the understanding of actual systems. This criterion is not met by colloids or emulsions in which the particles are of size or shape outside the range found in the real food system. The effects of particles ten times the size of a typical dairy cream droplet may not be relevant to the texture of dairy creams.[15] The effects of plate-like particles are unlikely to relate to those of globular particles.[16] The effects of cellulose granules may be quite different from those of dextrin granules.[17]

This point about closeness to reality applies equally to measurement rigs. Frictional effects between wet surfaces in the mouth are not well modelled, for instance, by dry leather.[18,19] Even now, there is no instrumental model for the smoothness of creams.[20]

So-called 'fat replacers' are generally thickeners that totally fail to mimic the smoothness aspect of creaminess. Indeed some starches have a graininess, and some gelling agents an elasticity, which bears no relation to a real milk product. Consumers looking only for a fruit flavour and a slightly acid dessert may not care about the absence of creaminess in a yoghurt, and consumer quality standards are undoubtedly changed (some might say degraded) by the products they get used to eating.

9 Real Foods and their Simulation

The sets of samples investigated in published sensory profiling studies are typically a rather heterogeneous collection of foods, rather than variants of a single type. With such sample sets, it is fallacious to assume that a single descriptor has a single physical source. There may be many different forms of roughness and smoothness (and indeed of 'thickness' or 'body'). Furthermore, if two or more physical parameters contribute to a single descriptive term or conscious sensation, then (as mentioned in section 7) the 'balance' amongst them might vary with examples of smoothness or creaminess.[11] This would show in different cognitive models of microstructural factors for the creaminess of, say, whole milk as compared with single cream—let alone yoghurt or crème caramel.

Yet a diversity of products is needed in order to get interesting pictures out of the statistical modelling that continues to dominate food sensory research. This is another feature of sensorimetrics that prevents that approach from contributing to scientific understanding of the perception of food. The only way that psychological science can begin to investigate the sensed physico-chemical basis of food quality is by focusing studies on ranges of products without qualitative divergences in character.

The practice of testing diverse foods can lead to totally erroneous conclusions. A key piece of evidence for the tribological theory of creamy smoothness[18,19] is plainly an artefact of including sherbert and frozen orange juice: these serve as outliers, and are solely responsible for the claimed relation between normalized frictional force (from lubrication in a leather-surfaced rig) and panel-judged smoothness. There is clearly no correlation at all between instrumental values and panel scores among the ten samples of various types of dairy emulsion.[18] (In any case, it is difficult to see how oil-in-water emulsions could vary in their lubricating effect in the mouth. The tissue surfaces are all thoroughly wetted already. Furthermore, saliva contains mucins that make it extremely slippery.)

The moral is that samples must be selected so that they each exemplify one of at least two different values of a hypothetically sensed material factor, unconfounded by correlated variation in any other potentially sensed factor. Such a sub-set of samples may not exist in the market. Indeed, it is in principle likely that sensorily influential factors are varied together during formulation because they arise from the same constituent, *e.g.*, increases in oil-soluble aroma release and apparent viscosity as the oil fraction increases in a dairy emulsion. Hence, advances in scientific understanding depend on designing at least some artificial samples with formulations that break up the confounding of factors in the set to be tested.

Yet the nature of the scientific problem requires that these artificial samples are perceived as natural, or are sensed to be close enough to familiar products to permit accurate quantitative judgements of differences from normal in overall quality or in described characteristics. This is not a requirement that the samples be 'palatable'. It is, however, a consequence of this requirement that none of the samples presented to an individual consumer should go outside that person's range of tolerance for deviations from preferred levels of all the sensed characteristics—and not just the factor(s) being investigated. It is crucial to the linearity of each psychophysical function that all tested samples are recognized as the same food, as might normally be expected from marketed variants.

10 A Microstructural Source of Texture

A relatively rigid macrostructure is obviously important to the texture of a solid food, together with its destruction in the mouth by processes such as compression between the teeth and wetting by saliva. Particles at the millimetre scale within semi-solid materials are important to the character and quality of a food; in size, shape and/or hardness, they range from the lumpiness of cereal grains, through the graininess of potato granules, to the smoothness of effectively milled cocoa solids in chocolate.

It is widely thought that particles have to be tens of micrometres in diameter in order to be felt. Nevertheless, undetectably small particles may help to make chocolate feel smooth. Furthermore, a clearly gritty feel is produced by particles of only ~10 μm in diameter if they have sharp points on them, such as the mineral particles serving as the abrasive in toothpaste. Nevertheless, the fingertips are capable of detecting particles as small as one micrometre as asperities on otherwise smooth plates.[21] It is therefore theoretically conceivable that colloidal particles and emulsion droplets as small as a half a micrometre in diameter could influence texture if the fluid were pressed sufficiently thin between tactually sensitive tissue surfaces for the two surfaces to press the particles in the monolayer into each other.

This microstructural hypothesis of a creamy 'smoothness', distinct from thickness, was supported by an increase in consumers' 'creamy' ratings halfway from 'high-fat milk' to 'single cream' of dairy emulsions when the fat content was raised, the emulsion homogenized, and thickener added.[13] Not only did the viscosity have to be raised, but also the number of fat droplets increased, and their sizes reduced and narrowed in distribution. Homogenization of fat droplets and size control of gas bubbles is now widely used to improve the creamy smoothness of dairy products and brewed stouts, respectively. Preliminary tests of samples in which dairy fat and vegetable oil emulsions were varied independently in viscosity and in sizes and spacings of droplets supports the hypothesis that droplet size and/or distance between droplets as well as viscosity contributed to rated creaminess.[22] That pilot work is now being replicated and extended in an attempt to assess the contributions of droplet size and interdroplet distance to the rated smoothness and creaminess of artificial creams, and possibly also to differences in shear-thinning behaviour and/or low-shear-rate viscosity or yield stress, as distinct from the contribution of apparent viscosity at high rates of shear, in relation to rated thickness and creaminess.

Only two of the possible types of cognitive interactions were calculated in our initial publication.[22] The same data have now been analysed for all cognitive models, the best of which is what the evidence indicates was the perceptual integration (Tables 2 and 3). Viscosity contributed to creaminess for 5 of the 6 assessors, but it was a dominant factor for only 3. The width of the distribution of droplet sizes contributed more often for the 5 assessors who were influenced by it, and the spacing between droplets occasionally contributed when viscosity did not.

A packet creamer consists of uniformly sized globules of vegetable fat suspended in maltodextrin: the dilution of this carbohydrate and the dispersion of the lipid particles may leave a rather slight rheological effect, and so the intensely creamy mouthfeel of the coffee (if it can be distinguished from creamy aroma and slightly sweet taste) may depend on the size and spacing of these minute 'ballbearings'. Presumably microparticulated egg protein[23] works in the same way.

Table 2 *Perceptual interactions in ratings of dairy emulsions relative to 'Gold Top' (high-fat) milk or single cream. This is a full analysis of data preliminarily presented in ref. 22. The physical parameters varied among the milk samples are: V=viscosity at shear-rate of 50 s⁻¹, representing rheological factors; S=size (mean diameter) of lipid droplets, fitting into tissue anisotropies; W=width ('span') of droplet sizes, the (un)evenness of the monolayer; and I=interdroplet distance (mean), the spacing (packing) of the monolayer. The models differentiate the sum of distances from the norm (+) from the square root of the sum of squares of distances from the norm—the hypotenuse of a right-angled triangle (¬).*

	Anchor term for ratings of 'creamy'	
Assessor number	*'Gold Top' milk*	*'Single cream'*
1	W ¬ S ¬ (S + V)	W ¬ S
2	V + W	(V + W) ¬ (W + S)
3	W ¬ I	W
4	(S + I + W) ¬ (S + I)	V ¬ W
5	I	V ¬ I
6	W	V ¬ W

Table 3 *Percentage contributions of rheology and microstructure to the perceptual interactions of Table 2.*

Assessor number code	\Anchor term for ratings of 'creamy'\							
	'Gold Top' milk				'Single cream'			
	Droplet properties				Droplet properties			
	Viscosity	Size	Width	Intercentre distance	Viscosity	Size	Width	Intercentre distance
1	32	0	38	30	0	2	97	0
2	59	0	40	0	33	22	32	13
3	0	0	60	39	0	14	50	37
4	0	39	6	55	44	28	28	0
5	55	0	0	45	66	0	0	33
6	0	0	100	0	51	0	48	0
Mean %	*24*	*7*	*41*	*28*	*32*	*11*	*43*	*14*
k	*3*	*1*	*5*	*4*	*4*	*4*	*5*	*3*

Other groups have found very small effects of fat droplets. However, weaknesses in experimental design can be adduced. An early study[14] estimated the contribution of fat to be only about 3%, but only droplet spacing was varied (over an unspecified range), and not droplet size. More recent studies[24,25] may show an effect of homogenization when viscosity is appropriately high; however, the critical data are hidden in the wide ranges used. Another study[26] showed a small effect of homogenization, but whitener was needed to bring it out.

A microstructural theory of texture falls at the first post if consumers cannot sense differences in droplet sizes or spacing. Preliminary work confirms that some consumers can discriminate distances between droplet centres, even in the mouth (Table 4). However, differential acuity ('suprathreshold' sensitivity)

Table 4 *Tactile sensing of rheology and microstructure by fingers, lips and tongue-on-palate: panel median acuities (50% discriminable differences) of cream descriptors for high shear-rate (50 s⁻¹) viscosity (V), correlated with low shear-rate viscosity, or mean intercentre distance between droplets (I). About 0.1 ml of butter fat emulsion was pipetted onto the lower tissue surface (unseen), the assessor wiped it once against the upper surface, and then immediately made the three ratings.*

| | Fingertips (dry) | | Dry lips | | Wet lips | | Palate/tongue | |
	V	I	V	I	V	I	V	I
'thick'	$>10^{-9}$	0.13	26	0.16	$>10^2$	0.7	14	0.20
'smooth'	$>10^{-2}$	0.44	69	0.27	$>10^2$	0.3	$>10^2$	0.53
'creamy'	$>10^{-2}$	1.18	17	0.42	$>10^4$	0.5	$>>10^2$	0.63

Table 5 *Relative specificity of 'thick' to rheology and of 'smooth' to microstructure: an illustration of evidence from double dissociation of the thickness/viscosity function from the smoothness/spacing function in perceptual measurements of butter oil emulsions during eating (assessor 57 4/03). Acuity (50% discriminable difference) for uncorrelated variations in viscosity and droplet spacing in cream-like samples.*

	Rating relative to 'double' cream		
	'thick'	'smooth'	'creamy'
Apparent viscosity (50 s⁻¹)	<u>0.08</u>	2.02	0.36
Oil fraction (inter-centre distance)	1.5×10^3	<u>1.01</u>	7.1

is poor at best. Also, descriptive vocabulary is problematic. Even 'thick' may not be totally unambiguous in its use to describe a dairy emulsion: ratings on this term are more sensitive to droplet spacing than are smoothness ratings (Table 4). The term 'smooth' may be even more ambiguous. Some assessors rate closer spacing of droplets as smoother (as predicted by the 'ball-race' model). Others use the word in the opposite sense. Aggregation of droplets may be evaluated variously as 'oily' or as 'lumpy' or 'rough'. Nonetheless, there are signs that viscosity strongly controls thickness while droplet spacing more strongly controls smoothness. Table 5 illustrates the characteristics of such a double dissociation from one assessor's data. This pattern of findings will have to be replicated within and across assessors before we have real evidence that perceived thickness is rheological while perceived smoothness is structural.

The microstructural theory of creamy smoothness faces a conundrum. It is conceivable that droplet sizes and spacings are sensed through the dynamic spatial distributions of forces generated by complex variations in shear. The closeness of droplets and similarity in size could affect the streaming of the emulsion as its bolus is squeezed into a thin layer between rounded (and rough) surfaces of sensing tissues. This is a challenging task to be tackled by computational rheology.

Experimentalists face a severe challenge in attempting to vary rheological properties independently of each other in sets of realistic emulsions. The asymptotic value of apparent viscosity that is approached with increasing rates of shear (such as at 50 s⁻¹) has to be dissociated from viscosity at very low rates that may approximate a yield stress, and both these parameters unconfounded from the rate of shear-thinning measured for example as the difference in apparent viscosity between a very low rate and a modest rate. To test microstructural against rheological theory, each of these rheological parameters must be dissociated in its effect on creaminess of a familiar product from the effect of distances between droplets, diversity in size of droplets, and the central tendency of droplet size.

Mental integration of these two sorts of information can be investigated by cognitive modelling of interactions among percepts based on either rheological or microstructural sources of tactile stimulation. However, only the investigation of the physical chemistry can determine if complex rheological interactions mediate the effects of microstructure.

11 Concluding Remark

It is beginning to be appreciated that the applicable scientific understanding of food properties requires hard quantitative evidence on how consumers build knowledge through the senses. It has long been realized that a deep understanding is needed of the forces arising from molecular interaction within the material as it moves through the mouth. An implication of the present work is that neither of these two fields of science can be fully effective in food research without close collaboration. Each science will always retain large areas completely separate, but where they overlap, in psychophysics, then some theoretical unification may be necessary.

Acknowledgement

Current work is being carried out in collaboration with the Procter Department of Food Science, University of Leeds, on funds from BBSRC (UK).

References

1. D. A. Booth, *Appetite*, 1986, **7**, 236.
2. D. A. Booth and M. T. Conner, *Food Qual. Pref.*, 1991, **2**, 75.
3. R. P. J. Freeman, N. J. Richardson, M. S. Kendal-Reed, and D. A. Booth, *Brit. Food J.*, 1993, **95**, 37.
4. D. A. Booth, S. Mobini, T. Earl, and C. J. Wainwright, *J. Food Sci.*, 2003, **68**, 382.
5. D. A. Booth, T. Earl, and S. Mobini, *Appetite*, 2003, **40**, 69.
6. D. A. Booth, S. Mobini, and T. Earl, *J. Food Qual.*, 2003, **26**, 425.
7. R. L. McBride and D. A. Booth, *J. Food Technol.*, 1986, **21**, 775.
8. D. A. Booth, A. L. Thompson, and B. Shahedian, *Appetite*, 1983, **4**, 301.
9. A. S. Szczesniak, M. A. Brandt, and H. H. Friedman, *J. Food Sci.*, 1963, **28**, 397.
10. D. Laming, 'Sensory Processes', Academic Press, London, 1986.
11. C. A. Yackinous and J.-X. Guinard, *J. Food Sci.*, 2000, **65**, 909.
12. D. A. Booth and R. P. J. Freeman, *Acta Psychol.*, 1993, **84**, 1.
13. N. J. Richardson, D. A. Booth, and N. L. Stanley, *J. Sensory Stud.*, 1993, **8**, 133.
14. N. L. Stanley, 'The physical and sensory quality of low fat dairy emulsions', Ph.D. Thesis, Silsoe College, Cranfield University, 1989.
15. D. Kilcast and S. Clegg, *Food Qual. Pref.*, 2002, **13**, 609.
16. P. Tyle, *Acta Psychol.*, 1993, **84**, 111.
17. I. Ramirez, *Physiol. Behavior*, 1992, **52**, 535.
18. J. L. Kokini, *Food Technol.*, 1985, **39**, 86.
19. J. L. Kokini, *J. Food Eng.*, 1987, **6**, 51.
20. R. B. Pereira, H. Singh, P. A. Munro, and M. S. Luckman, *Int. Dairy J.*, 2003, **13**, 655.

21. M. A. Srinivasan, J. M. Whitehouse, and R. H. LaMotte, *J. Neurophysiol.*, 1990, **63**, 1312.
22. N. J. Richardson and D. A. Booth, *Acta Psychol.*, 1993, **84**, 93.
23. N. S. Singer, in 'Handbook of Fat Replacers', eds S. Roller and S. A. Jones, CRC Press, Boca Raton, FL, 1996.
24. D. J. Mela, *Appetite*, 1988, **10**, 37.
25. D. J. Mela, K. R. Langley, and A. Martin, *Appetite*, 1994, **22**, 67.
26. M. B. Frost, G. Dijksterhuis, and M. Martens, *Food Qual. Pref.*, 2001, **12**, 327.

Colloidal Behaviour of Food Emulsions under Oral Conditions

By George A. van Aken,[1,2] Monique H. Vingerhoeds,[1,3] and Els H. A. de Hoog[1,2]

[1]WAGENINGEN CENTRE FOR FOOD SCIENCES (WCFS), P.O. BOX 557, 6700 AN WAGENINGEN, THE NETHERLANDS
[2]NIZO FOOD RESEARCH, P.O. BOX 20, 6710 BA EDE, THE NETHERLANDS
[3]AGROTECHNOLOGY & FOOD INNOVATIONS, P.O. BOX 17, 6700 AA WAGENINGEN, THE NETHERLANDS

1 Introduction

One of the key aspects of the consumer preference of food is its sensory quality. This explains the interest from the food industry in sensory evaluation and the numerous studies on sensory perception of food ingredients and compound food products. Sensory science includes the human perception of food attributes and the function of sensory receptors. The food industry takes an interest in sensory science for the purpose of product evaluation, and also to design new products based on an understanding of consumer preference. However, this connection between food preference on the consumer side and product composition on the technology side is extremely complicated. Figure 1 schematically shows some of the main aspects involved in making this connection. Starting at the top of the diagram, consumer preference is related to the hedonic reaction of a consumer to the sensory perception of the product, but it is also strongly influenced by factors such as cost, expected effect on health, reliability, culture and personal habits.

Sensory perception partially takes place inside the mouth, and, by the retronasal route, in the nose. But before the food is taken into the mouth, the food is judged by its visual appearance (*e.g.*, colour, homogeneity, smoothness/ shine of the surface, and wetting of the container), its rheological behaviour, and its fragrance. In the mouth, the food is masticated and tasted. Some understanding of these processes can be gained from the biological function of the mouth,[1] which is to detect nutritional and harmful aspects of a food

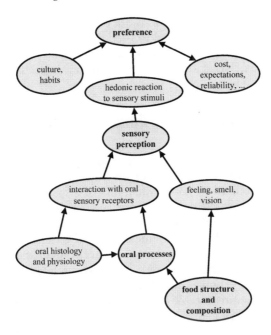

Figure 1 *Diagrammatic representation of the main aspects involved in the connection between food preference and product technology.*

product, and to process the food into a digestible form—often a slippery bolus—that can be swallowed safely. Sensory perception in the mouth is directly related to the stimulation of tactile, taste and flavour receptors, which interact with the various components and structure of the food *via* a multitude of oral processes. These processes include product deformation and flow, mixing with saliva, interaction with saliva components, friction, and transport of ingredients such as flavour compounds.

An important gap in the knowledge needed to make the connection between food preference and food technology lies in our understanding of the processes that take place in the mouth.[2] For some systems, sensory attributes such as thickness and creaminess can be related to the rheological behaviour of the food product measured extra-orally, as long as the appropriate flow regime (in terms of shear-rate, extensional *versus* shear deformation) is applied.[3] However, the oral environment is much more complicated than a rheometer. During mastication the product is exposed to a large variety of flow rates, with a change in temperature and interaction with saliva altering the composition and texture of the product. Moreover, the sensory receptors are located at, or buried inside, the oral surfaces; therefore the interaction of the food material with these surfaces (*e.g.*, friction, adhesion, and coating formation) must play an important role.

Recently, we have started an investigation into the physico-chemical aspects of the effect of emulsion droplets and foam bubbles on the oral perception of compound food systems. Our aim is to obtain a better understanding of common sensorial attributes related to food emulsions ('creaminess', 'smoothness', 'fatty', 'greasy', 'mouth-coating' and 'soothing'), and the release of fat-soluble flavours, based on the physico-chemical mechanisms that occur during oral processing.

2 Oral Behaviour of Emulsions

An important sensory aspect of emulsion droplets is that they provide the highly desired sense of 'creaminess' to a product. It is likely that the hedonic preference for creaminess is related to the expectation of a high nutritional value in a product caused by good availability of dispersed fat. This preference may be inherited or based on the learned connection between the sense of creaminess and the resulting satisfaction related to a high energy intake. Fat replacers in low-fat products may mimic some of these aspects, inducing the desired experience of creaminess, but in reality fooling the receptors in the mouth. However, if the experience of creaminess is related to a learned expectation of a high energy intake, the effect may decline. In any case, it is interesting to know how the presence of emulsion droplets and the easy release of fat from these droplets are identified in the mouth. In some way, the emulsion droplets must change the rheological properties of the food product as experienced in the mouth, or adhere to the oral surfaces and change lubrication, or destabilize by aggregation and coalescence.

The oral behaviour of emulsions must be related to a large number of rheological parameters and colloidal processes, some of which can be measured outside the mouth, but some of which are necessarily determined by the specific conditions in the oral cavity. During mastication the emulsion is heated or cooled to body temperature, mixed with saliva (involving dilution, agitation, and interaction with saliva components such as mucins and enzymes), brought into contact with interfaces (mucosa, teeth, air bubbles), and squeezed and sheared between the palate and tongue. These processes will lead to several structural changes in the food product.

The type and intensity of these oral processes may vary enormously. The first few times a food product is encountered, the product is carefully tasted, and the experienced appearance, rheological behaviour, flavour, and oral sensory profile are stored in the consumer's memory, together with the experience of well-being afterwards. It has been shown[4] that, during the sensory panel testing of mayonnaises and custards, the scoring of sensory attributes is the highest for natural oral manipulations ('normal') in comparison with various types of controlled oral manipulations (see Figure 2). This suggests that, during careful tasting, the food material is explored sensorially by varying and optimizing the oral manipulations to get a maximum sensory response. It is likely that this is the way the consumer learns which foods are pleasant, healthy, nutritious or noxious, and how they can be recognized by the appearance, rheological behaviour, flavour and taste. Once a product is known to a consumer, the tasting becomes more superficial and more focused on differences compared to previous experiences.

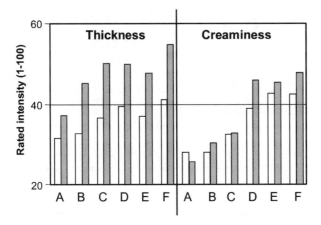

Figure 2 *Effect of oral movements on texture sensations of commercially available semi-solid vanilla custard dessert (3.0% fat, white bars) and mayonnaise (72% fat, grey bars). The various oral processes performed for 5 seconds before swallowing are: A, 'still', i.e., placed on the tongue; B, 'up', i.e., placed on the tongue, which is then pressed against the hard palate; C, 'up & down', i.e., placed on the tongue, which is pressed against the hard palate, and then lowered; D, 'suck', i.e., placed on the tongue, and the bottom of the tongue is raised and lowered 10 times; E, 'smear', i.e., placed on the tongue, and then the tip of the tongue is moved (10 times) in a figure of 8 pattern against the hard palate; F, 'normal', i.e., normal mouth movements.*
(Redrawn from ref. 4.)

3 Evidence for Structural Change of Emulsions in the Mouth

In this section we show and discuss some preliminary evidence of processes that alter the structure and behaviour of emulsions in the mouth. These changes may be directly or indirectly related to the perception of the emulsion.

Saliva-Induced Aggregation of Droplets

Whole saliva is the mixture of excretions from the various salivary glands in the mouth. The mixing of whole saliva into liquid emulsions leads to droplet flocculation and also to droplet adhesion to epithelial cells. This effect is immediately evident from Figure 3, which shows three images of emulsions mixed with saliva from which cellular debris had been removed by centrifugation.

An important component in saliva is the high-molecular-weight mucin (MUC5B).[5] Because of its large molecular weight and relatively high concentration in saliva, we suspect that this component may be important for the aggregation process. Therefore, we have studied the interaction of emulsions with a similar type of mucin, pig gastric mucin (PGM), which can be obtained relatively pure in large quantities. Also PGM induces flocculation of emulsions, and many of characteristics of PGM-induced flocculation resemble those of dextran-induced depletion flocculation,[6,7] as caused by dextran of a

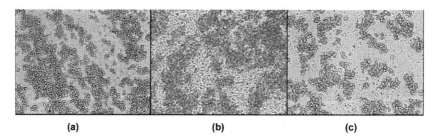

(a)   (b)   (c)

Figure 3 *Flocculation after mixing with saliva of 10% oil-in-water emulsions stabilized by 1% of (a) whey protein isolate, (b) sodium caseinate, and (c) Tween 20.*

Table 1 *Characteristics of the polymers, dextran and pig gastric mucin (PGM), which induce depletion flocculation of emulsions. (Data from refs. 6 and 7.)*

Polymer	M_w (kDa)	R_g (nm)
Dextran	2000	32
PGM	2000	49

similar molecular weight (Table 1). Firstly, the shape and density of the flocs are similar, and they are formed on a timescale roughly corresponding to transport-limited ('fast') flocculation kinetics, *i.e.*, for an unstirred 10% emulsion the flocculation proceeds within seconds. This also indicates that the flocculation is sufficiently fast to occur during passage through the mouth. Secondly, the flocculated emulsion slowly secretes a serum phase, separated by a sharp 'phase boundary' from the emulsion-rich phase. Although this process is too slow to be of importance during mastication, it indicates that the droplets form an aggregated network bound by relatively weak and flexible 'secondary' bonds. Thirdly, a sharply defined minimum PGM concentration was needed to induce the aggregation. This concentration, roughly corresponding to the theoretical value calculated according to the procedure described by Blijdenstein *et al.*,[8] does not strongly depend on the nature of the emulsifier but it decreases for smaller droplet radii (not shown). Finally, the observed aggregation was found to be reversible upon dilution with water, indicating that the aggregation process does not involve the formation of covalent bonds between droplets. Taken all together, these aspects strongly suggest that PGM-induced flocculation is caused by the depletion mechanism in the presence of mucins.

Although depletion may be an important mechanism in saliva-induced flocculation, the mucin concentration in saliva is considerably lower than the minimum PGM concentration required to induce flocculation. This may be related to the high degree of self-assembly of the salivary mucins, leading to a much higher effective molecular weight for the salivary mucins. Molecular weights exceeding 10 MDa are reported in the literature[9] for assemblies of salivary mucins. In addition, even though saliva samples collected from several volunteers at different times of the day always showed flocculation, there was some variation in structure of the flocs and in the degree of reversibility on dilution between samples from different individuals.[7] Therefore it can be inferred that possibly other mechanisms, such as bridging flocculation or specific coagulation reactions, could also be important in saliva.

To the best of our knowledge, the observation that mixing emulsions with saliva or mucins can induce droplet aggregation was not previously reported in the literature. We anticipate that the observed behaviour affects the texture and rheological properties of emulsions, which consequently affects the sensory perception of emulsions. For example, saliva-induced droplet flocculation leads to an increase in low-shear viscosity,[7] which should be taken into account when trying to correlate product viscosity with mouthfeel. We expect that sensory descriptors such as 'thickness', 'creaminess' and 'sliminess' are especially affected by this behaviour. In addition, it may also lead to droplet adhesion to the oral epithelia.

Formation of Slimy Structures

To study the relevance of the observed aggregation upon addition of PGM or saliva to emulsions, explorative experiments were carried out with milk (1.5% and 3% fat) and cream (40% fat). The emulsions were taken in the mouth for one minute to ensure interaction with the saliva and then spat out. The emulsions that had been spat out were clearly 'slimy' and contained loose droplet aggregates (Figure 4). The aggregates were often elongated in structure, and the droplets seemed to be bound to slimy 'strings' instead of actually touching each other. These slimy strings were often seen to incorporate loose epithelial cells. The sliminess of the spat-out material appeared to increase with the fat content: it was somewhat visible for the two milks and very apparent for the cream. For comparison, some sliminess was also visible in the spat-out sample

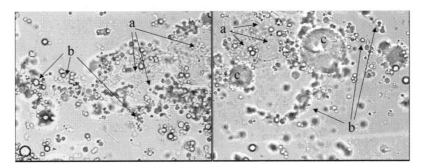

Figure 4 *Slimy 'stringy' structures incorporating emulsion droplets, observed in a spat-out sample of dairy cream kept in the mouth for 1 minute (left) and 4 minutes (right). Visible are (a) epithelial cells, (b) stringy structures and (c) large emulsion droplets formed by coalescence. Note that the epithelial cells are covered by emulsion droplets.*

of a 4% sodium caseinate solution, although to a much lesser degree. Sensory analysis revealed differences in 'melting', 'creamy afterfeel', 'slime formation' and 'fatty coating' between emulsions stabilized by whey protein isolate and sodium caseinate. It is hypothesised that the addition of mucin to casein, either from PGM or saliva, produces slimy complex coacervates, entrapping the emulsion droplets and binding them to particles and oral tissues. Possibly, the presence of fat globules enhances the coherence of the slimy matter by forming multi-point anchoring places for the slimy threads.

Shear-Induced Coalescence

Food emulsions are exposed to vigorous mechanical disturbance during their residence in the oral cavity. One important aspect that may affect the stability of the emulsion in the mouth is the rubbing of tongue and palate. Being confined to a small gap between the two mucosal surfaces, which may be of the order of the size of the emulsion droplets or even smaller, the emulsion droplets and their aggregates are subjected to severe shear stresses and will tend to interact intensely with the surfaces. For example, for a gap size of the order of the typical droplet size (1 μm) and a lateral velocity between the tongue and palate of typically 1 cm s^{-1}, the local shear-rate is estimated to be 10^4 s^{-1}. These hydrodynamic conditions can lead to flow-induced coalescence in many emulsion systems.

In a first explorative experiment, the behaviour of an emulsion between tongue and palate was simulated by rubbing a rubber pestle on the surface of microscopical cover slip in a cuvette filled with an emulsion (40% sunflower oil, stabilized by β-lactoglobulin, 0.1 M NaCl, average droplet diameter 1 μm). Rubber was used because it emulates realistic physical parameters (such as elasticity and roughness) of tribological contacts between soft, deformable surfaces like the epithelial surfaces in the oral cavity. Figure 5 shows a sequence of subsequent movements made with the pestle. We see that considerable coalescence has occurred, especially when the pestle was rotated, which corresponds to a simple shear movement between surfaces at a distance of approximately 5 μm. Estimating a pestle velocity of the order of 1 cm s^{-1} and a distance between the surfaces of the order of 5 μm (from CSLM images of the *xz*-plane, the plane parallel to the field of gravity), it can be calculated that the shear-rate would have been of the order of 2000 s^{-1}. Similar results were obtained by using a piece of a pig's tongue, instead of the rubber pestle. The tongue tissue was rubbed by hand over the underlying glass coverslip, with the emulsion in between, applying a slight pressure. Coalescence was indicated by the formation of regions of large emulsion droplets (observable as the black areas in Figure 6).

These experiments show that β-lactoglobulin-stabilized emulsions can become unstable under shear when confined to a small gap, even though the emulsions are otherwise very stable against shear. The precise mechanism by which coalescence occurred in these experiments is not yet known. It may be

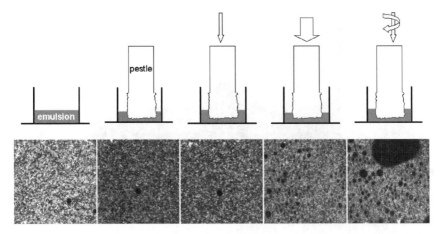

Figure 5 *Confocal microscopy images of the spacing between a rubber pestle and a glass surface in a cuvette filled with emulsion, after a variety of actions as schematically drawn above. The fluorescent labelling is by rhodamine, which shows the protein-rich areas as the bright areas. The black areas correspond to oil drops released by coalescence. The image sizes are 160 μm × 160 μm.*

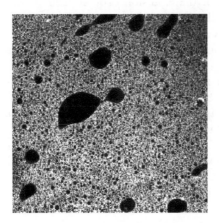

Figure 6 *Confocal microscopy image of an emulsion rubbed between a piece of a pig's tongue and a glass coverslip. The fluorescent labelling is by rhodamine, which shows the protein-rich areas as bright. The black areas consist of oil released by coalescence. The image size is 160 μm × 160 μm.*

caused directly by the shear flow, but it is also likely that the surfaces themselves are also involved *via* a hetero-coalescence process.[10] According to this latter explanation, the properties of the surfaces would be of great importance. The roughness of the surface might enhance or inhibit coalescence of emulsion droplets; the presence of a mucus layer on an epithelial surface would interact with the emulsion droplets and might induce aggregation or coalescence. Finally, the hydrophobicity of the surface determines the wetting properties, which will have an effect on the interaction with the droplets.

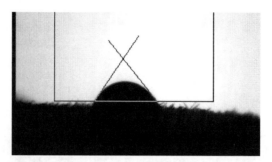

Figure 7 *Typical image obtained with the automated drop tensiometer of a water droplet on a pig's tongue surface pre-treated by rubbing with oil. The fitted lines are used in the determination of the contact angle. Since this surface is not well defined, the accuracy of the measurement is limited. The observed roughness is caused by the morphology of the papillae.*

Although the tongue surface is naturally hydrophilic, probably because of the adhering mucous coating, the surface can be made hydrophobic by rubbing with oil. Figure 7 shows that a droplet of water has a finite contact angle at a tongue surface that has been rubbed with oil.

The coalescence of the emulsion droplets and the associated release of oil will tend to affect the oral coating present during eating and that remaining after swallowing. We anticipate that the coating will strongly affect the perception of various attributes, such as astringency, fattiness, roughness, slipperiness and smoothness.

Another step in studying emulsion behaviour confined between shearing surfaces was made by measuring the friction coefficient. This is a measure of the lubricating behaviour of the emulsions. In several studies a correlation has been reported[11] between the instrumentally measured friction and various sensory attributes. We have carried out preliminary experiments which show that the lubricating properties of the emulsions can be tuned by varying the emulsion properties. The effect of the type of emulisfier was found to be especially great. Further investigations will be carried out to interpret the effects on the friction by studying the emulsion behaviour at the droplet level.

Fat Spreading at Air Bubble Surfaces

When a food emulsion is taken into the mouth, some air is inevitably included. The emulsion droplets, almost always based on triglyceride oils or fats, can release liquid oil by spreading it onto the air–water surface, and so forming a triglyceride monolayer. This process has been studied in detail by Hotrum *et al.*,[12] mainly focusing on the formation of a coating of fat globules on air bubbles in whipped emulsions. However, the same process can be expected to occur in the mouth during tasting and mastication of a food emulsion, finally leading to the release of liquid fat, droplet coalescence, and a dramatically enhanced release of fat-soluble flavours by the direct exposure of the droplet content to the air at a large area of oil–water interface.

A key factor in this process is the spreading coefficient S of the oil at the air–water interface, defined by

$$S = \gamma_{AW} - (\gamma_{OW} + \gamma_{OA}), \tag{1}$$

where γ_{AW}, γ_{OW} and γ_{AO} are the interfacial tensions at the air–water, oil–water and air–oil interfaces, respectively. For model food emulsions, composed of triglyceride oil dispersed in water at pH $= 7$ and stabilized by protein, the equilibrium spreading coefficient is negative, and therefore the emulsion droplets do not spread at the air–water interface. However, if the air–water interface is expanded, *e.g.*, by expansion and deformation of air bubbles during sucking while tasting the emulsion, the spreading coefficient can temporarily become positive because of the relatively low rate of adsorption and surface tension lowering by the protein solution. A triglyceride monolayer will spread out of droplets that are accidentally present at the oil–water interface, and this vigorous process will lead to a local convective motion of emulsion droplets toward the spreading oil patch, inducing a self-enhancing process of spreading emulsion droplets. This behaviour has been shown to be enhanced in the presence of low-molecular-weight surfactants, and it is especially effective in natural cream.

The first evidence for this process can be seen in Figure 8, which shows a spat-out sample of natural dairy cream masticated for 1 minute. The picture shows one of several large fat globules in the sample, which usually appeared in connection with air bubbles. We believe that these structures are formed by fat globule insertion and spreading at the bubble interface during expansion of the bubble surface area during flow in the mouth, leading to the condition $S > 0$. After the flow has ended, the bubble shrinks to its equilibrium area,

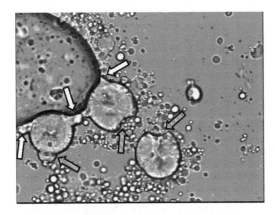

Figure 8 *Whipping cream, spat out after it was in the mouth for 1 minute, during which time some air was incorporated. Large, partially solid fat droplets are observed in contact with air bubbles. White arrows indicate places where liquid fat partially wets the bubble surface. Shaded arrows show places where liquid fat has been squeezed out of a droplet, leaving behind an indentation in the droplet.*

pushing the inserted and spread droplets together at the surface. When the surface has come to rest, the coefficient S becomes negative again, and as a consequence the adsorbed protein at the bubble surface pushes the spread monolayer back into the adhering emulsion droplets. In the final situation, a large droplet remains attached to the air bubble and its fatty content is partly solidified following cooling outside the mouth.

4 Conclusions

The perception of emulsions in the mouth cannot be directly related to the texture before consumption because several processes in the mouth change the emulsion properties. These processes include the aggregation of emulsion droplets due to mixing with saliva, the incorporation of droplets into slimy strings of mucous structures in saliva, the coalescence of droplets upon rubbing an emulsion between the tongue and palate, and the spreading of liquid triglycerides onto air bubbles. In addition, the lubricating properties of emulsions can be tuned by changing the composition at the oil–water interface.

We believe that knowledge and control of these processes will become an important tool for controlling the sensory quality of food emulsions and reduced-fat foods.

References

1. P. W. Lucas, J. F. Prinz, K. R. Agrawal, and I. C. Bruce, *Food Qual. Pref.*, 2002, **13**, 203.
2. H. T. Lawless, in 'Sensory Science: Theory and Applications in Foods', eds H. T. Lawless and B. P. Klein, Marcel Dekker, New York, 1991, p. 1.
3. T. van Vliet, *Food Qual. Pref.*, 2002, **13**, 227.
4. R. A. de Wijk, J. F. Prinz, and L. Engelen, *Appetite*, 2003, **40**, 1.
5. R. F. Troxler, I. Iontcheva, F. G. Oppenheim, D. P. Nunes, and G. D. Offner, *Glycobiology*, 1997, **7**, 965.
6. T. B. J. Blijdenstein, T. van Vliet, E. van der Linden, and G. A. van Aken, *Food Hydrocoll.*, 2003, **17**, 661.
7. M. H. Vingerhoeds, T. B. J. Blijdenstein, F. D. Zoet, and G. A. van Aken, submitted for publication.
8. T. B. J. Blijdenstein, W. P. G. Hendriks, E. van der Linden, T. van Vliet, and G. A. van Aken, *Langmuir*, 2003, **19**, 6657.
9. P. O. Glantz, *Colloids Surf. A*, 1997, **123–124**, 657.
10. G. A. van Aken, in 'Food Emulsions', 4th edn, eds S. E. Friberg, K. Larsson, and J. Sjöblom, Marcel Dekker, New York, 2004, p. 299.
11. M. E. Malone, I. A. M. Appelqvist, and I. T. Norton, *Food Hydrocoll.*, 2003, **17**, 763.
12. N. E. Hotrum, T. van Vliet, M. A. Cohen Stuart, and G. A. van Aken, *J. Colloid Interface Sci.*, 2002, **247**, 125.

Sensory Perception of Salad Dressings with Varying Fat Content, Oil Droplet Size, and Degree of Aggregation

By Eva Tornberg, Nicolas Carlier,[1] Ene Pilman Willers,[2] and Per Muhrbeck[2]

FOOD ENGINEERING, LUND UNIVERSITY, P.O. BOX 124, S-22100 LUND, SWEDEN
[1]ENSBANA, 1 ESPLANADE ERASME, 21000 DIJON, FRANCE
[2]ORKLA FOOD A/S, 241 81 ESLÖV, SWEDEN

1 Introduction

Salad dressings, belonging to the family of oil-in-water emulsions, are made out of oil, water, sugar, salt, mustard, egg-yolk emulsifier, hydrocolloids, and acid.[1] In order to meet the consumer demands, dressings with a lowered fat content are being produced. On lowering the oil content to make light salad dressings, more hydrocolloids are commonly added to maintain texture perception and physical stability.

The hydrocolloids considered in this investigation are xanthan gum and alginate. Xanthan gum has a glucose backbone with a trisaccharide side-chain on every other glucose at C-3, giving rise to a branched polysaccharide. Therefore, xanthan gum solutions are pseudoplastic and they form a highly ordered network of entangled, stiff molecules.[2] The alginates are a family of unbranched binary copolymers of linked mannuronic acid and guluronic acid.[2] A disadvantage of adding such hydrocolloids from the sensory point of view is that it decreases the intensity of flavours.[3]

The nature of the emulsifier is also important since it affects many of the emulsion properties. Egg yolk is mainly used as the emulsifying agent to produce dressings and mayonnaise.[1] All the constituents of yolk—low-density lipoprotein (LDL), high-density lipoprotein (HDL), phosvitin and livetin—have a strong tendency to adsorb on the oil droplets.[4] Yolk can be separated into plasma (85% LDL and 15% livetin) and granules (70% HDL, 16% phosvitin and 12% LDL). The plasma fraction is very rich in lipids (about 73% of dry matter) of which 25% are phospholipids. These phospholipids are mostly associated with the LDL, which is thought to become disrupted at the oil–water interface when the surface-active constituents adsorb there. The LDL is well represented at the oil–water interface (about 60% of the total proteins).[4]

Other food proteins also commonly used as emulsifying agents are whey proteins (WP) and caseinates. The protein molecules in caseinate have little secondary or tertiary structure, *i.e.*, they are mostly random and flexible and are therefore good emulsifiers.[5] Above pH 5.5, sodium caseinate is completely soluble or dispersible, but the casein molecules have a strong tendency to associate at lower pH and high ionic strengths (>2.0 M NaCl). The whey proteins possess more secondary and tertiary structure than the caseinates, but still they are almost as good emulsifiers.[5]

With proteins as emulsifiers, protein bridging between oil droplets can occur. This can lead to extensive network formation, which could be favourable for structure formation in salad dressings. Protein concentration per unit surface area, molecular weight, degree of association, and the conformation and surface hydrophobicity of the protein—all these are important factors governing the degree of protein bridging.[5] Darling and Butcher[6] have observed bridging flocculation with whey proteins and sodium caseinate at protein concentrations of 0.2–0.4% w/v in a 30% butter fat emulsion, corresponding to a range of oil/protein ratios between 75 and 150.

The product requirements for a salad dressing are physical stability and the perception of satisfactory texture, flavour and taste. The main physical instability phenomena are creaming, flocculation and droplet coalescence, which are mainly governed by the oil droplet size and the nature of the stabilizing film. The properties of the latter are deterimined by the emulsifying agent used. The main stabilizing action of the polysaccharide is *via* viscosity increase or gelation in the aqueous continous phase. However, hydrocolloids present in the aqueous phase can also cause destabilization of the emulsion by the mechanism of depletion flocculation.[7] Non-adsorbing hydrocolloids with extended and stiff polysaccharide backbones, like the xanthan gum used in this investigation, are especially effective at inducing depletion flocculation at low concentrations. For example, Cao *et al.* have shown[8] that the amount of xanthan has to exceed 0.125 wt% in a 10 wt% oil sodium caseinate-stabilized emulsion to overcome the effects of severe depletion flocculation and the resulting creaming. The inhibition of creaming at the higher concentrations of xanthan can be attributed to the formation of a weak gel-like network, in which the protein-coated oil droplets are entrapped.

One problem in low-fat foods is that the initial taste and flavour appear more intense than in high-fat foods; thereafter, taste/flavour perception quickly disappears, so that the desirable fullness of the texture is rapidly lost.[9] The oil also plays an important role by imparting a desired mouthfeel. It is believed[1] that this special mouthfeel of mayonnaise, which has a higher content of oil than most salad dressings, is derived from the loosely aggregated network of oil droplets. A salad dressing can be considered as a composite gel, *i.e.*, oil droplets embedded in a gel matrix. The textural properties of such a composite gel depends on the oil phase volume fraction and on the interaction between the gel matrix and the droplets. The size of the dispersed particles may also effect the rheological and fracture properties of the gel.[10]

In this investigation we explore some different ways of improving the texture perception of salad dressings—changes in the droplet-size distribution and in the degree of aggregation of the oil droplets. On lowering the oil content, it may be possible to maintain the textural perception by increasing the fat surface area, *i.e.*, by forming smaller oil droplets. An alternative approach is to use other proteins (apart from egg-yolk proteins) which may be capable of substantial protein bridging between droplets, thereby forming aggregates into a three-dimensional aggregated particle network that can contribute to the development of a textured emulsion.

2 Materials and Methods

Experimental Plan

An overview of the experimental plan is given in Figure 1. To vary the oil droplet size, a pressure drop of 100 or 25 bar was used in 10 passes (4 minutes) through a small-scale valve homogenizer (VH) as described by Tornberg and Lundh.[11] Another factor varied was the width of the oil droplet-size distribution. As well as the valve homogenizer, giving narrow size distributions, we have also used an ultrasonic device (S) (Branson Sonifier Cell Disruptor, model B-12) to prepare some emulsions with a wide droplet-size distribution.[12] The power input of the sonicator was 50 W for 2 minutes, which was equivalent in emulsifying time to 4 minutes using the VH. For comparison, some other salad dressings of 23% and 11% oil were made in the factory (samples N and O) using a large-scale homogenizer. All dressings, except for the samples N and O, were prepared in small batches of 100 ml.

We have considered four different hydrocolloid concentrations (0.2%, 0.3%, 0.4% and 0.5%), four fat contents (11%, 17%, 23% and 27%), and four kinds of emulsifier (whey protein at low and high concentrations, caseinate, and egg yolk). A variation in the degree of aggregation of the emulsion droplets was

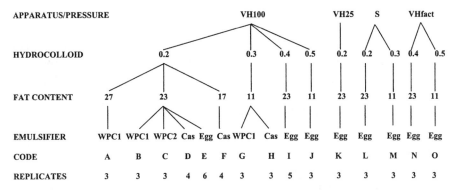

Figure 1 *Experimental plan of salad dressing formulation. VH100 = valve homogenizer at 100 bar; VH25 = valve homogenizer at 25 bar; VHfact = valve homogenizer in the factory; S = sonicator. WPC1 and WPC2 = whey protein concentrate at oil/ protein ratios 1 and 2; Cas = caseinate.*

achieved by choosing different oil/protein ratios with whey protein concentrate (WPC). The low concentration is WPC1 (ratio oil/WPC = 160), which was the concentration to obtain an optimum degree of aggregation, and the other is WPC2 (ratio oil/WPC = 60), which did not lead to any aggregation. For the caseinate, a single oil/protein ratio of 65 was used: at the pH of the salad dressing (pH ≈ 3.2) the caseinate was highly aggregated, and this led to a high degree of aggregation of the emulsion droplets.

Salad Dressing Preparation

All the ingredients came from the Orkla Food Company, except for the sodium caseinate (EM7, from the DMV Swedish distributor, ABR, Lundberg). The egg-yolk concentrations used were 1.6% and 2.3% at the oil concentrations of 11% and 23%, respectively. The concentrations of sugar (5%), mustard (1%), salt (1.5%) and acetic acid (3.5%) were kept the same for all the formulations.

The sugar and the two hydrocolloids (equal amounts of xanthan gum and alginates, each at the concentration given in the experimental plan) were first mixed together. When whey protein concentrate (WPC) or caseinate was used as emulsifier, it was added at this point. Water was then introduced during three consecutive additions, each time stirring at 10^3 rpm for three minutes at room temperature, in order to dissolve the hydrocolloids in an optimum way. Thereafter, each ingredient except salt was added, stirring for one minute between each addition. After mixing in the mustard and the egg yolk (if it was being used as emulsifier), acetic acid at a concentration of 12% was poured into the water phase, followed by rapeseed oil. This preparation was stored for half an hour before the salt was added. Emulsions were then made using either the valve homogenizer or the ultrasonic device. The pH values of the emulsions were checked for the different formulations using an electronic pH meter (pH M62 standard, Bergman & Beving Radiometer, Copenhagen).

Light Microscopy

Samples diluted 100 times were observed under a light microscope (Nikon Optiphot) with a magnification of × 400. Pictures taken for each sample were ocularly compared to a scale 0–5 based on the set of six pictures in Figure 2. This scale was used to give an arbitrary score of the degree of aggregation between 0 and 5 for the different emulsion samples.

Droplet-Size Distributions

A Coulter LS 130 laser light-scattering apparatus (Coulter Electronics, Luton, England), using polarization intensity differential scattering at wavelength 720 nm, was used to measure the size distributions of the oil droplets in the dressings. The samples were diluted 10 times in a 5% w/v solution of sodium dodecyl sulfate in order to dissociate the aggregated droplets. The geometric

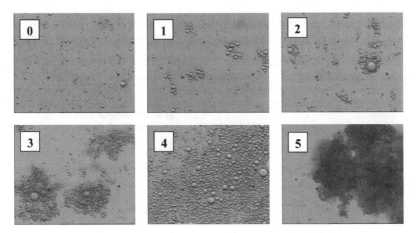

Figure 2 *Arbitrary flocculation scale (0 to 5) of the oil droplets in the salad dressing (magnification ×400).*

mean and the standard deviation of the volume-weighted distribution, d_v and SD-d_v, were calculated.

Rheology

Stress-sweep rheological analysis was carried out on a Stress Tech controlledstress rheometer (Reologica, Lund, Sweden) using a concentric cylinder measuring system (CC 25). An oscillatory stress sweep was generated, with changing stress amplitude from 0.05 to 20 Pa at a constant frequency of 1 Hz. We define G'_{max} as the storage modulus in the linear viscoelastic region, *i.e.*, where G' is independent of the stress. The critical shear stress $G'_{max}/2$ was defined as the shear stress applied at 0.5 $G'_{max.}$

Sensory Evaluation

A quantitative descriptive analysis, using a scale from 1 to 9, of the different salad dressings has been performed with an untrained panel composed of 5 females and 3 males. A list of the most relevant attributes has been established, and a definition for each can be found in Table 1.

Statistical Analysis

The statistical analysis was performed with Minitab Statistical Software (Minitab, USA). For the physical and sensory measurements, a principal component analysis (PCA) was used to determine how well the evaluated parameters were correlated.

To elucidate the effect of each of the studied factors, one-way ANOVA tests and a t-test were performed on a restricted number of samples containing formulations changing only by one factor. Although the ANOVA tests were

Table 1 *Sensory attributes and descriptions used for the sensory analysis. Abbreviations assigned to these attributes in the statistical analysis are also listed.*

Perception type	Attributes	Abbreviation	Description
Texture before consumption	Spreading	SPREAD	Pour the liquid into a dish to look if it spreads quickly or slowly
	Elasticity	ELAS	Stickiness of the dressing to a spoon when it is lifted from a plate
	Grainy texture	GRAINY	Spread the product to see if granules occur
Visual appearance	Yellow colour	YELLOW	Scale varying from white to yellow
	Smooth surface	SMOOTH	Surface smooth or rough
	Air bubbles	AIR	Few or many bubbles
Texture after consumption	Thick texture	THICK	Thick or liquid texture based on pressing the dressing on the palate
Taste and flavour	Fat taste	OILY	Scale varies from not fatty to very fatty
	Acid taste	ACID	Sour taste of vinegar and/or acetic acid
Overall acceptance		OVAC	Evaluate if the dressing is liked or not

based on a relatively small number of samples, they were able to show the influence of one factor without any interactions. The effect was deemed significant for $p < 0.05$. In the sensory analysis, the attributes used to describe the products supposed a certain consensus between the assessors on these attributes. However, for some sensory attributes the effect of assessor variability was higher than any effect studied. (General Linear Model ANOVAs were used ($p < 0.05$).) In those cases the attribute was deleted from the evaluation, as was the case for the attribute 'elasticity'. When the assessor effect was due to one or two panellists being extreme, they were removed to take out the assessor effect.

3 Results and Discussion

Effect of Oil Droplet Size and Distribution Width

To evaluate the importance oil droplet size and distribution width on the physical attributes of the salad dressings, samples with the following emulsion preparation conditions were compared: sample E for the valve homogenizer operating at 100 bar, sample K for the homogenizer at 25 bar, and sample L for the sonicator. For all these emulsions the emulsifying agent used was egg yolk and the hydrocolloid and fat contents were 0.2% and 23%, respectively. The results are shown in Table 2. We can see that, as expected,[12] both the mean size of the fat droplets (d_v) and the size distribution width (SD-d_v) were found

Table 2 *Physico-chemical parameters significantly affected by the emulsion preparation method.*

	Sample (Preparation method)			
	p-*value*	E (VH100)	K (VH25)	L (S)
d_v (μm)	*	1.57[b]	4.16[a]	4.1[a]
SD-d_v (μm)	*	0.52[b]	1.95[a]	2.42[a]
Aggregation degree	NS	1.50[a]	1.00[a]	0.30[a]
G' (Pa)	0.002	8.87[b]	4.18[a]	6.73[a]
Shear stress at $G'/2$ (Pa)	NS	1.71[a]	2.13[a]	1.78[a]

* p-value <0.001. NS = not significant.
[a,b] Different superscripts in one row are significantly different.

to be significantly higher for the sample made with the ultrasonic apparatus and the valve homogenizer at 25 bar than for the sample made with the valve homogenizer operating at 100 bar.

It is interesting to note in Table 2 that the elasticity of the emulsion, G', was significantly higher for VH100 as compared with the two other ways of making the emulsion. In a PCA analysis for all the emulsions (results not shown here), the degree of aggregation and not the oil droplet size was found to be more important with respect to the formation of a network, as expressed by the value of G'.

Smooth appearance was the only sensory attribute influenced by the emulsion preparation method (Table 3). It was significantly higher for VH100 as compared to S and VH25, *i.e.*, smoother surfaces were obtained when the oil droplets were smaller. It could be that the larger total interfacial area of oil, due to the presence of smaller droplets, produces a surface that is more homogeneous and lower in roughness.

The change in oil droplet size was not important enough to affect any of the other sensory properties. The absence of any effect on the other sensory attributes could be because the mean size of droplets was <5 μm. On varying the droplet size in the range 0.5–6.5 μm, Druaux *et al.* have reported[13] no significant effect on flavour release of the mean droplet size or the protein concentration at the interface.

Table 3 *The sensory attribute significantly affected by the emulsion preparation method.*

	Sample (Preparation method)			
	p-*value*	E (VH100)	K (VH25)	L (S)
Smooth appearance	*	7.37[b]	6.00[a]	5.25[a]

* p-value <0.001.
[a,b] Different superscripts in one row are significantly different.

Table 4 *Physico-chemical parameters significantly affected by the fat content.*

		Sample (Fat content)				
	p	A(27%)	B(23%)	p	D(23%)	F(17%)
d$_v$ (μm)	NS	3.94	3.53	NS	3.28	2.85
SD-d$_v$ (μm)	NS	1.75	1.81	NS	1.64	1.58
Aggregation degree	NS	1.33	2.00	NS	4.50	3.75
G' (Pa)	0.02	19.9	10.7	*	200	79.75
Shear stress at G'/2 (Pa)	NS	197	1.54	*	14.3	5.65

* *p*-value <0.001. NS = not significant.

Effect of Fat Content

Two sets of samples have been made with emulsification procedure VH100 to evaluate the effect of fat content: formulations A/B with 27% and 23% fat, and formulations D/F with 23 and 17% fat. WPC1 and caseinate were used as emulsifiers in formulations A/B and D/F, respectively, and the hydrcolloid content was set at 0.2%. The results are shown in Table 4.

For all comparisons, the size parameters and the aggregation state were not significantly different, *i.e.*, the fat content did not have any large impact on the oil droplet size or the degree of aggregation. However, when using an emulsifying agent causing a large degree of aggregation, like the caseinate, the value of G', reflecting the degree of network formation, was significantly higher when the fat content was increased. This indicates that the textural properties of highly aggregated gels depend sensitively on the volume fraction of oil droplets. The fracture properties of the emulsion gels, as reflected in shear stress at $G'/2$, were found to be significantly influenced by fat content when the emulsion was highly aggregated (D/F). Comparing the two pairs of samples, A with B and D with F, with regard to the sensory evaluated attributes, no significant effect of fat content could be observed.

Effect of Hydrocolloids

Formulations of salad dressings with hydrocolloid contents of 0.2% and 0.3% (L/M) and 0.2% and 0.4% (E/I) were compared (see Table 5). Sonication was used to make the salad dressings L and M with egg yolk as the emulsifier and the fat content varying from 11% to 23%. As the latter variables were already shown above not to affect significantly the droplet-size distribution, a satisfactory comparison can therefore be made. Egg-yolk emulsifier was also used to prepare samples E and I, but in this case the procedure VH100 was utilized for the emulsification and the fat content was kept at 23%.

We can see from Table 5 that increasing the amount of hydrocolloid by 0.1% did not affect any of the physico-chemical parameters, apart from the shear stress at $G'/2$, which increased. However, the larger change in the hydrocolloid content from 0.2% to 0.4% in the samples E and I was more

Table 5 *Comparison of physico-chemical parameters having different hydro-colloid contents using one way ANOVA.*

	p-*value*	L (0.2%)	M (0.3%)	p-*value*	E (0.2%)	I (0.4%)
		Sample (Hydrocolloid content)				
d_v (µm)	NS	4.10	4.60	0.006	1.57	1.69
SD-d_v (µm)	NS	2.43	2.93	NS	0.52	0.59
Aggregation degree	NS	0.33	0.00	0.03	1.50	0.40
G' (Pa)	NS	6.73	8.5	*	8.87	31.0
Shear stress at $G'/2$ (Pa)	0.02	1.78	3.57	*	1.71	10.0

* p-value <0.001. NS = not significant.

Table 6 *The sensory attribute significantly affected by the hydrocolloid content.*

	p-*value*	E (0.2%)	I (0.4%)
		Sample (Hydrocolloid content)	
Spreading	<0.001	7.62	1.69

significant: there were somewhat larger oil droplets, a reduced degree of aggregation, and an increased G' and shear stress at $G'/2$. Comparing the same samples as for the physical measurements, the only sensory attribute to be significantly influenced by the hydrocolloid content was 'spreading'; the value of this parameter was assessed as being significantly lower when the hydrocolloid content was higher, as shown in Table 6.

Effect of Emulsifying Agent

Table 7 compares the properties of a set of formulations containing the four different emulsifying agents. The emulsion making procedure here was VH100,

Table 7 *Physico-chemical parameters significantly affected by the type of emulsifier.*

	p-*value*	B (WPC1)	C (WPC2)	D (caseinate)	E (egg yolk)
		Sample (Emulsifier type)			
d_v (µm)	*	3.53[a]	1.96[b]	3.28[a]	1.57[c]
SD-d_v (µm)	*	1.81[a]	0.91[b]	1.64[a]	0.52[c]
Aggregation degree	*	2.00[a]	0.00[b]	4.50[c]	1.50[a]
G' (Pa)	*	10.70[a]	1.57[b]	200[c]	8.88[a]
Shear stress at $G'/2$ (Pa)	*	1.54[a]	0.92[a]	14.27[b]	1.71[a]

* p-value <0.001.
[a,b,c] Different superscripts in one row are significantly different.

and the oil and hydrocolloid contents were 23% and 0.2%, respectively. The largest amount of aggregation of the emulsions was obtained when using caseinate as the emulsifying agent. The level of aggregation was much lower (≤ 2) for the other emulsifiers. The pH of the dressings (3.1–3.2) and the high ionic strength reinforced this capacity for aggregation by the caseinate. Whey protein at the optimum concentration for aggregation (WPC1, oil/protein ratio $= 160$) was found to give a similar response to egg yolk. But when the oil/protein ratio was 60 (WPC2), no aggregation at all could be observed. When the emulsions were substantially aggregated, as for samples B and D, the oil droplet size and its width were significantly larger. The smallest droplets with the most narrow size distribution was achieved with egg yolk. With caseinate as emulsifier, the value of G' was very high. WPC1 and egg yolk gave emulsions with similar values of G', whereas WPC2 gave rise to substantially lower values.

Most of the sensory attributes of samples B–E were influenced by the nature of the emulsifier. This was mainly due to the caseinate, which had a large influence (Table 8). Using the caseinate as the emulsifying agent was found to reduce the spreading, while the other emulsifiers gave rise to substantial spreading. The concentration of air bubbles was higher with the whey protein than with egg yolk or caseinate. The thickness was higher with caseinate, whereas the other emulsifiers gave similar values. The dressings made with caseinate had a more grainy texture and the visual appearance with respect to the degree of smoothness of the surface was lower, but still the dressings prepared with the caseinate and egg yolk had the highest scores of overall acceptance.

Exchange of Hydrocolloid for Protein-Bridging Emulsifier

Using caseinate and WPC instead of egg yolk, and so producing highly aggregated emulsions, might be a useful way to lower the amount of hydrocolloid

Table 8 *Sensory attributes significantly affected by the type of emulsifier. (Assessors removed from the analysis are identified by the numbers in brackets.)*

		Sample (Emulsifier type)			
	p-value	B (WPC1)	C (WPC2)	D (caseinate)	E (egg yolk)
SPREAD	*	7.75[a]	8.37[a]	1.28[b]	7.62[a]
AIR (1,6)	0.03	6.33[bc]	7.33[bc]	3.50[a]	5.16[ab]
YELLOW	NS	1.37[a]	1.25[a]	1.37[a]	1.87[a]
THICK (1,6)	*	3.33[a]	2.33[a]	6.83[b]	3.83[a]
OVAC	*	3.87[ab]	3.00[a]	5.12[bc]	5.00[c]
GRAINY	*	1.12[a]	1.00[a]	6.87[b]	1.00[a]
SMOOTH	*	7.60[a]	6.28[a]	2.87[b]	7.37[a]
OILY	NS	3.71[a]	2.28[a]	2.42[a]	3.14[a]

* *p*-value <0.001.
[a,b,c] Different superscripts in one row are significantly different.

Table 9 *Means of physical and sensory attributes of samples G, H, J and O.*

Attribute	Sample (Emulsifier type, Hydrocolloid content)			
	G (WPC1, 0.3%)	H (Cas, 0.3%)	J (Egg yolk, 0.5%)	O (Egg yolk, 0.5%)
d_v (μm)	2.52	1.69	1.28	1.79
SD-d_v (μm)	1.57	1.41	0.58	1.99
Aggregation	1.00	1.00	0.33	0.00
G' (Pa)	3.72	5.72	26.50	20.57
Shear stress at $G'/2$ (Pa)	2.98	2.30	12.87	9.97
THICK	3.00	2.88	4.75	5.13
SPREAD	7.88	7.50	2.88	4.75
OVAC	3.88	4.00	5.25	6.00

used in a light salad dressings (low oil content) in order to improve the delivery of flavours. This has been done for samples G and H, where the concentration of hydrocolloid has been lowered to 0.3% in a salad dressing of 11% oil. These emulsions should be compared to samples J and O, where the oil content is the same (11%) and the hydrocolloid and emulsifier contents are the ones commonly used in light salad dressings, *i.e.*, 0.5% hydrocolloid and 1.6% egg yolk, respectively. Sample J was made using the laboratory valve homogenizer; sample O was made with a large-scale factory homogenizer.

According to the results in Table 9, the degree of aggregation was lowest for the egg-yolk emulsions. However, at this much lower concentration of oil (11%), the degree of aggregation of the caseinate-stabilized emulsions is down to a value of 1.0, as compared to the values of 3.75 and 4.50 at 17% and 23% oil, respectively. Evidently, the volume concentration of oil is very important to be able to form a network in these highly aggregated emulsions. This trend also shows up in the substantially lower values of G' and the shear stress at $G'/2$ for the two aggregated emulsions as compared to the emulsions made using a high concentration of hydrocolloid instead. The sensory attributes were also found to be influenced, and especially the spreading of these aggregated emulsions was greatly increased, at these low concentrations of oil. None of the samples based on WPC1 or caseinate, however, was able to reach the high overall acceptance levels of 5.3–6.0 obtained for the light salad dressings with the high content of hydrocolloid (0.5%).

Relationship between Physico-Chemical and Sensory Measurements

The two principal components of the averages of the physico-chemical and sensory measurements are able to explain 75% of the variation in the samples (Figure 3). In the first axis, two opposite groups were evaluated, one containing thickness, overall acceptance, yellow colour and shear stress at $G'/2$, and the second one opposed to air and spreadability. The second axis opposed G', the grainy texture, and the aggregation to a smooth surface.

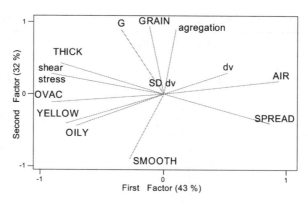

Figure 3 *Representation of various factors in the PCA analysis of the physico-chemical and sensory data of the emulsions.*

The results in Figure 3 tell us that, for those salad dressings investigated here, the overall acceptance was mainly governed positively in texture by the gel fracture property (shear stress at $G'/2$) more than the network strength *per se* (G'). The latter property, though, was mostly determined by the degree of aggregation, which caused a more grainy and less smooth appearance of the salad dressing. The texture sensed in the mouth as 'thickness' was positively related to the fracture behaviour (shear stress at $G'/2$) of the emulsion gel, whereas the visual appearance factors, such as the occurrence of air bubbles and the degree of spreading, were negatively correlated.

Conclusions

The variation in droplet-size distribution parameters was found not to influence the physico-chemical and sensory properties of the different salad dressings as much as an increase in the hydrocolloid content and in the degree of aggregation of the oil droplets.

The overall acceptability of the salad dressings from sensory evaluation was found to be mainly governed by the resistance of the emulsions to spread and to fracture with increasing shear stress (shear stress at $G'/2$), and by the perception of 'thick in the mouth'. This was achieved for salad dressings of 23% oil either by adding hydrocolloids to a level of 0.4% or by using highly aggregated caseinate as the emulsifying agent (instead of egg yolk) with just 0.2% hydrocolloid added. Though the latter emulsion also had a grainy texture, that did not substantially reduce the overall impression of it. However, when the concentration of oil was lowered to 11%, there was observed to be a substantially lower degree of aggregation for the caseinate-stabilized emulsions.

None of the new formulations studied reached the high overall acceptance level of 5.3–6.0 obtained with 'light' salad dressings having a high hydrocolloid content (0.5%). Further research seems justified to explore the possibility of using highly aggregated emulsions, in combination with the optimum amount of hydrocolloid, to maintain texture of light salad dressings.

References

1. J. A. Depree and G. P. Savage, *Trends Food Sci. Technol.*, 2001, **12**, 157.
2. G. O. Phillips and P. A. Williams (eds), 'Handbook of Hydrocolloids', CRC Press, Cambridge, 2000, pp. 104, 108, 381.
3. R. M. Pangborn, Z. M. Gibbs, and C. Tassan, *J. Texture Stud.*, 1978, **9**, 415.
4. M. Anton and G. Gandemer, *Colloids Surf. B*, 1999, **12**, 351.
5. E. Tornberg, A. Olsson, and K. Persson, in 'Food Emulsions', 2nd edn, eds K. Larsson and S. Friberg, Marcel Dekker, New York, 1990, p. 247.
6. D. F. Darling and D. W. Butcher, *J. Dairy Res.*, 1978, **45**, 197.
7. E. Dickinson, in 'Food Polysaccharides and their Applications', ed. A. M. Stephen, Marcel Dekker, New York, 1995, p. 501.
8. Y. Cao, E. Dickinson, and D. J. Wedlock, *Food Hydrocoll.*, 1990, **4**, 185.
9. K. B. de Roos, *Food Technol.*, 1997, **51**(1), 60.
10. T. van Vliet, *Colloid Polym. Sci.*, 1988, **266**, 518.
11. E. Tornberg and G. Lundh, *J. Food Sci.*, 1978, **43**, 1559.
12. E. Tornberg, *J. Food Sci.*, 1980, **45**, 1662.
13. C. Draux, J.-L. Courthaudon, and A. Voillet, in 'Hutieme Rencontres Scientifiques et Technologiques des Industrie Alimentaires: Production Industrielles et Qualité Sensorielle', Lavoisier, Tec & Doc, Paris, France, 1996, p. 255.

Acoustic Emission from Crispy/Crunchy Foods to Link Mechanical Properties and Sensory Perception

By Hannemieke Luyten,[1,2] Wim Lichtendonk,[1,3]
Eva M. Castro,[1,4] Jendo Visser,[1,4] and Ton van Vliet[1,4]

[1]WAGENINGEN CENTRE OF FOOD SCIENCES (WCFS),
P.O. BOX 557, 6700 AN WAGENINGEN, THE NETHERLANDS
[2]AGROTECHNOLOGY AND FOOD INNOVATIONS BV (A&F),
P.O. BOX 17, 6700 AA WAGENINGEN, THE NETHERLANDS
[3]TNO NUTRITION AND FOOD RESEARCH, P.O. BOX 360,
3700 AJ ZEIST, THE NETHERLANDS
[4]WAGENINGEN UNIVERSITY AND RESEARCH CENTRE (WUR),
P.O. BOX 9101, 6700 HB WAGENINGEN, THE NETHERLANDS

1 Introduction

For many food products the crispy and crunchy character is important for sensory perception, because it contributes to the appreciation of the food.[1,2] Although the terms 'crispy' and 'crunchy' are common, no generally accepted definitions are available. From the different descriptions of these sensory attributes given in the public literature, it is clear that to be considered as crisp or crunchy a food product has to be difficult to deform (it has to be stiff), but it should fracture at small deformations thereby producing a typical sound.[3] Likely, the fracture process must consist of several successive fracture events. The latter is especially of importance for crunchy foods.[2,3]

The sound emission from crispy/crunchy food products during deformation or eating originates from the fracturing process. This special position of the sound contribution is shown schematically in Figure 1.

To obtain acoustic emission, the crack has to proceed so fast through the material that shock waves are formed with the speed of sound. Thus a prerequisite for crispy/crunchy behaviour is that the fracture process is fast. It is also important to consider that sound is a type of energy: emission of sound during fracture thus implies that energy is dissipated as sound energy.[4,5] Although the acoustic energy is only a small part of the total deformation

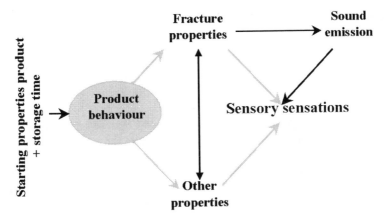

Figure 1 *Schematic overview of the different stimuli affecting crispy/crunchy sensory sensations.*

energy, other energy dissipative processes are likely to affect the amount of energy available for acoustic emission and thus to affect the sensory perception of crispy/crunchy.[3,6] Alchakra *et al.* showed[7] a clear decrease in acoustic energy in the audible region with increasing water content of dried pasta.

The amount of sound energy is related to the amount of damage. The frequency of the sound is related to the type of fracturing process and the (local) material properties of the material involved in the fracture.[5] One can imagine that the sizes of the broken pieces affect the sound due to vibrations. The amount of sound in the audible region is related to the perception of the loudness, and the frequency to the perception of the pitch.[8] The sound produced and its properties, and hence the sound heard while eating crispy/crunchy foods, depends amongst other things on the mechanical and morphological properties of these food products.

The acoustic emission during the fracturing of crispy food products has been studied here for two reasons. Firstly, we wish to investigate the sensory perception of crispy foods with a crust—including the role of sound within it. Secondly, as sound emission is determined by the fracture process and gives information about this process, an acoustic emission test has the advantage that it enables investigation of damage during the fracturing process.[9,10] In this way it may be even possible to detect irreversible processes before they can be noticed in the loading curve.

So, in principle, the studying of sound emission simultaneously with mechanical testing may help us to get a better understanding about which material properties determine the crispy or crunchy character of food. By linking mechanical and acoustic emission (AE) tests, we aim to understand what happens when crispy/crunchy food products fracture, and which properties cause the changes in these sensory attributes during deterioration. However, one should realise that acoustic emission results are difficult to record and interpret properly: the data may depend on the detection system, on the variable material structure, and on the propagation before recording.[5]

2 Materials and Methods

A biscuit named 'Knappertje' (Verkade) was chosen as a model for a food product that is simple, crisp, and completely dry. The biscuit dimensions were *ca.* 50 × 50 × 5 mm. The product was kept in its original packet under environmental conditions, *i.e.*, a temperature of 23 °C and a relative humidity of about 28%. For each experimental session a new packet was opened, and the first and last biscuits of each packet were discarded. Each biscuit was positioned in such a way that fracture took place in the middle, where the structure was assumed to be relatively homogeneous.

As a model for crispy/crunchy foods with a crust and a wetter/softer interior, two different model products were developed: a model bread and a deep-fried battered snack.[11,12] Examples are shown in Figure 2. These model products were used for the sensory tests (they are food-grade), for the mechanical tests (they have a convenient flat shape and are large compared to the inherent inhomogeneities), and for tests involving the transport of water (they are flat). In addition, using these model products, we were able to manipulate crust and core properties separately, and to prepare the products reproducibly.

A sensory QDA analysis was performed by a sensory panel of eight female members aged 20 to 52 years (average 41 years). They were trained to assess the different attributes that were generated by experts, food technologists and consumers when judging crispy foods with a crust.[13] The attributes were presented to the panellists using the Fizz computer program for automated sensory analysis.[14]

Sound emission during biting and eating was detected using an acoustic sensor (Brüel & Kjær, ½ inch, type 4189 free-field Deltatron microphone) with a frequency band of 6.3 Hz to 20 kHz and a sensitivity of 50 mV Pa^{-1}. The analogue sound signal was converted to a digital signal using a Brüel & Kjær 'front-end' system (type number 2827) with a sampling rate of 65 kHz. The mouth of the test person was open during biting, whereas during further

A   **B**

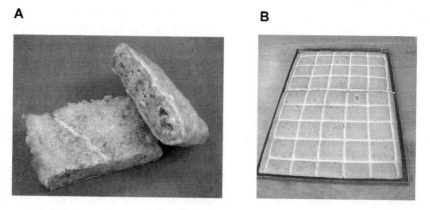

Figure 2 *Model products for crispy/crunchy foods with a crust and a softer interior: A, model for a deep-fried battered snack; B, model for bread.*

mastication the mouth was closed. The distance between the mouth and the microphone was set at 10 cm. The test person was free to choose her own biting size and frequency. Recording, replay and basic signal analysis were performed using Brüel & Kjær Pulse Labshop software (version 7.0). More detailed signal analysis was carried out using Brüel & Kjær Sound Quality type 7698 software (version 3.5).

A Texture Analyser (Stable Micro Systems (SMS) TA.XT.plus) was used to deform the food products in a reproducible way and to record simultaneously the force–deformation curves. The foods were fractured using a wedge penetration test,[15] with a wedge angle of 30°.[16] Analogue data were sent to the Brüel & Kjær Pulse front-end system, where they were converted into a digital signal (65 kHz). Sound emission during mechanical testing was detected using the same system as described above. Mechanical data and acoustic emission data were recorded synchronously. To reduce external noise, the tests were performed in an acoustically insulated room; the reduction in the audible region was around 50 dB.

We tried to follow the fracture of a crisp biscuit by recording it with a digital high-speed video camera. Chosen for testing was the best available camera regarding the combination of the speed of registration and the spatial resolution—a prerequisite as we do not know in advance the exact localization of the fracture or the fracture path. This camera from KSV had a memory of 32 MB and was able to collect 2200 frames of 256×128 pixels each second (0.45 ms between two frames). A test was performed with a three-point breaking test of the commercial biscuit 'Knappertje' using a rounded wedge. Fracture was followed by video recording at the above-mentioned speed.

3 Results and Discussion

Sensory Perception

A sensory vocabulary was constructed[13] to describe the sensations during the eating of crispy/crunchy foods with a crust. Attributes spontaneously used by the panellists could be distinguished based on the different stages in the eating process: 'appearance', 'first bite', 'chewing' and 'after taste'. The sound-related attributes used to describe the sensory perception of the crispy/crunchy breads and the deep-fried battered snacks are given in Table 1. These sound attributes were perceived both during the first bite and during chewing, and they were rated by the panellists separately.

Results obtained for different variants of a crispy/crunchy food product (Table 2) or for the effect of ageing under deterioration conditions (Table 2 and Figure 3) show that the different sound-related attributes describe different aspects of the sensory perception. Especially the sound intensity and the type of sound were affected in different ways by the food variant and the ageing time. It can be concluded from this that sound-related attributes have a special place in the vocabulary for describing the sensory perception of crispy or crunchy food products with a crust, and are very useful for describing differences between products or changes due to ageing.[13]

Table 1 *Sound-related attributes used to describe the sensory perception of crispy/crunchy breads and deep-fried battered snacks (ref. 13). Shown are the Dutch terms as used by the panellists, the English translation of these terms, and the definitions as produced by the members of the panel.*

Attribute (Dutch)	Attribute (English)	Definition
Geluid	Sound	Amount and loudness of sound
Type geluid	Type of sound	From low to high pitch
Krakend	Snappy	Loud, sharp, short sound
Knisperend	Crunchy	High pitched sound, light and longer

Table 2 *Results for ageing of two kinds of deep-fried battered snacks—a standard one and one with a dry interior. All sound-related attributes are rated at the first bite.*

Analysis	Standard Snack					Dry Interior Snack				
Time after frying (minutes)	1	2	3	5	10	1	2	3	5	10
Sound	53.1	45.7	45.1	40.9	36.2	46.3	44.5	40.3	41.1	40.3
	A[a]	AB	AB	BC	C	AB	B	BC	BC	BC
Type of sound	43.0	40.7	46.7	44.2	36.0	47.8	43.6	40.1	45.9	38.1
	ABC	BCD	AB	ABC	D	A	ABC	BCD	AB	CD
Snappy	46.7	43.5	38.9	31.3	31.7	39.5	42.5	32.6	31.5	33.9
	A	AB	ABCD	D	D	ABCD	ABC	CD	D	BCD
Crunchy	50.2	43.1	46.9	47.7	34.1	45.7	43.9	39.9	45.6	34.4
	A	BC	AB	AB	D	ABC	BC	CD	ABC	D

[a] Numbers labelled with the same letters A–D are not significantly different.

Fracture Path and Speed

In order to determine the exact fracture path and speed, we attempted to record the fracturing of a crispy dry biscuit with a high-speed video camera. Successive frames taken from the film shot are shown in Figure 4. These pictures reveal that fracture of the biscuit occurs within a single time frame, *i.e.*, within 0.45 ms. This shows that the fracture speed in crispy products is very fast—at least 100 m s^{-1} and probably faster.[17] This surprisingly high speed was not at all expected. To the best of our knowledge no such data have been published before reporting a similar observation for a food product. The speed of audible sound through air is ~ 340 m s^{-1}. As this is the same order of magnitude as the speed of the proceeding crack, this implies possible interference of the acoustic emission of the different fracture events that take place when fracturing a crispy/crunchy crust with a porous structure. Because the number of fracture events depends amongst other things on the morphology of the crust,

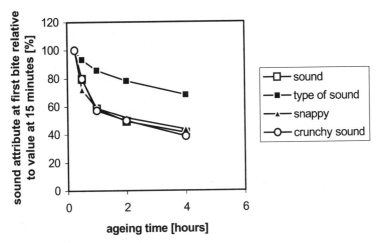

Figure 3 *Effect of ageing time after baking on the perception of some sound-related attributes during eating of a model bread. The values obtained after 15 minutes were set to 100%.*

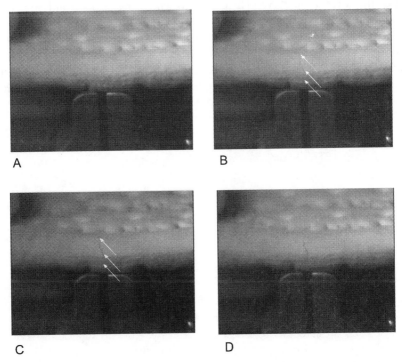

Figure 4 *Successive frames taken from a film shot with a high-speed camera of a fracturing dry biscuit. These figures show that fracture of the biscuit occurred within a single time-frame, i.e., within 0.45 ms. (A) At start of experiment. (B) After 1.35 ms from start of experiment: fracture through whole biscuit slightly visible. (C) After 1.8 ms from start of experiment: fracture through whole biscuit clearly visible. (D) After 2.25 ms from start of experiment: further crack opening.*

a relationship between this morphology and the audible sound as recorded outside the fracturing food is expected. The high fracture speed also makes the development of stress waves in a crispy/crunchy material that travel through the material with the speed of sound more understandable and also the emission of a characteristic sound. This, again, indicates that there is a relation between the acoustic emission and different mechanical properties, especially the fracture behaviour. The fast speed of fracture also implies that data acquisition during tests used to study the fracture of crispy/crunchy foods have to be very fast.

Mechanical Behaviour and Acoustic Emission

Based on a literature search and some preliminary tests,[3,16,18] we decided to choose a wedge test[15] for studying the mechanical properties. The penetration speed was set at 40 mm s[-1]. With this set-up we wanted to mimic the eating process, and especially the first bite. In Figure 5 an example of a force–time curve is shown. The wedge deforms the food product and initially especially the crust, thereby fracturing the crust. Some peaks are visible in the increasing part of the force–time curve. The extent to which the peaks could be resolved was improved by moving to a higher data acquisition rate.

After ∼200 ms, the force in Figure 5 decreases slowly. This part of the curve represents mainly the deformation and fracture of the interior part of the snack. By using a whole snack, and so including the crust and the softer interior, it was not possible to study just one structural part of the product. Hence the results are dependent on the properties of both the crust and the interior.

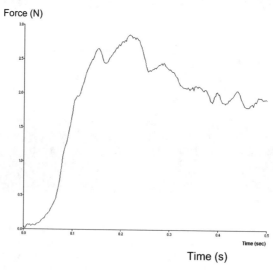

Time (s)

Figure 5 *Example of a force–deformation curve of a wedge test on a standard model snack with speed=40 mm s⁻¹ and wedge angle=30°.*

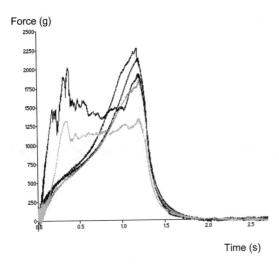

Figure 6 *Different examples of the force–deformation curves of a fresh model bread. Penetration test with a wedge of 30° at speed of 10 mm s⁻¹.*

The exact importance of the two contributions will depend on the size, shape and mechanical properties of the crust and the interior. Even very small differences in these factors can give a large deviation in the results, as is shown in Figure 6. Here five different force–deformation curves measured on the same bread and with the same test set-up are presented. In two cases the wedge indeed fractured the bread crust, while the three other tests showed a more predominant compression of the interior part, and a less obvious crust fracture. Probably the latter three samples had a somewhat less stiff interior or a less brittle crust—the exact reason is yet unknown. This type of problem is actually more important during the testing of bread than of our model snack, because the crust of the latter is more hard and brittle, and so it fractures easier.

The acoustic emission was recorded simultaneously with the mechanical test. An example of the results is shown in Figure 7. The contribution of the sound emission from the snack is distinguishable from the background sound and the sound produced by the texture analyser. It is remarkable that the sound emission from the snack is especially intense and distinguishable during the first 200 ms of the deformation. This is during the time that the force necessary to deform the snack is still increasing and has not yet reached its maximum value. The acoustic emission from the snack was observed to start almost immediately after the start of the deformation by the wedge, and it continued until the wedge mainly deformed the interior. In contrast to this, for a dry crispy biscuit like 'Knappertje', there is a distinct region in the deformation curve where no acoustic emission could be recorded.[16] The first acoustic emission peak is recorded simultaneously with the first peak in the force–time curve, as are the subsequent force and sound peaks. Both the force and sound peaks measured in the test with the 'Knappertje' were more clearly distinguishable, and limited

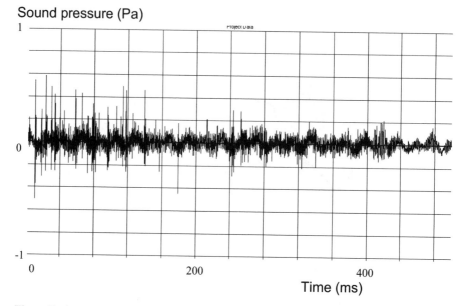

Figure 7 *Acoustic emission from a wedge test with the Texture Analyser on a standard model snack. Penetration speed=40 mm s⁻¹, wedge angle=30°.*

in their amount than the sound and force peaks emitted by the fracturing snack or bread crust.

On performing a Fourier analysis on the results, it becomes apparent that the whole audible frequency region, from 20 Hz to 20 kHz,[8] is involved in the acoustic emission. There are no clear distinguishable frequency peaks originating from the snack sound, although maybe the frequencies above 10–12 kHz are somewhat less intense in the sound produced during a mechanical test of a snack. This broad region indicates that it is very likely that also sound outside the audible region is emitted during the deformation of these crispy snacks. This feature of the acoustic emission can be used for a better understanding of the mechanism of fracturing,[5,9] and changes therein due to deterioration. Also this part of the sound energy has to be taken into account when establishing an energy balance for the fracturing process of crispy/crunchy food products,[3,6] which may be used as a tool to understand changes in crispy behaviour during product ageing.

In the lower frequency region, especially below ~2 kHz, the sound originating from the snack is not distinguishable from the sound of the Texture Analyser. A 'hand-deformation test' was performed with the same wedge to study the importance of the frequencies in this lower region where the sound of the Texture Analyser drowns out the sound of the snack. An example of the acoustic emission obtained by 'hand fracturing' can be seen in Figure 8. The sound emitted from the hand test is clearly comparable with the sound from the Texture Analyser test. The duration of both sounds was about 200 ms, and

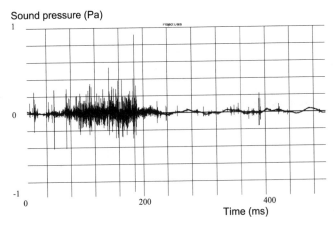

Figure 8 *Acoustic emission from a wedge test performed with 'hand deformation' on a standard model snack. The wedge had an angle of 30°.*

in both cases different sound pulses could be distinguished over this time interval. The whole frequency range (20 Hz to 20 kHz) was involved with again a higher intensity of frequencies below 10–12 kHz (as obtained by Fourier analysis). The hand test also shows a remarkable amount of acoustic emission in the region drowned out by the Texture Analyser, *i.e.*, between 20 Hz and 2 kHz. For aged snacks that had become less crispy, a remarkably lower sound pressure in the frequency range of 2–12 kHz was observed.[18] This indicates the possible importance of this type of sound, or of the fracturing process causing this type of sound, for the perception of crispiness. The changes at frequencies <2 kHz, and their relation to crispy perception, still have to be sorted out. The importance of sound within the frequency range 2–12 kHz is in accordance with literature data. Dacremont found[19] that frequencies of 5–12.8 kHz were typical for different crispy products. Lee *et al.* reported[20] a good relationship with the crispness of potato crisps ('chips') for frequencies of 3–4 kHz and of ∼6 kHz. Srisawas and Jindal found[21] that both the amplitude of the peaks and the relative importance of the sound below 7 kHz decreased with increasing moisture content of dry snacks. In contrast to this, De Belie *et al.* found[22] that different dry crisp snacks could best be distinguished based on the sound emission at frequencies of 200 and 900 Hz. In this context it should be noted that our QDA panel observed that there is a change in both the loudness and the type of sound perceived on eating when the model snacks had been aged (Figure 3).

The sound produced on biting the same type of snack is shown in Figure 9. The duration of this sound was ∼200 ms, similar to that obtained with the other two types of test. This indicates the closely comparable speeds of deformation produced by the Texture Analyser, by hand deformation, and by biting. Also the type of the sound was comparable: the whole audible frequency region was involved, and different sound pulses could be distinguished. The size of the sound pulses was roughly between 0.1 and 0.5 Pa. In Table 3 the concentration

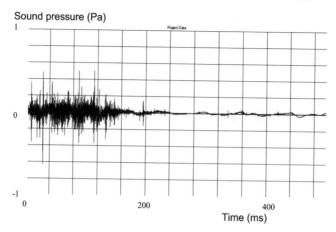

Figure 9 *Acoustic emission from a biting experiment on a standard model snack.*

Table 3 *Number of acoustic peaks with a size larger that 0.2 Pa during the first 200 ms of deformation of a deep-fried battered model snack.*

Type of test	Number of peaks >0.2 Pa
Biting	51
Wedge with Texture Analyser	55
Wedge by hand	51

of pulses larger than 0.2 Pa is given for all three types of test. The density of pulses was of the same order of magnitude, and did not clearly depend on the type of test. This again indicates that the deformations were roughly comparable, and that the wedge experiment mimics biting in a satisfactory way, as was found for apples and cheese previously.[15] The duration of a single sound peak was less than 0.25 ms, the same order of magnitude as the duration of the fracturing process of the dry brittle biscuits.

4 Discussion and Conclusions

It is generally accepted that the crispy/crunchy character of foods derives from the fracture properties and the accompanying acoustic emission. Here we have found that sound-related attributes play a special role in the vocabulary used to describe the sensory perception of crispy/crunchy foods with a crust. These attributes have been shown to be important for the description of differences between products or changes in one product due to ageing under deteriorating conditions.

The acoustic emission originates from the fracture process, and it depends on the amount of newly formed surface and on the mechanical properties of the (local) material that fractures. The fracture speed in crispy/crunchy foods is very fast, at least faster than ~ 100 m s^{-1}, and thus of the same order of magnitude as the speed of sound in air. This indicates not only that stress and sound waves travel through the material at a similar speed, but also that interference of these waves is very likely to occur. Thus the sound recorded outside the material will be considerably affected by the morphology as well.

The sound emitted during a wedge mechanical test at 40 mm s^{-1} was found to be comparable to the sound emitted while biting the same food. Both types of test showed acoustic emissions of the same duration, a similar amount of sound events and acoustic power, and comparable frequencies. This implies that the fracture process during the mechanical test is comparable to the fracture process during the first bite. In both types of test, high deformation speeds and fast fracture processes are involved, implying an important contribution of mixing and interference of the individual sound signals in the sound heard during eating or mechanical testing. The morphology of the crust is therefore important for acoustic emission and sound perception during eating. So it is important that, when mechanical tests (and simultaneous acoustic tests) are to be related to the sensory perception of the foods, the experiments are done at comparable speeds of deformation. On the other hand, when the distinct fracture events are intended to be studied, a slower rate of deformation, and a test set-up that gives a controlled fracture, is more favourable. Acoustic emission then may be less affected by interference, and it possibly can be better related to the local fracture processes and to the components involved. In both cases, sound recorded simultaneously with the mechanical test may give additional information on the process of fracturing and on the mechanical and morphological properties involved.

Acknowledgement

This study was funded by the Wageningen Centre for Food Sciences, an alliance of major Dutch food industries, TNO Nutrition and Food Research, and Wageningen University and Research Centre, with financial support by the Dutch government. The authors thank Garmt Dijksterhuis and René de Wijk for their work on the sensory properties of food products with a crust.

References

1. A. S. Szczesniak, *Food Qual. Pref.*, 2002, **13**, 215.
2. G. Roudaut, C. Dacremont, B. Vales Pamies, B. Colas, and M. Le Meste, *Trends Food Sci. Technol.*, 2002, **13**, 217.
3. H. Luyten, J. Plijter, and T. van Vliet, *J. Texture Stud.*, submitted for publication.
4. R. V. Williams, 'Acoustic Emission', Adam Hilger, Bristol, UK, 1980.
5. M. Wevers, *NDT&E International*, 1997, **30**, 99.
6. H. Luyten, J. J. Plijter, and T. van Vliet, in 'Proceedings of the 3rd International Symposium on Food Rheology and Structure', eds P. Fisher, I. Marti and E. J. Windhab, ETH Zürich, Switzerland, 2003, p. 379.
7. W. Alchakra, K. Allaf, and J. M. Ville, *Appl. Acoust.*, 1997, **52**, 53.

8. E. Zwicker and H. Fastl, 'Psychoacoustics: Facts and Models', Springer, Germany, 1999.
9. P. J. de Groot, P. A. M. Wijnen, and R. B. F. Janssen, *Composites Sci. Technol.*, 1995, **55**, 405.
10. J. Bohse, *Composites Sci. Technol.*, 2000, **60**, 1213.
11. M. Kootstra and H. Luyten, 'Development of a snack model system', Internal Report, WCFS, Wageningen, Netherlands, 2000.
12. M. W. L. J. van Son, 'Validatie modelsysteem en variëren van krokantheid binnen modelsysteem', Internal Research Note, WCFS, Wageningen, Netherlands, 2002.
13. G. Dijksterhuis, H. Luyten, and R. de Wijk, *Food Qual. Pref.*, submitted for publication.
14. Fizz Software for Sensory Analsyis, v2.01d., Biosystème, Dijon, France, 2002.
15. J. F. V. Vincent, G. Jeronimidis, A. A. Kahn, and H. Luyten, *J. Texture Stud.*, 1991, **22**, 45.
16. E. M. Castro, H. Luyten, and T. van Vliet, 'Simultaneous mechanical and acoustic emission measurement', Poster Presentation, Food Colloids Conference, Harrogate, UK, 2004.
17. J. Plijter, H. Luyten, and T. van Vliet, 'High speed fracture of crispy food products', Poster Presentation, Food Colloids Conference, Harrogate, UK, 2004.
18. J. E. Visser, unpublished results.
19. C. Dacremont, *J. Texture Stud.*, 1995, **26**, 27.
20. W. E. Lee III, A. E. Deibel, C. T. Glembin, and E. G. Munday, *J. Texture Stud.*, 1988, **19**, 27.
21. W. Srisawas and V. K. Jindal, *J. Texture Stud.*, 2003, **34**, 401.
22. N. de Belie, M. Sivertsvik, and J. de Baerdemaeker, *J. Sound Vibration*, 2003, **266**, 625.

Structure Control and Processing

Solubilization and Bioavailability of Nutraceuticals by New Self-Assembled Nanosized Liquid Structures in Food Systems

By Nissim Garti, Idit Amar-Yuli, Aviram Spernath, and Roy E. Hoffman[1]

CASALI INSTITUTE OF APPLIED CHEMISTRY, THE HEBREW UNIVERSITY OF JERUSALEM, JERUSALEM 91904, ISRAEL
[1]DEPARTMENT OF ORGANIC CHEMISTRY, THE HEBREW UNIVERSITY OF JERUSALEM, JERUSALEM 91904, ISRAEL

1 Introduction

As a result of new healthy diet patterns, motivated by the concept of prevention rather than cure, functional foods have shown promise at a time when consumer interest in diet and health is at an all-time high (10% growth per year).[1,2] The new health pattern emphasizes the role of food components as essential nutrients for preventing or delaying the premature onset of chronic diseases later in life. Bioactive phytochemicals, known as nutraceuticals,[3,4] providing medical or health benefits including the prevention and treatment of disease, have attracted much interest in the last decade as constituents of functional foods. These include several major categories of strategic nutraceuticals:[5] vitamins (*e.g.*, vitamin E), phytosterols, carotenoids (*e.g.*, β-carotene, lutein and lycopene), fibres (insulin, fenugreek, guar, flaxseed lignans), isoflavones (from soya), citrus limonoids, omega fatty acids, CoQ10, organosulfur compounds (garlic), polyphenols (such as tea catechins), phospholipids (such as phosphatidylserine), and others.

Some nutraceuticals are not soluble in water, and so their transport *via* the digestive tract membrane is very limited and their bioavailability is restricted. Therefore, attempts have been made to find new vehicles that can enhance solubility, transport the nutraceuticals, and improve their bioavailability. There is a growing interest[6-8] in using microemulsions as vehicles for nutraceuticals and pharmaceutical formulations due to their intrinsic physico-chemical properties, such as spontaneous formation, clear appearance, low viscosity,

Table 1 *Some characteristics of microemulsions, emulsions, and liposomes.*

	Microemulsions (U-type)	Emulsions	Liposomes
appearance	clear	turbid-milky	turbid-clear
size (nm)	8–15	100–10000	150–500
thermodynamic stability	stable	unstable	unstable
surface area	large	small	medium
curvature	high	low	medium
dispersity	monodisperse	polydisperse	good
solubilization capacity	high	low	low
pH sensitivity	insensitive	variable	sensitive
cost	low	low	high

thermodynamic stability, and high solubilization capacity.[6–9] Table 1 lists the major advantages of microemulsions in terms of stability as compared to emulsions and liposomes. Microemulsions (composed of water, oil and surfactant) are capable of solubilizing large amounts of lipophilic and hydrophilic food additives or drugs, thereby enhancing reaction efficiency, and allowing selective extraction.[6–10] Moreover, microemulsions have the potential to enhance nutraceutical absorption in the digestive system, as promoted by surfactant-induced permeability changes.[11,12] Following oral administration, the solubilized drug is widely distributed in the gastrointestinal tract by the small droplets.[13]

Three major microemulsion categories are known: water-in-oil (W/O), bicontinuous, and oil-in-water (O/W). Figure 1 illustrates these structures schematically.

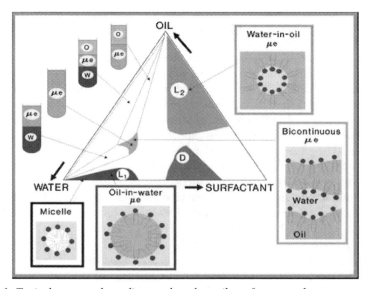

Figure 1 *Typical ternary phase diagram based on oil, surfactant and water.*

2 Microemulsion Formation and Phase Behaviour

The free energy of microemulsion formation is considered to depend on the extent to which the surfactant lowers the interfacial tension between the oil and the water. The change in free energy is derived from

$$\Delta G_f = \gamma \Delta A - T \Delta S, \tag{1}$$

where ΔG_f is the free energy of formation, γ is the tension of the oil–water interface, ΔA is the change in interfacial area upon microemulsification, T is the temperature, and ΔS is the change in entropy of the system. One should note that, when a microemulsion is formed, the change in the area is very large due to the many small droplets that are formed. However, γ is very small ($\ll 1$ mN m^{-1}) and so the term $\gamma \Delta A$ is offset by the entropic component. The dominant favorable entropic contribution is the large dispersion entropy arising from the mixing of one phase with the other in the form of a large number of small droplets. Expected entropic contributions arise from other processes such as surfactant mobility in the interfacial layer. Thus, a negative free energy of formation is achieved, with the consequence that the microemulsion forms spontaneously and remains thermodynamically stable.[14]

The phase behaviour of simple microemulsion systems containing oil, water and surfactant can be presented in the form of ternary phase diagrams in which each corner of the diagram represents 100% of one particular component. Figure 2 is a typical phase diagram of a non-food microemulsion based on a mixture of water, a non-ionic surfactant ($C_{18}E_{10}$, ethoxylated stearyl alcohol) and a hydrocarbon. In the case of microemulsions for food applications, the systems must be composed of food-grade surfactants and permitted solvents (that do not properly mutually dissolve); and therefore they must contain additional components such as a co-surfactant, and/or a co-solvent (and/or

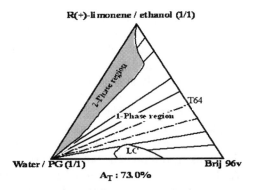

Figure 2 *Phase diagram of water/propylene glycol + R(+)-limonene/ethanol + Brij 96v ($C_{18:1}EO_{10}$) at 25 °C with constant weight ratio of water/propylene glycol (1:1) and a constant weight ratio of R(+)-limonene/EtOH (1:1). The grey area represents the two-phase region. A_T is the area of the total isotropic region as a percentage of the total area of the phase diagram.*

other solubilized additives). The co-surfactants, and in some cases also the co-solvents, are somewhat amphiphilic with an affinity for both the oil and the aqueous phases. The co-surfactant partitions between the phases, *i.e.*, at the oil–water interface, while the co-solvent dissolves mostly in the continuous phase and alters its hydrophilic properties. Medium-chain alcohols, which are commonly added as co-surfactants, further reduce the interfacial tension and increase the fluidity of the interface thereby increasing the entropy of the system.[15] But medium-chain alcohols are not permitted in food systems, and they must be replaced by ethanol.

Figure 1 gives a schematic representation of three types of microstructures that can be formed, depending on the microemulsion composition. O/W droplets are likely to be formed at low volume fractions of the oil phase. Conversely, W/O droplets tend to be formed when the volume fraction of the aqueous phase is low. In systems where the amounts of the aqueous and oil phases are roughly similar, a bicontinuous microemulsion may be formed.[6,7]

3 Microemulsions in Food Applications

The coexistence of oil and water within many foods has made the study of microemulsions important for food scientists and food producers over several decades.[6] The equilibrium nature of microemulsions means that they can remain stable indefinitely as long as the conditions around them are unchanged. Thus they are unaffected by the standard emulsion destabilization mechanisms (creaming, flocculation, *etc.*). On the other hand, by changing the temperature or composition to values at which microemulsion formation is no longer thermodynamically favourable, one may destabilize the system in a controlled fashion. Hence, microemulsions have the potential to bring unique advantages to foods that need long-term stability, clarity, *etc.*, and also to foods that require destabilization at a certain controlled stage or in a certain situation.

Food microemulsions (especially of the O/W type) are, however, difficult to formulate for two reasons. Firstly, there is a limited number of available food-grade emulsifiers that can create regions of solubilized oil phases in water, and these compositions are restricted to the water corner of the phase diagram, which permits solubilization of only small amounts of oil and solubilizates.[16] The second difficulty arises from the particular chemical nature of the oil phase commonly used in food applications.

Food emulsions typically contain triglycerides of unsaturated long-chain fatty acids (LCT)—such as the oils from soya, corn, cotton, or sunflower seeds. In some applications medium-chain fatty acid triglycerides (MCT) are used, mainly because they are liquids, fully saturated, and stable to oxidation. In a few food applications palm oil or coconut oil is used, but since this type of oil is solid at room temperature their microemulsion formulations are restricted to elevated temperatures (above the melting points of the triglycerides). Even if liquid at room temperature, triglycerides are difficult to solubilize in water/surfactant phases because of their large molecular volume and limited solubility in food-grade emulsifiers. Moreover, it has recently become a trend with food formulators to explore the use of liquid fractions of non-saponifable natural products. Some of these 'solvent-like' fractions, known as essential oils, contain various hydrocarbon-like groups (such as terpenes) along with some triglycerides and other lipophilic or lipid components. The monoterpene hydrocarbon, R(+)-limonene, the major constituent of citrus essential oils, is often present in foods; but, being essentially insoluble in water, it requires the addition of a surfactant to be solubilized or emulsified in an aqueous environment.[17] Other essential oils (perfume oils) involved in food applications are lemon oil, menthol oil, peppermint oil, spearmint oil, *etc.*

Food-Grade Microemulsion Characterization

Microemulsions based on five-component mixtures (R(+)-limonene, ethanol, ethoxylated sorbitan ester, polyol and water) have been studied extensively in our laboratory. Some typical five-component phase diagrams[15,16] are shown in Figures 2–4. Microemulsions containing R(+)-limonene/ethanol (1/1), Tween 60, and water/propylene glycol (PG) (1/1) (Figure 3), have a total monophasic area of $A_T = 64\%$, corresponding to a relatively high water-solubilization capacity. Replacing some of the water by glycerol (water/glycerol weight ratio = 3) and substituting the Tween 60 by Tween 80 reduces the total monophasic area to 52% (see Figure 4). According to Shiao *et al.*,[18] the mutual miscibility of the hydrophobic part of the surfactant and the oil affects the degree of oil penetration into the amphiphilic film. The solubilization of R(+)-limonene by non-ionic surfactants has been studied by Tokuoka *et al.*[17] and Kanei *et al.*[19] They found that 'perfume' molecules tend to be solubilized in the

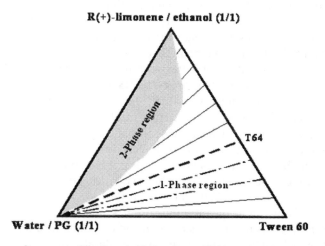

Figure 3 *Phase diagram at 25 °C and dilution lines (T64 means 60 wt% surfactant and 40 wt% oil phase in the micellar solution, a so-called 'oil-concentrate'), for a system composed of R(+)-limonene/EtOH (1:1 w/w) as the oil phase, Tween 60 as the emulsifier, and water/propylene glycol (1:1 w/w) as the aqueous phase.*

R(+)-limonene/ethanol (1/2)

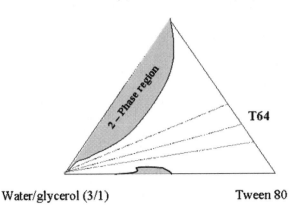

Water/glycerol (3/1)   Tween 80

Figure 4 *Phase diagram at 25 °C and dilution lines for a system composed of R(+)-limonene/EtOH (1:2 w/w) as the oil phase, Tween 80 as the emulsifier, and water/glycerol (3:1 w/w) as the aqueous phase. The dilution line 64 corresponds to 60 wt% surfactant and 40 wt% oil phase.*

vicinity of the interface between the water and the hydrocarbon moieties of the surfactant or at the surfactant palisade layer. The results are a good indication of improved water solubilization in R(+)-limonene because of the oil compatibility with the surfactant and its partial penetration into the surfactant film.

Many studies have shown that the maximum water solubilization in a W/O microemulsion is dictated by the chain length of the surfactant (N_S), the chain length of the alcohol (N_A), and the chain length of the oil (N_O). In our previous studies we discussed the effect of normal alcohols in non-ionic quaternary model systems; the observed data were presented in terms of the Bansal–Shah–O'Connell (BSO)[20] chain-length compatibility equation ($N_S = N_A + N_O$ for maximum solubilization of water) and more profoundly *via* the Hou and Shah phenomenological mechanism.[21] The BSO equation should be used with caution. Its predictive potential has been shown to be limited only to properly designed systems. In cases where the concentration of one component (such as the alcohol) is too high, or where the hydrophlic head group is too short or too long, and where the system contains branched molecules (triglyceride or perfume oils), the BSO equation is not obeyed or it has to be modified. The effect of the alcohol on the water solubilization is critical. If ethanol is used, it will play a different role when W/O microemulsions are formed from when they become transformed into bicontinuous or O/W microemulsions.

At present, there is no good theory that can correlate the amount of ethanol needed at the oil–water interface to the water solubilization, or any theory that can evaluate the role of the ethanol at the oil–water interface in terms of the amount of solubilized oil. Most of the work so far has been empirical, and only a few guidelines are presently available (see below).

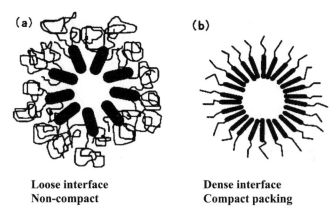

Loose interface **Dense interface**
Non-compact **Compact packing**

Figure 5 *Schematic structures of (a) a loose interface (e.g., a Tween 80 microemulsion droplet) and (b) a dense interface (e.g., a $C_{12}E_9$ microemulsion droplet).*

The spontaneous curvature R^0 and the interfacial elasticity R^C are two key additional parameters that must be considered when dealing with solubilization within surfactant + oil mixtures. Maximum solubilization is obtained when the alcohol chain length is optimized to allow R^0 and R^C to coincide into single crossing point (see Figure 5). The spontaneous curvature and elasticity parameters have not yet been examined in five-component mixtures or in microemulsions stabilized by a hydrophilic surfactant and an alcohol (ethanol).

The effects of polyols (propylene glycol and glycerol) and a short-chain alcohol (ethanol) on the phase behaviour of non-ionic surfactants and food-grade oils, and on the solubilization capacity, have been examined by Garti *et al.*[15,16] The aim was to find a mixture of ingredients that (i) imparts a positive interfacial curvature to the surfactant film in order to form an O/W microemulsion, (ii) inhibits liquid-crystal formation, and (iii) maximizes mutual dissolution of the ingredients. We established, empirically, that in order to formulate a food-grade O/W microemulsion with a hydrophilic surfactant, the addition of polyols and short-chain alcohols is essential. The formulation of such a system was visualized as a multi-stage preparation: formation of the solvated surfactant, formation of a hemi-micelle, closure of the micelle, and solubilization of water in the core of the micelle. The phase behaviour of the system composed of R(+)-limonene + ethanol + water/PG (1/1) + polyoxyethylene sorbitan monostearate (Tween 60), containing a 1:1:3 R(+)-limonene/ethanol/Tween weight ratio, is characterized by a single continuous microemulsion region starting from a pseudo-binary solution (surfactant/oil phase) towards the microemulsion/water PG (1:1) corner. Due to the addition of the short-chain alcohol and polyol, this dilution line inverts progressively from W/O to O/W *via* a bicontinuous region without any phase separation.

The effect of the co-solvent (glycerol) or co-surfactant (alcohol) that is not incorporated in the surfactant film has been explored. Figure 6 shows that, as the water/glycerol weight ratio decreases from 3:1 (Figure 6a) to 1:1 (Figure 6b), and as the alcohol content increases from an R(+)-limonene/

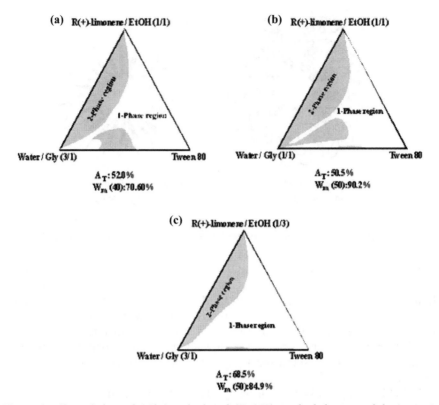

Figure 6 *Effect of glycerol (Gly) and ethanol (EtOH) on the behaviour of the isotropic regions of the system water/polyol + R(+)-limonene/EtOH (1:1) + Tween 80 at 25 °C for different ratios of water/polyol: (a) water/Gly (3:1), (b) water/Gly (1:1), and (c) water/Gly (3:1) + R(+)-limonene/EtOH (1:3). The composition of the oil phase in these phase diagrams is given in the upper corner.*

ethanol weight ratio of 1:1 to 1:3 (Figure 6c), larger isotropic areas are formed with broader dilution channels.

Microemulsions based on glycerol and polyoxyethylene sorbitan mono-oleate (Tween 80) show behaviour similar to that for micromulsions made from propylene glycol and Tween 60. However, one can note (Figure 3) that microemulsions based on Tween 60 and propylene glycol have larger isotropic areas, and broader dilution channels. It is possible that the propylene glycol, considered to be a co-solvent, is also behaving as co-emulsifier and is propagating into the interface, while glycerol remains mostly in the continuous phase.

4 Solubilization of Nutraceuticals within Microemulsions

One of the most obvious applications for microemulsions in foods is to facilitate the incorporation of food supplements or food additives into the final products. Solubilization may be defined[2] as the spontaneous 'dissolution' (molecular incorporation) of an insoluble (or only slightly soluble) substance into a given solvent, by reversible interaction with micelles in the solvent, to form a thermodynamically stable isotropic solution.

Flavours, preservatives, and nutrients that are poorly soluble in water can be incorporated into water-based foods by solubilizing the component within surfactant aggregates. Similarly, W/O microemulsions can be used to solubilize water-soluble substances. In both cases microemulsions can provide a well-controlled and highly stable medium for the incorporation of these ingredients.

The transparency of microemulsions provides new options for altering the appearance of food, as well as being a useful separation tool when employing these phases in food analysis. The small droplet size in microemulsions provides excellent contact between the lipid and aqueous phases which is particularly advantageous when solubilizing preservatives within the droplets. There is also some indication that microemulsions or micelles may protect solubilized components from unwanted degradative reactions.[22,23] Finally, the stability characteristics of microemulsions offer a real benefit during food processing and storage.

Flavours and Aromas

The hydrophobic nature of many flavour and aroma molecules makes them excellent candidates for incorporation within O/W microemulsions. However, there have been few studies of the solubilization of flavours within these surfactant systems. The solubilization of typical synthetic perfumes in anionic + non-ionic mixed surfactant solutions (micelles) has been investigated by Tokuoka *et al.*[17] In mixtures of sodium dodecyl sulfate (SDS) and ethoxylated palmitoyl alcohol ($C_{16}EO_{20}$), they found a positive synergistic effect on the maximum additive concentration of perfumes such as α-ionone, α-hexyl-cinnamaldehyde and R(+)-limonene. The concentrations of perfumes in the micelles are shown in Figure 7 as a fraction of the $C_{16}EO_{20}$ mole fraction in an SDS + $C_{16}EO_{20}$ system.

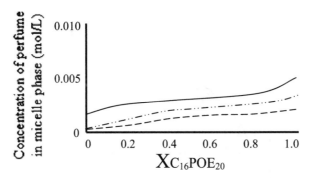

Figure 7 *Concentration of three different perfume molecules in the micelle phase versus mole fraction X of $C_{16}POE_{20}(C_{16}E_{20})$ in the SDS + $C_{16}POE_{20}$ mixed system at 30 °C: —, α-ionone, —·—, α-hexylcinnamaldehyde, and – – –, R(+)-limonene.*
(Taken from ref. 17.)

Slocum *et al.* found[24] that greatly enhanced solubility of a range of flavour elements (including alcohols, 2-ketones, 1-alcohols, and ethyl esters) could be achieved through solubilization in a variety of food-grade surfactants. Tween 20 was able to solubilize relatively large quantities of flavours—up to 3 moles of flavour compound per mole of surfactant—a quite attractive result considering the very small quantities of flavour components that are generally needed in foods. Several of the microemulsions discussed by Wolf and Havekotte[25] contained essential oils as the organic phase. In most of these studies, it was inferred that the flavour molecule is incorporated directly into the micelle, at the interface, or at the micelle's core.[26] It should be stressed also that flavour incorporation is not limited to oil-soluble flavours. El-Nokaly *et al.* reported[27] the incorporation of water-soluble flavours within food-grade W/O microemulsions. It should be also noted that, in the early reports, solubilization was achieved by incorporating the nutraceuticals into the oil core, and in none of these studies could the systems be diluted to infinity with aqueous phase. In contrast, we have been able to form dilutable systems within which the additives remain incorporated and solubilized to infinite dilution.[26]

Vitamins and Nutraceuticals

Microemulsions have the potential for enhancing the solubility of hydrophobic vitamins and other nutraceuticals within water-based foods, or water-soluble nutraceuticals within oil-based foods. The pharmaceutical and medical literature is replete with studies of the enhanced micellar delivery of vitamins—in particular, vitamin E, vitamin K_1 and β-carotene (see Figure 8).[28-30]

Chiu and Yang have claimed[28] that solubilization within non-ionic surfactants protects vitamin E from oxidative degradation. Studies in our laboratory have shown[30] that vitamin E (α-tocopherol and α-tocopheryl acetate) can be solubilized in W/O microemulsions and remain solubilized even after progressive dilution and structural conversion to O/W microemulsions. The

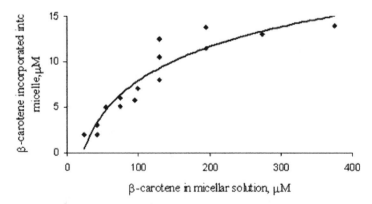

Figure 8 *Concentration of β-carotene incorporated into a mixed micellar solution as a function of β-carotene added.*
(Taken from ref. 29.)

vitamin E serves as the oil phase (partially or entirely replacing the R(+)-limonene), the co-emulsifier, and the solubilizate. The basic system used to solubilize the nutraceuticals consists of R(+)-limonene and ethanol as the oil phase, water/PG (1/1) as the aqueous phase, and Tween 60 as the surfactant (Figure 3). The difference in solubilization behaviour of tocopheryl acetate and tocopherol arises from differences in their relative lipophilicity, which leads to them partitioning more to the interface or to the core, respectively. Tocopheryl acetate seems to decrease the solubilization capacity of water and oil, whilst tocopherol enhances the solubilization of water in the oil-rich region, and reduces oil solubilization in water to an even greater extent than tocopheryl acetate.[30]

Recently, other oil-soluble nutraceutical solubilization capacities have been tested[31] within the same food-grade microemulsion. Lycopene, a non-soluble crystalline nutraceutical antioxidant with a solubility of only 700 ppm in R(+)-limonene, has been solubilized in the core and at the interface of W/O, bicontinuous and O/W microemulsions. Spernath *et al.*[31] found that lycopene cannot serve as an oil phase; yet it can be solubilized up to 10 times more than its dissolution capacity in R(+)-limonene or any other edible oil (Figure 9). The incorporation of molecular phytosterols (a cholesterol-lowering agent) within the microemulsion (Figure 10) was also tested. Phytosterols were solubilized up to 12 times more than their dissolution capacity in the R(+)-limonene.[32]

The synergistic effect of solubilization in non-ionic mixed surfactant micro-emulsions[31] is shown in Figure 11. The solubilization capacity of lycopene was found to be increased by *ca.* 40% when the microemulsions were stabilized in

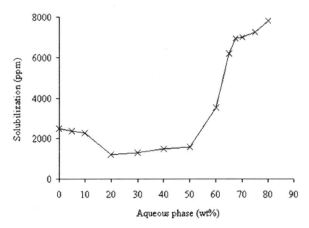

Figure 9 *Solubilization capacity of lycopene normalized to the amount of oil (solubilization efficiency) as a function of the water content at each aqueous dilution point. The solubilization at 25 °C is plotted against the aqueous phase content along the dilution line 64 in a microemulsion composed of R(+)-limonene/EtOH (1:1 w/w) as the oil phase, Tween 60 as the emulsifier, and water/propylene glycol (1:1 w/w) as the aqueous phase.*

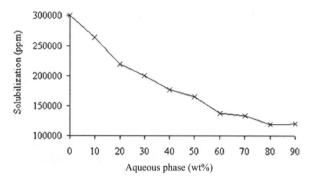

Figure 10 *The normalized solubilization capacity (solubilization efficiency) of phytosterols (ppm) as a function of the amount of oil phase at each dilution point. The solubilization at 25 °C is plotted against the aqueous phase content along the dilution line 64 in a microemulsion composed of R(+)-limonene/EtOH (1:1 w/w) as the oil phase, Tween 60 as the emulsifier, and water/propylene glycol (1:1 w/w) as the aqueous phase.*

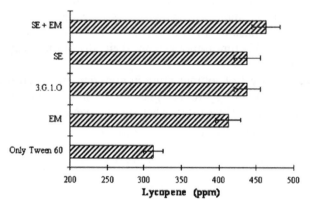

Figure 11 *Maximum solubilization capacity of lycopene (in ppm) in microemulsions stabilized by mixed surfactants at 25 °C. The composition of the microemulsion is R(+)-limonene + ethanol + Tween 60 (1:1:3) + 80% aqueous phase. The additional surfactants are: EM = ethoxylated monodiglyceride, 3.G.1.O = triglycerol monooleate, and SE = sucrose ester, O-1570.*

a mixture of non-ionic surfactant such as Tween 60 and 3.G.1.O (triglycerol monooleate) or Tween 60 and SE (sucrose monooleate O1570) in the ratio of 9:1. A mixture of three non-ionic surfactants, Tween 60 + SE + EM (ethoxylated monodiglyceride), increased the solubilization by 48%. The synergism was attributed to a better interfacial organization (orientation) of the mixed surfactants around the oil droplets, thereby allowing better interfacial solubilization.

We have examined the solubilization capacity of another microemulsion (see Figure 4) composed of water, Tween 80, R(+)-limonene, ethanol and polyol

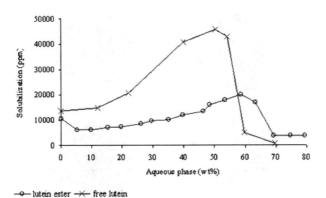

Figure 12 *The normalized solubilization capacity (ppm) of lutein ester (O) and free lutein (×) related to the amount of oil phase at each dilution point tested. The solubilization at 25 °C is plotted against the aqueous phase content along dilution line 64 in a microemulsion composed of R(+)-limonene/EtOH (1:2 w/w) as the oil phase, Tween 80 as the emulsifier, and water/glycerol (3:1 w/w) as the aqueous phase.*

(propylene glycol or glycerol) at somewhat different compositions. The differences are in the oil/alcohol and water/polyol weight ratios. The solubilization of both lutein and lutein mono-/di-esters was examined, and the differences in behaviour between the two solubilizates were determined.[26] Lutein is a naturally occurring carotenoid, which is claimed to reduce the risk of cataracts and age-related macular degeneration (AMD). Both forms of lutein—free and esterified—are practically insoluble in aqueous systems, and their solubility in food-grade oils such as R(+)-limonene is very limited (1200 and 7000 ppm, respectively). Free lutein and the lutein ester behave somewhat like phytosterols and lycopene, respectively. The luteins cannot serve as the oil phase. Both kinds of lutein require high alcohol ratios in the oil phase, and their solubilization capacity has been found to be surfactant-dependent in all the isotropic regions and structures. From Figure 12, one can note that the solubilization is highest in the bicontinuous phase. Free and esterified lutein can be solubilized, respectively, up to 10 and 8 times more than their dissolution capacity in the oil phase (R(+)-limonene/ethanol in the ratio 1:2).

5 Relationship of Guest Molecule to Microemulsion Microstructure

Locus of Solubilization

The location of a solubilized molecule (a guest molecule) in the microemulsion relative to the structural components of the surfactant is determined primarily by the chemical structure of the additive (Figure 13). Data on the locus of solubilization is obtained mainly from studies on the solubilizate and the other components of the microemulsion, before and after solubilization, using

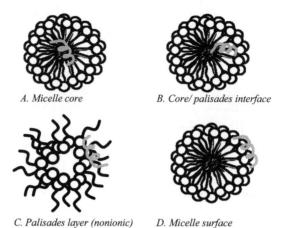

A. Micelle core   B. Core/ palisades interface

C. Palisades layer (nonionic)   D. Micelle surface

Figure 13 *Possible locations for the solubilizate in surfactant-based micelles.*

small-angle X-ray scattering (SAXS) and NMR spectrometry (pulsed-gradient spin-echo self-diffusion, PGSE-SD-NMR).[33] The SAXS studies measure changes in micellar dimensions on solubilization, whereas PGSE-SD-NMR indicates changes in the environment of the components in the microemulsion. Based on these combined techniques, solubilization is believed to occur at a number of different sites in the micelle.[34]

In aqueous solution (O/W microemulsion), non-polar additives such as hydrocarbons are intimately associated with the core of the micelle (Figure 13A), while slightly polar guest molecules such as fatty acids, alcohols and esters are usually located in what is termed the palisade layer (Figure 13B). The orientation of such molecules is probably more or less radial, with the hydrocarbon tail remaining closely associated with the micellar core. Guest molecules may also be found entirely in the palisade region of a water-in-oil micelle (Figure 13C), or on the micellar surface (Figure 13D).[34]

Gerhardt and Dungan suggest[35] two possible solubilization sites for IgG protein in W/O microemulsions depending on the salt concentration in the initial aqueous phase. At low salt concentration, where the droplets tend to be larger than IgG, the protein will be incorporated into one single droplet (Figure 14a). At higher salt concentrations, in the absence of protein, the droplet size is smaller than the protein molecule; hence the IgG is associated with a cluster of at least two or three droplets, as shown in Figure 14b.[35]

Even though the chemical structure may dictate the average preferred location for an additive, one should note that solubilized systems are dynamic, as are the micelles, and so the location of specific molecules will fluctuate rapidly with time.[34]

Microstructural Effects due to Solubilization

The effect of solubilized additives on the micellar/microemulsion properties varies according to the structure of the guest molecules. The solubilization of

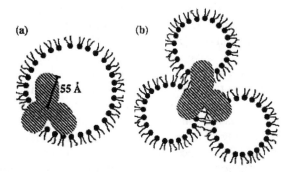

Figure 14 *Possible solubilization sites for IgG in W/O microemulsion droplets: (a) large droplets, (b) small droplets.*
(Based on ref. 35.)

hydrocarbon (decane) guest molecules into ternary surfactant systems has been examined,[36] and it was found that increasing the level of decane caused formation of the lamellar phase instead of a microemulsion phase. The phase transition was attributed to the increased rigidity of the bilayer interface of the liquid-crystalline lamellar phase. Other authors[37] have shown that a guest protein (human serum albumin, HSA) can induce bicontinuous microstructure formation in typical W/O AOT microemulsions. Figure 15 shows the relative diffusion coefficient of water (D/D_0) *versus* the water/surfactant (W/S) molar ratio. In the presence of protein at W/S <20, the relative self-diffusion coefficient of water is significantly higher than for the surfactant-free system. The observed values of the relative diffusion coefficient are more typical of bicontinuous domains than those expected for diffusion in restricted domains such as W/O microemulsions. The possibility of protein groups (in HSA)

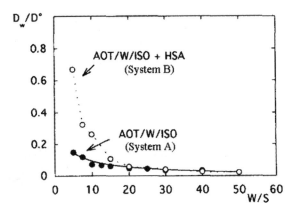

Figure 15 *Reduced ^1H self-diffusion coefficient of water $D_w/D°$ versus W/S (water/ surfactant molar ratio) at 25 °C, where $D°=2.29$ m^2 s^{-1}: •, system A, sodium bis-(2-ethylhexyl) sulfosuccinate (AOT)+water+isooctane; o, system B, AOT+water+isooctane and human serum albumin (HAS) (ref. 37).*

acting as co-surfactants allows a decrease in inverse curvature of the interface to produce a water-connected network (*i.e.*, a bicontinuous microemulsion).

In our U-type microemulsions (as described above), the L-phase, which is characterized by a progressive and uninterrupted isotropic phase, undergoes two major transitions along the dilution line 64 (60 wt% surfactant, 40 wt% oil phase). These structural transitions, W/O → bicontinuous and bicontinuous → O/W, were detected by conductivity and viscosity measurements and confirmed by SD-NMR measurements. The effect on the transitions of various guest molecules differing in their molecular structures at the structural transitions was examined. The expectation was that guest molecules solubilized within the microemulsion core, or at its interface, would contribute to the interfacial area, curvature, thickness and rigidity.

The presence of planar molecules such as phytosterols or lycopene (Figures 16a and 16b) at the oil–water interface leads to the interface becoming more rigid; the change in the bending elasticity of the interface results in an increase in the persistence length of the interfacial domains. However, lutein is a non-planar molecule (Figure 16c), with a 40° bending of the end ring out of the skeletal plane, and so having a limited tendency to penetrate into the interface. Therefore, it is solubilized in smaller amounts, at a lower molar guest

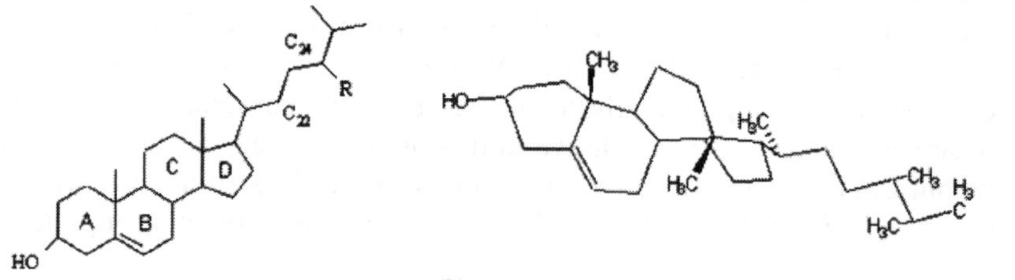

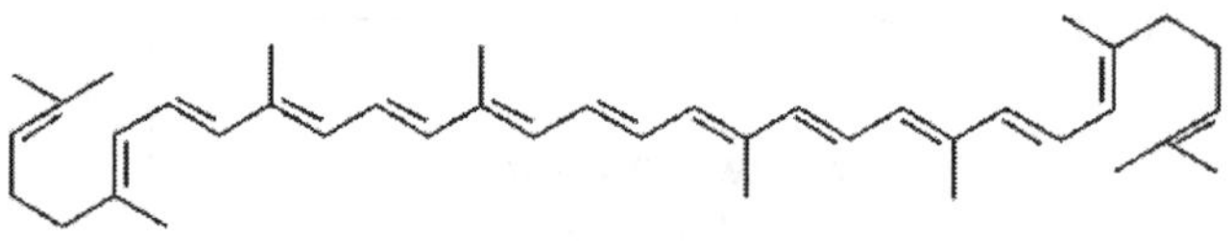

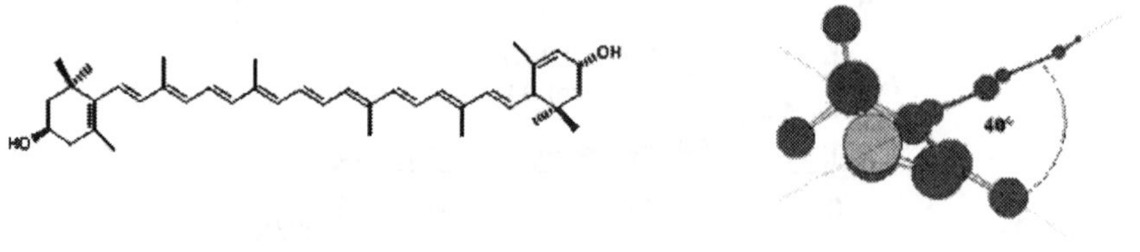

Figure 16 *Molecular structures of (a) free phytosterols (R=H, cholesterol; R= CH$_2$CH$_3$, β-sitosterol; R=CH$_2$CH$_3$ and additional double bond at C$_{22}$, stigamsterol; R=CH$_3$ campasterol; R=CH$_3$ and additional double bond at C$_{22}$, brassicasterol); (b) lycopene; (c) free lutein.*

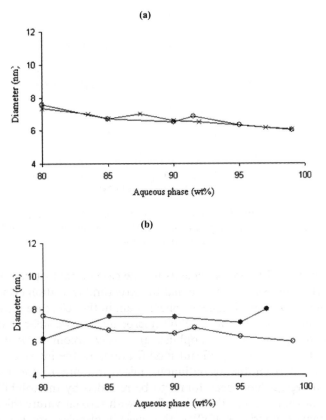

Figure 17 *O/W droplet diameter (nm) as a function of the aqueous phase content along dilution line 64 of (a) microemulsion (○) and microemulsion containing free lutein (×); (b) microemulsion (○) and microemulsion containing lutein ester (●).*

molecule/surfactant ratio, and therefore likely to less affect the nature of the interface.

Figure 17 shows the variation in droplet diameter for three systems studied by dynamic light scattering and SAXS. With increasing aqueous phase content (80–99 wt%), the droplets were observed to shrink in size from 7.5 ± 0.1 nm in the basic microemulsion to 6.0 ± 0.1 nm in the microemulsion containing free lutein. By contrast, solubilizing lutein ester in the O/W microemulsion swelled the droplets and increased their volume by up to 30%. These results suggest that, in the presence of lutein ester, additional hydrophilic surfactant is required at the interface in order to accommodate the lipophilic solubilizate (swelled droplets). No excess surfactant is required at the interface if free lutein is solubilized, and the opposite effect occurs: some of the hydrophilic surfactant is displaced by the guest molecules and the droplets shrink.[38]

Structural transitions from W/O to bicontinuous microstructures can be induced by any guest molecules such as tocopherol, phytosterols, lycopene and

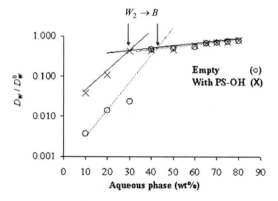

Figure 18 *Relative self-diffusion coefficient of water (D_w/D_w^0) derived from SD-NMR as a function of aqueous phase content along dilution line 64: ○, microemulsion; ×, microemulsion containing phytosterols (PS-OH).*

lutein. From SD-NMR measurements for the case of phytosterols as the guest molecules (Figure 18), one can see that the structural transition occurs at lower water content (20–30 wt% aqueous phase) than in the reference microemulsion (30 wt% aqueous phase). In the W/O microemulsion we infer that the guest molecules, being practically lipophilic, are easily accommodated at convex water interfaces without significant modifications to the interfacial structure. The transitions from the bicontinuous microstructure to the O/W micro-emulsion, however, have been found to be retarded by the solubilizates (*i.e.*, lutein), as shown in Figure 19. At interfaces with zero curvature (bicontinuous) and for convex interfaces (O/W), the guest molecules are accommodated mostly in the palisade layer, and they contribute to the flattened curvature and

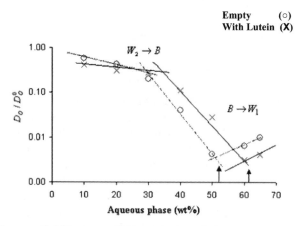

Figure 19 *Relative self-diffusion coefficient of oil (D_o/D_o^0) derived from SD-NMR as a function of aqueous phase content along dilution line 64: ○, microemulsion; ×, microemulsion containing lutein.*

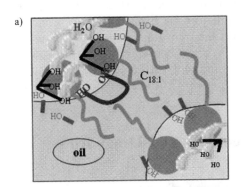

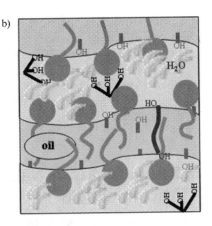

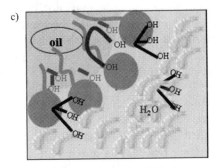

Figure 20 *Schematic representation of (a) W/O, (b) bicontinuous and (c) O/W microemulsions containing lutein.*

to its rigidity; as a result, the bicontinuous isotropic region becomes much broader. Figure 20 gives a schematic presentation, summarizing the behaviour of the solubilized guest molecules in (a) W/O, (b) bicontinuous and (c) O/W microemulsions, as indicated by the SD-NMR measurements.

Some guest molecules have been found not to affect the microstructure of microemulsions, as discussed earlier. Kreilgaard *et al.*[39] solubilized the hydrophobic drug, lidocaine, and the hydrophilic drug, prilocaine hydrochloride, in a microemulsion composed of Labrasol, Plurol-Isostearic, isostearylic isostearate and water. (Labrasol is a mixture consisting of 30% mono-, di- and triglycerides of C_8 and C_{10} fatty acids, 50% of mono- and di-esters of poly(ethylene glycol) (PEG 400) and 20% of free PEG 400), Plurol-isostearic is isostearic acid ester of polyglycerol, containing 30–35% of diglycerol, 20–25% of triglycerol, 15–20% of tetraglycerol, and 10% of pentaglycerol and higher oligomers). The authors claim[39] that the addition of the hydrophilic/hydrophobic guest molecules had no effect on the microemulsion microstructure or its phase transition. Similarly, in our laboratory we found that, when cholesterol was solubilized into a microemulsion, the mobility of the components observed by SD-NMR did not change significantly (unpublished results).

Our conclusion is, therefore, that loading microemulsions with guest molecules can affect the elasticity and curvature of the interface if the structure of the molecules is very different from that of the surfactant. As a result the presence of the guest molecule can affect the transitions from one structure to another. However, if the solubilizate is at a low concentration (well below its maximum solubilization) or has a similar molecular structure to the surfactant, so that it fits in well at the interface, the presence of the guest molecule will not affect the transitions.

6 Bioavailability of Nutraceuticals and Drugs from Microemulsions

When individuals are given the same amount of a nutraceutical, the degree of physiological/biochemical response may not be the same. The length of duration of the response may vary greatly because the physical form of the food may affect the partial absorption of the nutrients.

Some scientists define 'bioavailability' as the proportion of a drug or nutrient that is digested, absorbed, and utilized in normal metabolism. However, a more practical definition estimates the amount of nutrient absorbed into the serum. Some scientists have also coined 'bioactivity' as a term signifying the "journey of a compound from food to its functional site".[40]

The pharmacological definition of bioavailability means the rate and extent to which a drug substance reacts at its site of action. This parameter is difficult to determine because of ethnicity differences between individuals. The pharmacological definition calls for the fraction of an oral dose from a particular preparation that reaches the systemic circulation.[40] The most accurate definition refers to bioavailability as a sum of events that include absorption and bioactivity, or tissue distribution, and the functional consequences of absorption.

For our purposes here, bioavailability is defined as (i) the fraction of an ingested nutrient that is available for utilization in normal physiological functions or for storage, or, alternatively, as (ii) the amount of drug transported through the digestive membranes to the serum. Published information on the bioavailability of nutraceuticals (*i.e.*, carotenoids, vitamins) is based mainly on measurements of the nutrient in serum or plasma after ingestion.[41,42] Figure 21 shows a schematic representation of neutraceutical absorption in the small intestine. The nutraceuticals are associated with bile salts, and they are solubilized and absorbed at the brush border membrane of the mucosa by passive diffusion, along with other lipids in the micelles formed there. This absorbed material is then incorporated into a lipoprotein known as chylomicron, and secreted into the intestinal lymph on its way to the different organs and tissues *via* the blood-stream.[43]

Factors that may interfere with the rates of each absorption step will affect the overall bioavailability of the nutrient. Some of these factors are as follows.

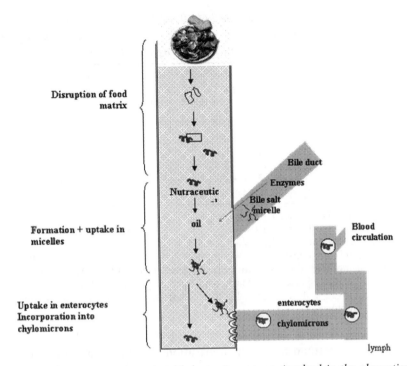

Figure 21 *Schematic representation of the various stages involved in the absorption of oil-soluble nutraceuticals.*

(1) *The molecular configuration (cis/trans) of the nutraceutical molecules.* The bioavailability of the *cis*-isomer is higher than the bioavailability of the *trans*-isomer. This may be the result of the greater solubility of *cis*-isomers in mixed micelles and the lower tendency of *cis*-isomers to aggregate.[41,42]

(2) *The particle size.* Conversion of tomatoes into tomato paste has been found[44] to enhance the bioavailability of lycopene, as the processing includes particle-size reduction and heat treatment.

(3) *The matrix in which a nutraceutical is incorporated.* The relative bioavailability of lutein appears to be lower from leafy green vegetables (*e.g.*, 45% from spinach) than from a range of other vegetables (67%).[43] This may be attributed to entrapment of lutein and its complexation to proteins in chloroplasts and within cell structures.

(4) *The presence of triglycerides or other oils.* Studies have shown[43,45] that intake together with vegetable oils improves bioavailability. As bile salts are required in nutraceutical absorption, any nutraceuticals that are administered as a dietary supplement (such as lycopene, lutein and vitamin E) should preferably be taken with food that induces bile secretion,[46] and not on an empty stomach.

(5) *The presence of surfactants.* Fat-soluble vitamins, including vitamin E, are known to be absorbed better in the presence of surfactants from emulsified

Table 2 *Steady-state flux J over period 20–28 h of lidocaine (4.8% w/w) and prilocaine hydrochloride (2.4% w/w), through rat skin from different microemulsion systems with equal drug load (ref. 37).*

Water (wt%)	Lidocaine J (mg h^{-1} cm^{-2})	Prilocaine hydrochloride J (mg h^{-1} cm^{-2})
7	14.6 ± 7.8	–
11	8.3 ± 0.8	6.2 ± 0.7
20[a]	7.1 ± 0.3	6.2 ± 1.7
20[b]	10.3 ± 1.1	10.7 ± 2.3
20[c]	8.2 ± 2.3	10.1 ± 3.5
55	25.1 ± 2.0	16.3 ± 4.9
65	26.3 ± 6.5	6.1 ± 1.2

[a] Labrasol–Plurol:Isostearic, 1:1 (w/w).
[b] Labrasol–Plurol:Isostearic, 2:1 (w/w).
[c] Labrasol–Plurol:Isostearic, 3:1 (w/w).
(Labrasol is a mixture consisting of 30% mono-, di- and triglycerides of C_8 and C_{10} fatty acids, 50% of mono- and di-esters of poly(ethylene glycol) (PEG 400) and 20% of free PEG 400), Plurol-isostearic is isostearic acid ester of polyglycerol, containing 30–35% of diglycerol, 20–25% of triglycerol, 15–20% of tetraglycerol, and 10% of pentaglycerol and higher oligomers).

vehicles rather than from oily preparations.[47] Emulsification of vitamin E for improved bioavailability has been reported,[48] but (macro)emulsions do not possess thermodynamic stability and have to be shaken before use. Hence, a thermodynamically stable microemulsion is a more appropriate delivery system. When β-itosterol—an antipyretic, cholesterol-reducing agent—was prepared in the form of O/W microemulsions using lecithin as the carrier, the bioavailability was found[49] to be enhanced to 24%, while from suspension in water the value was 18%.

Not much is yet known on the bioavailability of nutraceuticals from microemulsions. However, the bioavailability of drugs has been extensively explored. In particular, a correlation was reported by Kreilgaard *et al.*[39] between the bioavailability of two drugs and the structure of the delivery microemulsions. It was shown that the transdermal permeation rate of lidocaine and prilocaine hydrochloride (hydrophobic and hydrophilic drugs, respectively) from the Labrasol/Plurol-Isostearic/isostearylic-isostearate/water microemulsion system depends on the microemulsion composition (Table 2). The authors postulated[39] that the variation in mean transdermal flux in different microemulsion compositions provides confirmation that the drug delivery potential is greatly dependent on the internal structure and/or the fractional composition of the phases. The mechanism of this enhancement is still not clear, but it is probably due to structural and fluidity changes caused by these surface-active materials upon insertion into the intestinal membrane.[11]

It is worth commenting that, although there is presently, to the best of our knowledge, only one microemulsion formulation on the market (*i.e.*, Neoral®), it is widely anticipated that there will be many more microemulsion-based nutraceutical delivery systems commercially available in the not too distant future.

7 Conclusions

This article has attempted to review the current state of knowledge on self-assembled surfactant systems used as vehicles for nutraceuticals in food matrices.

It can be seen that efforts have been made to find the correct mixture of microemulsion ingredients and surfactants to inhibit liquid-crystal formation and phase separation, and to maximize dissolution of mutual ingredients, in order to structure food-grade microemulsions. The formation of progressively dilutable microemulsions (with oil and/or water) has been made possible by the addition of polyols and short-chain alcohols (ethanol).

The effect of guest molecules such as phytosterols, vitamin E, β-carotene, lycopene, lutein, proteins, *etc.*, on the phase behaviour and microstructure of microemulsions has been partially elucidated. A guest molecule that is considerably lipophilic, such as lutein ester or tocopherol, can be readily accommodated into the oil–water interface and can be easily solubilized at high capacities; but, with further water dilution, the solubilizate is 'pushed out' of the interface and released (phase separation). Thus, when O/W microemulsions are formed, the maximum solubilization will be strongly reduced. Molecules that are less lipophilic (*e.g.*, free lutein, phytosterols) are less effectively solubilized at any of the interfaces. Molecules that are totally insoluble in the oil phase such as lycopene exhibit the lowest solubilization capacity; yet if the calculated solubilization capacity is normalized with respect to the oil content, the solubilization can be considered to be greatly enhanced, and it is certainly many times higher than the oil solubility capacity.

The presence of guest molecules, at their maximum degree of solubilization, can affect the transition between one phase and another. Some nutraceuticals induce the transition from W/O morphology to the bicontinuous microstructure, while others affect the transition from bicontinuous to O/W microstructure. In most systems the guest molecules serve as triggering agents for microstructural transitions. Poorly water-soluble nutraceuticals such as free lutein and phytosterols can trigger the transition from bicontinuous to O/W microstructure. Similarly, water-soluble proteins have been found to trigger the transition from the W/O morphology to the bicontinuous microemulsion.

References

1. W. R. Bidlack and W. Wang, in 'Phytochemicals as Bioactive Agents', eds W. R. Bidlack, S. T. Omaye, M. S. Meskin, and D. K. M. Topham, Technomic, Lancaster, PA, 2000, p. 241.

2. W. R. Bidlack, S. T. Omaye, M. S. Meskin, and D. Jahner, in 'Phytochemicals: A New Paradigm', eds W. R. Bidlack, S. T. Omaye, M. S. Meskin, and D. K. M. Topham, Technomic, Lancaster, PA, 1998, p. 1.

3. R. Cheruvanky, in 'Phytochemicals as Bioactive Agents', eds W. R. Bidlack, S. T. Omaye, M. S. Meskin, and D. K. M. Topham, Technomic, Lancaster, PA, 2000, p. 213.

4. S. L. DeFelice, *Trends Food Sci. Technol.*, 1995, **6**, 59.

5. C. J. Dillard and J. B. German, *J. Sci. Food Agric.*, 2000, **80**, 1744.

6. S. R. Dungan, in 'Industrial Applications of Microemulsions', eds C. Solans and H. Kunieda, Marcel Dekker, New York, 1997, p. 147.

7. C. Solans, R. Pons, and H. Kunieda, in 'Industrial Applications of Microemulsions', eds C. Solans and H. Kunieda, Marcel Dekker, New York, 1997, p. 1.

8. R. Zana, *Heterogen. Chem. Rev.*, 1994, **1**, 145.

9. K. Kawakami, T. Yoshikawa, Y. Moroto, E. Kanaoka, K. Takahashi, Y. Nishihara, and K. Masuda, *J. Control. Release*, 2002, **81**, 65.

10. K. Holmberg, in 'Micelles, Microemulsions and Monolayers', ed. D. O. Shah, Marcel Dekker, New York, 1998, p. 161.

11. P. P. Constantinides, *Pharm. Res.*, 1995, **12**, 1561.

12. N. H. Shah, M. T. Carvajal, C. I. Patel, M. H. Infeld, and A. W. Malick. *Int. J. Pharm.*, 1994, **106**, 15.

13. C. K. Kim, Y. J. Cho, and Z. G. Gao, *J. Control. Release*, 2001, **70**, 149.

14. M. J. Lawrence and G. D. Rees, *Adv. Drug Delivery Rev.*, 2000, **45**, 89.

15. A. Yaghmur, A. Aserin, and N. Garti, *Colloids Surf. A*, 2002, **209**, 71.

16. N. Garti, A. Yaghmur, M. E. Leser, V. Clement, and H. J. Watzke, *J. Agric. Food Chem.*, 2001, **49**, 2552

17. Y. Tokuoka, H. Uchiyama, M. Abe, and S. D. Christian, *Langmuir*, 1995, **11**, 725.

18. S. Y. Shiao, V. Chhabra, A. Patist, M. L. Free, P. D. T. Huibers, A. Gregory, S. Patel, and D. O. Shah, *Adv. Colloid. Interface Sci.*, 1998, **74**, 1.

19. N. Kanei, Y. Tamura, and H. Kunieda, *J. Colloid Interface Sci.*, 1999, **218**, 13.

20. V. K. Bansal, D. O. Shah, and J. P. O'Connell, *J. Colloid Interface Sci.*, 1980, **75**, 462.

21. M. J. Hou and D. O. Shah, *Langmuir*, 1987, **3**, 1086.

22. H. Krasowska, *Int. J. Pharm.*, 1979, **4**, 89.

23. M. E. Moro, J. Novillofertrell, M. M. Velazquez, and L. J. Rodriguez, *J. Pharm. Sci.*, 1991, **80**, 459.

24. S. A. Slocum, A. Kilara, and R. Nagarajan, in 'Flavors and Off-flavors', ed. G. Charalambous, Elsevier, Amsterdam, 1990.

25. P. A. Wolf and M. J. Havekotte, US Patent 4835002, 1989.

26. I. Amar, A. Aerin, and N. Garti, *J. Agric. Food Sci.*, 2003, **51**, 4775.

27. M. El-Nokaly, G. Hiler, and J. McGrady, in 'Microemulsions and Emulsions in Foods', eds M. El-Nokaly and D. Cornell, ACS Symposium Series 448, American Chemical Society, Washington, DC, 1991, p. 26.

28. Y. C. Chiu and W. L. Yang, *Colloids Surf.*, 1992, **63**, 311.

29. L. M. Canfield, T. A. Fritz, and T. E. Tarara, *Methods Enzymol.*, 1990, **189**, 418.

30. N. Garti, I. Zakharia, A. Spernath, A. Yaghmur, A. Aserin, and R. E. Hoffman, *Prog. Colloid Polym. Sci.*, submitted for publication.

31. A. Spernath, A. Yaghmur, A. Aserin, R. E. Hoffman, and N. Garti, *J. Agric. Food Chem.*, 2003, **50**, 6917.

32. A. Spernath, A. Yaghmur, A. Aserin, R. E. Hoffman, and N. Garti, *J. Agric. Food Chem.*, 2003, **51**, 2359.
33. M. J. Rosen, 'Surfactants and Interfacial Phenomena', 2nd edn, Wiley, New York, 1989.
34. D. Myers, 'Surfaces, Interfaces and Colloids: Principles and Applications', 2nd edn, Wiley, New York, 1999, p. 333.
35. N. I. Gerhardt and S. R. Dungan, *Biotechnol. Bioeng.*, 2002, **78**, 60.
36. A. Hufnagl and M. Gradzielski, *Colloid Surf. A*, 2001, **183**, 227.
37. M. Monduzzi, F. Caboi, and C. Moriconi, *Colloid Surf. A*, 1997, **130**, 327.
38. I. Amar, A. Aserin, and N. Garti, *Colloids Surf. B*, 2004, **33**, 143.
39. M. Kreilgaard, E. J. Pedersen, and J. W. Jaroszewski, *J. Control. Release*, 2000, **69**, 421.
40. K. Schumann, H. G. Classen, M. Hages, R. PrinzLangenohl, K. Pietrzik, and H. K. Biesalski, *Arzneimittel-Forsch.*, 1997, **47**, 369.
41. J. J. M. Castenmille and C. E. West, *Ann. Rev. Nutr.*, 1998, **118**, 19.
42. P. M. Bramley, I. Elmadfa, A. Kafatos, F. J. Kelly, Y. Manios, H. E. Roxborough, W. Schuch, P. J. A. Sheehy, and K. H. Wagner, *J. Sci. Food Agric.*, 2000, **80**, 913.
43. K. H. van het Hof, C. Gärtner, C. E. West, and L. B. M. Tijburg, *Int. J. Vitamin Nutr. Res.*, 1998, **68**, 366.
44. S. A. R. Paiva and R. M. Russell, *J. Am. College Nutr.*, 1999, **18**, 426.
45. K. H. van het Hof, C. E. West, J. A. Weststrate, and G. A. J. Hautvast, *J. Nutrition*, 2000, **130**, 503.
46. H. E. G. Torres, *Lipids*, 1970, **5**, 379.
47. N. E. Bateman and D. A. Uccellini, *J. Pharm. Pharmacol.*, 1984, **36**, 461.
48. T. Julianto, K. H. Yuen, and A. M. Noor, *Int. J. Pharm.*, 2000, **200**, 53.
49. Patent DE4038385, 1992.

Particle Dynamics in a Transient Gel Network—Ageing of a Monodisperse Emulsion

By A. D. Watson, G. C. Barker, D. J. Hibberd, B. P. Hills,
A. R. Mackie, G. K. Moates, R. A. Penfold, and M. M. Robins

INSTITUTE OF FOOD RESEARCH, NORWICH RESEARCH PARK,
COLNEY, NORWICH NR4 7UA, UK

1 Introduction

Emulsions evolve over time in ways that are not yet fully understood. In this respect it seems important to appreciate the physics of particle dynamics in close-packed configurations, as well as the chemistry of the mixture of oil, water and surfactant. In particular, aggregated emulsion systems can form transient gel states where the droplets occupy positions in a relatively immobile network that spans the system. The properties of such states are of interest to the food industry as they can be used to control product texture and quality. The slow evolution and potential collapse of transient gel states is of considerable interest because of its implications for shelf-life and storage stability.

In this work we consider a transient gel based on a model emulsion system that is weakly aggregated *via* depletion flocculation through the addition of a non-adsorbing polymer. In order to access the explicit structure and dynamics at the level of droplets and clusters of droplets, we are obliged to go beyond the more traditional techniques of light-scattering and rheology, since these are averaging techniques. Instead, we rely on confocal microscopy and magnetic resonance imaging (MRI), combined with image analysis and individual droplet tracking. The confocal microscopy, coupled with the novel composition of the emulsion system, allows us to study structure and dynamics at the individual droplet length scale. The MRI allows us to expand from the droplet length-scale up to the larger length scale of aggregates and voids. The ultimate focus of our study is towards the dynamics of such voids and multi-droplet clusters, and their role in the macroscopic changes occurring in the network structure of the bulk system.

2 The Emulsion System

The twin demands of the confocal microscopy technique and the desire for density matching of the continuous and dispersed phases makes for a challenging emulsion system design. Effective confocal microscopy requires the oil and continuous phases to be refractive index matched, since otherwise the sample is too turbid at industrially relevant volume fractions. Effective control over gravity, and in particular the power to 'switch off' gravity—either to monitor its role or to increase the observation time—demands that the density difference between phases be continuously variable and also adjustable to zero. Matching either the optical properties alone or the buoyancy alone is not difficult, and indeed it has been achieved by many groups. But to fulfil both conditions together is a much more challenging task.

We have devised an emulsion system that meets these objectives. The oil phase is a mixture of *n*-hexane and tetradecafluorohexane. The use of a perfluorinated oil appears essential, since this type of oil has a refractive index lower than that of water, and a density greater than water. Tetradecafluorohexane is also known as the specialist refrigerant FC-72; its properties are well documented, and it is readily available at reasonable cost. There already exists a considerable literature on fluorocarbon emulsions, which have found a number of biomedical uses such as blood substitutes.[1] Hexane and FC-72 are miscible, although the consolute point is close to room temperature for compositions between 20% and 60% FC-72. A diblock copolymer F6H6 (CH_3-$(CH_2)_5$-$(CF_2)_5$-CF_3) is necessary to reduce the consolute temperature and so to prevent accidental phase separation of the two oils at ambient laboratory temperature. A dispersed phase mixture of 57.23% w/w FC-72 + 39.77% w/w hexane + 3.00% w/w F6H6 results in an emulsion that is suitably matched for both refractive index and buoyancy. This emulsion is stabilized using the non-ionic surfactant Brij-35, and it contains sodium azide as a preservative. The emulsion system can be flocculated using polyethylene oxide (PEO) of molecular weight $M_w \approx 9 \times 10^5$ daltons at concentrations less than $C_p^* \approx 0.9\%$ w/w in the continuous phase.

This novel emulsion formulation has the advantage of not relying on the use of large quantities of continuous phase contrast agents such as sucrose or salt, which can modify system characteristics in other ways. This is a model system which is, nevertheless, an attempt to be representative of a real food emulsion. We are able to modify continuously the density difference between the two phases in order to monitor the impact of gravity, but without sacrificing the refractive index matching. For example, a 10% density difference between the phases can be achieved with the introduction of 5% w/w of sucrose into the continuous phase, coupled with a suitable adjustment of the oil composition.

Our use of the technique of particle tracking introduces an additional constraint. Particle tracking is a two-step process comprising (i) feature location and (ii) tracking of identified features from one instant to another.[2,3] The identification stage is much simpler if the significant features (*i.e.*, emulsion droplets) are all of the same size, thus imposing a further restriction on our emulsion system. To achieve this we have commissioned a cross-flow membrane emulsification (XME) device,[4] capable of producing large quantities of pseudo-monodisperse emulsions in batch mode. In our XME facility, the continuous phase circulates past the outer surface of a porous ceramic membrane, as shown in Figure 1. The dispersed phase is supplied to the inner surface of the membrane *via* a reservoir, and is driven through the membrane by compressed air. Droplets emerging from the outer membrane surface are swept away by the flowing continuous phase, within which they accumulate over time. Emulsions of dispersed phase volume fraction $\phi = 0.14$ can be

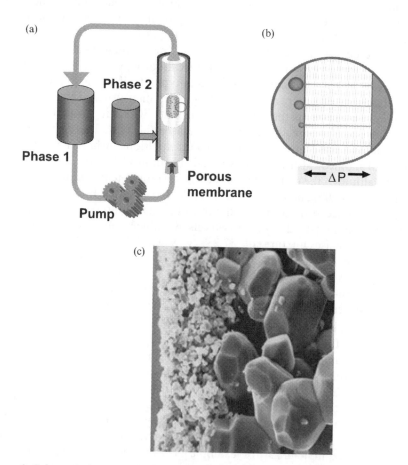

Figure 1 *Schematic diagram of the cross-flow emulsification set-up. (a) Continuous phase (1) from a reservoir on the left passes via a pump into the tall cylinder housing the porous ceramic membrane. The reservoir for the dispersed phase (2) is indicated in the centre. (b) Oil moves through the membrane (right to left) under a pressure drop, emerging from the membrane inner surface as droplets. (c) An electron micrograph of the membrane. The large 'boulders' are the supporting substrate. The fine surface structure controls droplet formation. (Courtesy of Emulsion Systems Ltd.).*

produced in this way. The flow-rate, the pressure drop across the membrane, and the membrane structure all impact systematically and reproducibly upon the droplet size.

Figure 2 shows the size distribution of emulsion droplets obtained using the XME apparatus as compared to the distribution obtained with a conventional blender. The conventional polydisperse emulsion spans two orders of magnitude in droplet diameter and has the expected log-normal distribution. The XME size distribution spans one order of magnitude, and it shows a slightly skewed distribution with a sharp cut-off at larger sizes. Although we cannot describe our present emulsion as properly monodisperse, we believe that the degree of monodispersity is compromised by poor control of the pressure drop across the membrane, and we have indications that sharper control of droplet size is indeed available with the technique. The currently achievable XME droplet-size distribution does, however, exhibit some useful features. In particular, few droplets are of diameter below 1 μm, which favours optical microscopy where smaller droplets are difficult to resolve. Moreover, the sharpening of the distribution compared with the conventional emulsification technique means there are fewer giant droplets to distort the structure and dynamics. Figure 3 reproduces a confocal image of a small region of the depletion-flocculated XME emulsion.

We have learnt that the XME apparatus is also beneficial for creating 'delicate emulsions'. Both *n*-hexane and FC-72 have high vapour pressures, boiling at just 69 °C and 55 °C, respectively. Attempting to emulsify these oils using a conventional laboratory blender causes them to vaporize. In contrast, the low-energy, low-shear XME apparatus encounters no such difficulty. This advantage may prove useful in pharmaceutical applications, *e.g.*, creating

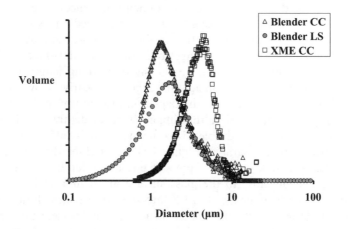

Figure 2 *The droplet-size distribution of an emulsion created using a blender compared to that of an XME emulsion. A combination of Coulter counter (CC) and light scattering (LS) was required to capture the full size range of the blender emulsion. The right-hand plot is the XME emulsion. The oil used was n-hexadecane.*

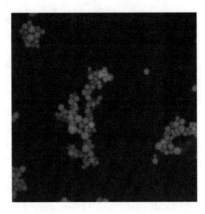

Figure 3 *Confocal image of an XME emulsion of nominal volume fraction 7% flocculated with PEO. This image slice is close to the glass container wall. 'Gravity up' is to the right. The viewing area is 65 μm × 65 μm.*

an emulsion of an active ingredient that is a volatile oil. Of more interest to the food industry might be the creation of emulsions stabilized by protein at the oil–water interface in a gentle hydrodynamic environment that does not cause shear-induced protein modification.

3 Confocal Imaging

We have used confocal scanning laser microscopy (CSLM) to study the emulsion system at the individual particle level, exploiting the ability to form a two-dimensional image slice of the sample interior. Three-dimensional images can then be created as a "stack" of two-dimensional slices taken at different distances into the sample. Unusually, we have imaged the system along a horizontal axis, to maximize resolution in the direction of gravity.

We have obtained preliminary images of samples in a flat-sided capillary, in order to minimize limitations due to the short working length of the high magnification objective. The image in Figure 3 clearly shows individual emulsion droplets, aggregated *via* the depletion mechanism in the presence of PEO, although the droplets in this image were evidently attached to the capillary surface. Individual droplets become increasingly difficult to resolve at larger distances into the sample. Figure 4 shows a three dimensional stack, corresponding to a view field of 65 μm × 65 μm and a depth of 75 μm into the sample. Regions too close to the glass surface are demonstrably dominated by wall effects, whereas images too far into the sample show little individual droplet detail. However, the flocculated structure remains clear throughout. A working region for particle tracking is likely to extend from ∼25 μm inside the sample, to avoid wall effects, to ∼40 μm into the sample, to give satisfactory image quality.[5]

The images shown here are raw data. They have not been filtered or enhanced in any way. Nevertheless, we have developed a full three-dimensional

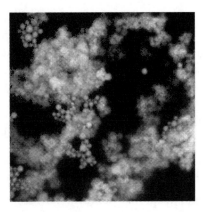

Figure 4 *A three-dimensional confocal image stack for which Figure 3 is the upper-most slice. The image depth is 75 μm.*

regularized deconvolution package that can significantly improve the image quality by mathematically correcting for the effects of non-ideal microscope optics.

With the microscope used here, a single stack of 32 slices takes approximately 2 minutes to acquire. Slow dynamic processes are probed by taking a time sequence of stacks. One qualitative way to visualize the evolution of the system is to display *differences* between stacks taken at successive times (so that stationary droplets and flocculated regions do not appear) and then project them into two dimensions. A preliminary version of such a sequence has been generated, and a still picture from this is shown in Figure 5. As the difference sequence unfolds, flocculated regions are seen to remain relatively intact. In this region there is an overall movement at roughly 45° to the

Figure 5 *The difference between two image stacks taken at different times, projected into two dimensions. A time sequence of such difference images shows that the large aggregated regions drift towards the lower right corner. Other regions exhibit a range of motions. 'Gravity up' is directed towards the right.*

direction of gravity. Other regions are seen to move in quite different directions and in non-trivial ways. The general impression is one of a surprisingly high level of quite complicated movement. It is difficult, of course, to draw unequivocal conclusions from a single image sequence, but we speculate that some degree of thermally driven bulk flow may be involved in the observed movement, since no attempt has been made to ensure isothermal conditions at the level of the particle clusters. In addition, other regions of the container may display compensating flows in different directions.

4 Particle Tracking

To quantify the images from the confocal microscopy, it is necessary to identify and locate specific features. At the most fundamental level this means determining the location of individual droplets in three dimensions. These coordinate data can then be used to measure structure, or, through successive image stacks, to generate droplet trajectories from which dynamical information may be extracted. In addition, large-scale features—at the level of regions of aggregated droplets or voids between aggregated regions—are amenable to dynamic three-dimensional analysis using a variety of space-analysing algorithms.

Droplet location and tracking has been performed using available code written in IDL (Interactive Data Language).[6] The feature-finding algorithm has been described in detail elsewhere.[2] The algorithm expects spheres of uniform radius. We have applied a two-dimensional implementation of the code to simulated data having known centre coordinates in which each 'droplet' has a Gaussian brightness pattern. We have demonstrated an ability to be able to locate discs of uniform size, and also discs varying in size by a factor of 4. We have made a preliminary attempt to locate droplet centres in the image shown in Figure 3. The result is shown in Figure 6, where the bright spots mark located drops. We believe this to be the first attempt to apply this

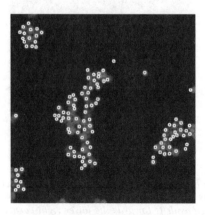

Figure 6 *Located droplets, shown as bright spots, from the confocal image of Figure 3.*

locating technique to individual emulsion droplets. Despite the polydispersity of the sample, it appears that a satisfactory level of droplet identification can be achieved in two dimensions. The extension to three dimensions is in progress. Continuing challenges are the image resolution, the degree of droplet monodispersity, and the ensemble statistics.

5 Simulation and Modelling

In line with the strategy of accommodating the particulate nature of the aggregated emulsion system explicitly, we make use of techniques developed in the context of granular systems. We are developing a three-dimensional dynamical extension to the granular q-model of Coppersmith et al.,[7] a two-dimensional model in which the gravitational stress on a single particle is partitioned randomly between those particles that support it. This model is a useful way of exploring the development of 'stress chains' that may play a role in the long-term structural evolution of transient gels. With a similar focus we are also developing a model based on the fabric tensor, in which a system is described as a material in terms of point-to-point contacts between its component entities, in our case oil droplets. The fabric tensor describes structures at the particulate level and has proved to be a powerful description of granular materials.[8]

Thermal drivers for the formation and evolution of transient gels are being studied by Brownian dynamics simulation at the Smoluchowski level of description (the 'position Langevin' equation).[9] In addition to strictly hard-sphere repulsion, the simplest appropriate potential form includes a short-ranged pairwise attractive term, as introduced phenomenologically by Asakura and Oosawa[10-12] to account for the osmotic pressure imbalance arising from depletion of small polymer particles in the vicinity of the relatively large emulsion droplets. The hard-sphere discontinuity of the corresponding effective force is not suitable, however, for a discretized dynamical simulation. After consulting the extensive literature on Brownian dynamics methods applied to depletion flocculation, we have incorporated an effective potential based on that proposed by Puertas and co-workers.[13] This comprises a smooth repulsive term of range r^{-36}, and it truncates the Asakura–Oosawa potential near particle contact to ensure that it does not contribute in a physically unjustifiable range (less than the particle size). At the nominal contact separation, a first-order smooth transition is affected by a harmonic term that fixes the total potential minimum at the particle diameter. The corresponding total potential is depicted in Figure 7.

The simulation parameters are specified to match the experimental system with a dispersed phase volume fraction of 10%, a polymer concentration (relative to the overlap concentration C_p^*) of $C_p/C_p^* \approx 95\%$, and a size ratio of $\xi = R_g/R \approx 6\%$ based on an estimated polymer radius of gyration of $R_g \approx 0.06\ \mu\text{m}$ and a droplet radius of $R = 1\ \mu\text{m}$. We have carried out a Brownian dynamics simulation of modest size (512 particles) for a period of several Brownian time units ($\tau_B = 4R^2/D_0$) in steps of 40 μs. Values appropriate for micron-sized spheres in aqueous solution give $\tau_B = 20$ s.

Total Potential

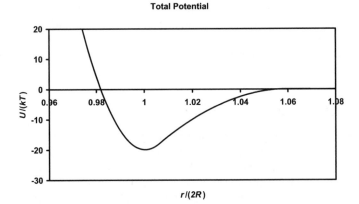

Figure 7 *The potential $U(r)$ used in the Brownian dynamics simulations. Note the symmetric form of the harmonic well, and the short range of the potential. The additional intermediate-range potential barrier term of Puertas et al. (ref. 13) is omitted.*

After equilibrating the purely repulsive system for a period of at least $5\tau_B$, the attractive forces were instantaneously 'switched on'. As expected, the values of the internal energy E and the long-time diffusion constant D were found to drop immediately, with a slower decrease beyond τ_B (see Figure 8), by

Internal Energy / Relative Diffusion

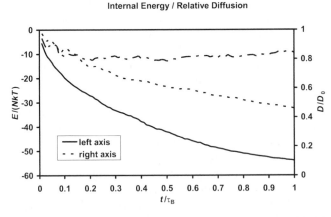

Figure 8 *The internal energy per particle, E, relative to the thermal background (left ordinate) and the mean-square displacement relative to the free-particle diffusion (right ordinate) as a function of time from Brownian dynamics trajectories. For the purely repulsive system, the energy (not shown) is static on the scale of the graph, while the mean-square displacement rapidly equilibrates $(\cdot - \cdot - \cdot)$. After instantaneously switching on the attractive depletion interaction (see Figure 7), the internal energy $(-)$ decreases dramatically and remains unequilibrated after one Brownian time unit. Similarly, the mean-square displacement $(\cdots\cdots)$ falls as particles become more confined to flocculated clusters.*

which time flocculated particle clusters have clearly formed and discernable local inhomogeneities ('voids') have developed (see Figure 9). We are presently implementing quantitative measures to characterize these structures using static and dynamic isotropic two-body and three-body density correlation functions.

6 NMR Imaging

To access microstructure with MRI on length scales of the order of tens of droplet diameters, we exploit the technique of nuclear magnetic resonance (NMR). Two relevant demonstrations of principle have been performed.

We have imaged a PTFE phantom in water, and demonstrated the ability to resolve features as small as 80 μm. This should permit us to use MRI to map features in the bulk in a way that complements the microscopic imaging field of CSLM. We note that MRI does not demand refractive index matching, and so the technique is suitable for microstructural imaging of 'off-the-shelf' food

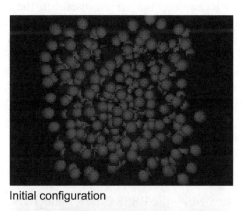

Initial configuration

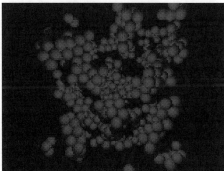

Final configuration

Figure 9 *The initial and final configurations from a Brownian dynamics simulation run of 512 particles over one Brownian time unit τ_B, after the attractive interactions were instantaneously switched on at time zero.*

systems. But some challenges still facing those using the technique include signal-to-noise issues for real samples and limitations of data acquisition rates.

We have demonstrated that our existing spectrometer can be tuned to perform F-NMR based on the fluorine signal from the FC-72 oil, rather than the more conventional hydrogen-based signal. This may offer an advantage over H-NMR, since the latter is complicated by the large number of different hydrogen environments that exist within the sample. In contrast, the FC-72 oil is the sole source of the fluorine signal.

7 Conclusions

We have here presented an overview of an ongoing study examining the ageing of weakly aggregated emulsions, focusing on the role of fractures in the collapse of transient gels. To date, we have formulated a novel emulsion system, based on a mixture of FC-72 and *n*-hexane, which is matched for both buoyancy and refractive index. We have used cross-flow membrane emulsification to create emulsions of restricted droplet-size distribution. We have noted that this technology may prove useful in the creation of delicate emulsions, such as those based on volatile oils or protein-stabilized droplets. Using confocal microscopy we have created a time series of three-dimensional image stacks, and have shown that these do reveal qualitative information on emulsion dynamics. Feature location has been successfully applied to emulsions in two dimensions. NMR microimaging and the use of F-NMR have been demonstrated in principle. Granular-inspired stress-chain simulations are being developed, and a Brownian dynamics simulation in support of the experimental programme has been implemented.

Acknowledgements

This work is being funded by the BBSRC (UK) under grant 218/D17326. We thank Grant Calder and the John Innes Centre for access to their confocal microscope suite. We are grateful to John Crocker, David Grier and Eric Weeks for making their IDL particle tracking code available to us.

References

1. M. P. Krafft, A. Chittofrati, and J. G. Riess, *Curr. Opin. Colloid Interface Sci.*, 2003, **8**, 251.
2. J. C. Crocker and D. G. Grier, *J. Colloid Interface Sci.*, 1996, **179**, 298.
3. A. D. Dinsmore, E. R. Weeks, V. Prasad, A. C. Levitt, and D. A. Weitz, *Appl. Optics*, 2001, **40**, 4152.
4. Emulsion Systems Ltd, UK, http://www.emulsionsystems.com.
5. E. R. Weeks, J. C. Crocker, A. C. Levitt, A. Schofield, and D.A. Weitz, *Science*, 2000, **287**, 627.
6. J. C. Crocker, D. G. Grier, and E. R. Weeks, http://www.physics.emory.edu/~weeks/idl/.

7. S. N. Coppersmith, C. Liu, S. Majumdar, O. Narayan, and T. A. Witten, *Phys. Rev. E*, 1996, **53**, 4673.
8. M. Tassopoulos and D. E. Rosner, *AIChE J.*, 1992, **38**, 15.
9. M. P. Allen and D. J. Tildesley, *Computer Simulation of Liquids*, Oxford University Press, Oxford, 1990.
10. S. Asakura and F. Oosawa, *J. Chem. Phys.*, 1954, **22**, 1255.
11. S. Asakura and F. Oosawa, *J. Polym. Sci.*, 1958, **33**, 183.
12. A. Vrij, *Pure Appl. Chem.*, 1976, **48**, 471.
13. A. M. Puertas, M. Fuchs, and M. E. Cates, *Phys. Rev. E*, 2003, **67**, 031406.

Gels, Particle Mobility, and Diffusing Wave Spectroscopy—A Cautionary Tale

By David S. Horne, Eric Dickinson,[1] Caroline Eliot,[1] and Yacine Hemar[2]

HANNAH RESEARCH INSTITUTE, AYR, SCOTLAND KA6 5HL, UK
[1]PROCTER DEPARTMENT OF FOOD SCIENCE, UNIVERSITY OF LEEDS, LEEDS LS2 9JT, UK
[2]FONTERRA RESEARCH, PALMERSTON NORTH, NEW ZEALAND

1 Introduction

The problem of defining the gel state is one that recurs throughout the literature of colloid science. Flory[1] recalled the quotation attributed to D. Jordan Lloyd in a 1920's text-book: "the colloidal condition, the gel, is one which is easier to recognize than to define". And as recently as the previous food colloids conference in Wageningen, the same issue of the relationship of colloidal aggregation to gel formation was assessed and reviewed.[2]

Bungenberg de Jong, in 1949,[3] defined a gel as a colloidal system of solid character in which the colloidal particles somehow constitute a coherent structure, the latter interpenetrated by a liquid system. Phenomenologically this definition has connotations of the concepts of percolation. Modern terminology describes the gel as a viscoelastic solid, *i.e.*, a system which, depending on the circumstances, can flow like a viscous liquid or behave as an elastic solid. It is the coherent structure or network that gives rise to this characteristic behaviour, the ability to respond to the applied mechanical force in the rheometer. The problem in the study of gelation kinetics is recognizing when this load-bearing network has been created. Implicitly, in considering gels resulting from an aggregation reaction beginning with monomers, this is tantamount to asking when a container-spanning cluster is first formed. The availability of such experimental information allows the testing of various models of the aggregation reactions leading to gel formation.[2]

The gel point is most easily defined by considering the rheological properties of the system. One common definition of the gel point[4] is the time in the progress of the reaction from sol to gel when the rheological phase angle δ falls below 45°, *i.e.*, when the storage modulus G' becomes greater than the loss modulus G''. More complicated definitions of the gel point have been proposed by other groups.[5,6] The definition $\delta < 45°$ considers the measurement to have been made at a single oscillation frequency. So Almdal and coworkers required[5] that the mechanical response should also be independent of oscillation frequency to demonstrate the gel point. Alternatively, Winter and Chambon suggested[6] a criterion for the gel point as the time when, in a frequency-sweep experiment, both G' and G'' show a power-law dependence on the frequency, each modulus having the same positive exponent.

Such complex criteria involving studies of frequency-dependent responses demand slowly changing systems, and so are not readily met. Moreover, the ability to detect the simple cross-over may depend on the sensitivity of the measuring instrument. In these circumstances, the gel point is often judged to be when the instrument response from the gelling system becomes greater than the noise level. In some concentrated systems, the simple criterion may not be met because, although still only very weakly elastic, the elastic modulus may already dominate. Then, a working definition of exceeding a particular value of the storage modulus is often adopted as the gel point criterion.

Diffusing wave spectroscopy (DWS) has emerged as a particularly useful technique[7–10] for studying aggregation reactions in opaque colloidal systems which eventually progress through to gel or curd formation. DWS is a light-scattering technique which extends dynamic light scattering into concentrated opaque systems by exploiting the diffusive nature of the propagation of light through such samples. A major advantage of light scattering over conventional rheology is that it is non-perturbing.[8,9] However low is the shear strain, the very act of applying a finite stress during a small-deformation rheological experiment can substantially disturb or even destroy the developing microstructure. In contrast, DWS is a 'zero stress' technique, since the signal originates from the changes in scattering behaviour driven by ever-present spontaneous thermal fluctuations. In this communication, we critically review some of the approaches that are currently utilizing the DWS technique in the study of (micro)rheology. In so doing, we explore some of its limitations and emphasize the need to use all of the information that the light-scattering behaviour reveals during the gelation process.

2 Diffusing Wave Spectroscopy

Diffusing wave spectroscopy (DWS) is a powerful tool for obtaining information about local dynamical properties in a suspension.[12] In a typical DWS experiment, coherent light from a laser impinges on one side of an opaque sample. The scattered light fluctuates in intensity due to the motion of the scattering entities. The DWS technique demands strong multiple scattering so that the propagation of light through the sample can be described by a diffusion model. Analytical solutions, subject to the boundary conditions set by sample and instrument geometry, can then be written down for the distribution of scattering paths, allowing the temporal autocorrelation functions of the intensity fluctuations to be derived.

Two geometrical arrangements using optical fibres are widely employed— namely transmission or backscattering—with either single launch and detector fibres or multiple arrangements. Commercial instrumentation is also available, sometimes aimed at specific end-users. In our own instrument, which operates in back-scattering mode with a plane wave-source and a point detector, the electric field correlation function $g_{(1)}(t)$ simplifies to a stretched exponential form with exponent ½, *i.e.*,

$$g_{(1)}(t) = \exp\left[-\frac{z_0}{l^*}\sqrt{\frac{6t}{\tau}}\right]. \tag{1}$$

Here, t is the decay time of the correlator, l^* is the photon diffusional mean free path, z_0 is the distance into the sample at which light transport can be considered diffusional ($\sim 2l^*$), and τ is the relaxation time of the scattering particle motion in the suspension. DWS therefore provides unique information on particle motion and the factors influencing or controlling that motion. In a conventional dynamic light-scattering scenario, we have

$$\tau = (2Dq^2)^{-1}, \tag{2}$$

where q is the scattering wave vector and D is the particle diffusion coefficient. From the Stokes–Einstein equation, D is related to the particle radius r by

$$D = kT/6\pi\eta r, \tag{3}$$

where T is temperature, k is Boltzmann's constant, and η is the viscosity of the background medium. Even here, we can see that an increase in relaxation time can arise not just from an increase in particle size, but also from a change in background viscosity—as would happen, for example, if temperature were lowered or solvent added. Moreover, when dealing with emulsion droplets, an increase in (effective) particle size can result from Ostwald ripening as well as from aggregation processes, and the kinetics of this droplet ripening process can be monitored by DWS.[14]

The dynamics of particle motion does not depend only on particle size, but also on the nature of the particle–particle interactions. The strength and effects of these, in turn, depend on the local particle concentration. So, even without any change in particle size, we see that the particle mobility decreases with increasing concentration, and again this can be followed by DWS.[8,9]

Mean-Square Displacements from DWS Measurements

In a fluid system, where particles are undergoing Brownian motion, the mean-square displacement $<\Delta r^2>$ increases linearly with time, the gradient being the diffusion coefficient D, as in

$$<\Delta r^2> = 6Dt. \tag{4}$$

The decaying correlation function can also be expressed as a function of mean-square displacement, thereby emphasizing its relationship to particle dynamics, and allowing the mean-square displacement to be determined simply and easily:

$$g_{(1)} = \exp\left[-\gamma(k^2 < \Delta r^2(t) >)^{1/2}\right]. \tag{5}$$

Schurtenberger *et al.*[11] have used the correlation functions measured during acid-induced gelation of skim (and whole) milk to show how the mean-square displacement changes from the simple linear dependence of the fluid, as in equation (4), to a more complex dependence, where $<\Delta r^2>$ reaches a plateau value at long times, and where its short-time increase follows a power-law dependence, $<\Delta r^2> \propto t^\delta$, with exponent $\delta = 0.7$. These authors have shown[11] that the transition in exponent value from unity to 0.7 is almost a step function, and that it takes place in the same reaction time zone as that for the marked growth in the shear moduli as measured in a rheometer. This has led to the suggestion[11] that the transition in the exponent value corresponds to the gel point. In practical terms, however, the width of the DWS transition zone is not negligible in terms of reaction time, and during that period the measured elastic modulus rises by 4 to 5 orders of magnitude, though admittedly from a particularly low level detected with a sensitive rheometer. The question can therefore be asked: what point in this transition should be considered as the gel point—the onset, the mid-point, or the end-point?

In theoretical terms, the use of this DWS transition to define the gel point can itself be questioned. In principle, the full frequency dependence of the viscoelastic moduli can be obtained from the time dependence of the mean-squared displacement using the generalized Stokes–Einstein equation,

$$\tilde{G} = \frac{kT}{\pi a s \langle \Delta \tilde{r}^2(s) \rangle}, \tag{6}$$

where a is the particle radius. In equation (6), the mean-square displacement in real space has been transformed into Laplace space to obtain $\tilde{G}(s)$. An arbitrary functional form is then fitted to $\tilde{G}(s)$ and the modulus is expressed as a complex function in Fourier space by substituting $s = i\omega$ into the functional form used in the fit of the experimental data. Thus $G'(\omega)$ and $G''(\omega)$ may be obtained from the DWS decay function, provided an appropriate analytical functional form can be found to fit $\tilde{G}(s)$. Newtonian viscosity behaviour provides a particularly simple form which transforms into the linear growth of mean-square displacement with time. The departure of $<\Delta r^2(t)>$ from linearity thus marks the onset of viscoelasticity in the system, constraining particle motion to reach a plateau value set by the elasticity of the medium or network. The onset of viscoelasticity does not of itself, however, imply the formation of a percolating network, and therefore it should not necessarily be taken to indicate the gel point. More tangible additional evidence is required to define this point properly.

The Correlation Function Amplitude

By considering only the time constant for the correlation function decay, or even the behaviour of the normalized decay function in terms of mean-square displacement, we are throwing away valuable physical information on the behaviour of our system. The DWS correlation function represents the decay of the variance of the fluctuations in light-scattering intensity. It represents the decay from the mean-squared intensity to the square of the mean intensity, and the correlation function amplitude (*CFA*) is therefore defined by

$$CFA = <I^2(t)> - <I(t)>^2 . \tag{7}$$

Intuitively, Horne *et al.* have argued[16] that the magnitude of the amplitude is related to the number of mobile scatterers, a number that would be expected to decline during gelation as these particles become incorporated into the developing gel network. The time of onset of this decline would then define the gel point. Conversely, unchanging behaviour in the correlation function amplitude, despite an increasing relaxation time, would indicate that particles were simply being slowed in their movements with no decrease in their number.

A lowering of the value of the correlation function amplitude has been identified previously[17] in studies of the sol–gel transition in latex-doped gelatin solutions. In a more extensive study of the kinetics of gel formation in acidified milks, the rate of decay of the correlation function amplitude was observed[16] to correlate inversely with the rate of increase in the elastic modulus measured in a rheometer. In addition, the onset of the decay in the *CFA* was found to correlate with the rheological gelation time,[16] whether this was taken as the time when the modulus rises from the baseline noise or the time at which tan δ achieves a value of unity. Both of these examples[16,17] of the usefulness of the correlation function amplitude relate to its application in situations where gel formation is clear-cut and unambiguous. Oil-in-water emulsions stabilized with (excess) sodium caseinate, and prepared in the presence of ionic calcium and the acidulant glucono-δ-lactone (GDL), show a wide range of temperature-sensitive viscoelastic behaviour depending on calcium content and pH.[18,19] Such behaviour provides a more stringent test of the usefulness of DWS, and in particular of the onset of the decline in correlation function amplitude in defining the gel point.

3 Temperature-Sensitive Caseinate Emulsions and DWS

Materials and Methods

The preparation of these temperature-sensitive emulsions has been described elsewhere.[18,19] It is important to recall that the appropriate amounts of $CaCl_2$

and GDL were added to the protein solution prior to homogenization. The oil-in-water emulsions (30 vol% oil) were prepared with an excess of sodium caseinate (4%).

DWS measurements were carried out using the apparatus described previously.[13] This operates in back-scattering geometry. It employs a bifurcated multi-fibre bundle as light guide. One leg of the Y-configuration, containing half of the optical fibres, carries light to the sample from a 5 mW He–Ne laser, and the other leg takes the back-scattered light back from the sample to the photomultiplier detector. The common end (with all of the fibres) dips directly into the emulsion sample together with a temperature probe (precision ± 0.3 °C). Temperature ramps (~1 °C min^{-1}) similar to those achieved previously in rheological experiments[19] were achieved with a Peltier plate heater (Gene-Tech, Stuart Scientific). The DWS correlation functions were accumulated for 60 s at approximately one minute intervals over the experimental duration using a Malvern 4700 photon correlation spectrometer, operated in multi-time-base mode and analyzed using software developed in-house. The two parameters of interest extracted from each decay function were the relaxation time τ (equation (1)) and the *CFA* (equation (7)).[20]

Parallel rheological measurements were performed on a controlled stress rheometer (Bohlin Instruments, Model CVO) in a stainless-steel Couette double-gap geometry. The instrument was operated in oscillation mode at a frequency of 0.1 Hz. After loading, the sample was subjected to rotational pre-shear tempering at 10 Pa for 10 min before application of the temperature gradient of 1 °C min^{-1} from 20 to 60 °C.

Results

The stability behaviour of these temperature-sensitive emulsions is summarized in Figure 1 as a function of emulsion pH and calcium/protein molar ratio. At high pH and low ionic calcium content (top left corner of the plot in Figure 1), the excess protein content located in the serum phase causes depletion flocculation. Raising the temperature in these particular instances lowers the extent of depletion flocculation and causes the viscosity to decrease by over an order of magnitude. This decline is faithfully mirrored by the drop in the value of the DWS relaxation time as temperature is increased (Figure 2). There is possibly a slight monotonic decrease in the correlation function amplitude of less than 10% over the temperature range encompassed, either due to changes in the scattering properties of the emulsion (loss of depletion-inducing entities) or a drift in laser power (data not shown).

When the pH is lowered or the calcium content increased, first the level of depletion flocculation is diminished. The emulsions are stable to temperature increase in this region. At low pH or high calcium content, flocculation is again induced, either by Ca^{2+} or H^+, or both. Compositions of emulsions in the bottom right corner of the plot of Figure 1 show a rich spectrum of rheological behaviour as the temperature gradient is applied. The shear moduli rise abruptly from noise level at a critical temperature that depends on the level of

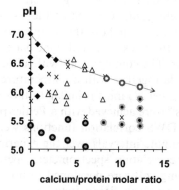

Figure 1 *Stability diagram for sodium caseinate-based emulsions (4 wt% protein, 30 vol% vegetable oil) containing various amounts of GDL and CaCl₂, incorporated before emulsification. Samples were stored for 2 days at 25 °C. Emulsion pH is plotted against the Ca²⁺/protein molar ratio (R_{Ca}).*

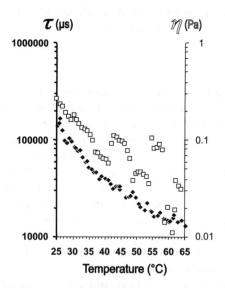

Figure 2 *Semi-logarithmic plots of the temperature dependence of the DWS relaxation time τ(◆) and viscosity η(□) of a sodium caseinate-based emulsion (4 wt% protein, 30 vol% vegetable oil) with no added CaCl₂ or GDL. The applied temperature gradient was 1 °C min⁻¹.*

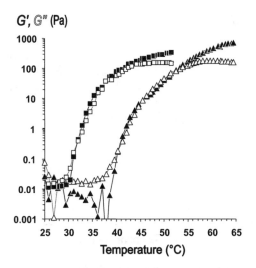

Figure 3 *Temperature dependence of the storage modulus G' (filled symbols) and the loss modulus G'' (open symbols) at 0.1 Hz for two sodium caseinate-based emulsions (4 wt% protein, 30 vol% vegetable oil, $R_{Ca}=8.3$) heated at 1 °C min^{-1} : ▲, △, $R_{GDL}=0$; ■, □, $R_{GDL}=5.8$.*

CaCl$_2$ or GDL included (Figure 3). Increasing the CaCl$_2$ or GDL content pushes the critical temperature to a lower value. Both storage and loss moduli rise in parallel over several orders of magnitude, often criss-crossing one another several times in the course of this rise. Equality of moduli ($G'=G''$) means that we have tan $\delta=1$, satisfying one of the gel point definitions—but making it difficult to define in circumstances of "now you see it, now you don't". At a particular temperature, however, the moduli do clearly separate, and then the system becomes more elastic, with the viscous modulus typically decaying slowly (Figure 3). Again this separation temperature (T_{sep}) is dependent on the contents of added emulsion components, CaCl$_2$ and GDL. If the temperature gradient is reversed at $T<T_{sep}$, the rheological behaviour is reversible. If heating continues beyond T_{sep}, then the temperature-sensitive behaviour becomes irreversible.[18]

At the highest Ca^{2+} level in these caseinate-stabilized emulsions, the emulsions as prepared show rheology with $G'>G''$ at 0.1 Hz. By another accepted definition, these are gels, albeit relatively weak. Applying the standard temperature ramping to such an emulsion finds both the elastic and viscous moduli increasing in parallel, with G' still always greater than G'', as indicated in Figure 4. Also shown in Figure 4 are the changes in DWS relaxation time over the same temperature range. These follow the increases in shear moduli at least until the separation temperature, where the relaxation time falls, as the moving entities giving rise to the fluctuating intensity apparently again move faster. Such behaviour has been observed previously for strong gels,[8,17] where it has been ascribed to some contributors being frozen out, whilst the remainder rattle around more freely, although trapped in a network cage.

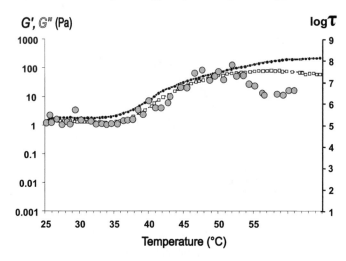

Figure 4 *Temperature dependence of the elastic modulus G' (small •) and viscous modulus G" (small ■) (both left-hand axis) at 0.1 Hz and the DWS relaxation time τ (large circle, right-hand axis) for a sodium caseinate-based emulsion (4 wt% protein, 30 vol% vegetable oil, $R_{GDL} = 0$; $R_{Ca} = 12.4$) heated at 1 °C min^{-1}.*

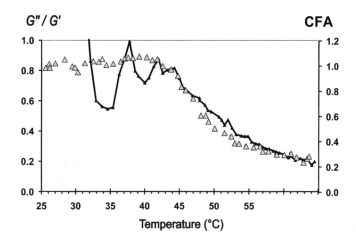

Figure 5 *Temperature dependence of the tangent of the phase angle, tan δ = G"/G', at 0.1 Hz (small ▲, left-hand axis) and the correlation function amplitude, normalized to its value at 25 °C (large shaded △, right-hand axis) for a sodium caseinate-based emulsion (4 wt% protein, 30 vol% vegetable oil, $R_{GDL} = 5.8$; $R_{Ca} = 8.3$) heated at 1 °C min^{-1}.*

The separation temperature, where G' begins to dominate over G'', is more readily apparent in a plot of tan δ ($= G''/G'$) as a function of temperature (Figure 5). Also shown on this plot is the behaviour of the correlation function amplitude, normalized to its value at the starting temperature. The value of

CFA remains constant until T_{sep} is reached, when it then falls in line with the decreasing value of tan δ as the system becomes more elastic. This means that the number of mobile scatterers remains essentially constant through the temperature range where the shear moduli and the DWS relaxation time are increasing by orders of magnitude.[20] This implies that such changes should be interpreted as thickening behaviour rather than gelation. It also implies that true gelation, in the sense of a container-spanning percolating network, begins only above the temperature where the shear moduli separate and tan δ begins to fall, detected in DWS by the onset of the decline in correlation function amplitude.[20] This ability to distinguish simultaneously various aspects of the rheological behaviour illustrates the power of the DWS technique.

4 Conclusions

We have demonstrated that changes in relaxation time alone do not indicate that a gelation mechanism is operating during a growth in relaxation time. We have also argued that, when the correlation function has been transformed into a mean-squared displacement with time, a change in its power-law exponent does not mark the onset of gelation, merely a change to viscoelastic behaviour from Newtonian flow. We have shown that, when all the aspects of DWS behaviour are considered, including both the relaxation time and the correlation function amplitude, we can recognize the gel point more clearly. Moreover, for the subtle and interesting case of a thermally labile caseinate-stabilized emulsion, the process of gel formation can be distinguished from strong thickening behaviour.

Acknowledgements

We thank Mrs Celia Davidson and Ms Jo-Ann Smith for skilled technical assistance with the DWS and rheology experiments, respectively. C.E. acknowledges receipt of a BBSRC Research Studentship and financial support from Unilever Research, Colworth Laboratory (UK). Core funding for the Hannah Research Institute is provided by the Scottish Executive Environment and Rural Affairs Department.

References

1. P. J. Flory, *Faraday Discuss. Chem. Soc.*, 1974, **57**, 7.
2. E. Dickinson, in 'Food Colloids, Biopolymers and Materials', eds E. Dickinson and T. van Vliet, Royal Society of Chemistry, Cambridge, 2003, p. 68.
3. H. G. Bungenberg de Jong, in 'Colloid Science', ed. H. R. Kruyt, Elsevier, Amsterdam, 1949, vol. 2, p. 1.
4. S. B. Ross-Murphy, *J. Texture Stud.*, 1995, **26**, 391.
5. K. Almdal, J. Dyre, S. Hvidt, and O. Kramer, *Polymer Gels and Networks*, 1993, **1**, 5.
6. H. H. Winter and F. Chambon, *J. Rheol.*, 1986, **30**, 367.

7. D. S. Horne, in 'Photon Correlation Spectroscopy: Multicomponent Systems', ed. K. S. Schmitz, SPIE Proceedings Vol. 1439, SPIE, Bellingham, USA, 1991, p. 166.
8. D. S. Horne, in 'New Physico-Chemical Techniques for the Characterization of Complex Food Systems' ed. E. Dickinson, Blackie, Glasgow, 1995, p. 240.
9. D. S. Horne, in 'Gums and Stabilisers for the Food Industry', eds G. O. Phillips and P. A. Williams, Royal Society of Chemistry, Cambridge, 2004, vol. 4, p. 368.
10. A. J. Vasbinder, P. J. J. M van Mil, A. Bot, and C. G. de Kruif. *Colloids Surf. B.*, 2001, **21**, 245.
11. P. Schurtenberger, A. Stradner, S. Romer, C. Urban, and F. Scheffold, *Chimia*, 2001, **55**, 155.
12. D. A. Weitz and D. J. Pine, in 'Dynamic Light Scattering: The Method and Some of its Applications', ed. W. Brown, Clarendon Press, Oxford, 1993, p. 652.
13. D. S. Horne, *J. Phys. D*, 1989, **22**, 1257.
14. Y. Hemar and D. S. Horne, *Colloids Surf. B*, 1999, **12**, 239.
15. B. V. Dasgupta, S. Y. Tee, J. C. Crocker, B. J. Frisken, and D. A. Weitz, *Phys. Rev. E*, 2002, **65**, 051505.
16. D. S. Horne, Y. Hemar, and C. M. Davidson, in 'Food Colloids, Biopolymers and Materials', eds E. Dickinson and T. van Vliet, Royal Society of Chemistry, Cambridge, 2003, p. 17.
17. D. S. Horne and I. R. McKinnon, *Prog. Colloid Polym. Sci.*, 1997, **104**, 163.
18. E. Dickinson and C. Eliot, *Colloids Surf. B*, 2003, **29**, 89.
19. C. Eliot and E. Dickinson, *Int. Dairy J.*, 2003, **13**, 679.
20. C. Eliot, D. S. Horne, and E. Dickinson, *Food Hydrocoll.*, 2005, **19**, 279.

Small-angle Static Light-Scattering Study of Associative Phase Separation Kinetics in β-Lactoglobulin + Xanthan Gum Mixtures under Shear

By S. I. Laneuville, C. Sanchez,[1] S. L. Turgeon, J. Hardy,[1] and P. Paquin

DAIRY RESEARCH CENTER STELA, FACULTY OF AGRICULTURE AND FOOD SCIENCE, UNIVERSITÉ LAVAL, PAVILLON PAUL – COMTOIS, QUÉBEC G1K 7P4, CANADA
[1]LABORATOIRE DE PHYSICO-CHIMIE ET GÉNIE ALIMENTAIRES, INPL-ENSAIA, 2 AVENUE DE LA FORÊT DE HAYES, BP172, 54505 VANDOEUVRE-LÈS-NANCY CEDEX, FRANCE

1 Introduction

Several studies have shown that the associative or segregative interaction between proteins and polysaccharides can determine the texture, structure and viscoelastic properties of processed foods. As a result, diverse protein + polysaccharide systems have been studied for their practical applications in food technology for the production of functional ingredients such as texturizers,[1] fat replacers,[2] meat and caviar analogues,[1,3,4] edible films,[5] microencapsulated vitamins and flavours,[1,6] and emulsion and foam stabilizers.[7-9] However, one of the most appealing aspects of this field is the challenge of understanding at a fundamental level the forces and mechanisms behind protein/polyelectrolyte phase separation. The potential benefits that may arise from mastering this subject are vast, including the improvement of current processes, and the development of novel textured ingredients or products with specific particle size, structure, texture, rheology, and superior shelf-life.

Abundant literature[1,10-14] shows that the character of protein–polyelectrolyte interactions is affected by environmental factors, i.e., pH, ionic strength, temperature, macromolecular ratio, total solids content, etc., and by internal factors, i.e., the molecular characteristics—molecular weight, chain rigidity, net charge, and relative charge density.[2,10] However, only a small number of studies[15-19] have dealt with the underlying mechanisms and the kinetics of associative/segregative phase separation.

According to the theory of Tainaka,[20] which is based on earlier studies by Veis and Aranyi,[21] protein–polyelectrolyte associative phase separation kinetics involves two stages delimited by two critical pH values, designated pHc and pHφ.[22,23] The onset of interaction occurs at pHc. At this stage, the molecules of the different species begin to interact, principally due to the opposite charges they carry, to form primary (or intrapolymeric) complexes. These primary complexes are composed of a single polymer chain and multiple bound proteins, and thus they remain soluble since they still carry a net charge.[24,25] Prior to reaching pHc, no interaction is possible due to repulsive Coulombic forces predominating, as both molecules are negatively charged. With further pH decrease below pHc, and as charge neutralization of the primary complex is approached, the second critical pH value (pHφ) is attained. At this pH the soluble complexes interact to form interpolymeric complexes that undergo extensive higher order aggregation leading on to macroscopic phase separation.[22,23]

Light scattering is the technique of choice for studying phase separation kinetics in colloids because it is very sensitive to concentration fluctuations and to the presence of aggregated structures appearing in the sample.[19,26,27] Furthermore, a number of theoretical studies and computer simulations that predict the characteristics of the light-scattering profiles during the early and late stages of phase separation are available.[28–30] In a static light-scattering experiment, the time evolution of the scattered light intensity, $I(q)$, measured at different scattering angles (related to q) can provide useful information about the phase separation mechanism of the system. Additionally, the size and structure of the developing aggregates can be determined.

The initial separation of a homogenous mixture into two phases may proceed through two distinct processes: spinodal decomposition or nucleation and growth. Spinodal decomposition (SD) occurs under conditions far from the binodal, *i.e.*, deep into the two-phase region, where concentration fluctuations are unstable and readily develop into phase separation. In SD, the morphology is characterized by the presence of one particular length scale, Λ, correlated to the prevailing domain size in the system. The characteristic length scale is associated with the presence of an intensity peak I_{max} at a constant scattering angle q_{max}, the wave vector of the fastest growing fluctuation; the quantities are related by $\Lambda = 2\pi/q_{max}$. The course of the process can be separated into three regimes—the early, intermediate, and late stages—giving different and well-defined patterns in light-scattering measurements. In the early stage, the intensity of scattered light increases exponentially due to the rapid increase in the number of structural domains of length scale Λ. The intermediate stage is characterized by a shift in the intensity peak towards lower q values due to the coarsening of the system. In the late stage, when the system has attained equilibrium, only one length scale is important, and dynamic scaling applies.[28]

Nucleation and growth (NG) occurs when a mixture is taken close to the binodal and the system is still stable to small concentration fluctuations. Under these conditions, phase separation will occur only if a certain amount of energy is input into the system to overcome the energy barrier necessary for the formation of a nucleus of a determined critical size. Once this critical size is attained, the development of a polydisperse array of particles will follow. Generally, the NG mechanism is characterized by the absence of a scattering peak and a monotonic decrease in the scattered light intensity with the measuring wave vector.[31] However, a peak related to NG has been reported in numerical simulations, and in systems undergoing polymerization-induced phase separation processes.[26,32–35] Based on numerical simulations and practical studies of aggregating colloidal particles,[32,36,37] the peak in NG has been attributed to the presence of a depletion layer surrounding the formed nuclei, or at high concentrations due to the correlations in the locations of individual scatterers.[33,34] Mie theory has shown, however, that the two mechanisms (NG and SD) can be differentiated,[17,33,35] even if both profiles present a peak in the scattering function, by the different behaviour of the peak position over time.

The macromolecules selected for this study, β-lactoglobulin and xanthan gum, are well known biomacromolecules, and are widely used in the food industry. β-Lactoglobulin (β-lg) is a globular protein with a molecular mass of about 18.3 kDa. Its tertiary structure consists of eight antiparallel β-sheets that fold to form a β-barrel with a hydrophobic core, some β-turns, and one α-helix.[38,39] The quaternary structure of β-lg depends on pH, ionic strength and temperature. It exists as an equilibrium between monomeric and dimeric forms in the pH range from the isoelectric point ($pI = 5.1$) to pH = 6.6. Xanthan gum is an anionic polyelectrolyte which undergoes an order–disorder transition with temperature. The ordered conformation exists as a double-stranded chain with a persistence length of *ca.* 150 nm at high ionic strengths.[40,41] It is well-known nowadays that xanthan can exist in several aggregated states—single, double or multistranded helices—depending on the primary structure of the sample (pyruvate and acetate content),[42] on the solution conditions (ionic strength, temperature, solvent quality),[40,41] and on its treatment history during and after fermentation, *i.e.*, drying, filtration, pasteurization, centrifugation and purification.

Previous studies on the β-lg + xanthan system have shown that the protein/polysaccharide ratio *r* has a major effect on the characteristics of the resulting complexes, namely, the complex size, composition, and viscosity in solution.[43] It was also found[43,44] that a microfluidization step (a dynamic high-pressure treatment) of xanthan gum prior to the complexation process allows formation of particulated complexes, instead of fibrous ones, from the beginning of production. In this paper, the mixtures of β-lg + xanthan (native or microfluidized), in aqueous dispersion and under shear, are further characterized by determining the associative phase separation and coarsening kinetics by turbidimetry and static light-scattering experiments.

2 Experimental

Materials

Whey protein isolate (WPI) (lot no. JE 002-8-922, 98.2 wt% protein, of which 85% is β-lg, 1.8% minerals, 4% moisture) was obtained from Davisco Foods

(Eden Prairie, MN). Due to the high content of β-lg in this WPI powder, it was assumed that the β-lg governed its behaviour;[45] it is therefore referred to simply as β-lg in the rest of this article.

Xanthan gum (Keltrol F, lot no. 9D2192K, 96.36% total sugar, 3.02% protein) was purchased from Kelco (San Diego, CA). Microfluidized xanthan samples (treated with 1, 4 or 12 passes at 750 bar) were prepared and dialysed as described previously.[43] Microfluidization is a dynamic high-pressure treatment that decreases the molecular weight (M_w) and the aggregated state of xanthan to different extents, depending on the operating conditions (pressure, number of passes, temperature, solvent, *etc.*). The principal characteristics of the xanthan samples tested in this study are presented in Table 1.

The protein content in the β-lg and xanthan samples was determined by measuring total nitrogen content by combustion according to the Dumas principle (IDF standard 185:2002) using Leco equipment (FP-528, Leco Corporation, St. Joseph, MI) and a conversion factor of 6.38. Total carbohydrate content was determined by the phenol–sulphuric acid method.[46] All chemicals used were of reagent-grade quality.

Preparation of Biopolymer Dispersions

Dispersions containing 0.1 wt% total biopolymer concentration of protein and xanthan were prepared in filtered (0.2 μm) deionized water (Milli-Q, Millipore, US) under gentle stirring for 1 h (protein) or strong stirring, and avoiding air incorporation, for 2 h (polysaccharide). Solutions were kept overnight at 4 °C to allow complete hydration. Subsequently, the dispersions were centrifuged for 30 min at $10^4 g$ (20 °C) and filtered through a 0.2 μm filter (protein) or a 0.8 μm filter (polysaccharide) (Sartorius Minisart filters, Gottingen, Germany). The β-lg + xanthan dispersions (0.1 wt%) were prepared at the desired protein/ polysaccharide ratios r, namely 2:1, 5:1; 10:1 or 15:1 ($r = 2$, 5, 10 or 15, respectively), using the appropriate quantities of stock solutions, and mixed gently for 30 min. Tests with microfluidized xanthan were carried out at $r = 5$. The initial pH of the mixed dispersions was 6.58 ± 0.08.

Associative phase separation of the mixed dispersions of β-lg + xanthan (native or microfluidized) was induced by slow acidification using 0.015% w/w glucono-δ-lactone (GDL) (Merck, Darmstadt, Germany) to a final pH ≈ 4.5.

Table 1 *Molecular weight (M_w) and intrinsic viscosity [η] of the xanthan gum samples used in this study.*

Sample	Code	$M_w \times 10^6$ (Da)	[η] (dL g^{-1})
Native xanthan gum	XG	5.08	26.9
Microfluidized 1 pass[a]	X1	4.46	20.1
Microfluidized 4 passes[a]	X4	3.94	15.2
Microfluidized 12 passes[a]	X12	3.22	11.3

[a] Number of passes through the microfluidizer at 750 bar.

After the addition of GDL, dispersions were slowly stirred for an additional 15 min before starting the light-scattering experiments. The pH of the dispersion was followed in parallel in order to determine the acidification curve for each system. Pure β-lg and xanthan dispersions were also tested separately for comparison. All tests were carried out at least in duplicate. Protein blanks presented some minor aggregation starting at pH=5.25, which is close to the β-lg pI. The maximum intensity reached was ∼8 arbitrary units, very much less than the intensity obtained for the mixed system (3500 arbitrary units). This slight aggregation was possibly due to the presence of some denatured protein produced during WPI manufacture. The xanthan blanks presented a low intensity profile that remained almost constant over the whole acidification course.

Time-Resolved Small-angle Static Light-Scattering

The associative phase separation, the coarsening kinetics, and the temporal evolution of turbidity of β-lg+xanthan dispersions were followed through acidification using a Mastersizer S equipped with a 5 mW He–Ne laser of wavelength λ=632.8 nm (Malvern Instruments, UK). The sample was maintained under constant circulation with the stirred sample unit operating at 2000 rpm (laminar flow). The optical cell had a path length d=2.4 mm and the range of scattering angle q covered was 0.01–10.4 μm^{-1}, where $q=4\pi n_s \sin(\theta/2)/\lambda$, with θ the scattering angle of observation and n_s the refractive index of the solvent. This q range allowed the study of particle sizes in the range between 0.05 and 900 μm.

The background intensity was recorded on filtered (0.2 μm) Milli-Q water with a recording time of 5 s, before starting the measurements on the mixed dispersions. The instrument was programmed to measure the scattered light intensity from the sample every minute over a period of 10 to 15 h (recording time 10 s). The experimental scattering intensity was then corrected with respect to the angle of the detectors. Finally, corrections were applied to take into account the turbidity development in the sample over time according to the relationship:[47,48]

$$I_c(\theta) = I_m(\theta)\left\{\frac{\exp[\tau d/\cos\theta]\tau d[\cos(\theta-1)]^{-1}}{\exp(\tau d[\cos(\theta-1)]^{-1})-1}\right\}. \tag{1}$$

Here, I_m and I_c are the measured and corrected scattered intensities, respectively, and d is the sample thickness. The turbidity τ of the specimen is calculated from

$$\tau = \frac{\ln(I_0/I_t)}{d}, \tag{2}$$

$$I_t = 1 - O(t), \tag{3}$$

where I_0 is the incident intensity, I_t is the transmitted intensity, and $O(t)$ the obscuration value (%) of the sample at a given time t, which is automatically measured by the Mastersizer S equipment. The corrected scattered intensity I_c will henceforth be referred to as $I(q)$.

The evolution of the internal structure of the complexes was determined by measuring the apparent fractal dimension d_f from the slope over one decade at large q values using the power law $I(q) \sim q^{-d_f}$. The presence of a fractal structure implies that many aggregate properties will show a power-law scaling. The scattering intensity profile obeys power-law behaviour at large q values, allowing the data to collapse to master curves for the aggregation kinetics data. The d_f value indicates the degree of compactness of the structure: if particles aggregate in linear arrays, then the value of d_f would tend to approach 1; but if they form compact spherical aggregates, d_f would tend to approach 3.[49]

Temporal Evolution of the Size of Complexes

To gain deeper insight into the evolution and coarsening of the developing complexes, we have determined the averaged scattered light intensity $I_a(q)$, according to three ranges of q, 0.0124–0.042 μm^{-1}, 0.05–0.67 μm^{-1} and 0.8–10.4 μm^{-1}. The three ranges were assumed to represent contributions to the scattering attributable to large ($> 130\,\mu m$), medium (10–130 μm) and small particles (~ 1–10 μm). The final values of d_{43} and $d(v, 0.5)$ for the formed complexes, as calculated automatically by the Mastersizer software, were also recorded.

Results were analysed in terms of the temporal evolution of the turbidity, the scattered light intensity $I(q)$, and the apparent fractal dimension d_f of the formed complexes.

3 Results and Discussion

Temporal Evolution of Turbidity and Reaction Kinetics

Figure 1 presents the turbidity evolution upon acidification for $r = 5$ and 15. The plot shows that the β-lg + xanthan system follows a two-step associative phase separation process as predicted by the theory of Tainaka[20] for complex coacervation. The measured values of pHc and pHφ are presented in Table 2. The pHc value denotes the onset of the interaction that leads to the formation of soluble intrapolymeric complexes,[22,23] and it was taken as the point where a subtle, but distinct, increase in turbidity was detectable (inset to Figure 1), indicating an increase in size and/or concentration of the forming complexes. We find pHc $= 5.76 \pm 0.03$ for all the systems studied at low ionic strength, indicating that it is independent of the macromolecular ratio r (Table 2). It has been suggested that such complexation is controlled by the interaction of a single protein molecule and a single sequence of polymer segments. Consequently, the process does not follow a mass-action law, and pHc should not be affected by r or M_w since it only depends on local charge densities and the intrinsic stiffness of the polymer chains, which are in turn controlled only by conditions such as pH and ionic strength.[24,50]

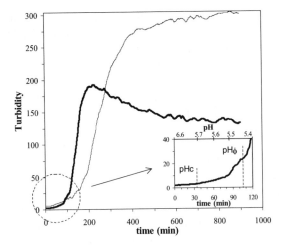

Figure 1 *Temporal evolution of turbidity for r = 5 (—) and r = 15 (——). The inset shows an enlargement of the initial region for r = 15, indicating the positions of pHc and pHϕ. Average standard deviations are 5.3 and 2.9 for r = 5 and r = 15, respectively.*

Table 2 *Critical pH values determined from turbidity measurements for all the studied systems.*

Composition (β-lg/xanthan ratio)	pHc	pHϕ	EEP
2:1	5.78[a]	5.10[a]	N/A[b]
5:1	5.79	5.36	4.46
10:1	5.78	5.43	4.87 ± 0.04
15:1	5.75	5.47	4.92
5:1 X1	5.75	5.31	4.49
5:1 X4	5.74 ± 0.03	5.23	4.45
5:1 X12	5.79 ± 0.03	5.19	4.50

[a]Standard error ± 0.02, unless otherwise stated.
[b]Not attained under the current conditions.

At pHc, the acid has yielded its protons to enough carboxylate and amino groups on the protein to decrease the repulsive forces between the two molecules, so that electrostatic attraction is spontaneously supported. Since we have pHc ≫ 5.1, the p*I* of β-lg, it could be inferred that the electrostatic interaction starts with positively charged patches on the protein molecule surface providing enough binding sites for the xanthan. Similar effects have been observed in other protein + polyelectrolyte systems[15,22,24,45,51] and have been predicted by Monte Carlo computer simulations.[29,30] Moreover, Girard *et al.*[15] found that β-lg possesses several charged patches (including the 132/148 peptide, which is part of the α-helix) susceptible to taking part in complexation with polyanions above p*I*.[15] Incidentally, it was found[52] by dichroism measurements that, indeed, the β-lg helix is at least partially lost during coacervation with the anionic acacia gum.

With further pH decrease and as charge neutralization of the primary complex is approached, pHφ is attained (inset Figure 1, Table 2). This is characterized by an almost exponential increase in turbidity caused by extensive interpolymeric complexation and macroscopic phase separation.[22,23] At this stage, the protein charge density has increased enough for mass action effects to be observed.[24] Accordingly, it was found that pHφ increases with r from 5.10 to 5.47 for $r = 2$ to 15 (Table 2). Complexes being neutralized sooner at higher r (more protein available for interaction) may explain this. Also, more protein can be bound onto the xanthan, and thus the charge per protein required to achieve complex neutralization decreases, resulting in an increase of pHφ.[24]

The sharp increase in turbidity after pHφ (Figure 1) continued up to a peak for high protein systems ($r = 10$ and 15) or to a plateau for low protein systems ($r = 2$ and 5) (see Figure 1). This peak or plateau is attributed to the isoelectric point of the β-lg–xanthan complex, also known as the stoichiometric electrical charge equivalence pH (EEP), where molecules carry equal but opposite charges, and complexation production is at its maximum.[24,53] The decrease in turbidity for $r = 10$ and 15 is related to a reduced density of scattering entities, possibly as a result of flocculation due to protein over-aggregation.[43] It cannot be attributed to sedimentation since the dispersions were under continual stirring. For $r = 2$, the turbidity still seemed to be increasing at pH = 4.6, indicating that this system may have not yet fully attained its EEP. Despite the protein's pI seeming to have no special significance for pHc, we found that the EEP condition always took place after passing the pI, and that it lay very close to it for high protein containing systems ($r = 10$ and 15; see Table 2). These results reveal that protein aggregation is present and forms an important part of the process. The effect of a high protein content on complex characteristics was previously detected[43] for the same system using chemical and microscopic analysis.

Xanthan is known to form aggregates in solution by linkages at the end of the chains, and to form a tenuous network that has been compared to a weak gel;[40,54,55] such structuring breaks up readily upon shearing.[40] Previous studies have shown[43] that microfluidization of xanthan dispersions results in the break-up of aggregates and in a net reduction in the value of M_w, the thickening power, and the capacity to re-aggregate to different extents, depending on the severity of the treatment (microfluidizer pressure and number of passes). The principal characteristics of the xanthan samples tested in this study are presented in Table 1.

The associative phase separation process in systems containing microfluidized xanthan was found to start at the same pHc (Table 2), indicating that the M_w of the polyelectrolyte had no noticeable influence on the initial step of complexation. This result is in agreement with previous studies, where it was proposed that the interaction is initiated at small patches on the protein surface involving short segments of the polyelectrolyte chain. Therefore, the value of M_w for the polyelectrolyte should have no effect on pHc.[24,50,56]

On the other hand, pHφ was found to decrease significantly with increasing severity of microfluidization treatment, from 5.36 ± 0.01 with native xanthan to 5.19 ± 0.02 with xanthan treated 12 times (Table 2). This result shows that the transition at pHϕ depends strongly on polyelectrolyte M_w,[56] which may be explained by the formation of smaller primary complexes, resulting from the lower M_w of the xanthan (Table 1) that serves as the backbone for the forming complexes. The formation of smaller intrapolymeric complexes would result in a weaker turbidity increase upon interpolymeric aggregation, thereby shifting the pHϕ to lower values. Accordingly, at the same r, final turbidity values were lower for samples containing microfluidized xanthan (*e.g.*, $\sim 14\%$ turbidity reduction for sample X12). Moreover, the formation of smaller intrapolymeric complexes was observed from the beginning of the complexation process when microfluidized xanthan was used (see below). It has been proposed[57] that, when a high M_w polyelectrolyte is used, the critical pH values are attained sooner due to a higher local concentration of protein molecules bound to a single chain.

Complex Size Evolution and Coarsening Kinetics

The evolution of the averaged scattered light intensity of small (~ 1–$10\,\mu m$), medium (10–$130\,\mu m$) and large ($>130\,\mu m$) particles for $r=2$, 5 and 15 is presented in Figure 2a–c. This approach allows us to gain deeper insight into the evolution and coarsening of the developing complexes. For all the studied values of r, it could be observed that, once pHc was attained, the number of small and medium-sized particles, assumed to be intrapolymeric complexes, increased rapidly while the number of large particles decreased. The reduction in the scattered intensity due to large particles at the early times (Figure 2) may be in part ascribed to the formation of more compact primary complexes. As more protein binds to the xanthan molecule, a reduction of the net internal repulsion between the polyelectrolyte segments occurs resulting in the collapse of the polyelectrolyte chains. This loss in conformational entropy is partly compensated for by the release of counter-ions and occluded water generating compact structures.[10,45,56,58] The reduction in size of the complexes has also been noticed through viscosity measurements.[43,45,59] Additionally, a decrease in number of large particles at early times was observed in the protein and xanthan control samples (results not shown); this can be attributed to dissolution of GDL crystals.

The production of primary complexes was found to increase up to a peak occurring around pHϕ (indicated by a dashed line in Figure 2a–c). At this point large interpolymeric complexes begin to form (sharp increase in large particles for $r=5$, 10 and 15) at the expense of the primary complexes (decrease in smaller particles). For the lowest protein content system ($r=2$), the increase in the number of large particles was less steep, indicating a slower reaction rate. Subsequently, the number of interpolymeric complexes continue to increase slowly (Figure 2a) down to the lowest pH attained for this sample (pH $=4.6$), and thus the complexation process for this ratio may have not been completed. For higher protein content systems ($r=5$ and 15 in Figures 2b and 2c, respectively) the coarsening process followed a more complex path: after

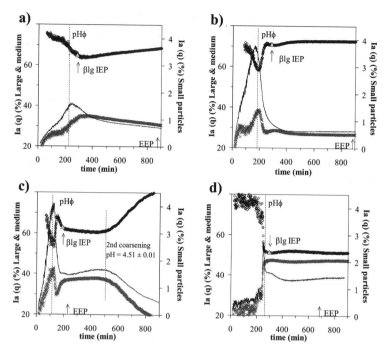

Figure 2 *Evolution of the averaged scattered light intensity $I_a(q)$ from large ($>130\ \mu m$)*
($\Diamond$), medium-sized (10–130 μm) ($\times$) and small (~ 1–10 μm) (—) particles
for (a) r=2, (b) r=5, (c) r=15, and (d) r=5 (xanthan X12). Critical pHϕ
values are indicated by the dashed vertical lines. The isoelectric points of β-lg
(IEP) and of the complexes (EEP) are indicated by the arrows. Mean
standard deviations are ≤ 4.0 for large and medium-sized particles, and ≤ 0.1
for small particles.

pHϕ the number of interpolymeric complexes increased abruptly up to peak,
soon after which the protein's pI was attained. For $r=5$, a plateau was attained
at pH$=4.52\pm0.02$ (~ 650 min) (Figure 2b) indicating the stabilization of
domain growth. This is probably due to the saturation of reactive sites, caused
by a stoichiometric composition of the neutralized complex (electrostatic equilibrium) which does not allow any further number of protein molecules to bind
to the polysaccharide and so forces the complexes to reach a maximum size.[24]
The results presented in Figure 2a–c are in accordance with the mechanism
of a two-step complexation, where interpolymeric complexes (large particles)
grow at the expense of smaller ones (intrapolymeric complexes).

For $r=15$, a second coarsening phase was observed at pH$=4.51\pm0.01$ (time
~ 500 min) (Figure 2c) in which very large particles formed while small and
medium-sized particles were almost completely depleted from the system.
This time interval corresponds to the decrease in turbidity after the peak
corresponding to the EEP for this system (Figure 1). In the β-lg+xanthan
system the neutralized complexes obtained at non-stoichiometric ratios are
destabilized by the excess component causing a secondary aggregation. This is

contrary to previous studies where it was found[59] that the excess component stabilized the neutralized core against secondary aggregation. We argue that the major component can be incorporated in excess into the complex only up to a certain level, after which secondary aggregation is induced. This explanation is supported by the results found for $r=10$ and 15, where a plateau precedes the second coarsening (Figure 2c). This plateau may correspond to the time where excess protein is binding to the complex, before being completely destabilized due to the reduction in repulsive forces caused by protein over-aggregation onto the complex.[43] The ratio 10:1 presents intermediate behaviour between $r=5$ and 15, with a less dramatic second coarsening (results not shown).

From Figures 2a–c it can be deduced that the process of complex coarsening occurs faster at high protein content. This is caused by at least two driving forces: (1) mass-action equilibrium since at high r more protein is available to interact and saturate the xanthan gum; and (2) the higher viscosity in systems at lower r (a higher proportion of xanthan) may have reduced the collision rate between molecules and therefore hindered faster growth.[22,24] Accordingly, pHϕ is reached sooner for high r (Table 2), indicating that the critical amount of bounded protein needed to initiate interpolymeric complexation is achieved sooner.[22]

Figure 2d presents the complex size evolution for a system with xanthan microfluidized 12 times (sample X12). In general, when microfluidized xanthan samples were used, the increase of scattered light from small particles (intrapolymeric complexes) upon arrival to pHc was lower, specially for X4 and X12, compared to the system where native xanthan was used, presumably as a result of smaller scattering entities. The transition to form interpolymeric complexes at pHϕ also occurred after the peak in small complex production. However, large particles continued to disappear while medium-sized complexes were produced, and both fractions tended towards the same final equilibrium size (Figure 2d). This reveals that the intrapolymeric and interpolymeric complexes were much smaller when microfluidized xanthan was used. Accordingly, the final d_{43} and $d(v,0.5)$ values (Table 3) were reduced with the number of passes through the microfluidizer.

Table 3 *Characteristics of the intermolecular complexes for all the studied systems.*

β-Lg/xanthan ratio	d_{43} (μm)	$d(v,0.5)$ (μm)	Λ (μm)	Final $d_f{}^a$	β
2:1	18.9±0.7	13.5±0.6	360±2	2.25	2.5±0.3
5:1	24.5±0.5	19.0±0.4	233±20	2.23	10.2±0.1
10:1	17.3±2.8	10.7±0.3	467±40	2.30	12.8±0.4[b]
15:1	70.3±20	19.8±5.7	730±42	2.37	11.8±0.8[b]
5:1 X1	14.3±0.4	11.8±0.5	213.0±18	2.26	15.1±0.2
5:1 X4	13.4±0.7	11.5±0.6	106.5±20	2.46	14.5±0.8
5:1 X12	10.5±0.2	9.2±0.1	81.5±6	2.53	30.1±0.8

[a] Standard error ±0.02, unless otherwise stated.
[b] Values before the second coarsening process.

Previous studies have also shown[56,57,60] that an increase in the polymer M_w results in the formation of larger primary complexes that aggregate more readily into interpolymer complexes, presumably due to the higher gain in entropy[20] or to bridging effects.[57]

Therefore, the final size of the complexes can be controlled by adjusting the size of the polyelectrolyte. Accordingly, the production of smaller complexes when microfluidized samples are used shows that the severity of the micro-fluidization treatment prior to complexation allows us to control the size of the complexes from the beginning of production. Nevertheless, it should be noted that the acidification procedure also influences the structure and size distribution of protein–xanthan gum complexes. In this study, the effect of M_w could be detected due to the use of GDL, which provides a slow and uniform acidification. As a result, more homogeneous complexes are formed compared to complexes obtained in previous work where HCl was used for acidification.[43] The former results had led us to hypothesize that the initial r value had an overwhelmingly high effect on the size of the complexes, masking the effect of M_w.[43] However, it is now evident that the acidification method used also had an impact on our results.

Furthermore, the use here of GDL allows the formation of particulated complexes even when native xanthan is used, instead of the fibrous complexes obtained at the same concentration when HCl is used for acidification.[2,44] Therefore, to develop further the previously presented hypothesis for the formation of fibrous complexes of protein with native xanthan,[43] we propose that in addition to the association pattern of xanthan (*i.e.*, weak associations at the end of chains) acting as the guide for fibre formation, a fast complexation with proteins should occur to fix in place the fibrous structure. In this context, a rapid interaction and not a gradual one is essential. Such rapid interactions may be induced by an abrupt local decrease in pH, *e.g.* by adding a sufficiently concentrated acid, even if it is slowly incorporated (as shown by the results of Chen and Soucie[3]), in the presence of an adequate protein quantity, otherwise the forming fibres would be subsequently torn apart due to insufficient joining material. This scenario presupposes the existence of a certain range of r at which fibre formation would be optimal, as previously found.[2,3] Evidently, a large chain length of the polyelectrolyte is essential to form fibrous complexes, and low M_w molecules will tend to form particulated complexes.[43,61] Another important parameter to consider is the total solids concentration: below a certain critical macromolecular concentration (c_r), fibre formation is suppressed since xanthan molecules are too far apart for bridging to occur. In the β-lg + xanthan system, we have $c_r < 0.1\%$ w/w. The effect of other factors influencing fibre formation, such as degree of shearing, temperature and ionic strength, are consistent with this model.

Temporal Evolution of the Scattering Profiles: the Phase Separation Mechanism

Time-resolved light-scattering profiles are shown for $r = 2$ and $r = 15$ in Figures 3 and 4, respectively. In each plot the phase separation progress is

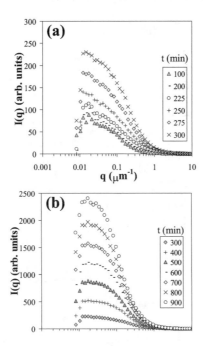

Figure 3 *Time evolution of light-scattering profile I(q) for r=2 after onset of interaction (pHc ∼65 min, pHφ ∼225 min): (a) 100–300 min; (b) 300–900.*

presented over time in the following order: (a) early (before pHφ), (b) intermediate and late (after pHφ) stages of complexation, and (c) second coarsening for $r=15$. The scattered light intensity for all r presents a similar profile at the early and intermediate times of measurement. Conversely, in the late stages of coarsening, systems with a low protein content ($r=2$ and 5) seem to approach or reach equilibrium, while systems with a high protein content ($r=10$ and 15) become destabilized as the second coarsening is reached.

Initially, all samples present a correlation peak (I_{max}) located at a constant q_{max} value between 0.0175 and 0.0147 μm^{-1} (Figures 3a and 4a). On attaining pHc, as determined by turbidity measurements, the $I(q)$ was found to increase faintly, in the intermediate q region between 0.0295 and 4.159 μm^{-1}, after an induction time that increased from ∼30 min to ∼90 min as r decreased from 15 to 2. Then, around pHφ, I_{max} increased almost exponentially as phase separation progressed, indicating an increase in the total number and size of structural domains, whereas q_{max} shifted very slightly to smaller q values, only shortly after pHφ, and remained constant at $q=0.0124\ \mu m^{-1}$, for ∼150 min (Figures 3b and 4b).

The evolution of I_{max} as a function of time for $r=5$ and 15 is shown in Figure 5. In all cases, the I_{max} increase after pHφ can be expressed approximately by a power relation of the type $I_{max}(t) \propto t^{\beta}$. The estimated values of β are presented in Table 3. These values are well above the 3 or 1 values predicted

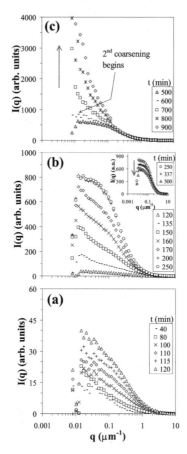

Figure 4 *Time evolution of light-scattering profile $I(q)$ for $r=15$ after onset of interaction ($pHc \sim 50$ min, $pH\phi \sim 115$ min): (a) 40–120 min; (b) 120–500 min; (c) 500–900 min.*

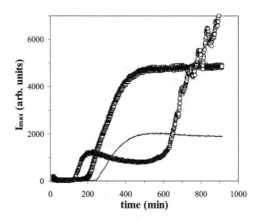

Figure 5 *Temporal evolution of I_{max} for $r=5$ (□), $r=15$ (○), and $r=5$ with X12 (—).*

for classical SD or NG in the intermediate and late times of coarsening.[28] Higher β values indicate a higher reaction rate. It was found that the complexation rate was proportional to the amount of protein in the system. In other words, coarsening appears slower in systems with a high proportion of xanthan. This is also evident from the lower structuring rate for $r=2$. High values of β have been also found[26] in systems undergoing polymerization-induced phase separation.

In the intermediate stages of coarsening (Figures 3b and 4b), the region between $q=0.0295$ and 0.334 μm^{-1} became stronger and formed a shoulder, while I_{max} continued to increase. Subsequently, q_{max} shifted to larger values starting at 500 and 250 min, for $r=2$ and 15 respectively (Figures 3b and 4b). At this stage, the systems contained a sort of bimodal cluster distribution. The presence of a shoulder is consistent with the formation of smaller primary complexes, detected at shorter wavelengths, which act as nuclei for the formation of larger interpolymeric complexes detected at larger wavelengths. A similar pattern was found in all the studied systems. The q_{max} is related to a characteristic length scale ($\Lambda = 2\pi/q_{max}$), which corresponds to the main domain size (periodic structure) being formed. The final Λ values increase with protein content (Table 3).

A major difference was evident in the late stages of coarsening between systems with low ($r=2$ and 5) or high ($r=10$ and 15) protein content. For low protein content systems, the I_{max} growth levelled off at times >550 min for $r=5$ (Figure 5), indicating stabilization of domain growth at the EEP $\sim 4.46 \pm 0.02$. Accordingly, a plateau in the turbidity (Figure 1) and the complex size (Figure 2b) was found at the late times of interaction for this $r=5$. The same pattern was followed at $r=2$, but for the latter system the equilibrium plateau had not been fully attained at the end pH of this study. On the other hand, for high protein content systems, after a shorter shift of q_{max} to higher values, I_{max} effectively stopped its increase between 230 and 250 min (Figure 4b). This may be due to the attainment of a transient electrostatic equilibrium in this mixture, which coincides with the short plateau attained in the complexes sizes (Figure 2c, 220 to 250 min, pH$=4.8$ to 4.9), just before the beginning of the second coarsening. As pH decreased further, the interpolymeric complexes continued to incorporate the excess protein in solution until a certain limit, after which massive aggregation was induced. The second coarsening began with a decrease in the scattered light intensity in the small q range, between 250 and 500 minutes for $r=15$ (inset of Figure 4), suggesting a clustering of the complexes into larger structural domains. Then, I_{max} resumed its increase at ~ 550 min, while q_{max} shifted towards smaller angles until it moved out of the measurement window (Figure 4c). This coincides with the sharp increase in turbidity (Figure 1) and the depletion of small particles (Figure 2c) at times beyond 500 min. A similar decrease in $I(q)$ was observed in a system of β-lg + acacia gum prior to a second growth phase.[19] Conversely, other studies have found[62] that in complexes of protein and strong polyacids, the domain size was independent of protein concentration.

The profiles presented in Figures 3 and 4 are different from the classical NG mechanism where a monotonic decrease in $I(q)$ with q is expected. However, the presence of a peak during NG has been reported in experimental[33,63] and numerical simulation studies,[35] where its development at high volume fractions has been attributed to the formation of a depletion layer around the forming nuclei and/or to multiple scattering. A peak related to NG has also been reported for systems undergoing polymerization-induced phase separation.[26,34,64] Moreover, the presence of a shoulder or a secondary peak, with the concomitant shift of q_{max} toward large values, is present for some of these systems. It was proposed that the formation of newer domains in the interdomain region would cause the shift of the scattering function to higher scattering angles due to a decrease of the average spacing between particles.[17,26,33,34,63,64] Development of a shoulder has been recently reported[16] during β-lg–pectin complexation and attributed to the formation of smaller assemblies resulting from the NG mechanism.

Several characteristics have led us to classify the initial phase as being associated with the NG mechanism, *i.e.*, the lag time between pHc and the start of $I(q)$ increase, and the stability of q_{max} at the intermediate times followed by the later shift to larger values. The presence of a correlation length scale is related to the formation of primary complexes, possibly xanthan-rich nuclei, surrounded by a layer depleted in protein. The shoulder is assumed to result from the continual formation of smaller scattering bodies (primary complexes) as nucleation proceeds. Correspondingly, the production of intrapolymeric complexes is around its maximum when the shoulder begins to gain strength.

The $I(q)$ profiles for samples at $r = 5$ with microfluidized xanthan present similar behaviour to the system with native xanthan except that the reaction proceeded more rapidly, *i.e.*, the β values are higher (see Table 3). It was found that the maximum $I(q)$ attained was lower with lower M_w (Figure 6, with different scale on the axis). Interestingly, the region of q at which the shoulder was found to develop appears at larger q values with decreasing xanthan M_w (see Figure 6). It has been proposed[33] that during NG the formed nuclei will have a correlation length of the order of the largest species in the system. Furthermore, the final Λ values decrease with decreasing xanthan M_w (Table 3). Therefore, the xanthan molecule acts as the support for the formation of primary complexes and determines the final size of the interpolymeric complexes, *i.e.*, the final size is proportional to the molecular weight of the polyelectrolyte.

Dynamical scaling of the scattering function was checked by plotting $I(q)/I_{max}$ against q/q_{max}. When dynamical scaling holds, all the scattering data collapse onto a single master curve. Scaling was found only for low r, and at long enough complexation times, when complexes were close to their equilibrium composition. Similar effects were found in polymerization-induced phase separation.[63] The inset in Figure 7 shows the collapse of the data for $r = 5$ at times well beyond pHφ. For high r, a transient scaling, with a pattern similar to that obtained at low r, was found before the second coarsening process. Afterwards, the continuous changes in the coarseness of the system hindered dynamic scaling.

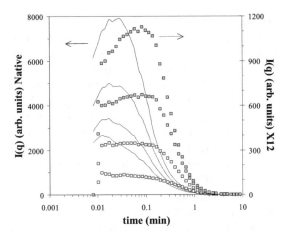

Figure 6 *Temporal evolution of $I(q)$ for $r=5$: —, native xanthan (XG); □, microfluidized xanthan (X12). Times depicted are from bottom to top: 290, 310, 360 and 800 min (pH $\approx$ 5.1 to 4.5).*

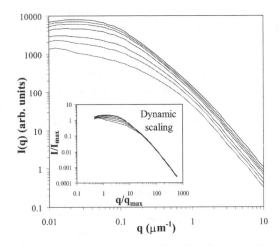

Figure 7 *Time evolution of light-scattering profile $I(q)$ for $r=5$ after pHϕ in a double logarithmic scale. Inset: dynamic scaling of the same data. The deviation at low q reveals breakage due to shear forces on large length scales. Times depicted are from bottom to top: 250, 275, 300, 350, 500 and 900 min.*

Apparent Fractal Dimension and Cluster Morphology

Figure 7 shows similar data to that presented in Figures 3 and 4, but in a double logarithmic scale for $r=5$ at times beyond pHϕ. The scattering curves display an increasingly linear relationship over more than a decade in the q range $>0.65\ \mu m^{-1}$, which is a signature of the scattering from a mass fractal object.[27] As the complexes grow in size, the curved transition from the Guinier regime (plateau at small q values) to fractal scattering migrates to lower q, while the scattered light intensity in the fractal regime increases as the mass of the average aggregate increases, showing that the final structure has not yet been fully developed.[27,49] At sufficiently long complexation times, the $I(q)$ in the power-law region remains unchanged indicating a conservation of mass over this range of length scales. At this stage, information about the internal structure (compactness) of the aggregates, in this case β-lg–xanthan complexes, can be determined by measuring the apparent fractal dimension d_f from the slope at large q values using the power law $I(q) \sim q^{-d_f}$. At the beginning of complexation the measured values are not a real fractal dimension, as given in the strict definition of a fractal structure,[27] since the complexes are still evolving and have not yet attained equilibrium. Nevertheless, over small periods of time (*i.e.*, of the order of 5–10 minutes in this work, due to the very slow acidification), the formed structures are relatively stable, and it is possible to measure an 'apparent' d_f from the slope at large q values after pHc, in order to estimate the morphology of the intrapolymeric complexes.

The evolution of the apparent d_f over the course of acidification for different r is presented in Figure 8. The complexes underwent a restructuring process in several stages, starting with the formation of diffuse aggregates (aggregation of nucleated primary complexes) with $d_f = 1.65 \pm 0.02$ for $r=2$ and $d_f = 1.90 \pm 0.05$ for $r=5$, 10 and 15, indicating a diffusion-limited cluster aggregation mechanism (DLCA). This type of mechanism occurs when repulsive forces are negligible, and the model assumes that collision always results in irreversible sticking, causing particles to stick upon contact. Loose and highly tenuous structures are formed and the aggregation proceeds very rapidly.[65] At pHφ a rapid compaction of the structure occurs, showing that the interpolymeric complexes are denser than the primary complexes. Then, a slight loosening of

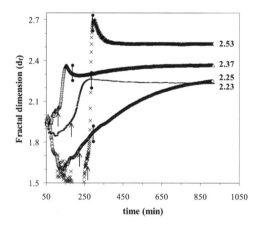

Figure 8 *Temporal evolution of the apparent fractal dimension d_f during associative phase separation and coarsening for $r=2$ (○), $r=5$ (-), $r=15$ (△), and $r=5$ with xanthan X12 (×). The pHφ values are indicated by arrows, and the protein pI values by bars. The error bars are smaller than data points after ~100 min.*

the structure occurs around the protein's pI (especially at high r) (Figure 8), followed by a gradual reorganization towards more compact structures.

The final structures of the interpolymeric complexes (Table 3) appear more compact than the values predicted by the classical DLCA and RLCA regimes ($d_f = 1.78$ and 2.1, respectively).[65] These models assume irreversible aggregation with an infinite interparticle attraction. However, several studies of polyelectrolyte complexes with proteins[19,59] and mixed micelles,[56] as well as aggregation of biological molecules,[66] produce clusters of higher d_f. It has been suggested that compact structure formation is driven by excluded volume effects and entropy loss factors, since binding induces a loss of conformational entropy on the polyelectrolyte chains. This loss is compensated for to some extent by the release of occluded water and bound counter-ions from both the protein and polyelectrolyte.[53,56] This also compels the xanthan molecules to try to retain their original overall conformation, forcing protein molecules to occupy the space between segments. As a result, compactness of the structures will tend to increase with increasing r, especially in pure water, due to the screening of the Coulombic interactions, leading to rearrangements as found in soluble complexes.[56,59]

Accordingly, d_f was found to be affected by r in this way (Table 2). The formation of dense complexes at high r is a geometric effect. As the number of large negative particles increases ($r = 2$), the packing density and the fractal dimension decrease because relatively few smaller positive particles are available to fill in the spaces between the negative particles. Similar trends have been found during the aggregation of biomacromolecules[59] and colloidal particles.[27] For $r = 15$, the form of $I(q)$ is still a power law, but it decreases in amplitude after reaching pH $= 4.80$, as very large complexes are produced during the secondary aggregation. The effective d_f value remains essentially unchanged during the second coarsening process, indicating that the changes detected by turbidimetry (Figure 1) and scattering profiles (Figure 4c) occurred at large length scales, i.e., at the cluster–cluster level, with no change on the basic structure of the complexes.

The compaction of the structures during complexation may be explained by results from experimental and numerical studies of rapid aggregation processes, where aggregate structures with an initial lower d_f can restructure into more compact clusters.[27,66] Initially, when the interparticle interactions are weak (e.g., the protein has a low surface charge density), as is the case at the beginning of the complexation process, for pH $\gg$ pHϕ, the salt linkages formed between protein and polyelectrolyte are very loose,[58,67] and the particle clustering may be considered reversible.[66] As a result, soluble complexes may be able to loosen and re-form repeatedly after the first collision, allowing protein rearrangements to take place and leading to denser clusters through a restructuring process. As the pH decreases, an increase of the interparticle attractive energy, due to an increase of opposite charges with acidification, may favour the progressive compactation of the structure.[66] This effect may be supported by the fact that the value of d_f increased more rapidly for pH < 5.2, where both β-lg and xanthan carry net opposite charges. Accordingly, recent studies on protein–polysaccharide complexes[68] have found that spontaneous rearrangement occurs over time (without any external forces) to produce the most favorable state of protein co-operative binding.[10,24]

The restructuring process may be amplified by shear, which provides a supplementary driving force for reorganization. Studies made on various molecules and colloidal materials have shown that shearing strongly influences the packing geometry and the tendency for aggregates to restructure.[27,49,69–71] This effect has been attributed to a selective break-up that removes the weaker and more porous parts of the aggregates, followed by re-aggregation and internal rearrangement processes which lead to denser structures and lower final equilibrium sizes.[27,69–71] The break-up due to hydrodynamic forces tends to occur at larger length scales, *i.e.*, once the aggregates are big enough, and at sufficiently high shear-rates.[19,69–72] On small length scales, the differential velocity of the fluid is low and no restructuring can occur.[27,70]

A similar effect could be observed in the β-lg + xanthan system under shear. As can be seen in Figure 2b for $r = 5$, after pHφ has been attained, large complexes begin to form up to a point where the critical size at which shear forces have an effect is reached, and selective break-up of the structures occurs (decrease in large particles at ~ 250 min). This effect is more noticeable at higher r. In the scattering intensity profile this effect can be seen in the inset of Figure 7. The timescale presented in this plot ($t = 250$–350 min) coincides with the decrease in the scattered light intensity from large particles after the peak following pHφ (Figure 2b). A slight deviation from scaling is observed at the larger length scales, *i.e.*, changes are occurring closer to the edge of the complexes. At large q values, the d_f value remains unchanged. Several authors have reported this deviation in the scattering profiles of aggregates subjected to shear, and have attributed it to a restructuring process induced by hydrodynamic forces.[49,69,71] This shows that the size and structure of interpolymeric complexes are determined by restructuring processes driven by a competition between attractive electrostatic forces and the rupture forces caused by the flow.

In systems where microfluidized xanthan was present, the initial intrapolymeric complexes formed at pHc were more tenuous, with d_f ranging from 1.8 ± 0.02 to 1.5 ± 0.03 for xanthan treated from one to twelve passes, respectively (Figure 8). The subsequent restructuring yielded interpolymeric complexes with a more compact structure compared to the complexes obtained with native xanthan (Table 3). As discussed previously, the xanthan molecules of lower M_w will tend to form more compact structures, since their entropy loss is lower.

4 Conclusions

Associative phase-separation kinetics and cluster morphology of β-lg + xanthan complexes produced under steady shear conditions were studied by static light scattering. The complexation kinetics for this system followed a nucleation and growth mechanism resulting in the formation of fractal structures. The correlation length observed in the scattering profiles may be the result of the formation of a xanthan-rich nucleus surrounded by a layer depleted in protein. The interaction between β-lg and xanthan starts at positively charged patches on the protein surface, before reaching the pI of β-lg. Primary complexes with a diffuse structure form initially, and they subsequently aggregate into more dense interpolymeric complexes. The compactation of the complexes over the course of interaction may reflect their susceptibility to restructuring due to external forces (pH and shear) and internal processes (co-operative protein binding). The β-lg/xanthan ratio has an important effect on the reaction rate, and the internal structure and average size of the formed complexes, principally as the result of mass-action equilibrium. Moreover, the complexes could bind excess protein only to certain extent, after which the system was destabilized due to the decline of repulsive forces caused by protein over-aggregation onto the complex. Finally, varying the M_w of xanthan also permits control of the characteristics of the resulting complexes. Notably, the mean size of the complexes was found to be proportional to the molecular weight of the polysaccharide.

Acknowledgements

The authors acknowledge financial support from the NSERC industrial chair, and from the industrial partners—Agropur, Novalait, and Parmalat.

References

1. V. B. Tolstoguzov, *Food Hydrocoll.*, 2003, **17**, 1.
2. W. S. Chen *et al.*, European Patent Application 0.340.035, 1989.
3. W. S. Chen and W. G. Soucie, U.S. Patent 4.559.233, 1985.
4. W. G. Soucie, W.-S. Chen, V. C. Witte, G. A. Henry, and W. D. Drehkoff, U.S. Patent 4.762.726, 1988.
5. H. Zaleska, S. G. Ring, and P. Tomasik, *Food Hydrocoll.*, 2000, **14**, 377.
6. T. V. Burova, N. V. Grinberg, I. A. Golubeva, A. Y. Mashkevich, V. Y. Grinberg, and V. B. Tolstoguzov, *Food Hydrocoll.*, 1999, **13**, 7.
7. M. Girard, S. L. Turgeon, and P. Paquin, *J. Food Sci.*, 2002, **67**, 113.
8. S. Laplante, 'Étude des propriétés stabilisantes d'émulsion du chitosane en présence d'isolat de protéines de lactosérum', Ph.D. thesis, Université Laval, Canada, 2004.
9. E. Dickinson, *Colloids Surf. B*, 1999, **15**, 161.
10. V. B. Tolstoguzov, *ACS Symp Ser.*, 1996, **650**, 2.
11. D. A. Ledward, in 'Protein Functionality in Food Systems', eds N. S. Hettierachchy and G. R. Ziegler, Marcel Dekker, New York, 1996, p. 225.
12. J. Xia and P. L. Dubin, in 'Macromolecular Complexes in Chemistry and Biology', eds P. Dubin, J. Bock, R. M. Davies, D. N. Schulz, and C. Thies, Springer, Berlin, 1994, p. 247.
13. J. C. Cheftel and E. Dumay, *Food Rev. Int.*, 1993, **9**, 473.
14. S. K. Samant, R. S. Singhal, P. R. Kulkarni, and D. V. Rege, *Int. J. Food Sci. Technol.*, 1993, **28**, 547.
15. M. Girard, S. L. Turgeon, and S. F. Gauthier, *J. Agric. Food Chem.*, 2003, **51**, 6043.

16. M. Girard, C. Sanchez, S. I. Laneuville, S. L. Turgeon, and S. F. Gauthier, *Colloids Surf. B*, 2004, **35**, 15.
17. M. F. Butler and M. Heppenstall-Butler, *Food Hydrocoll.*, 2003, **17**, 815.
18. R. Tuinier, J. K. G. Dhont, and C. G. de Kruif, *Langmuir*, 2000, **16**, 1497.
19. C. Sanchez, G. Mekhloufi, C. Schmitt, D. Renard, P. Robert, C.-M. Lehr, A. Lamprecht, and J. Hardy, *Langmuir*, 2002, **18**, 10323.
20. K. Tainaka, *J. Phys. Soc. Jpn*, 1979, **46**, 1899.
21. A. Veis and C. Aranyi, *J. Phys. Chem.*, 1960, **64**, 1203.
22. J. M. Park, B. B. Muhoberac, P. L. Dubin, and J. Xia, *Macromolecules*, 1992, **25**, 290.
23. P. L. Dubin and J. M. Murrell, *Macromolecules*, 1988, **21**, 2291.
24. K. W. Mattison, Y. Wang, K. Grymonpré, and P. L. Dubin, *Macromol. Symp.*, 1999, **104**, 53.
25. J. Xia, P. L. Dubin, Y. Kim, B. B. Muhoberac, and V. J. Klimkowski, *J. Phys. Chem.*, 1993, **97**, 4528.
26. W. Chen, X. Li, and M. Jiang,. *Macromol. Chem. Phys.*, 1998, **199**, 319.
27. A. Y. Kim and J. C. Berg,. *J. Colloid Interface Sci.*, 2000, **229**, 607.
28. K. Binder, in 'Materials Science and Technology. A Comprehensive Treatment', eds R.W. Cahn, P. Haasen, and E. J. Kramer, VCH, Weinheim, 1991, vol. 5, p. 405.
29. R. J. Allen and P. B. Warren, *Langmuir*, 2004, **20**, 1997.
30. F. Carlsson, M. Malmsten, and P. Linse, *J. Am. Chem. Soc.*, 2003, **125**, 3140.
31. R. Wagner and R. Kampmann, in 'Materials Science and Technology—A Comprehensive Treatment', eds R. W. Cahn, P. Haasen, and E. J. Kramer, VCH, Weinheim, 1991, vol. 5, p. 213.
32. M. Carpineti and M. Giglio, *Phys. Rev. Lett.*, 1992, **68**, 3327.
33. G. E. Eliçabe, H. A. Larrondo, and R. J. J. Williams, *Macromolecules*, 1997, **30**, 6550.
34. G. E. Eliçabe, H. A. Larrondo, and R. J. J. Williams, *Macromolecules*, 1998, **31**, 8173.
35. J. Maugey, T. van Nuland, and P. Navard, *Polymer*, 2001, **42**, 4353.
36. W. C. K. Poon, A. D. Pirie, M. D. Haw, and P. N. Pusey, *Physica A*, 1997, **235**, 110.
37. G. Ramírez-Santiago and A. E. González, *Physica A*, 1997, **236**, 75.
38. M. Verheul, J. S. Pedersen, S. P. F. M. Roefs, and C. G. de Kruif, *Biopoymers*, 1999, **49**, 11.
39. D. W. S. Wong, W. M. Camirand, and A. E. Pavlath, *CRC Crit. Rev. Food Sci. Nutr.*, 1996, **36**, 807.
40. V. J. Morris, in 'Food Polysaccharides and their Applications', ed. A. M. Stephen, Marcel Dekker, New York, 1995, p. 341.
41. T. Sato, T. Norisuye, and H. Fujita, *Polymer J.*, 1984, **16**, 341.
42. J. Lecourtier, G. Chauveteau, and G. Muller, *Int. J. Biol. Macromol.*, 1986, **8**, 306.
43. S. I. Laneuville, P. Paquin, and S. L. Turgeon, *Food Hydrocoll.*, 2000, **14**, 305.
44. S. Le Hénaff, 'Microparticules de complexes de protéines de lactosérum et de xanthane comme substitut de matière grasse'. M.Sc. thesis, Université Laval, Canada, 1996.
45. F. Weinbreck, R. de Vries, P. Schrooyen, and C. G. de Kruif, *Biomacromoecules*, 2003, **4**, 293.
46. M. F. Chaplin, in 'Carbohydrate Analysis: A Practical Approach', eds M. F. Chaplin and J. F. Kennedy, IRL Press, Oxford, 1986, p. 1.

47. T. Hashimoto, M. Itakura, and H. Hasegawa. *J. Chem. Phys.*, 1986, **85**, 6118.
48. R. S. Stein and J. J. Keane, *Polym. Sci.*, 1955, **17**, 21.
49. G. C. Bushell, Y. D. Yan, D. Woodfield, J. Raper, and R. Amal, *Adv. Colloid Interface Sci.*, 2002, **95**, 1.
50. K. W. Mattison, I. J. Brittain, and P. L. Dubin, *Biotechnol. Prog.*, 1995, **11**, 632.
51. C. Schmidt, C. Sanchez, F. Thomas, and J. Hardy, *Food Hydrocoll.*, 1999, **13**, 483.
52. C. Schmitt, C. Sanchez, S. Despond, D. Renard, P. Robert, and J. Hardy, in 'Food Colloids: Fundamentals of Formulation', eds E. Dickinson and R. Miller, Royal Society of Chemistry, Cambridge, 2001, p. 323.
53. D. J. Burgess, in 'Macromolecular Complexes in Chemistry and Biology', eds P. Dubin, J. Bock, R. M. Davies, D. N. Schulz, and C. Thies, Springer, Berlin, 1994, p. 285.
54. K. S. Kang and D. J. Pettitt, in 'Industrial Gums: Polysaccharides and their Derivatives', 3rd edn, eds R. L. Whistler and J. N. BeMiller, Elsevier, Amsterdam, 1992, p. 341.
55. I. Capron, G. Brigand, and G. Muller, *Polymer*, 1997, **38**, 5289.
56. Y. Li, J. Xia, and P. L. Dubin, *Macromolecules*, 1994, **27**, 7049.
57. J.-Y. Shieh and Ch. E. Glatz, in 'Macromolecular Complexes in Chemistry and Biology', eds P. Dubin, J. Bock, R. M. Davies, D. N. Schulz, and C. Thies, Springer, Berlin, 1994, p. 273.
58. A. Tsuboi, T. Izumi, M. Hirata, J. Xia, P. L. Dubin, and E. Kokufuta, *Langmuir*, 1996, **12**, 6295.
59. H. Dautzenberg, *Macromol. Symp.*, 2000, **162**, 1.
60. Y. Wang, K. Kimura, P. L. Dubin, and W. Jaeger, *Macromolecules*, 2000, **33**, 3324.
61. E. Tsuchida, K. Abe, and M. Honma, *Macromolecules*, 1976, **9**, 112.
62. J. H. E. Hone, A. M. Howe, and T. Cosgrove, *Macromolecules*, 2000, **33**, 1206.
63. M. F. Butler, *Biomacromolecules*, 2002, **3**, 676.
64. J. Zhang, H. Zhang, and Y. Yang, *J. Appl. Polym. Sci.*, 1999, **72**, 59.
65. S. Tang, J. M. Preece, C. M. McFarlane, and Z. Zhang, *J. Colloid Interface Sci.*, 2000, **221**, 114.
66. J. A. Molina-Bolívar, F. Galisteo-González, and R. Hidalgo-Álvarez, *J. Colloid Interface Sci.*, 1998, **208**, 445.
67. E. Kokufuta, in 'Macromolecular Complexes in Chemistry and Biology', eds P. Dubin, J. Bock, R. M. Davies, D. N. Schulz, and C. Thies, Springer, Berlin, 1994, p. 301.
68. F. Weinbreck, H. Nieuwenhuijse, G. W. Robijn, and C. G. de Kruif, *Langmuir*, 2003, **19**, 9404.
69. P. Marsh, G. Bushell, and R. Amal, *J. Colloid Interface Sci.*, 2001, **241**, 286.
70. G. Bushell and R. Amal, *J. Colloid Interface Sci.*, 2000, **221**, 186.
71. C. Selomulya, R. Amal, G. Bushell, and T. D. Waite, *J. Colloid Interface Sci.*, 2001, **236**, 67.
72. C. Sanchez, S. Despond, C. Schmitt, and J. Hardy, in 'Food Colloids: Fundamentals of Formulation', eds E. Dickinson and R. Miller, Royal Society of Chemistry, Cambridge, 2001, p. 332.

Break-up and Coalescence of Bubbles in Agitated Protein Solutions at High Air Volume Fraction

By B. J. Hu, A. W. Pacek, and A. W. Nienow

SCHOOL OF ENGINEERING (CHEMICAL ENGINEERING),
UNIVERSITY OF BIRMINGHAM, EDGBASTON,
BIRMINGHAM B15 2TT, UK

1 Introduction

A common situation in the chemical/biochemical processing industries is the flow of gas (air) in and out of a reactor containing relatively pure liquids or through solid particle suspensions. An extensive investigation over a wide range of hydrodynamic conditions and physical properties of the latter systems has been carried out.[1] Most research on gas/liquid dispersion has been carried out with flow-through systems at relatively low gas hold-up.[2] On the other hand, in many aerated food manufacturing processes, a high level of air dispersion is often required, e.g., the proportion of air in ice-cream can be up to 60%.[3] Yet studies of bubble behaviour in batch air/liquid food systems at high volume fraction under dynamic agitation conditions are rather limited.[4] This paper is a contribution towards filling that gap.

Three proteins, which are commonly used to stabilize dispersions in the food industry, have been studied here under a variety of batch processing conditions at high air volume fractions. Solutions of each protein were characterized by time-dependent measurements of surface tension. Bubble-size distributions and mean sizes under dynamic and steady-state conditions have been measured by a microscope–video–computer technique. Two main aspects are addressed in this paper. These are the long-term stability of proteins and the effect of pressure and agitation intensity on bubble size in agitated air/protein solution dispersions. Bubble sizes are correlated with processing parameters (pressure, impeller speed, air volume fraction and equilibrium surface tension) and the difference between static and dynamic systems are analysed and discussed.

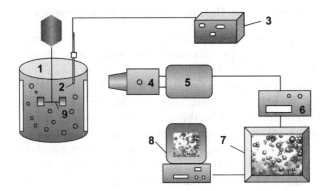

Figure 1 *Experimental rig for measurement of bubble-size distribution: (1) stirred pressure vessel, (2) strobe lamp, (3) strobe flash, (4) microscope, (5) video camera, (6) video recorder, (7) monitor, (8) computer, and (9) impeller.*

2 Materials and Methods

The Equipment

Mean bubble sizes and bubble-size distributions were measured in a pressurized stainless-steel vessel of diameter $T = 0.015$ m using the experimental rig shown in Figure 1. The vessel was fitted with four baffles, each of width $T/10$, and the liquids were agitated with a Rushton turbine of diameter $D = T/2$ (power number Po $= 5.0$). The bubble-size distributions were measured with an updated stroboscope–camera–microscope–computer system, involving a high-energy fibre-optic strobe light synchronized to a digital camera with shutter frequency linked to a microscope. Both frequencies could be adjusted to produce sharp images of air bubbles, the diameters of which were measured using the image analysis software Bubble Pro.

Preparation of Protein Solutions

The proteins were separately dissolved in 0.1 M imidazole hydrochloride buffer solution (pH $= 7$) at concentrations of 0.01% and 0.05% w/w. The protein samples were: (1) commercial sodium caseinate with $> 82\%$ dry protein, 6% moisture, $< 6\%$ fat and ash, supplied by DMV International (Veghel, Netherlands); (2) whey protein isolate (WPI) with 95% dry protein, $< 5\%$ moisture, $< 4\%$ fat and ash, supplied by Davisco Foods International; (3) β-lactoglobulin (β-lg) supplied by Sigma (containing bovine β-lg variants A and B, $3 \times$ crystallized).

Surface Tension Measurement

Surface tension as a function of time was measured using the Wilhelmy plate technique on a surface tensiometer, model CDCA-100, supplied by Camtel Ltd. The apparatus was controlled using Windows™ '95 software.

Bubble Generation and Bubble Stability Experiments

Air (40% v/v) was introduced into the closed vessel and dispersed by agitation at 25 °C into one of the protein solutions. Speeds were used at and above the minimum required to fully disperse the air. These speeds corresponded to values of the energy dissipation rate $\bar{\varepsilon}_T$ ranging from ~ 15 W kg^{-1} to 50 W kg^{-1}. Pressures of 1, 7, and 13 bar were applied, with step increases between them, and transient and steady-state bubble-size distributions were measured. Also, in order to simulate industrial processing, some experiments were carried out with reduction of pressure from 13 bar to 7 bar in steps of 2 bar, controlled by a needle valve fitted on the top lid of the stirred vessel. In both these cases of changing pressure, the agitation was continuous at a fixed speed, and with increasing pressure, air was rapidly admitted each time until the new pressure was reached. Since the liquid was essentially incompressible, the volume of air remained constant at the original level of 40%. However, on reducing the pressure *via* the needle valve, a gas/liquid mixture was discharged, and the exit from the needle valve was connected to an empty volumetric flask in order to collect the liquid. Thus, the volume discharged could be measured and the new increased volume of air could be estimated. Clearly, in the pressure reduction case, both the pressure and air volume fraction were being changed.

At each fixed pressure and for each protein solution, the effects of the hydrodynamic conditions on bubble size and coalescence rate were also investigated using the impeller-speed step-change technique.[5] Dispersions were stirred at constant impeller speed until the mean bubble size reached dynamic equilibrium between break-up and coalescence. After this time, the impeller speed was lowered, leading to an increase in d_{32}, initially at a rate controlled by the coalescence. Thus, a new 'equilibrium' size was 'approached *via* coalescence'. Alternatively, the speed was increased, leading to a decrease in d_{32}, and the new 'equilibrium' size was 'approached *via* breakage'. In all cases, the speed was sufficient to ensure that the air remained well dispersed.

Overall, this experimental study is essentially similar to that previously reported by us,[6] but extended to include additional proteins, higher air volume fractions, and coalescence due to incremental speed reductions under elevated pressure. Some data from our previous work are included as appropriate.

3 Results and Discussion

Effect of Prolonged Agitation on Protein Stability in Aerated Systems

It has often been considered that shear stresses associated with intense agitation may damage proteins, eventually causing them to lose activity. For instance, Pearce and Kinsella observed[7] that over-processing (prolonged agitation) of oil-in-water emulsions in the presence of protein resulted in an increase in mean droplet size. The review of the literature by Tirrell[8] suggests that protein damage is very complex and often depends on the solvent type. Thus, a protein which is found to be damaged in one condition may not be damaged at all in another.

Clearly, the effect of agitation on protein stability is still not resolved. Recently, Walstra[9] reviewed the literature, and also compared the stability of globular proteins under various operating conditions, taking into account both the hydrodynamic forces and the surface effects. By calculating the specific energy dissipation rate in turbulent flow, the shear stresses in laminar flow, the surface forces, and the energy required for protein denaturation, he concluded[9] that, in general, the energy dissipation associated with hydrodynamic processes alone is insufficient to cause denaturation even in intense turbulent flow. On the other hand, he proposed that protein is denatured mainly due to its adsorption onto fresh surfaces, e.g., air bubbles or solid surfaces.

We showed previously[6] that the steady-state bubble size in 0.05–0.5 wt% sodium caseinate solutions of air volume fraction $\Phi \leq 0.2$ remains the same under the same agitation conditions—even after 3 hours of intense agitation. From this observation, it was concluded, in agreement with Walstra,[9] that the protein was relatively stable with respect to agitation. On the other hand, since during all this time, fresh air–liquid interfaces were being formed under the action of bubble coalescence and break-up, the result may be considered surprising in the light of Walstra's other conclusion concerning the effect of interfaces.

The uncertainty surrounding the above matter clearly meant that it was worthy of further investigation. It was felt that one of the reasons for the extended protein stability might have been the relatively high protein concentration used. It was also considered that the result might have been specific to sodium caseinate. Hence the above question has also been addressed here using whey protein isolate (WPI). Firstly, the dynamic surface tension was measured as a function of WPI concentration. The equilibrium surface tension σ for each of the three proteins is given in Figure 2. It can be seen that initially

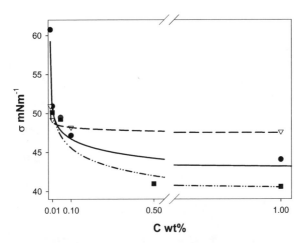

Figure 2 *Equilibrium surface tension σ as a function of protein concentration C:* ●, *sodium caseinate;* ∇, *β-lg;* ■, *WPI.*

there is a rapid fall in σ up to about 0.01 wt% WPI, and then the value remains
almost constant. It was decided to work within the range where σ is a strong
function of concentration, so that, if the protein became denatured, it would be
seen as a change in σ and in mean bubble size (because of its sensitivity to σ).
Hence experiments were undertaken at 0.01 wt% WPI; and in order to maximize the interfacial area between the phases, a value of $\Phi = 0.4$ was used. The
system was stirred vigorously for 7 hours at a constant mean specific energy
dissipation rate of 28 W kg^{-1}. Bubble sizes were measured initially after
1 minute, and then after 10 minutes, and subsequently at hourly intervals; the
results are shown in Figure 3. It can be seen that the bubble size falls quite
quickly, but after ∼60 minutes the size stabilizes. This result again suggests
that a dynamic breakage–coalescence equilibrium is achieved, which also
implies that the interfacial activity of the protein solution is also unchanged.
The amount of protein present, if it were *all* adsorbed, corresponds to a surface
coverage of *ca.* 3.2 mg m^{-2} (calculated assuming all bubbles have the same size,
d_{32}). This value is close to, but probably slightly higher than, the likely required
minimum amount of WPI for saturation coverage (about 2–3 mg m^{-2}). This
slight excess of non-adsorbed protein is perhaps the reason that the protein
still remains effective as a bubble stabilizer, but, as pointed out earlier, the concentration was chosen so that any shear-induced loss of protein functionality
should have been exhibited as a change in surface tension.

Figure 4 shows the time-dependent surface tension of three WPI solutions.
One data set relates to measurements made at the time the aeration experiment
was started; one to measurements after a day of quiescent ageing; and the
remaining one was a measurement made after stirring for 7 hours and then
waiting for a day. (It was not possible to measure σ earlier than this following
agitation because of the time it took for all of the bubbles to disappear.) All
the data sets in Figure 4 are are essentially the same. Thus, neither the intense

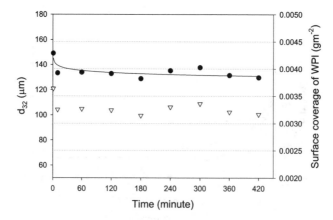

Figure 3 *Mean bubble size d_{32} (●, left axis) and effective maximum surface coverage (∇,
right axis) as a function of time for 0.01 wt% WPI solution at $\Phi = 0.4$ and
$N = 1000$ rpm.*

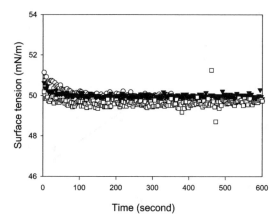

Figure 4 *Surface tension as a function of time for 0.01 wt% WPI solution before and after agitation (N = 1000 rpm): O, after 7 h of agitation, and waiting for 24 h; ▼, with no stirring, and waiting for 23 h; □, fresh protein solution.*

agitation nor the interaction of the protein with the new surface appears to denature or change substantially the structure of the protein. Or, if it does, it is not detectable, either directly by changes in the time-dependent surface tension or indirectly by changes in the mean bubble size.

Overall, both this study and our previous one[6] indicate less impact on protein stability during the turbulent dispersion of air in protein solutions than might have been expected based on the previous literature. A particular surprise is that this conclusion does not seem to agree with a study of such proteins at a quiescent planar interface, where it has been suggested[10] that irreversible adsorption causes protein to change structure, and that repeated adsorption/desorption events eventually lead to extensive protein aggregation and loss of interfacial activity.

Effect of Protein Type and Energy Dissipation Rate

Over the past decades, because of the complexity and importance of proteins in food processing, much research has been carried out[10] on how protein structure changes under different conditions and, more importantly, on the nature of the interactions of proteins at air–water and oil–water interfaces. However, most of the published results are based on laboratory studies at quiescent planar interfaces. Whether such studies can give results that are applicable to the dynamic turbulent flow conditions common in food processing is far from certain; and our results outlined above increase the need to answer this question. In addition, most of the investigations under dynamic conditions have been limited to the study of emulsions, *i.e.*, involving oil–water interfaces rather than air–water interfaces. Such work that has been reported, for instance, has found that casein-ates are more efficient emulsifiers,[11,12] insofar as they often lead to emulsions of higher stability (assessed under static conditions) than whey proteins.[13] However, such a conclusion may not be appropriate for air/water dispersions.

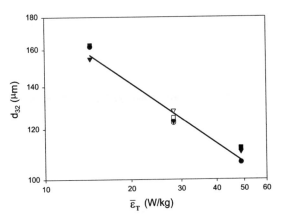

Figure 5 *Mean bubble diameter d_{32} as a function of mean specific energy dissipation rate $\bar{\varepsilon}_T$ for $\Phi = 0.4$ and protein concentration $= 0.01$ wt%: ●, β-lg; O, β-lg, return to 1000 rpm; ▼, WPI; ▽, WPI, return to 1000 rpm; ■, caseinate; □, caseinate, return to 1000 rpm.*

Sodium caseinate, whey protein isolate and β-lg have been investigated here at a concentration of 0.01 wt%, initially at atmospheric pressure, and with $\Phi = 0.4$. Figure 5 indicates the steady-state bubble size at mean specific energy dissipation rates of $\bar{\varepsilon}_T = 14$, 28 and 49 W kg⁻¹. At each agitation intensity, the mean bubble size is the same for each protein solution. Thus, again, whilst interfacial studies under quiescent planar conditions have shown differences in behaviour, we find under dynamic turbulent conditions that the behaviour (*i.e.*, the mean bubble size) is the same for the three proteins.

From Figure 5, the relationship between the Sauter mean diameter d_{32} and the mean specific energy dissipation rate $\bar{\varepsilon}_T$ is

$$(d_{32} / \mu m) = 361(\bar{\varepsilon}_T / \mathrm{W\ kg^{-1}})^{-0.31}. \tag{1}$$

If the value of the negative exponent in equation (1) is compared to the value of 0.35 for a low air volume fraction of $\Phi = 0.05$ for 0.05 wt% sodium caseinate found in our previous work,[6] we can conclude that the two correlations are similar, even though the $\bar{\varepsilon}_T$ values are much higher here (in order to ensure that a homogenous dispersion is achieved at the higher air volume fraction). On the other hand, the larger Φ value gives larger bubbles, presumably because of greater coalescence. In order to get further insight, the dispersion was continuously agitated at 1000 rpm, then at 800 rpm, and finally returning to 1000 rpm. The agitation time at each speed was two hours to ensure that the steady-state bubble size was reached at each stage. We can see from Figure 5 that the steady-state bubble size at 1000 rpm at the different times was essentially the same, which provides further evidence for the stability of the protein under severe agitation.

The impact of an adsorbed protein on the bubble size is very much related to the rate of mass transfer from the bulk solution to the interface. In a quiescent environment, protein is transported by diffusion, which may be different from protein to protein. But under dynamic turbulent conditions, protein is transported by turbulent diffusion and convection.[14] Equivalently, in making liquid/liquid dispersions, Walstra comments[14] that protein emulsifiers are transported to the oil–water interface by convection. Given the similarity between emulsification and our experimental conditions, convection may be the controlling mechanism responsible for transporting the different proteins to the air–water interface, and so one would not expect that mechanism to depend significantly on the type of protein.

Correlating Bubble Sizes through the Weber Number

Bubble (and drop) sizes in turbulent agitated dispersions have often been correlated in terms of the Weber number (We $= \rho_c N^2 D^3 / \sigma$), which represents the balance between inertial forces causing break-up and surface/interfacial tension stabilizing the bubble (drop). Such a correlation has been used even when coalescence is also involved. Indeed, this approach was used successfully together with the equilibrium surface tension in our previous work.[6] The dynamic surface tensions for the three proteins at a concentration of 0.01 wt% were determined over a period of an hour, and the equilibrium σ values were found to be very close as shown in Table 1. With this approach, since the equilibrium σ values are so similar, it is not surprising that the bubble sizes are also similar for the three protein solutions in Figure 5. The results are also correlated well by the Weber number, as can be seen in Figure 6. Thus, we have fitted the relationship

$$d_{32}/D = 0.064 \, \text{We}^{-0.47}, \tag{2}$$

where D is the turbine diameter. The equation for the low volume fraction reported previously[6] is

$$d_{32}/D = 0.034 \, \text{We}^{-0.44}. \tag{3}$$

Clearly, the Weber number approach, based on the application of Kolmogoroff turbulence concepts, which was found previously[6] to work well for sodium caseinate solutions, seems also to be able to correlate the data well for the other protein solutions based on their steady-state surface tensions.

Table 1 *Equilibrium surface tension for the three protein samples at a bulk aqueous phase concentration of 0.01 wt%.*

Protein	Surface tension (mN m^{-1})
β-lactoglobulin	49.2
whey protein isolate	50.1
sodium caseinate	51.0

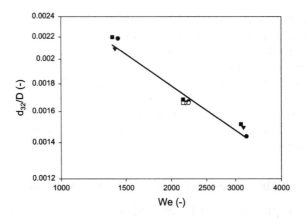

Figure 6 *Normalized mean bubble diameter d_{32}/D as a function of Weber number (We) for $\Phi=0.4$ and protein concentration $= 0.01$ wt% : •, β-lg; O, β-lg, return to 1000 rpm; ▼, WPI; ▽, WPI, return to 1000 rpm; ■, caseinate; □, caseinate, return to 1000 rpm.*

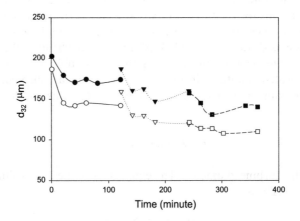

Figure 7 *Mean bubble diameter d_{32} as a function of time for 0.05 wt% caseinate solution at elevated pressure at constant air volume fraction $\Phi=0.4$ and different agitation speeds: •, 1 bar, 800 rpm; ▼, 7 bar, 800 rpm; ■, 13 bar, 800 rpm; O, 1 bar, 1000 rpm; ▽, 7 bar, 1000 rpm; □, 13 bar, 1000 rpm.*

Effect on Bubble Behaviour of Raising and Lowering the Pressure

For these experiments, a number of different protein concentrations and Φ values were tested. The results for 0.05 wt% sodium caseinate and $\Phi=0.4$ are shown in Figure 7. The bubble size is plotted as a function of time for three pressures, from atmospheric ($p_a=1$ bar) to 13 bar, while keeping a constant impeller speed (800 or 1000 rpm). The initial increase in bubble size, each time the pressure is raised, is associated with the extra air that has to be introduced into the vessel in order to keep Φ constant. Similar results were found at $\Phi=0.05$,[6] and at both Φ values the steady-state d_{32} was found to drop by

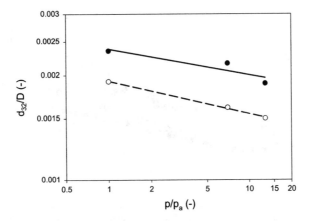

Figure 8 *Correlation of normalized mean bubble size (d_{32}/D) and normalized pressure (p/p_a) for 0.05 wt% caseinate solution at $\Phi = 0.4$ and different energy dissipation rates: O, 1000 rpm; ●, 800 rpm.*

~20% for the same pressure increase, even though the agitation speeds and the $\bar{\varepsilon}_T$ values were very different. The steady-state sizes as a function of pressure at the two $\bar{\varepsilon}_T$ values are also shown in Figure 8. These results at $\Phi = 0.4$, and the similarity in behaviour with the earlier low volume fraction work, clearly indicate the existence of a pressure effect, albeit a rather small one, *i.e.*, a 13-fold increase in pressure gives a 20% reduction in bubble size.

Similar results have also been reported for bubble columns,[15–17] but so far a satisfactory explanation has not been proposed. For example, it has been suggested[16,17] that the increasing pressure has an equivalent effect to increasing gas density, and that it is the latter that is responsible for the decrease in bubble size. On the other hand, work with different gases (from helium to xenon) in a sparged stirred vessel has shown[18] that changing the gas density has no significant influence on bubble size.

To gain more insight into the effect of pressure, the technique of making a step change in speed has been used to separate the breakage and coalescence processes. For a short time after reducing the impeller speed (say, up to 5 minutes), the increase in size will give an indication of the rate of coalescence, as everywhere the local ε_T is insufficient to break up any of the bubbles. Therefore the coalescence rate can be determined using a simplified coalescence model[19] from the initial gradient of the Sauter mean diameter against time. Figure 9 shows the transient mean bubble size for a sequence of impeller speed reductions from 1000 rpm to 500 rpm at three pressures (1, 7 and 13 bar) for 0.05 wt% sodium caseinate and $\Phi = 0.05$. Initially, the bubble size increases much faster at atmospheric pressure than at 7 bar or 13 bar, indicating that coalescence is slower at the elevated pressures. An estimate of the initial gradient of d_{32} *versus* time is shown in Table 2, which provides further evidence for the reduction in coalescence rate with increasing pressure. The reason for the smaller bubble sizes at the elevated pressure can therefore be explained by the change in dynamic equilibrium between break-up and coalescence. If the

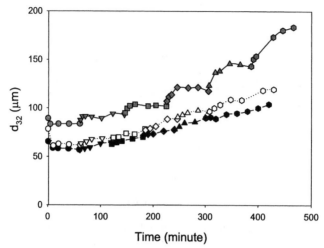

Figure 9 *Mean bubble size d_{32} as a function of time with impeller speed reduction at different constant pressures for 0.05 wt% caseinate solution at $\Phi=0.05$: •, 1 bar, 1000 rpm; ▼, 1 bar, 900 rpm; ■, 1 bar, 800 rpm; ♦, 1 bar, 700 rpm; ▲, 1 bar, 600 rpm; ●, 1 bar, 500 rpm; O, 7 bar, 1000 rpm; ∇, 7 bar, 900 rpm; □, 7 bar, 800 rpm; ◊, 7 bar, 700 rpm; Δ, 7 bar, 600 rpm; ●(open), 7 bar, 500 rpm; •, 13 bar, 1000 rpm; ▼, 13 bar, 900 rpm; ■, 13 bar, 800 rpm; ♦, 13 bar, 700 rpm; ▲, 13 bar, 600 rpm; ●, 13 bar, 500 rpm.*

Table 2 *Coalescence rate (dd_{32}/dt) over a period of 10 minutes from the data in Figure 9 for a step decrease in impeller speed starting from the steady state.*

Pressure(bar)	Coalescence rate $(\mu m\ min^{-1})$			
	Impeller speed			
	900 rpm	800 rpm	700 rpm	600 rpm
13	0.144	0.178	0.156	0.290
7	0.380	0.220	0.189	0.478
1	0.925	1.175	1.150	1.067

coalescence rate decreases, the equilibrium size should be smaller. On the other hand, the underlying reason why coalescence is slower is far from obvious. Maybe the higher pressure affects the protein structure at the two surfaces of the interfacial film separating a pair of neighbouring bubbles, thereby slowing down film drainage and hence coalescence. Certainly, as discussed below, the effect of the volume fraction Φ on d_{32} is relatively weak, which is a clear indication that the presence of protein is inhibiting coalescence even though it does not prevent it completely.

The effect of pressure reduction on bubble size is relevant to many aerated food products. Figure 10 shows the transient bubble size as a function of time for 2 bar incremental lowering of pressure from 13 bar to 7 bar. Because the degree of agitation has to be kept constant while reducing the pressure to

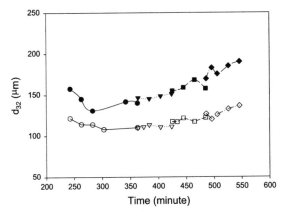

Figure 10 *Mean bubble size d_{32} as a function of time with pressure reduction at different constant impeller speed:* O, *13 bar,* $\Phi=0.4$, *1000 rpm;* ∇, *11 bar,* $\Phi=0.433$, *1000 rpm;* □, *9 bar,* $\Phi=0.482$, *1000 rpm;* ◇, *7 bar,* $\Phi=0.545$, *1000 rpm;* •, *13 bar,* $\Phi=0.4$, *800 rpm;* ▼, *11 bar,* $\Phi=0.441$, *800 rpm;* ■, *9 bar,* $\Phi=0.487$, *800 rpm;* ◆, *7 bar,* $\Phi=0.555$, *800 rpm.*

approximate to a steady hydrodynamic state, it is inevitable that some of the liquid phase is lost. The air volume fraction, which can be estimated as described earlier, increases up to 0.55. As a result, a much bigger mean bubble size is obtained—an increase of between 24% to 36% for 1000 rpm and 800 rpm, respectively—due to the greater degree of coalescence with increasing Φ and pressure.

Overall Correlation for the Mean Bubble Size

Considering all the steady-state bubble sizes at different pressures, but otherwise identical conditions (protein concentration, Φ value, and hydrodynamic conditions), we find the relationship $d_{32}/D \propto (p/p_a)^{-0.10}$, i.e., a rather weak effect. Assuming that d_{32}/D is related to the Weber number *via* the exponent in equation (2), then the effect of the change in air volume fraction from 0.05 to 0.55 can be quantified by plotting the group $(d_{32}/D)We^{0.47}(p/p_a)^{0.1}$ against $(1+C\Phi)$ where C is an empirical dimensionless constant.[6] Such a plot is shown in Figure 11. This manipulation leads to the fitted equation

$$d_{32}/D = 0.036(1+2.20\Phi)We^{-0.47}(p/p_a)^{-0.10} \tag{4}$$

with a correlation coefficient of 0.87. The fitted value of $C=2.2$, which characterizes the degree of coalescence of the dispersion, is quite close to the one obtained in our previous work,[6] which only included lower air volume fractions up to $\Phi=0.2$. This relationship, based on all the available data, suggests that even at air volume fraction up to 0.55, the dispersion can still be classified as a low-coalescence system. That is, the value of $C=2.2$ is close to the value of $C=3$ considered to represent non-coalescing liquid/liquid dispersions.[20] Figure 12 shows the experimental d_{32} data plotted against the values calculated from

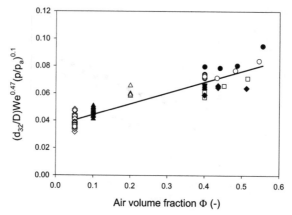

Figure 11 *The quantity $(d_{32}/D)We^{0.47}(p/p_a)^{0.1}$ as a function of air volume fraction Φ: •, $\Phi \geq 0.4$, 0.05 wt% caseinate, 800 rpm, 1–13 bar; O, $\Phi \geq 0.4$, 0.05 wt% casein-ate, 1000 rpm, 1–13 bar; ▼, $\Phi = 0.4$, 0.01 wt% caseinate, 1 bar; ∇, $\Phi = 0.4$, 0.01 wt% β-lg, 1 bar; ■, $\Phi = 0.4$, 0.01 wt% WPI, 1 bar; □, $\Phi \geq 0.4$, 0.01 wt% WPI, 7–13 bar; ♦, $\Phi \geq 0.4$, 0.05 wt% WPI, 7–13 bar; ◇, $\Phi = 0.05$, caseinate; ▲, $\Phi = 0.1$, caseinate; △, $\Phi = 0.2$, caseinate; ◕, $\Phi = 0.05$, 0.05 wt% caseinate, 7 bar; ◕(open), $\Phi = 0.05$, 0.05 wt% caseinate, 13 bar.*

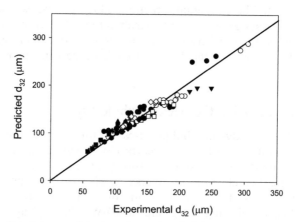

Figure 12 *Predicted d_{32} as a function of experimental d_{32}: •, $\Phi = 0.05$, 1 bar; O, $\Phi = 0.10$, 1 bar; ▼, $\Phi = 0.20$, 1 bar; ∇, $\Phi = 0.05$, 0.05 wt% caseinate, 7 bar; ■, $\Phi = 0.05$, 0.05 wt% caeinate, 13 bar; □, $\Phi \geq 0.40$, 0.05 wt% caseinate, 1–13 bar, 800 rpm; ♦, $\Phi \geq 0.40$, 0.05 wt% caseinate, 1–13 bar, 1000 rpm; ◇, $\Phi = 0.40$, 0.01 wt% β-lg , caseinate, WPI; ▲, $\Phi \geq 0.40$, WPI, 1–13 bar, 1000 rpm.*

equation (4). Given the difficulty of the measurement and other experimental problems, the correlation appears very good.

4 Conclusions

Bubble behaviour in agitated aqueous solutions of three different proteins at high air volume fractions (up to ~0.55) has been studied. We have examined effects on the transient and steady-state bubble sizes of prolonged periods of agitation, of stepwise increases and decreases in pressure at constant agitator speed, and similarly of stepwise speed changes at constant pressure.

Agitation for 7 hours at a high-energy dissipation rate, a 0.4 volume fraction air/solution dispersion containing 0.01 wt% WPI did not give any significant change in mean bubble size or equilibrium surface tension of the aqueous phase. This finding suggests that the surface activity of the protein remained unchanged even at this relatively low concentration of WPI. This is in accordance with our previous paper,[6] where the protein concentration was significantly higher, and any protein damage might have gone undetected. Other aspects of the present study associated with the step changes in speed are also consistent with good protein stability. This conclusion seems to agree well with other studies in relation to the impact of agitation alone, but not with respect to the denaturing effect of proteins through interaction with interfaces.[9,21]

The three food proteins investigated here (sodium caseinate, WPI and β-lg) at 0.01 wt% concentration in air/solution dispersion all gave essentially the same steady-state bubble size for the same $\bar{\varepsilon}_T$. They also have similar equilibrium surface tensions. Combining these two points suggests that, although their structure and their behaviour at quiescent surfaces may be different,[22] under dynamic conditions the equilibrium surface tension is apparently sufficient to determine the bubble size. This conclusion is supported by the fact that all of the steady-state bubble size data can be correlated by equation (4), including results from both this work and our earlier study.[6] It shows that coalescence does occur, as confirmed by the 'step-down-in-speed' experiments. The constant C has been called the coalescence tendency parameter,[20] and its numerical value here of 2.2 is low in relation to a range of values previously reported for liquid/liquid dispersion. This suggests that the bubble coalescence is actually a rather slow process, presumably because of the presence of protein at the interface reducing the film drainage rates. Overall then, in these protein solutions where coalescence is apparently rather slow, one parameter—the equilibrium surface tension—is sufficient to take into account the impact of the fluid interface and its associated protein on the mean bubble size.

This conclusion is quite different from that typically reported under quiescent conditions.[22] We postulate that the similar behaviour of all three proteins here under dynamic conditions is because the protein interaction with the air–water interface is dominated in our experiments by turbulent diffusion and convection. Thus, the system behaviour is controlled by macroscopic processes, unaffected by the precise protein molecular structure. On the other hand, under quiescent conditions, the interaction is determined by diffusion occurring on the molecular level, which may depend significantly on the nature of the specific protein.

This work has also shown that, under otherwise similar conditions, the steady-state mean bubble size is slightly smaller at elevated pressures (up to 13 bar). The reduction in bubble size is linked to a reduction in the coalescence rate that affects the break-up/coalescence equilibrium. While the reason for the reduction in coalescence rate is not totally clear, we speculate that the protein stabilizes the film between bubbles better under pressure, and thus lowers the film drainage rate. Reduction in pressure appears to increase the bubble size; however, because of the difficulty of undertaking these experiments, this result should not be treated as conclusive.

It is interesting to compare the results for these air dispersions in protein solutions with our recent work on air/alcohol dispersions.[23] In all the present work with proteins at concentrations ≥ 0.01 wt%, and for Φ values from 0.05 to 0.55, the bubbles were observed to be spherical. The equilibrium σ values were never less than $\sim 40\%$ mN m^{-1}. With the alcohols, however, σ is generally less than 30 mN m^{-1} and it can even be as low as ~ 20 mN m^{-1}.[23] And the bubbles are, under broadly equivalent hydrodynamic conditions, an order of magnitude bigger. Even at $\Phi = 0.01$, many of them are non-spherical. Therefore equation (4) grossly underpredicts bubble sizes in such systems. This comparison, taken together with the points made earlier in this paper, clearly shows that the adsorbed proteins are indeed causing a dramatic reduction in the rate of coalescence, even though they do not prevent it entirely.

Acknowledgements

This work forms part of a project sponsored by BBSRC (UK) and Unilever plc. The authors acknowledge helpful discussions with Eric Dickinson and Brent Murray (University of Leeds) and Iain Cambell and his colleagues at Unilever Research (Colworth Laboratory).

References

1. J. C. Middleton, *Mixing in the Process Industries*, 2nd edn, eds N. Harnby, M. F. Edwards, and A. W. Nienow, Butterworth-Heinemann, London, 1992, p. 322.
2. V. Machon, A. W. Pacek, and A. W. Nienow, *Trans. Inst. Chem. Eng.*, 1997, **75A**, 339.
3. K. G. Berger, in 'Food Emulsions', ed. S. Friberg, Marcel Dekker, New York, 1976, p. 141.
4. K. Niranjan, 'Bubbles in Foods', eds G. M. Campbell, C. Webb, S. Pandiello, and K. Niranjan, Eagan Press, St Paul, MN, 1999, p. 3.
5. A. W. Pacek, C. C. Man, and A. W. Nienow, *Chem. Eng. Sci.*, 1998, **53**, 2005.
6. B. Hu, A. W. Nienow, and A. W. Pacek, *Colloids Surf. B*, 2003, **31**, 3.
7. K. N. Pearce and J. Kinsella, *J. Agric. Food Chem.*, 1978, **26**, 716.
8. M. V. Tirrell, *J. Bioeng.*, 1978, **2**, 183.
9. P. Walstra, in 'Food Colloids: Fundamentals of Formulation', eds E. Dickinson and R. Miller, Royal Society of Chemistry, Cambridge, 2001, p. 245.
10. E. Dickinson, B. S. Murray, and G. Stainsby, in 'Advances in Food Emulsions and Foams', eds E. Dickinson and G. Stainsby, Elsevier Applied Science, London, 1988, p. 123.
11. J. E. Kinsella, *CRC Crit. Rev. Food Sci. Nutr.*, 1984, **21**, 197.
12. E. Tornberg, A. Olsson, and K. Persson, in 'Food Emulsions', 2nd edn, eds K. Larsson and S. Friberg, Marcel Dekker, New York, 1990, p. 247.

13. A. Millqvist-Fureby, N. Burns, K. Landström, P. Fäldt, and B. Bergenståhl, in 'Food Emusions and Foams: Interfaces, Interactions and Stability', eds E. Dickinson and J. M. Rodríguez Patino, Royal Society of Chemistry, Cambridge, 1999, p. 236.
14. P. Walstra, in 'Encyclopedia of Emulsion Technology', ed. P. Becher, Marcel Dekker, New York, 1983, vol. 1, p. 57.
15. P. M. Wilkinson and L. L. V. Dierendonck, *Chem. Eng. Sci.*, 1990, **45**, 2309.
16. P. Jiang, T.-J. Lin, X. Luo, and L.-S. Fan, *Trans. Inst. Chem. Eng.*, 1995, **73A**, 269.
17. T.-J. Lin, K. Tsuchiya, and L.-S. Fan, *AIChE J.*, 1998, **44**, 545.
18. K. Takahashi and A. W. Nienow, *J. Chem. Eng. Jpn*, 1992, **25**, 432.
19. W. J. Howarth, *AIChE J.*, 1967, **13**, 1007.
20. A. W. Nienow, *Adv. Colloid Interface Sci.*, 2004, **108–109**, 95.
21. C. R. Thomas, A. W. Nienow, and P. Dunnill, *Biotechnol. Bioeng.*, 1979, **21**, 2263.
22. B. S. Murray, E. Dickinson, Z. Du, R. Ettelaie, K. Maisonneuve, and I. Söderberg, in 'Food Colloids, Biopolymers and Materials', eds E. Dickinson and T. van Vliet, Royal Society of Chemistry, Cambridge, 2003, p. 165.
23. A. W. Nienow, B. Hu, and A. W. Pacek, Annual Meeting of the American Institution of Chemical Engineers, San Francisco, CA, 2003, Paper 361a.

Subject Index